MICROFLUIDIC TECHNOLOGIES
FOR HUMAN HEALTH

MICROFLUIDIC TECHNOLOGIES
FOR HUMAN HEALTH

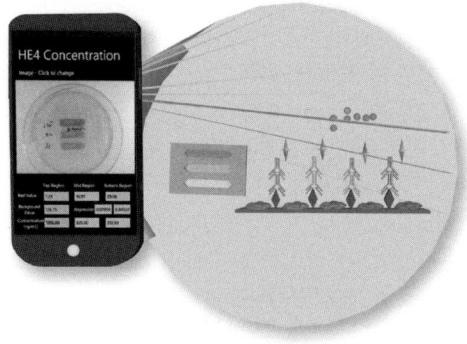

edited by

Utkan Demirci
Harvard Medical School, USA & Brigham and Women's Hospital, USA

Ali Khademhosseini
Harvard Medical School, USA & Brigham and Women's Hospital, USA

Robert Langer
Massachusetts Institute of Technology, USA

Jeffrey Blander
Harvard Medical School, USA & Brigham and Women's Hospital, USA

NEW JERSEY · LONDON · SINGAPORE · BEIJING · SHANGHAI · HONG KONG · TAIPEI · CHENNAI

Published by

World Scientific Publishing Co. Pte. Ltd.
5 Toh Tuck Link, Singapore 596224
USA office: 27 Warren Street, Suite 401-402, Hackensack, NJ 07601
UK office: 57 Shelton Street, Covent Garden, London WC2H 9HE

British Library Cataloguing-in-Publication Data
A catalogue record for this book is available from the British Library.

MICROFLUIDIC TECHNOLOGIES FOR HUMAN HEALTH

Copyright © 2012 by World Scientific Publishing Co. Pte. Ltd.

All rights reserved. This book, or parts thereof, may not be reproduced in any form or by any means, electronic or mechanical, including photocopying, recording or any information storage and retrieval system now known or to be invented, without written permission from the Publisher.

For photocopying of material in this volume, please pay a copying fee through the Copyright Clearance Center, Inc., 222 Rosewood Drive, Danvers, MA 01923, USA. In this case permission to photocopy is not required from the publisher.

ISBN 978-981-4405-51-5

Typeset by Stallion Press
Email: enquiries@stallionpress.com

Printed by FuIsland Offset Printing (S) Pte Ltd Singapore

Contents

Preface		vii
List of Contributors		ix
Chapter 1	A Microscale Bioinspired Cochlear-like Sensor *Robert D. White, Robert Littrell, and Karl Grosh*	1
Chapter 2	Systematic Evaluation of the Efficiencies of Proteins and Chemicals in Pharmaceutical Applications *Morgan Hamon and Jong Wook Hong*	21
Chapter 3	Microfluidic Glucose Sensors *Jithesh V. Veetil, Sruthi Ravindranathan, Sha Jin, and Kaiming Ye*	47
Chapter 4	Applications of Microfabrication and Microfluidic Techniques in Mesenchymal Stem Cell Research *Abhijit Majumder, Jyotsna Dhawan, Oren Levy, and Jeffrey M. Karp*	69
Chapter 5	Patient-Specific Modeling of Low-Density Lipoprotein Transport in Coronary Arteries *Ufuk Olgac*	97
Chapter 6	Point-of-Care Microdevices for Global Health Diagnostics of Infectious Diseases *Sau Yin Chin, Tassaneewan Laksanasopin, Curtis D. Chin, and Samuel K. Sia*	115
Chapter 7	Integrated Microfluidic Sample Preparation for Chip-based Molecular Diagnostics *Jane Y. Zhang, Qingqing Cao, Madhumita Mahalanabis, and Catherine Klapperich*	135
Chapter 8	Microfluidic Devices for Cellular Proteomic Studies *Yihong Zhan and Chang Lu*	161

Chapter 9	Microfluidics for Neuroscience: Novel Tools and Future Implications Vivian M. Hernandez and P. Hande Özdinler	185
Chapter 10	Microfluidics: On-Chip Platforms as *In Vitro* Disease Models Shan Gao, Erkin Şeker, and Martin L. Yarmush	213
Chapter 11	Application of Microfluidics in Stem Cell and Tissue Engineering Sasha H. Bakhru, Christopher Highley, and Stefan Zappe	241
Chapter 12	Microfluidic "On-the-Fly" Fabrication of Microstructures for Biomedical Applications Edward Kang, Sau Fung Wong, and Sang-Hoon Lee	293
Chapter 13	Microfluidics as a Promising Tool Toward Distributed Viral Detection Elodie Sollier and Dino Di Carlo	311
Chapter 14	Electrophoresis and Dielectrophoresis for Lab-on-a-Chip (LOC) Analyses Yağmur Demircan, Gürkan Yilmaz, and Haluk Külah	341
Chapter 15	Ultrasonic Embossing of Carbon Nanotubes for the Fabrication of Polymer Microfluidic Chips for DNA Sample Purification Puttachat Khuntontong, Min Gong, and Zhiping Wang	377
Chapter 16	Ferrofluidics A. Rezzan Kose and Hur Koser	391
Chapter 17	Antibody-based Blood Bioparticle Capture and Separation Using Microfluidics for Global Health ZhengYuan Luo, ShuQi Wang, Utkan Demirci, TianJian Lu, Feng Xu, and BoFeng Bai	417
Chapter 18	Applications of Quantum Dots for Fluorescence Imaging in Biomedical Research ShuQi Wang, Matin Esfahani, Dusan Sarenac, Bettina Cheung, Aishwarya Vasudevan, Fatih Inci, and Utkan Demirci	451

Biographies of the Editors   471
Index   475

Preface

The dramatic rise of health care costs in the developed world and the emerging double burden of chronic and infectious diseases in developing countries have created a great need for innovative solutions to reduce costs while improving quality and expanding access to care and treatment services. A key component of the health care delivery value chain is to enhance the clinician's ability to detect and diagnose medical conditions faster at the point of care to allocate scarce technical and human resources for the treatment and monitoring of patients.

Traditional models that have once influenced tertiary centers of excellence are now being extended and decentralized into the community, home, and rural villages through technology innovation. Remarkable advances in the integration of mobile communications, diagnostic imaging, and microfluidic technologies are transitioning from brick and mortar health care approaches into virtual health monitoring platforms that serve to enhance the global medical management of clients.

Microfluidic technologies which leverage best on the breed principles of mechanical, chemical, and biomedical engineering can be used across an array of groundbreaking medical applications that range from monitoring a patient's progress on chemotherapy or the recurrence of cancer to monitoring HIV-infected patients for antiretroviral treatment failure as well as measuring levels of blood glucose in diabetic patients. These technologies seek to minimize the need for invasive procedures for sampling blood or other bodily fluids such as urine and saliva to accurately analyze, diagnose, treat, and monitor conditions.

Examples of important applications of these technologies include the diagnosis and treatment of HIV/AIDS in pregnant women. Recently, microfluidic technologies that can be used to detect the level of $CD4^+$ cells and viral load assays have been developed, and have been envisioned by enabling the capture and quantification of cells or viruses from a small volume of a patient's whole blood sample without sample preprocessing steps. Thus, given their simplicity and cost-effectiveness, such technological platforms can be of high impact on the international stage as broadly applicable biotechnological tools, reducing the rate of mother-to-child transmission of infections.

Other applications extend beyond the diagnosis and treatment of disease. Such applications include the fields of medical research and drug delivery. For example, microfluidic devices can be used to test the effects of new drugs on human cells by directly simulating the pharmacokinetics and pharmacodynamics of medicine inside miniaturized systems located on a chip device. Such platforms may increase the efficiency of existing testing systems leading to faster approval times as well as the lowering of drug development costs.

This book provides a comprehensive overview of the myriad of microfluidic systems and their vast potential on human health. Topics covered in this book include: Sensor Devices (Chapters 1 and 3), Pharmaceutical Applications (Chapter 2), Stem Cell Research (Chapters 4 and 11), Chronic Diseases (Chapters 5 and 9), Global Health (Chapters 6 and 17), Molecular Diagnostics and Viral Detection (Chapters 7 and 13), Applied Research (Chapters 8, 10, 12, 14, and 18), Techniques for Microfluidic Chip Manufacturing (Chapter 15), and Ferrofluidics (Chapter 16).

It is our belief that future advances in medicine and human health will continue to be greatly influenced through the application of microfluidic technologies, reducing cost and increasing the quality of diagnosis, treatment, and monitoring of patients in the process.

Utkan Demirci
Ali Khademhosseini
Jeffrey Blander
Cambridge, Massachusetts
October 2012

List of Contributors

BoFeng Bai
State Key Laboratory of
  Multiphase Flow in Power
  Engineering *and*
Biomedical Engineering and
  Biomechanics Center
Xi'an Jiaotong University
Xi'an 710049, China
Chapter 17

Sasha H. Bakhru
Department of Molecular
  Pharmacology, Physiology, and
  Biotechnology
Brown University, Providence
RI 02912, USA
Chapter 11

Qingqing Cao
Departments of Biomedical and
  Mechanical Engineering
Boston University, Boston
MA 02215, USA
Chapter 7

Bettina Cheung
Demirci Bio-Acoustic-MEMS in
  Medicine (BAMM) Laboratory
Division of Biomedical Engineering
Department of Medicine

Brigham and Women's Hospital
Harvard Medical School
Boston, MA 02139, USA
Chapter 18

Curtis D. Chin
Department of Biomedical
  Engineering
Columbia University
351 Engineering Terrace
1210 Amsterdam Ave
New York City, NY 10027, USA
Chapter 6

Sau Yin Chin
Department of Biomedical
  Engineering
Columbia University
351 Engineering Terrace
1210 Amsterdam Ave
New York City, NY 10027, USA
Chapter 6

Yağmur Demircan
METU-MEMS Center *and*
Department of Electrical and
  Electronics Engineering
Middle East Technical University
  (METU)
Ankara, Turkey
Chapter 14

Utkan Demirci
Demirci Bio-Acoustic-MEMS in
  Medicine (BAMM) Laboratory
Division of Biomedical Engineering
Department of Medicine
Brigham and Women's Hospital
Harvard Medical School
Boston, MA 02139, USA and
Harvard–MIT Health Sciences and
  Technology
Cambridge, MA 02139, USA
Chapters 17 and 18

Jyotsna Dhawan
Institute for Stem Cell Biology and
  Regenerative Medicine (inStem)
Bangalore 560065, India and
Center for Cellular and Molecular
  Biology
Council of Scientific and Industrial
  Research
Hyderabad 500007, India
Chapter 4

Dino Di Carlo
Department of Bioengineering
Henry Samueli School of
  Engineering and Applied Science
University of California, Los Angeles
Los Angeles, CA 90095, USA
Chapter 13

Matin Esfahani
Demirci Bio-Acoustic-MEMS in
  Medicine (BAMM) Laboratory
Division of Biomedical Engineering
Department of Medicine
Brigham and Women's Hospital
Harvard Medical School
Boston, MA 02139, USA
Chapter 18

Shan Gao
Center for Engineering in Medicine
Massachusetts General Hospital
Harvard Medical School and
  Shriners Hospitals for Children
51 Blossom Street, Boston
MA 02114, USA
Chapter 10

Min Gong
Genome Institute of Singapore
60 Biopolis Street, Singapore 138672
Chapter 15

Karl Grosh
Department of Mechanical
  Engineering
2026 GGB (George G. Brown
  Laboratory)
University of Michigan
2350 Hayward Ave
Ann Arbor, MI 48109-2125, USA
Chapter 1

Morgan Hamon
Materials Research and Education
  Center
Department of Mechanical
  Engineering
275 Wilmore Lab, Auburn University
Auburn, AL 36849, USA
Chapter 2

Vivian M. Hernandez
Les Turner ALS Laboratory II
Department of Neurology
Feinberg School of Medicine
Northwestern University
Chicago, IL 60611, USA
Chapter 9

Christopher Highley
Materials Research and Education
 Center
Department of Biomedical
 Engineering
Carnegie Mellon University
Pittsburgh, PA 15213, USA
Chapter 11

Jong Wook Hong
Materials Research and Education
 Center
Department of Mechanical
 Engineering
275 Wilmore Lab, Auburn University
Auburn, AL 36849, USA
Chapter 2

Fatih Inci
Demirci Bio-Acoustic-MEMS in
 Medicine (BAMM) Laboratory
Division of Biomedical Engineering
Department of Medicine
Brigham and Women's Hospital
Harvard Medical School
Boston, MA 02139, USA
Chapter 18

Sha Jin
Biomedical Engineering Program
College of Engineering
203 Engineering Hall
University of Arkansas
Fayetteville, AR 72701, USA
Chapter 3

Edward Kang
Department of Biomedical
 Engineering
College of Health Science
Korea University
Jeongneung-dong, Seongbuk-gu
Seoul, 136-703, Republic of Korea
Chapter 12

Jeffrey M. Karp
Division of Biomedical Engineering
Department of Medicine
Center for Regenerative
 Therapeutics
Brigham and Women's Hospital
Harvard Medical School
Cambridge, MA 02139, USA,
Harvard Stem Cell Institute
Harvard University
Cambridge, MA 02138, USA, *and*
Harvard–MIT Division of Health
 Sciences and Technology
Massachusetts Institute of
 Technology
Cambridge, MA 02139, USA
Chapter 4

Puttachat Khuntontong
Microfluidics Manufacturing
 Programme
Singapore Institute of
 Manufacturing Technology
71 Nanyang Drive
Singapore 638075
Chapter 15

Catherine Klapperich
Departments of Biomedical and
 Mechanical Engineering
Boston University, Boston
MA 02215, USA
Chapter 7

A. Rezzan Kose
Department of Electrical
 Engineering
Yale University
15 Prospect Street, BCT 507
New Haven, CT 06511, USA
Chapter 16

Hur Koser
Department of Electrical
 Engineering
Yale University
15 Prospect Street, BCT 507
New Haven, CT 06511, USA
Chapter 16

Haluk Külah
METU-MEMS Center *and*
Department of Electrical and
 Electronics Engineering
Middle East Technical University
 (METU)
Ankara, Turkey
Chapter 14

Tassaneewan Laksanasopin
Department of Biomedical
 Engineering
Columbia University
351 Engineering Terrace
1210 Amsterdam Ave
New York City, NY 10027, USA
Chapter 6

Sang-Hoon Lee
Department of Biomedical
 Engineering
College of Health Science
Korea University
Jeongneung-dong, Seongbuk-gu
Seoul, 136-703, Republic of Korea
Chapter 12

Oren Levy
Division of Biomedical Engineering
Department of Medicine
Center for Regenerative
 Therapeutics
Brigham and Women's Hospital
Harvard Medical School
Cambridge, MA 02139, USA,
Harvard Stem Cell Institute
Harvard University
Cambridge, MA 02138, USA, *and*
Harvard–MIT Division of Health
 Sciences and Technology
Massachusetts Institute of
 Technology
Cambridge, MA 02139, USA
Chapter 4

Robert Littrell
Department of Mechanical
 Engineering
2026 GGB (George G. Brown
 Laboratory)
University of Michigan
2350 Hayward Ave
Ann Arbor, MI 48109-2125, USA
Chapter 1

Chang Lu
Department of Chemical
 Engineering, Virginia Tech

Blacksburg, VA 24061, USA
Chapter 8

TianJian Lu
Biomedical Engineering and
  Biomechanics Center
Xi'an Jiaotong University
Xi'an 710049, China
Chapter 17

Zheng Yuan Luo
State Key Laboratory of Multiphase
  Flow in Power Engineering *and*
Biomedical Engineering and
  Biomechanics Center
Xi'an Jiaotong University
Xi'an 710049, China
Chapter 17

Madhumita Mahalanabis
Departments of Biomedical and
  Mechanical Engineering
Boston University, Boston
MA 02215, USA
Chapter 7

Abhijit Majumder
Institute for Stem Cell Biology and
  Regenerative Medicine (inStem)
Bangalore 560065, India
Division of Biomedical Engineering
Department of Medicine
Center for Regenerative
  Therapeutics
Brigham and Women's Hospital
Harvard Medical School
Cambridge, MA 02139, USA,
Harvard Stem Cell Institute
Harvard University
Cambridge, MA 02138, USA,
and

Harvard–MIT Division of Health
  Sciences and Technology
Massachusetts Institute of
  Technology
Cambridge, MA 02139, USA
Chapter 4

Ufuk Olgac
Department of Mechanical
  Engineering
Koc University, Rumeli Feneri Yolu
34450 Sariyer, Istanbul, Turkey
Chapter 5

P. Hande Özdinler
Les Turner ALS Laboratory II
Department of Neurology,
Robert H. Lurie Comprehensive
  Cancer Center, *and*
Cognitive Neurology and
  Alzheimer's Disease Center
Feinberg School of Medicine
Northwestern University
Chicago, IL 60611, USA
Chapter 9

Sruthi Ravindranathan
Biomedical Engineering Program
College of Engineering
203 Engineering Hall
University of Arkansas
Fayetteville, AR 72701, USA
Chapter 3

Dusan Sarenac
Demirci Bio-Acoustic-MEMS in
  Medicine (BAMM) Laboratory
Division of Biomedical Engineering
Department of Medicine
Brigham and Women's Hospital
Harvard Medical School

Boston, MA 02139, USA
Chapter 18

Erkin Şeker
Department of Electrical and
   Computer Engineering
University of California, Davis
Davis, CA 95616, USA
Chapter 10

Samuel K. Sia
Department of Biomedical
   Engineering
Columbia University
351 Engineering Terrace
1210 Amsterdam Ave
New York City, NY 10027, USA
Chapter 6

Elodie Sollier
Department of Bioengineering
Henry Samueli School of
   Engineering and Applied Science
University of California
   Los Angeles
Los Angeles, CA 90095, USA
Chapter 13

Aishwarya Vasudevan
Demirci Bio-Acoustic-MEMS in
   Medicine (BAMM) Laboratory
Division of Biomedical Engineering
Department of Medicine
Brigham and Women's Hospital
Harvard Medical School
Boston, MA 02139, USA
Chapter 18

Jithesh V. Veetil
Section of Cellular Biophotonics
National Institute of Health
5625 Fishers Lane, Bethesda
MD 20892, USA *and*
Biomedical Engineering Program
College of Engineering
203 Engineering Hall
University of Arkansas
Fayetteville, AR 72701
USA
Chapter 3

ShuQi Wang
Demirci Bio-Acoustic-MEMS in
   Medicine (BAMM) Laboratory
Division of Biomedical Engineering
Department of Medicine
Brigham and Women's Hospital
Harvard Medical School
Boston, MA 02139, USA
Chapters 17 and 18

Zhiping Wang
Microfluidics Manufacturing
   Programme
Singapore Institute of
   Manufacturing Technology
71 Nanyang Drive
Singapore 638075
Chapter 15

Robert D. White
Department of Mechanical
   Engineering
301 Anderson Hall
Tufts University
Medford, MA 02155, USA *and*

Department of Mechanical
  Engineering
2026 GGB (George G. Brown
  Laboratory)
University of Michigan
2350 Hayward Ave
Ann Arbor, MI 48109-2125
USA
Chapter 1

Sau Fung Wong
Department of Biomedical
  Engineering
College of Health Science
Korea University
Jeongneung-dong, Seongbuk-gu
Seoul, 136-703, Republic of Korea
Chapter 12

Feng Xu
Biomedical Engineering and
  Biomechanics Center *and*
The Key Laboratory of Biomedical
  Information Engineering of
  Ministry of Education
School of Life Science and
  Technology
Xi'an Jiaotong University
Xi'an 710049, China
Chapter 17

Martin L. Yarmush
Center for Engineering in Medicine
Massachusetts General Hospital
Harvard Medical School and Shriners
  Hospital for Children
51 Blossom Street, Boston
MA 02114, USA *and*
Department of Biomedical
Engineering, Rutgers University
Piscataway, NJ 08854, USA
Chapter 10

Kaiming Ye
Biomedical Engineering Program
College of Engineering
203 Engineering Hall
University of Arkansas
Fayetteville, AR 72701, USA
Chapter 3

Gürkan Yilmaz
METU-MEMS Center *and*
Department of Electrical and
  Electronics Engineering
Middle East Technical University
  (METU)
Ankara, Turkey
Chapter 14

Stefan Zappe
Department of Biomedical
  Engineering
Carnegie Mellon University
Pittsburgh, PA 15213, USA
Chapter 11

Yihong Zhan
Birck Nanotechnology Center
Purdue University
West Lafayette, IN 47907, USA
Chapter 8

Jane Y. Zhang
Departments of Biomedical and
  Mechanical Engineering
Boston University, Boston
MA 02215, USA
Chapter 7

Chapter 1

A Microscale Bioinspired Cochlear-like Sensor

Robert D. White[*,‡], *Robert Littrell*[†], *and Karl Grosh*[†]

*Department of Mechanical Engineering, Tufts University
301 Anderson Hall, Medford, MA 02155, USA
†Department of Mechanical Engineering, University of Michigan,
2026 GGB (George G. Brown Laboratory)
2350 Hayward Ave, Ann Arbor, MI 48109-2125, USA

Introduction

Microfluidic systems have often been applied to human health directly in an attempt to deliver new diagnostic tools or treatment methods. In this chapter, we explore an alternative paradigm for the use of microscale systems in relation to biology and human health. Rather than attempting to create either diagnostic systems or treatment methods, we explore the interface between the auditory system and engineering microsystems as an example of two creative processes in sensor design and biomechanical modeling:

(1) *Drawing inspiration from biology in the design of new sensing and analysis systems.* In this creative process, termed biomimetic design, we draw inspiration from nature to design complex, high-performance engineered systems using natural architectures. These architectures may be innovative in comparison with conventional designs. It is certainly not true that biomimicry will always lead to better designs, but it does serve as an inspiration for novel systems that can be evaluated in comparison with existing designs.

[‡]Corresponding author. Email: r.white@tufts.edu

(2) *Using physical models to study electromechanical processes occurring in biological systems.* In this creative process, rather than relying entirely on mathematical models (with their many simplifications) and animal experiments (with their inherent variability, lack of control, and difficulty), we can take an intermediate step to something that is physically accurate (in particular, it is always three-dimensional and made of real materials), but can still be simplified and controlled in ways that animals cannot be.

The choice of the auditory system as a model case for these two processes is motivated both by the high performance of the system as a sensor, and also by the complex biomechanical processes that occur in the organ. We will focus on the cochlea, the organ in the inner ear which is responsible for transducing acoustical signals to neural signals. The mammalian cochlea is a marvel of micro-electro-mechano-fluidic design. At threshold, the sensory cells detect sub-nanometer displacements.[1] The mechanics of the fluid–structure interaction within the cochlea create a sensitive real-time frequency analyzer, delivering approximately 3,000 channels of frequency information. The human cochlea operates over a three-decade band in frequency, 120 dB of dynamic range, and can distinguish tones which differ by less than 0.5%. These characteristics are superb, particularly when one considers that the cochlea operates on 10–100 µW of power[2,3] and takes up only 1 cm^3 of volume.[4] Can we build an engineered version? There are compelling reasons to try. First, an engineered device that achieves the remarkable response characteristics of the cochlea would provide analog signal processing and sensitivity to acoustic fluctuations not yet possible. Second, building an artificial cochlea provides insight into the operation of the biological cochlea by enabling the testing of different hypotheses of cochlear function through hardware implementation.

Researchers in cochlear mechanics have produced a number of physical models of the cochlea, early versions of which were of an exaggerated size (e.g., Refs. 5 and 6). Lechner's work[6] is notable as it includes an active element for force feedback. In recent years, a number of life-sized, micromachined cochlear models have been reported.[7–10] These devices demonstrated the ability to achieve passive cochlear-like acoustic filtering in a life-sized engineered cochlea, incorporating both the variable impedance structure coupled with a heavy fluid (with a density similar to water), as in the cochlea. Cochlear mimicry has also inspired the design of micromachines, spatially distributed sensors with variable resonant frequencies configured in a xylophone-like arrangement.[11–13] However, these designs did not include any heavy fluid loading in their composition (only air loading). In this chapter, we describe the current progress of our group on the development of a fully micromachined cochlear analog transducer that incorporates both integrated sensing elements and heavy fluid loading. We used a capacitive sensing scheme, while Chen et al.[7] and Shintaku et al.[13] used laminated piezoelectric-based sensing (using PVDF). The current version of the cochlear-like sensor system has gone through a number of stages which are described in previous publications by the authors.[14–17] We refer to our device as a micro-cochlear analog transducer (μCAT).

What is the fundamental innovation of a cochlear-like sensor design? There are two ways to approach spectral analysis. One is to design a single sensor with a broad bandwidth and then take the single output signal and determine the spectral content in post-processing. The drawback to this scheme is that the sensor must be very broadband. An alternative approach is to use a group of sensors, each sensitive to a particular band. These sensors would each be narrowband, making them easier to design and producing lower noise floors and higher sensitivity. However, the large number of sensors would take up a large area. Most of the acoustic energy impinging on that area would be wasted; it would be reflected or passed by the out-of-band sensors. The cochlea solves this problem in a unique way. An array of narrowband sensors is used, but the sensors are arrayed along a one-dimensional variable impedance waveguide. Energy entering the system through its relatively large input window passes down the waveguide, meeting each narrowband sensor in turn, with the highest frequencies sensed at the base near the input. The energy is absorbed by the sensor tuned to that band of sound. Hence, little energy is wasted; the acoustic energy is passed to the sensor tuned to optimally transduce at that band. This is this idea that we are trying to replicate here.

From a signal-to-noise perspective, it is desirable to extract as much energy as possible out of the external acoustic field to present to the sensors. This requires a good impedance match to the environment. A standard microphone presents a large impedance mismatch to the environment at most frequencies, reflecting almost all of the incoming energy. However, the mammalian auditory system matches the acoustic input impedance across a large frequency band, thereby extracting almost all of the energy out of the incoming signal. In order to achieve this with structural components, a continuous, variable mechanical impedance coupled to an acoustic waveguide is required. This is exactly what the basic mechanical structure of the mammalian cochlea achieves. By extracting almost all of the energy out of the incoming signal, the maximum possible signal-to-noise ratio is achieved, resulting in maximum possible sensitivity.

For maximum sensitivity, it is also critical that all of the energy extracted from the mechanical signal be converted into electrical energy. This requires the dominant damping mechanism in the device to be an active electrically driven damping mechanism, rather than mechanical damping from fluid viscosity or material losses. One way to achieve this is with local transducers and control loops that respond to structural vibration over some limited portions of the device. By presenting a force in phase with velocity, energy is removed from the mechanical domain and transduced into the electrical domain. This system is exactly analogous to the electromotile outer hair cell (OHC) mechanics that are present in the organ of Corti (OoC) in the animal. The local active feedback mechanism also gives the opportunity to inject a compressive nonlinearity into the system, reducing the sensitivity at high sound pressure levels. This is also crucial; since the transducer will have a high sensitivity, a compressive nonlinearity must be used to achieve a high dynamic range.

Finally, the structure of the system gives us mechanical frequency analysis. The incoming signal is naturally filtered into separate bands by the mechanical process, giving information about the spectral content of the incoming sound. We therefore envision the artificial cochlea to facilitate a number of analog signal processing tasks where time-frequency analysis is required. The cochlear-like transducer could be applied for structural health monitoring applications as a combined sensor/analysis system for acoustic emissions (AE) in the structure.[18,19] There is also a potential natural synergy between the biomimetic sensory design of the cochlear-like sensor and the biomimetic neural net analysis methods currently under investigation in the structural health monitoring community.[20,21] We also envision the possibility of using this device for either underwater or airborne acoustic sensing, where real-time time-frequency analysis is desirable, including speech processing and target recognition. It would be most applicable for situations, like unattended sensors, in which the spectral content of the acoustic signal must be analyzed in an ultra-low power process. Although other tasks where ultra-low power is not important, such as musical transcription, they may also be made more facile with a μCAT. Naturally, another compelling application would be to use the μCAT as a front end to a cochlear implant, or in the future, implant a miniaturized version of the device. Significant advances in packaging, power supplies, communication, signal processing, and biocompatibility are, however, required before such direct human health applications could be considered.

In addition, although we will not explore this topic in detail here, it is compelling to consider the use of the μCAT as an alternative means for studying questions related to the function of the biological organ. The μCAT might be a useful third tool that complements mathematical models and animal studies. Mathematical models of the hydromechanics and active elements in the biological cochlea abound and lay the groundwork for any understanding of the function of the cochlea. Yet, it is sometimes difficult to assess the effectiveness of competing models at predicting the mechanics of the biological structure. This is due to the difficulty of animal experiments, the complexity of the physiology, and the fundamental variability of biological systems from individual to individual. It can therefore be difficult to distinguish the relative merits of competing models and hard to apply those models to the development of clinical diagnostic tools. In particular, the tuning of model parameters is often carried out. This can lead to a situation in which it is unclear whether the "cochlear-like" phenomena that emerge are a result of a good mathematical representation of the physiology, or simply due to tuning parameters to achieve a desired effect.

A physical life-sized model of the cochlea would provide a well-controlled, vastly simplified, and easily experimentally accessible tool for comparing the validity of mathematical models. The system would retain three-dimensional fluid chambers, be constructed of real materials with well-known material properties, and cannot be manipulated in physically impossible ways (unlike some mathematical models!). There is, naturally, significant concern that the engineered model may not be relevant to biological questions as it does not include the full complexity of the organ. However, we argue that this is not true. In fact, in a simplified system, we can state with confidence that we understand *everything* that is happening and therefore can demonstrate clear cause-and-effect relationships. When the same phenomena are observed in the biological system, we now have a clear understanding of their source. If only biological studies are conducted, the complexity of the response of the organ (not to mention its fragility and variability) may make it very difficult to identify the cause and effect relationships. The simplified nature of the model is therefore its strength; the simplicity that makes it tractable allows us to use it to understand the vastly more complex biology in a modular fashion.

Physiology and Design

The design of the cochlear-like sensor is derived directly from the physiology of the ear. (For an extensive review of cochlear physiology, experimental results, and models, see Refs. 1, 4, and 22.) Briefly, the outer ear captures sound from the air and transmits it as structural vibrations through the tympanic membrane and the middle ear bones. The innermost bone in the middle ear is the stapes (Fig. 1), a stiff bone embedded in a flexible support structure that is in contact with the cochlear fluid. It acts as a rigid piston to drive acoustic energy into the cochlear ducts.

The cochlea itself is a spiral-shaped organ consisting mainly of three ducts filled with endolymph and perilymph, two ion-rich water-like fluids. Figure 1 shows two schematics of the structure. The three ducts — the scala tympani, scala media, and scala vestibuli — are separated by Reissner's membrane and the

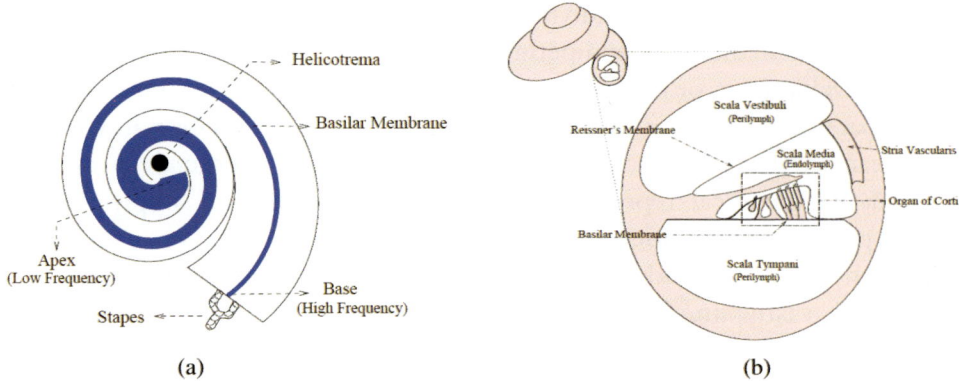

Fig. 1. Schematic of some of the major physiological features of the cochlea.
Source: Niranjan Deo, University of Michigan.

basilar membrane (BM), respectively. Acoustic energy is injected into the scala vestibuli via the stapes. A fluid–structure wave then travels down the length of the cochlear spiral, interacting with the microstructures in the OoC and the BM. The BM changes in width, thickness, and stiffness along the length of the cochlear spiral and is the major structural feature that gives rise to the spatial frequency selectivity map of the cochlea. Due to its changing acoustic impedance, high-frequency sounds excite amplified BM motion near the base of the cochlea, and low-frequency sounds excite motion near the apex. The sensory cells are housed in the OoC, which is perched atop the BM in the scala media.

The simplest analogy to the mechanical structure of the cochlea is a single, straight, fluid-filled duct (representing the fluid-filled scala) bounded along one side by a tapered membrane (representing the variable structural impedance BM and OoC). This mechanical structure forms the basis of the μCAT design. In addition, we incorporated capacitive sensing elements along the length of the membrane (representing the sensory inner hair cells). This was accomplished by using a conducting membrane (highly boron-doped polysilicon) and bonding on a glass die with patterned Cr/Pt electrodes to form a series of parallel plate capacitors along the length of the membrane. Each electrode is a separate channel of output, corresponding to vibration of the section of membrane under that electrode.

Figure 2 shows the dimensioned design. The length of the fluid-filled duct, 3.5 cm, is similar to the length of an uncoiled human cochlea (4 cm). The fluid chamber height, 0.475 mm, lies within the range of human scala height (0.1–2 mm). The volume compliance of the variable width membrane, 10^{-11}–10^{-14} m^4/N, is similar to the human BM (10^{-10}–10^{-14} m^4/N[23]).

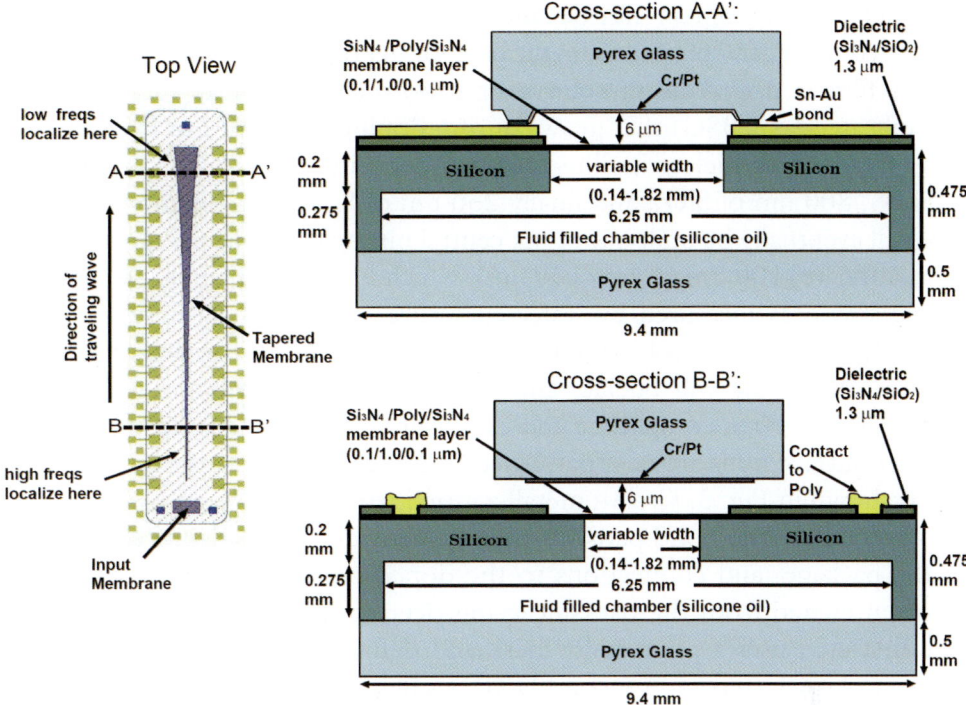

Fig. 2. Schematic of the μCAT design showing a top view (left) and two cross-sections (right). (Reprinted from Ref. 24.)

One notable difference between this device and the cochlea is the viscosity of the filling fluid. In this design, 200 cSt fluid silicone oil was used to add sufficient damping to prevent wave reflections. The endolymph and perilymph in the cochlea have viscosities similar to water (1 cSt). In addition, the width of the fluid duct, 6.25 mm, is greater than the physiological width (1–2 mm), in order to provide substantial frequency without varying the fluid duct geometry. It should also be noted that in the cochlea, the height and width of the fluid duct vary with longitudinal position down the spiral; in the μCAT, the duct height is constant due to fabrication limitations.

Microfabrication

The μCAT was fabricated at the University of Michigan's Lurie Nanofabrication Facility. The starting substrates were 100 mm in diameter, 475-μm thick, <100> oriented, p-type (boron) (1–10 ohm·cm) silicon wafers. An SiO_2 film, 2-μm thick, was grown by pyrogenic oxidation. Then, 100 nm of stoichiometric silicon nitride and 1.1 μm of low-stress polysilicon are deposited using low-pressure chemical vapor deposition (LPCVD). The polysilicon film is doped using solid source boron diffusion at 1,175°C. The borosilicate glass (BSG) that grows during doping is stripped in hydrofluoric acid (HF). Three dielectric films are then deposited by LPCVD — 100 nm of Si_3N_4, 800 nm of SiO_2, and finally 250 nm of Si_3N_4. The membrane structure will eventually be built out of the central nitride/polysilicon/nitride structural laminate (100 nm/1 μm/100 nm), which has approximately 40 MPa net tensile stress (as measured by wafer curvature) and has a sheet resistivity of 10–50 Ω/square. Control of the film stress is critical for device design. A compressive membrane will buckle, and a highly tensile membrane will exhibit much lower volume compliance and drastically shift device bandwidth.

The surface films are next patterned by reactive ion etching and wet etching as shown in Fig. 3. Cr/Au metallization is sputtered on and patterned via liftoff to define the bond pads for the top glass die, the pads for connection to the package, and connections to the doped polysilicon layer. The wafers are then etched from the backside using deep reactive ion etching (DRIE). The first etch uses a photoresist mask and defines the membrane shapes, as shown in Fig. 3, step 3a. A second DRIE etch is performed using the backside oxide as a hardmask, and stopping on the buried oxide etch stop. This etch

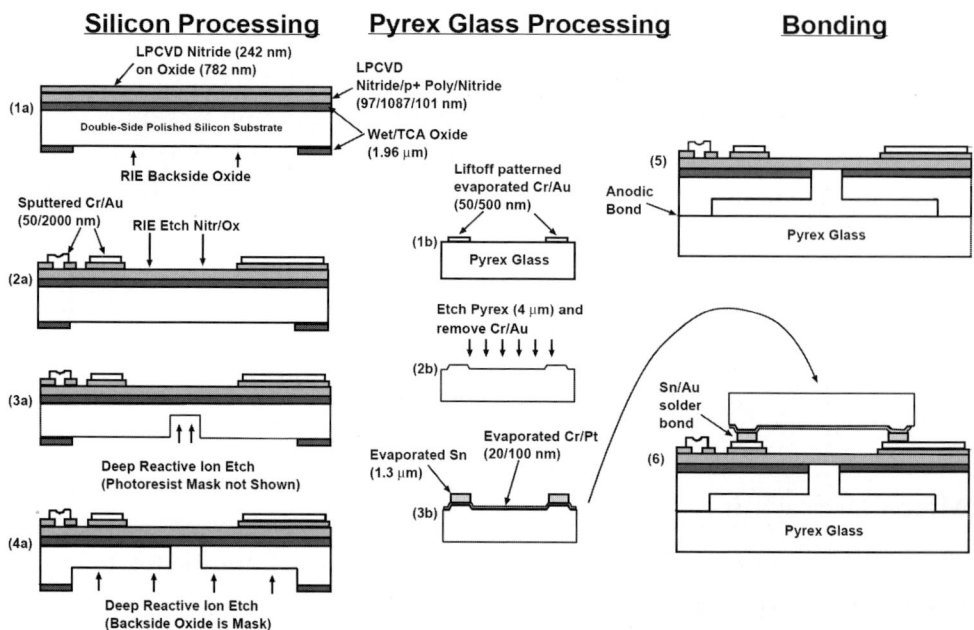

Fig. 3. Diagram of the microfabrication process. (Reprinted from Ref. 24.)

defined the fluid chamber shape. The situation at this point is seen in Fig. 3, step 4a. The buried oxide was then removed from the membrane regions in 1:1 HF, releasing the membranes.

In parallel with silicon processing, a Pyrex glass wafer (Corning type 7740), was processed to produce the top electrodes for capacitive sensing. First, 4-µm high legs were etched into the glass using 3:1 HF and an evaporated Cr/Au mask. Evaporated Cr/Pt electrodes were then evaporated on and patterned using liftoff, as shown in Fig. 3, step 2b. Sn bumps were evaporated on and patterned using liftoff. Diced Pyrex glass pieces were anodically bonded onto the backside of the silicon at 330°C and 700 V, sealing the fluid chambers. The Pyrex top pieces were finally bonded on using Sn–Au fluxless solder bonding at 350°C with 15 MPa applied clamping pressure. This completed the structure, making it ready to be packaged. Needles were epoxied into the filling ports and silicone oil was injected using a microinjection jig. The needles were cut and sealed off with epoxy. The finished chip was then mounted into a hybrid leadless ceramic chip carrier (LCCC) package with double-sided tape and wire bonded to make an electrical connection with Au wire bonds. Photographs of the finished device are shown in Figs. 4 and 5.

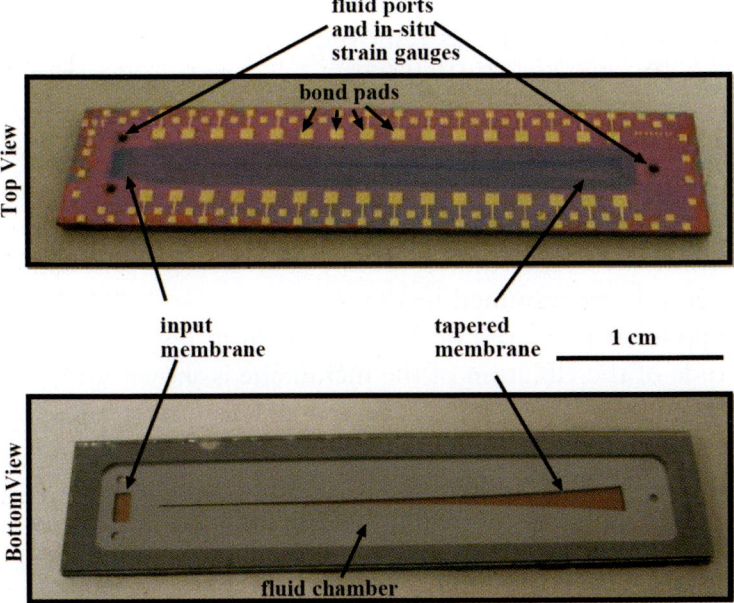

Fig. 4. Photographs of the device after anodic bonding, but before bonding on the top electrode plate (that is, after step 5 in Fig. 3). (Reprinted from Ref. 24.)

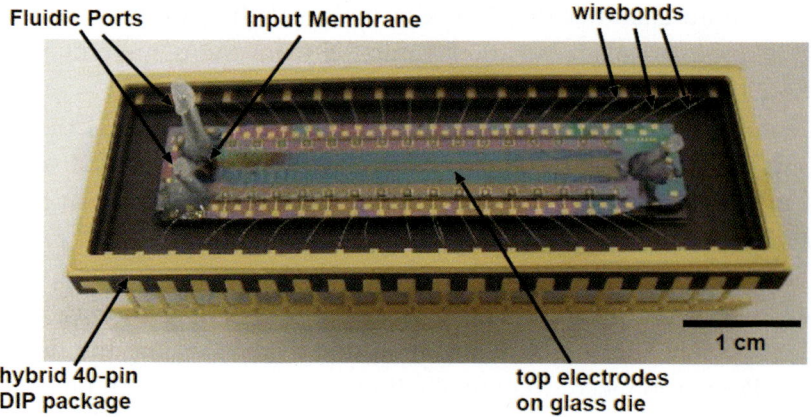

Fig. 5. Photograph of the device with top glass plate bonded on, filled with fluid and sealed, and wirebonded into a DIP hybrid package. (Reprinted from Ref. 24.)

Experimental Results

Measurement of the mechanical vibration of the tapered membrane structure in the μCAT was carried out using a laser Doppler velocimetry (LDV) system. The LDV system was composed of a Polytec OFV-301 sensor head, custom optics, and a computer-controlled micropositioning stage. The top plate of the μCAT was removed, 10/50 nm of Cr/Au was sputtered on for reflectivity, and the chip was mounted on the stage with rubberized clamps. An acoustically baffled piezoelectric tweeter was placed over the front end of the chip, delivering approximately 100 dB of SPL pure tone acoustic excitation in the 10–70 kHz band, where the device is most sensitive. A Larson–Davis 0.25" pressure microphone was used as a reference inside the baffle to measure the strength and phase of the excitation. The setup is diagrammed in Fig. 6.

The response as measured by LDV is shown in Figs. 7 and 8. In Fig. 7, the magnitude of the vibration of the membrane is shown for eight different driving frequencies. In Fig. 8, the centerline magnitude and phase of the membrane vibration are shown. In the 10–70-kHz band, maximum membrane vibration is observed at a frequency-dependent location. This frequency–position mapping is the primary cochlear-like phenomenon we hoped to reproduce. Displacements are on the order of 0.2 nm/Pa. The phase shows that traveling waves dominate, another cochlear-like feature. Phase accumulations in the 30-mm length of the membrane are between 10 and 40 radians.

Commercial off-the-shelf components were used to construct the signal conditioning electronics for the μCAT. The electronics were implemented on

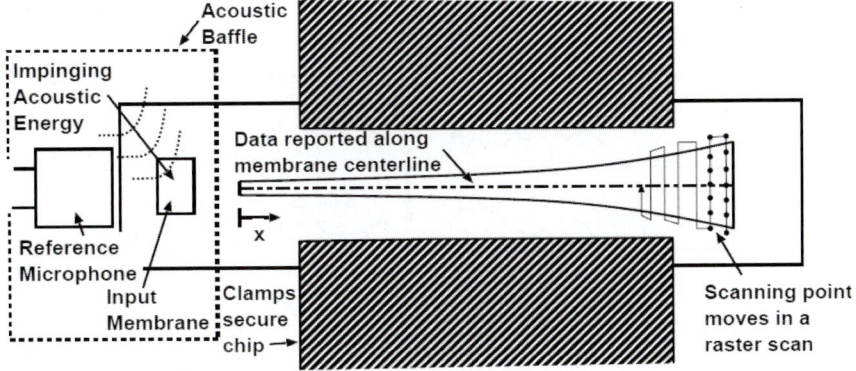

Fig. 6. Diagram of mechanical measurement setup. (Reprinted from Ref. 24.)

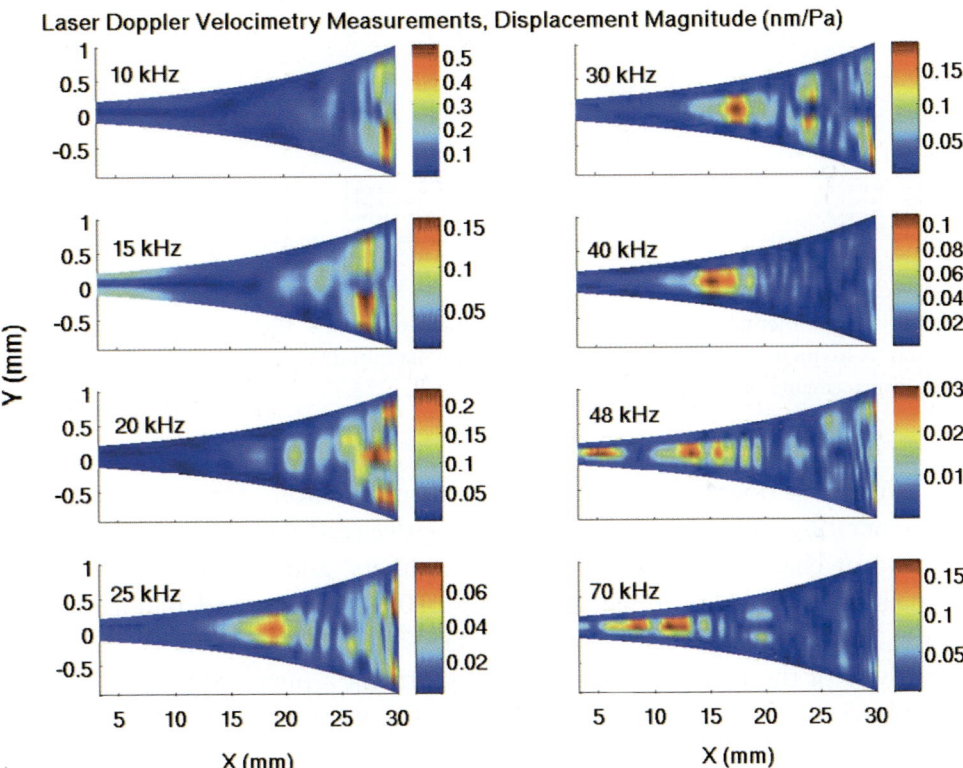

Fig. 7. LDV experimental results showing steady state membrane vibration at eight driving frequencies. A frequency–position map is evident in the magnitude results in the 10–70-kHz band with displacements on the order of 0.2 nm/Pa.

12 *Microfluidic Technologies for Human Health*

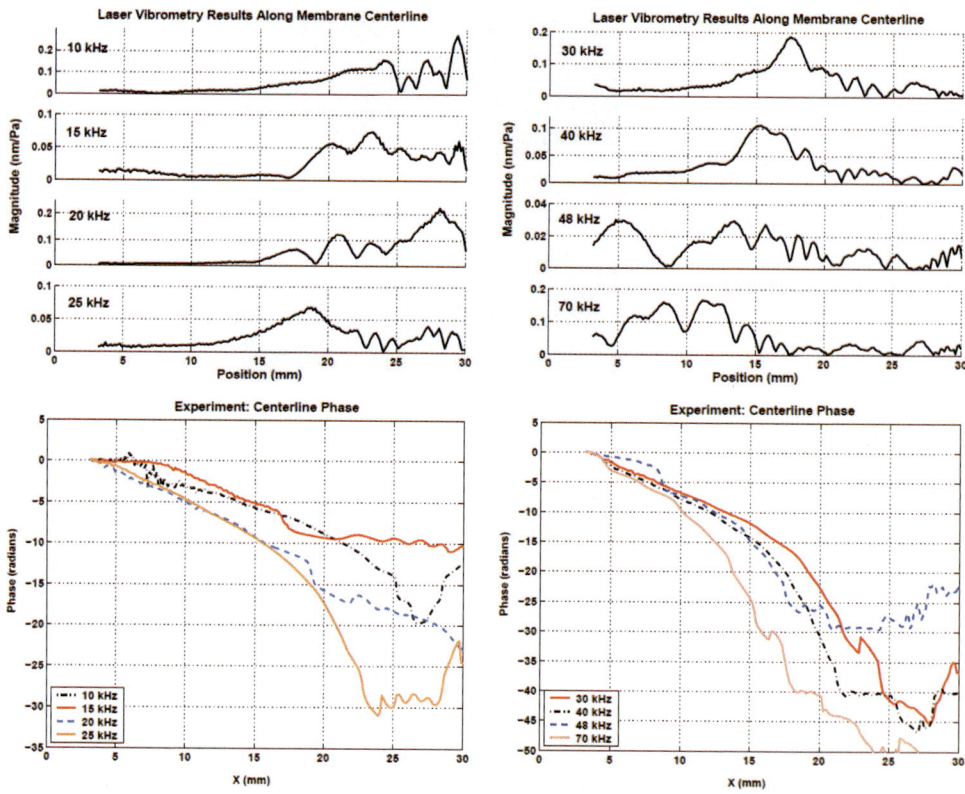

Fig. 8. Experimental results along the membrane centerline after subtracting outer edge motion. A frequency–position map is evident in the magnitude results in the 10–70-kHz band with displacements of 0.2 nm/Pa. (Reprinted from Ref. 24.)

a custom PC board. There are three main functions for the electronics: (1) generating a bias reference for the capacitive readout, (2) multiplexing the 32 output channels to one of two preamplifiers, and (3) preamplifying the capacitive signal using a charge amp (at 100 mV/pC) and bandpass filter (with 40 dB of passband gain, 70 Hz–70 kHz bandwidth). Electrical sensitivity of the µCAT was measured by driving a pure acoustic tone using the piezoelectric tweeter and measuring the output from all 32 channels. The measurements were conducted for 9 V applied bias and for 0 V applied bias. With 0 V bias, any measured output should be electrical cross-talk from the speaker. Typical results are shown in Fig. 9 for four driving frequencies. The measured sensitivities of the low frequency channels are 10–50 mV/Pa at the band pass filter output (0.1–0.5 mV/Pa at the preamp output). This is 100 times higher sensitivity than that predicted by mathematical models of the system. Response is only seen from the channels at the wide end of the device. Very low sensitivity is seen for the narrow end channels, at all frequencies tested (tests were conducted from 2–70 kHz). This also does not match model predictions or expectations based on the frequency–position map observed in LDV experiments.

The sensitivity reduces clearly on all channels when the bias is set to zero, indicating that the measurements are recording true sensitivity to sound, and not simply electrical cross-talk from the speaker drive. The circuitry used to achieve the multiplexing is known not to have perfect "off" isolation. Such a lack of isolation (cross-talk, or "ghosting" as it is sometimes called) will give rise to the sort of contamination seen in the results. With a different scheme, this can be overcome. It is not known why the pattern of vibration measured using capacitive sensing is very different from the model predictions. It seems likely that this is mainly due to cross-talk in the multiplexing electronics, but it could also be due to poor performance of the acoustic baffle, resulting in

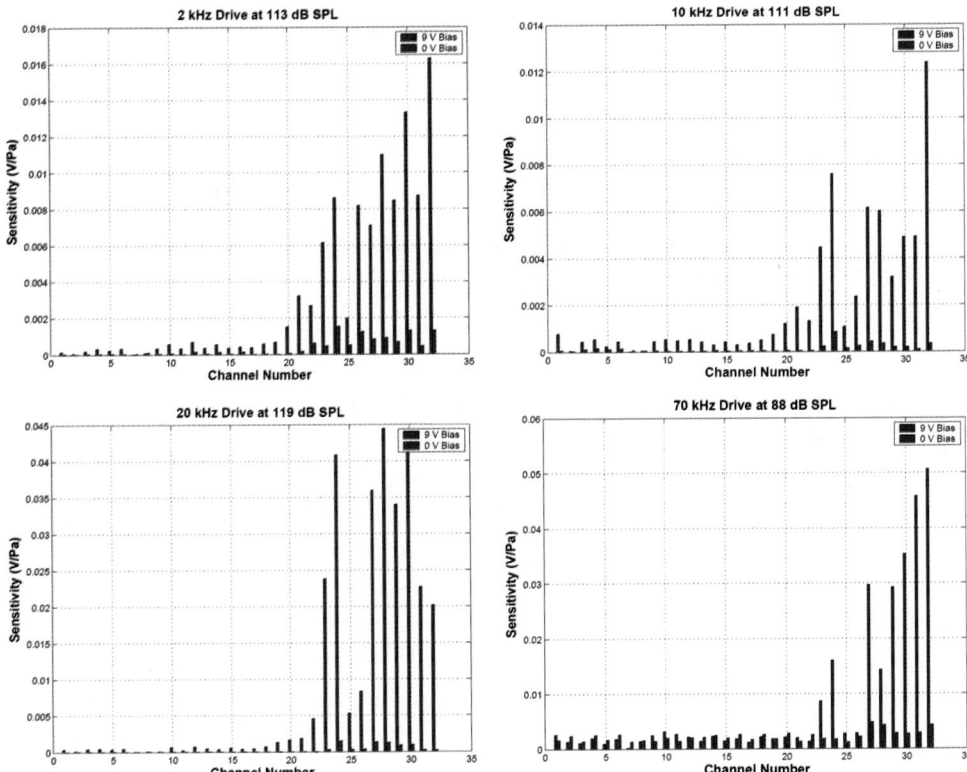

Fig. 9. Electrical sensitivity of the 32 channels to four pure tone acoustic signals, at two applied bias levels (9 V and 0 V). (Reprinted from Ref. 24.)

acoustic pressure directly forcing the tapered membrane rather then entering primarily through the input membrane as desired. Additional work on this respect is needed.

Mathematical Modeling

In our design process, mathematical modeling is the first step and upon which all fabrication is based. Furthermore, as there are some uncertainty and variability in the as-deposited material properties (especially the residual stress in each layer), modeling is a crucial post-processing step to interpret the experimental results. In this chapter, we will only summarize modeling efforts of our group and others for the cochlear analog transducer design, as these efforts have been described elsewhere and adequate treatment of this topic would require more space than possible in this chapter. Analytic mathematical approximations using the WKB asymptotic expansions have been co-opted from cochlear mechanics research and applied to cochlear analog design.[13,26] We too have used these WKB techniques as a way to quickly develop a design for the cochlear analog device.[14] Indeed, we recommend this semi-analytic method as an initial approach to sizing for the structural dimensions for microfabrication and channel dimensions for microfluidics. For designs where the fluid duct height is small or comparable to the viscous boundary layer (e.g., on the order of one to two hundred microns), we used another efficient technique for design[15] by applying an approximation from Beltman[27] in order to develop a two-dimensional finite element solution for the problem which approximates the fluid pressure loading and viscosity. In our studies, including the embodiment shown in Fig. 2, the geometry does not conform to the simplifying approximations of a WKB formulation or Beltman's approach. While those approximations can be used to obtain a "ballpark" design, a full three-dimensional finite element analysis is needed to make accurate predictions. Because of the numerical challenges associated with the locking seen in finite element plate formulations and the spurious modes in the fluid pressure elements for nearly incompressible fluids, we have developed a three-dimensional finite element approximation for viscous fluid–structure interaction seen in the micromachined cochlear analog devices[28] to overcome these problems.

In all three of these modeling schemes,[14,15,28] the geometry comprises a three-dimensional fluid domain interacting with a two-dimensional structural domain. The structural domain represents the flexible BM, which may be modeled as a pre-tensioned thin bending plate. The structure is very thin and can thus be modeled successfully as a two-dimensional orthotropic pretensioned Kirchhoff plate:

$$T_x \frac{\partial^2}{\partial x^2} u + T_y \frac{\partial^2}{\partial y^2} u - D_{xx} \frac{\partial^4}{\partial x^4} u$$
$$-2D_{xy} \frac{\partial^4}{\partial x^2 y^2} u - D_{yy} \frac{\partial^4}{\partial y^4} u + \omega^2 m_a u = P, \quad (1)$$

where T_x and T_y are the tensions in the two orthogonal directions, m_a is the structure mass per unit area, P is the complex pressure applied to the structure from the fluid domain, and u is the complex normal displacement of the structure. Both P and u are functions of the two Cartesian spatial dimensions in the plane of the structure, x and y. They are both assumed to take on a time dependence $e^{j\omega t}$. In all cases, pure harmonic motion is assumed; this is the steady state response.

The three orthotropic plate flexural rigidities are

$$D_{xx} = \frac{E_x h^3}{12(1 - \nu_{xy}\nu_{yx})}, \quad (2)$$

$$D_{yy} = \frac{E_y h^3}{12(1 - \nu_{xy}\nu_{yx})}, \quad \text{and} \quad (3)$$

$$D_{xy} = \frac{E_x h^3 \nu_{yx}}{12(1 - \nu_{xy}\nu_{yx})} + \frac{1}{6} G_{xy} h^3, \quad (4)$$

where E_x and E_y are the elastic moduli in the two Cartesian directions, G_{xy} is the in-plane shear modulus, ν_{xy} and ν_{yx} are the two Poisson ratios, which must satisfy $\nu_{xy}/\nu_{yx} = E_x/E_y$, and h is the structure thickness.

Note that there are four independent parameters for such a structure. In the case of an isotropic plate,

$$E = E_x = E_y, \quad \nu = \nu_{xy} = \nu_{yx} \quad \text{and} \quad G = E/(2(1+\nu)). \quad (5) \text{ and } (6)$$

This structural model can be applied to data from animal experiments as well as to engineered cochlear models. For example, Liu and White[29] analyzed probe deflection data for gerbil BM structures and arrive at orthotropic plate moduli estimates for the biological membrane. In this work, D_{yy} the flexural rigidity for the transverse direction, was found to vary between 5×10^{-9} and 7×10^{-10} Nm while moving from the base to the apex of the cochlea. The BM was approximately 200-µm wide. D_x and D_{xy} were between one and two orders of magnitude smaller than D_y. This indicates that the gerbil BM is mildly orthotropic, with orthotropy ratios (D_y/D_x) of between 10 and 65. This justifies the use of a mildly orthotropic structural model, rather than a strongly othotropic model or isotropic model.

In the fluid domain, fluid displacements are small, viscosity is important, the fluid is modeled as Newtonian, and the fluid is slightly compressible. Again, all displacements and pressures are assumed to be time harmonic as $e^{j\omega t}$. This leads to a linearized Navier–Stokes equation:

$$-\rho_f \omega^2 \bar{u} = -\nabla P + j\omega \mu \nabla^2 \bar{u} + \frac{1}{3} j\omega \mu (\nabla \cdot \bar{u}), \qquad (7)$$

where μ is the fluid dynamic viscosity, ρ_f is the fluid density, P is the complex pressure in the fluid domain, and $\bar{u}$ is the complex fluid displacement, a vector quantity. The coupling between the fluid and structural domains is easily handled. The pressure at the structure surface becomes the forcing pressure on the structural model, and the displacement of the structure becomes a displacement boundary condition on the fluid. In this manner, a fully coupled fluid–structure interaction problem is produced.

The solution of this coupled problem can be handled relatively easily for rectangular duct geometries, particularly if the height of the fluid domain is small compared to the viscous boundary layer and the wavelength of structure oscillations. In these situations, a two-dimensional formulation, as described by White and Grosh[15] and Beltman et al.,[27] is computationally efficient and produces accurate results.[15,18,27] However, the geometry shown here is fully three-dimensional in the fluid domain, which therefore requires a three-dimensional solution method. A mixed formulation finite element method has been produced to solve this problem, as described by Cheng et al.[28] A quantitative match between measured and predicted membrane vibration patterns, both magnitude and phase, have been achieved for predominantly two-dimensional geometries using thin film models.[15] Some discrepancies, mainly related to the predicted frequency response, remain to be resolved for the three-dimensional structure presented here.

Conclusions

The microscale cochlear analog transducer (μCAT) described here demonstrates the feasibility of a life-sized acoustic sensor, which mimics the dynamics of the cochlea. This is an innovative approach to acoustic sensing, differing significantly from the traditional designs for hydrophones and microphones.

Laser vibrometry measurements of µCAT dynamics demonstrated frequency–position mapping over the 10–100-kHz band with displacements on the order of 0.2 nm/Pa. Sensitivity measurements on the µCAT indicate electrical sensitivity of the lower frequency channels (close to the wide end of the device) of 0.1–0.5 mV/Pa (at the preamp output).

There is still work to be done on this topic. At this point for our device, no consistent frequency–position mapping has been demonstrated in the electrical measurements (Fig. 8) even though the structural vibrations measured using LDV has shown such a mapping. A lack of isolation between the channels in the multiplexed circuitry is the most likely culprit. Another possibility is that the top Pyrex layer is not a sufficient acoustic baffle, hence incoming sound pressure is directly exciting the motion of the tapered membrane, rather than only directly impinging on the device input. In other researchers' attempts at direct electrical sensing in an artificial cochlea when using heavy fluid loading in the microfluidic duct (e.g., Refs. 7 and 13), the noisefloor of the µCAT is dominated by the noise of the preamplifier circuit, mainly the Johnson noise of the DC-stabilizing feedback resistor and the voltage noise of the operational amplifier. Improving the noise characteristics of the device would require improvements to this preamplifier stage.

Additional work needs to be done on improving the sensor capabilities, particularly electrical sensing. Our recent analysis of the system indicates that a piezoelectric sensing scheme could both facilitate microfabrication (e.g., eliminating the need for a controlled small gap for capacitive sensing) and reduce the overall noise in the system. Other recent work indicates that a better design of the multiplexing circuitry can reduce channel to channel crosstalk to less than −80 dB. Further work on packaging is also needed in order to make the transition from a laboratory experiment to a more robust measurement device. Isolation of the BM from environmental disturbances will be an important part of this work. In addition, a "middle ear" analog is needed to couple the µCAT more efficiently with the external acoustic environment.

Finally, to achieve a truly cochlear-like filter, active feedback mechanisms will need to be added to the system to mimic the operation of OHC mechanics. In animal experiments, the active cochlea exhibits higher amplitudes (by at least an order of magnitude) and much better tuning than the passive cochlea. Neurological measurements of tuning curves at different sound pressure levels also show a clear improvement in frequency selective at low sound pressure levels where the active elements are most effective. Finally, the active elements produce a compressive nonlinearity, whereby the cochlea is more sensitive at low sound pressure levels than at high sound levels, allowing for large dynamic range. For all of these reasons, distributed active feedback elements should be explored in future versions of the system.

Acknowledgment

Portions of this article appeared in Ref. 24 and appear here with permission. We also acknowledge funding from the Office of Naval Research — Code 332 (KG and RDW), the National Science Foundation (KG, RJL, and RDW), U.S. Department of Homeland Security (RJL), and the University of Michigan.

References

1. Robles, L., Ruggero, M.A., Mechanics of the mammalian cochlea. *Physiol Rev*, 2001, **81**: p. 1305–1352.
2. Johnstone, B.M., Genesis of the cochlear endolymphatic potential. *Curr Topics Bioenergy*, 1967, **2**: p. 335–352.
3. Sarpeshkar, R., Lyon, R.F., Mead, C., A low-power wide-dynamic-range analog VLSI cochlea. *Analog Integr Circuits Signal Proc*, 1998, **16**(3): p. 245–274.
4. Dallos, P., Popper, A., Fay, R., *The Cochlea. Springer Handbook of Auditory Research*, Vol. 8. Springer, New York, 1996.
5. Chadwick, R.S., Fourney, M.E., Neiswander, P., Modes and waves in a cochlear model. *Hear Res*, 1980, **2**: p. 475–483.
6. Lechner, T.P., A hydromechanical model of the cochlea with nonlinear feedback using PVF_2 bending transducers. *Hear Res*, 1993, **66**: p. 202–212.
7. Chen, F.Y., Cohen, H.I., Bifano, T.G., Castle, J., Fortin, J., Kapusta, C., Mountain, D.C., Zosuls, A., Hubbard, A.E., A hydromechanical biomimetic cochlea: Experiments and models. *J Acoust Soc Am*, 2006, **119**(1): p. 394–405.
8. Hemmert, W.A. Life-sized hydrodynamical, micromechanical inner ear. In: Gummer, A.W., ed. *Biophysics of the Cochlea: From Molecules to Models*, World Scientific, Singapore, 2002, p. 409–416.
9. Wittbrodt, M.J., Steele, C.R., Puria, S., Developing a physical model of the cochlea using microfabrication methods. *Audiol Neuro-otol*, 2006, **11**: p. 104–112.
10. Zhou, G., Bintz, L., Anderson, D.Z., A life-sized physical model of the human cochlea with optical holographic readout. *J Acoust Soc Am*, 1993, **93**: p. 1516–1523.
11. Tanaka, K., Abe, M., Ando, S., Novel mechanical cochlea fishbone with dual sensor/actuator characteristics. *IEEE/ASME Trans Mechatron*, 1998, **3**: pp. 98–105.
12. Haronian, D., MacDonald, N.C., A microelectromechanics-based frequency-signature sensor, *Sens Actuators A*, 1996, **53**, pp. 288–298.
13. Shintaku, H., Nakagawa, T., Kitagawa, D., Tanujaya, H., Kawano, S., Ito, J., Development of piezoelectric sensor with frequency selectivity for artificial cochlea, *Sens Actuators A*, **158**: p. 183–192.
14. White, R.D., *Biomimetic Trapped-fluid Microsystems for Acoustic Sensing*, in *Mechanical Engineering*. University of Michigan, Ann Arbor, 2005.

15. White, R.D., Grosh, K., Microengineered hydromechanical cochlear model. *Proc Natl Acad Sci*, 2005, **102**: p. 1296–1301.
16. White, R.D., Grosh, K., Trapped-fluid traveling wave filters based on the mammalian cochlea. *Proc MicroTAS 2005, 9th Int Conf Miniaturized Syst Chem Life Sci*, 2005.
17. White, R.D., Grosh, K., Fully micromachined lifesize cochlear model. In Nuttall, A.L., ed. *Auditory Mechanisms: Processes and Models*, World Scientific, Singapore, 2005.
18. Finlayson, R.D., Friesel, M., Carlos, M., Cole, P., Lenain, J.C., Health monitoring of aerospace structures with acoustic emission and acousto-ultrasonics. *Insight*, 2001, **43**(3): p. 155–158.
19. Rizzo, P., di Scalea, F.L., Acoustic emission monitoring of carbon-fiber-reinforced-polymer bridge stay cables in large-scale testing. *Exp Mech*, 2001, **41**(3): p. 282–290.
20. Martin, W.N. et al., An artificial neural receptor system for structural health monitoring. *Int J Struct Health Monitoring*, 2005, **4**(3): p. 229–245.
21. Kirikera, G.R. et al., Damage localisation in composite and metallic structures using a structural neural system and simulated acoustic emissions. *Mech Syst Signal Proc*, 2007, **21**(1): p. 280–297.
22. Geisler, C.D., *From Sound to Synapse*, Oxford University Press, 1998.
23. Von Békésy, G., *Experiments in Hearing*, McGraw Hill, Oxford, 1960.
24. White, R.D., Littrell, R., Grosh, K., *Copying the Cochlea: Micromachined Biomimetic Acoustic Sensors*, in *Structural Health Monitoring* 2007 Quantification, Validation, and Implementation, *Proc 6th Int Workshop on Structural Health Monitoring*. DEStech Publications, Inc., Lancaster, PA, 2007.
25. Xu, T., Bachman, M., Zeng, F.G., Li, G.P., Polymeric micro-cantilever array for auditory front-end processing, *Sens Actuators A*, 2004, **114**(2–3): p. 176–182.
26. Lim, K.M., Fitzgerald, A.M., Steele, C.R., Building a physical cochlear model on a silicon chip. In: Wada H., Takasaka T., Ikeda K., Ohyama K., Koike T., eds. *Recent Developments in Auditory Mechanics*, World Scientific, Singapore, 1999, p. 223–229.
27. Beltman, W.M., van der Hoogt, P.J.M., Spiering, R.M.E.J., Tijdeman, H., Implementation and experimental validation of a new viscothermal acoustic finite element for acousto-elastic problems, *J Sound Vib*, 1998, **216**(1): p. 159–185.
28. Cheng, L., White, R.D., Grosh, K., Three-dimensional viscous finite element formulation for acoustic fluid-structure interaction. *Comput Methods Appl Mech Eng*, 2008, **197**(49–50): p. 4160–4172.
29. Liu, S., White, R.D., Orthotropic material properties of the gerbil basilar membrane. *J Acoust Soc Am*, 2008, **123**(4): p. 2160–2171.

Chapter 2

Systematic Evaluation of the Efficiencies of Proteins and Chemicals in Pharmaceutical Applications

*Morgan Hamon and Jong Wook Hong**

*Materials Research and Eduction Center
Department of Mechanical Engineering
Auburn University, 275 Wilmore Lab
Auburn, AL 36849, USA*

Introduction

The recent completion of human genome sequencing[1] coupled with the progress in genomics[2,3] and proteomics[4,5] has dramatically increased the number of potential drug targets.[6] In parallel, the number of new screenable drugs for those targets has also increased due to the use of combinatorial synthesis and the increased access to natural molecules,[7] because of combinatorial biosynthesis.[8,9] However, despite increasing costs for new drug development,[10] leading to an increase in new drug submission to the FDA,[11] the number of approved new drugs delivered to patient has not increased.[12] One reason for this is the low accuracy of the conventional system used for the quantitative characterization of drug candidates. Considering the multiple steps required for the realization of conventional drug discovery process, the addition of errors in each step could lead to results that are not representative of the true drug effects. Another reason is the high failure rate of drug candidate in the late and expensive part of the drug discovery process due to factors that are not considered in the early stage of the discovery process such as adsorption,

---

*Corresponding author. Email: hongjon@auburn.edu

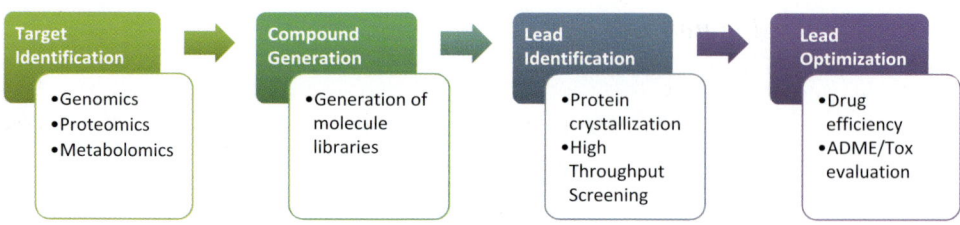

Fig. 1. Microfluidic applications in the four steps of the drug discovery process.

distribution, metabolism, excretion, and toxicity (ADME/Tox).[13] Thus, the evaluation of these parameters in the earliest part of the drug discovery process is the main interest of pharmaceutical companies. In addition, each step of drug discovery is a time-consuming process that slows down the appearance of new marketed drugs. Therefore, there is an opportunity to develop new tools to overcome those limitations in the discovery of new drug and thereby increase the accuracy of measurements and provide systems that can be used to evaluate ADME/Tox.

Drug discovery involves different steps that can be regrouped in four principal stages: target identification, compound generation, lead identification, and lead optimization (Fig. 1). The drug target identification process aims to identify a molecular structure that can interact with a drug candidate. In parallel, compound libraries that allow regrouping of thousands of small molecules are generated. The small libraries are then screened for potential drug targets and lead compounds are identified. These compounds will be characterized and optimized before they are tested and developed clinically. In this chapter, we will introduce different microfluidic systems that have been developed and have been useful for drug discovery. We will present different systems for the different steps of the drug discovery process.

Microbioreactors for Drug Target Selection

The first step in the drug discovery process is to identify a target, i.e., a particular biological molecule that may have therapeutic utility.[14] It can include proteins, polysaccharides, lipids, or nucleic acids. Polysaccharides, lipids, and nucleic acids are less frequently used in drug discovery because their physiopathological roles are less understood.[3,15] However, because of the recent sequencing of the human genome, the possibility of gene repair, alteration, or therapy hold promises for the future.[16] To identify the specific target, proteomic, genomic, and more recently metabolomics[17] studies have been performed. To understand the different pathways and interactions within a cell, various microfluidic devices have been developed.

Proteomics and genomics

To select a specific target, whole-cell proteins and genes must be understood.[3,18] However, mRNA and proteins often exist in very low copy numbers, which makes them difficult to be detected in single-cell analysis.[19–22] Microfluidic reactors have been developed to analyze the DNA, mRNA, or cell proteins from multiple cells or a single cell.[23–26] Based on single-molecule imaging technology, Taniguchi et al.[27] developed an automated imaging platform based on a microfluidic system to quantify the *Escherichia coli* proteome and transcriptome with single-molecule sensitivity in single cells, as shown in Fig. 2. The microfluidic platform was realized in polydimethylsiloxane (PDMS), using photolithography and soft lithography. In the microfluidic chip, 96 independent channels were designed in parallel to hold 96 kinds of cell samples in parallel channels on a single cover slip, each channel measuring 150 µm (width) × 10 mm (length) × 25 µm (height). Channels were pre-coated with poly-L-lysine to immobilize the bacteria in the microchannels.[28] After pre-coating, cells were injected into the channels and incubated for stable binding to the channel surface. Floating cells in the channels were washed out, resulting in a single-cell layer on the cover slip surface. In this device, 1,018 strains of an *E. coli* yellow fluorescent protein fusion bank were tested, simultaneously showing the expressions of specific proteins or mRNA in a single cell at a single-molecule level. Moreover, the authors could localize the proteins in different cell compartments (nucleus, cytoplasm, membrane).

Another technique to measure the protein expression in cells is flow cytometry, which is commonly used to count microscopic elements.[29] In the recent years, a growing interest for the miniaturization and automation of complex laboratory processes (lab-on-a-chip) provided different microfluidic devices to perform flow cytometry,[30] allowing the measurement of antibody staining, green fluorescent protein transfection efficiency, and cell characterization.[31,32] Chan et al.[32] described the use of a commercial lab-on-a-chip device (2100 Agilent bioanalyzer, Agilent Technologies GmbH) (Fig. 3) in analyzing protein expression on primary cells. The microfluidic chip is made of glass and glued into a plastic caddy. It is composed of six sample wells — two buffer wells, one for the reference dye, and one to create the vacuum interface that contributes to control the cell movement and collect fluid waste. Each sample channel is joined by a buffer channel to generate a single-file cell stream that will be analyzed. The cells will be led to the detection area which is able to detect two different fluorescent colors. The authors demonstrated the possibility of analyzing protein expressions and also the apoptosis of different types of mammalian cells — non-adherent (lymphocytes) or adherent (human umbilical vein endothelial cells (HUVECs) and normal human dermal fibroblasts).

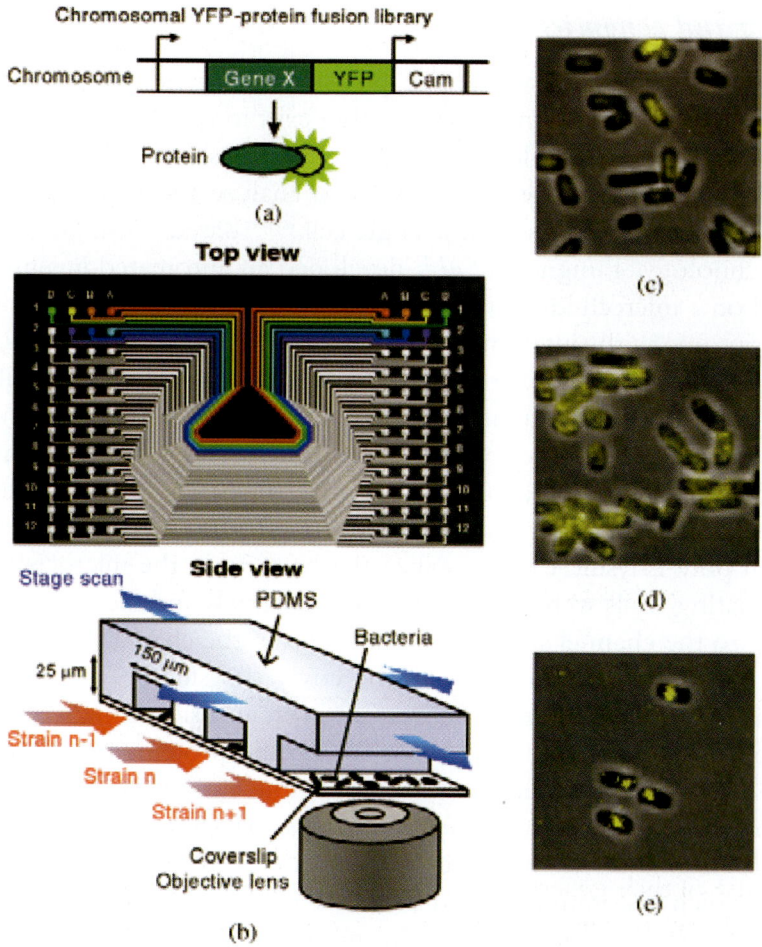

Fig. 2. Quantitative imaging of a yellow fluorescent protein (YFP) fusion library. (a) Chromosomal YFP fusion process. A gene X is coupled with the gene coding for YFP and inserted in the genome. Once produced, the protein will become fluorescent. (b) The microfluidic chip used for the experiment (top); schematics of the experiments representing three channels of the device containing three different strains (bottom). (c)–(e) Representative fluorescence images showing cytoplasmic proteins (c), membrane proteins (d), and DNA-binding proteins (e). (Reproduced with permission from Ref. 27.)

Metabolomics

Recently, metabolomics, a relatively new field among the "– omics" technologies, has been attracting the attention of various researchers.[33,34] Its first concern is the characterization of small-molecule metabolites, which are intermediates or products of metabolism, focusing on their physiological effects[35] such as the regulation of gene expression, cell signaling, cell–cell communications, and cell differentiation. Commonly, metabolomics studies are done using nuclear

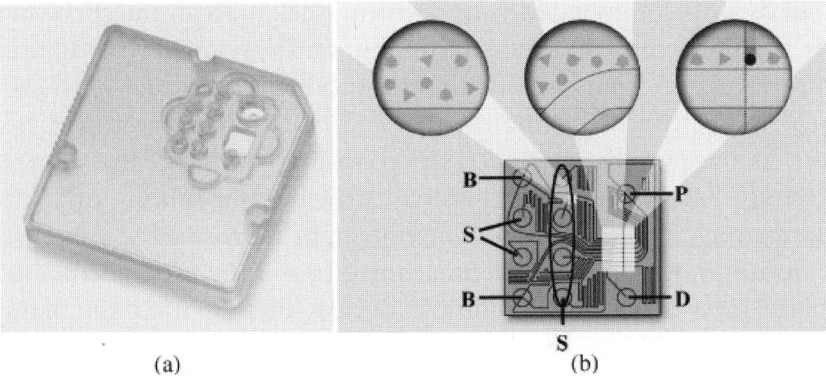

Fig. 3. Microfluidic chip for flow cytometry. (a) General view of the microfluidic chip in its plastic holder. (b) Layout of the microfluidic glass chip with indication of the sample wells (S), buffer wells (B), the well for the reference dye (D), and the priming well (P), which collects waste and acts as the interface to the vacuum source. Insert illustrates cells in the sample channel (left), the junction of buffer and sample channel (middle), and the cells while passing through detection spot (right). (Reproduced with permission from Ref. 32.)

magnetic resonance (NMR) spectroscopy[36] and mass spectrometry (MS)[37,38] coupled to a high-resolution separation technique like liquid chromatography (LC)[39] or gas chromatography (GC),[40] which are techniques that have the ability to detect many different species in a single sample.[41] The quantification of metabolites under physiological, *in vivo*, and dynamic conditions presents major challenges because of their generally very low concentration, high turnover rate, and chemical diversity. To this end, microfluidic systems can play an important role because of their characteristics for sample treatment and analysis, e.g., their highly spatial and temporal resolution. Despite progresses in microtechnology in recent years, many microfluidic studies or devices remain at the level of proof-of-concept and have been rarely applied to the real world of metabolomic analysis. Numerous devices have been developed to perform analytical methodologies applicable to metabolomics studies such as microfluidic chips to perform NMR[42] and high-performance liquid chromatography (HPLC)[43] coupled to MS.[44] So far, there is still a lack of an integrated microfluidic system allowing the culture of cells and analyses of their metabolisms on a single chip.

Microfluidic Devices for Compound Generation

One of the key processes of drug discovery is the identification of small-molecule hits and their subsequent optimization to lead compounds. The goal is to identify a lead compound whose properties make it a good drug candidate. To increase the chance of success, a high number of screenable compounds is recommended.[13] In conventional macroscale chemistry, this process is labor-intensive, time-consuming, and expensive. In addition, it is difficult to scale up from research to mass production because of the volume-to-surface ratio that affects the thermal and mass transport properties of the reaction. As a result, it is often necessary to readjust the process at each stage of the scale-up. Microreaction technology provides the means of addressing these limitations. Laboratory experiments can be performed in a single microfluidic device, reducing the cost, time, and labor of the experiments. In order to achieve mass production, multiple devices can be used simultaneously, preventing the problems due to scale-up. Moreover, these systems can be used to synthesize molecule libraries for drug screening.

To rapidly generate a large number of novel compounds that can be used as potential drug candidates, considerable attention has been given to combinatorial chemistry. This field quickly became a ubiquitous procedure in the pharmaceutical industry for generating potential drug candidates.[45] Several recent studies have reported advances in microfluidic combinatorial chemical synthesis, which combined the advantages of solution-phase combinatorial chemistry and microfluidic technology on conventional solid-phase combinatorial chemistry.[46-48] For example, a microreactor, in which Knorr's pyrazole synthesis[49] can be carried out, has been developed.[50] As shown in Fig. 4, in a single microchannel device, successions of reactions were realized to obtain a succession of products and to analyze the reaction product during chip operation or on-line, in opposition with other microdevices where the number of compounds made in one run was limited by the chip design and where the analysis of the reaction products was realized off-line, therefore requiring some manipulation.[48,51-53] The automated microreactor system was fabricated by mounting a borosilicate top plate onto a borosilicate bottom plate where the microchannels were designed. Most of the microreactors for combinatorial chemistry were made of hard material such as glass,[52,54,55] borosilicate,[51,56] or pyrex[57] because of the use of a solvent in the reaction process that might interact with PDMS. (The device was composed of the following: an auto-sampler, to introduce the reactants in the chip via capillaries; a pumping system, to mobilize reagents and the product through the chip using hydrodynamically driven flow; a dilution system, to dilute the reaction slug on-chip; a detection system, to detect the slug at the chip outlet; and an analysis system.) All components were on-line and configured in-house. A pyrazole compound library was prepared sequentially by 7×3 combinatorial series of simultaneous injections of solutions of 1,3-dicarbonyl and solutions of hydrazines, as summarized in Table 1. The reactant slugs reached the chip at the same time, ensuring their mixing on-chip. After an on-chip residence

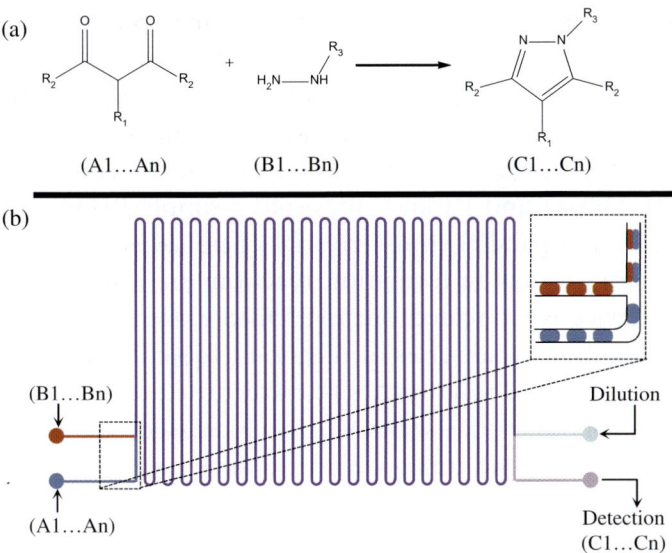

Fig. 4. 7 × 3 compound library generation in a simple microfluidic device. (a) General reaction scheme for Knorr's pyrazole (C1...Cn). (b) Schematic representation of the microfluidic device. 1,3-dicarbonyl compounds (A1...An) and hydrazines (B1...Bn) are sequentially introduced (blue and red channels, respectively) in the device, diluted, detected, and analyzed. (Reproduced from Ref. 50.)

Table 1. Pyrazole library obtained from the automated platform. Pyrazoles (C1...Cn) were obtained from the reaction of 1,3-dicarbonyl compounds (A1...An) with hydrazines (B1...Bn). (Reproduced from Ref. 50.)

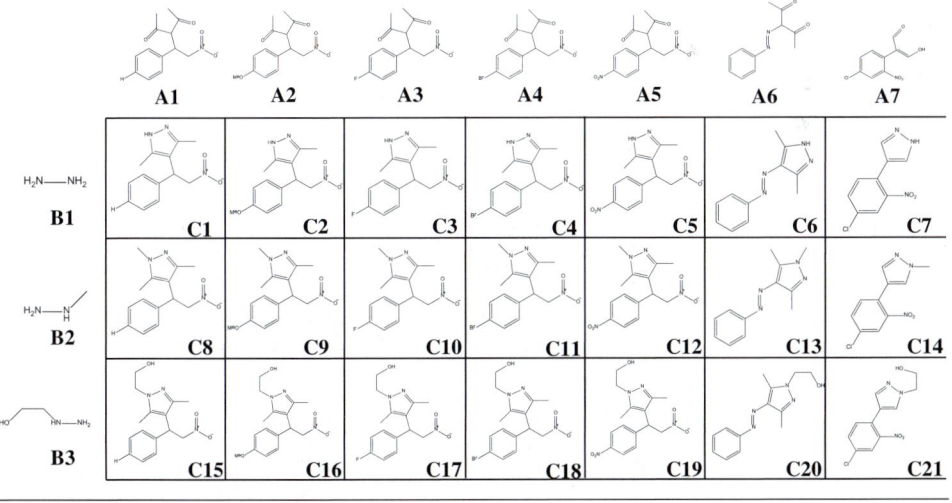

time of 210 s, the reaction slug was diluted on-chip and reaction products were identified on-line.

Although effective, microfluidic reactions are run in a sequential fashion by performing sequential reactions on a single microfluidic channel to obtain one compound, reducing the throughput of the combinatorial synthesis. In 2009, Dexter and Parker[58] reported a simple PDMS microfluidic device for performing parallel combinatorial chemical synthesis, realizing a combinatorial series of amide formation reactions on a single chip, shown in Fig. 5. The device is fabricated out of PDMS using soft lithography. To perform 2 × 2 combinatorial series of reaction (Table 2), four channels are created with five inlets and four outlets to collect the compounds produced. The number of combinatorial series can be increased (Table 3), but the number of channels and the total size of the device may limit the possible combinations.

Fig. 5. 2 × 2 simultaneous compound library generation in a simple microfluidic device. Schematic representation of the chip. Inlets for the reactants A (A1 and A2) and B (B1 and B2) and outlet for the library products A1B1, A2B1, A2B2, and A1B2 are indicated. (Reproduced with permission from Ref. 58.)

Table 2. Pyrazole library obtained from the automated platform. (Reproduced from Ref. 58.)

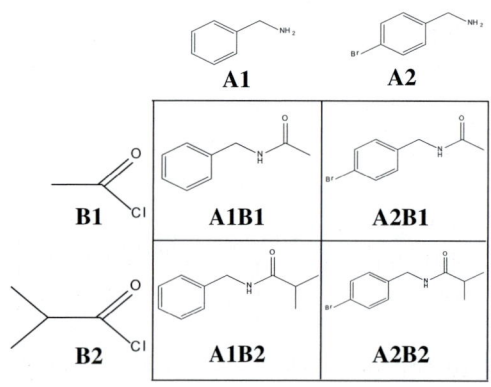

Table 3. Table of equations for calculating the number of inputs required for performing an arbitrary combinatorial synthesis. (Reproduced from Ref. 50.)

m	n	No. of Inputs
Even	Even	$mn + 1$
Even	Odd	$mn + m/2$
Odd	Even	$mn + n/2$
Odd	Odd	$mn + (m + n)/2$

Lead Identification in Microfluidic Systems

Protein crystallization

Protein crystallization is an important component of the drug discovery process.[59] It is performed to obtain the structure of macromolecules.[60] Investigated proteins are mixed with various combinations of reagents in an attempt to discover the recipe that will create conditions suitable for the formation of protein crystals, which can then be examined via X-ray diffraction.[59,61] The search for the optimum recipe is labor-intensive because the number of possible mixtures can be in millions. The use of microfluidic systems can greatly affect the field of protein crystallization in two ways: first, it allows scaling down of experiments that are conventionally performed, reducing the amount of proteins and reagents needed that are rare and/or expensive; second, the miniaturization of the process provides better control of experimental conditions, such as the diffusion of molecules or nucleation of crystals that are difficult or impossible to apply in large-scale experiments. Traditionally, there are four different techniques for protein crystallization: microbatch, vapor diffusion, dialysis, and free interface diffusion (FID).

In the microbatch technique, a small drop of protein is combined with the crystallization agent at a constant concentration to induce crystallization.[62] This technique requires the testing of various reagent concentrations to identify a good combination. In the vapor diffusion technique, a small drop containing the protein, buffer, and precipitant equilibrates with a large reservoir containing similar buffers and precipitants in a higher concentration.[63] The evaporation of water from the droplet increases protein and reagent concentrations. When the concentration is optimal, proteins start to crystallize. In the dialysis technique, pure solutions of the protein are dialyzed against a precipitant solution through a semi-permeable membrane.[64] The crystallization trial, on the side of the membrane containing the protein, starts when the concentration is optimal. In FID, pure solutions of the protein and precipitant are connected by an interface and both gradually diffuse under the influence of a concentration gradient.[65] Proteins start to crystallize when the concentration is optimal.[59] There are three microfluidic approaches covering three of the four techniques of crystallization: (1) the valve-based system, (2) the droplet-based system, and (3) the SlipChip-based system. All these systems are present nanoliter volume reactors for crystallization analysis; however, these volumes are created using different systems.

In the valve-based system, protein and precipitant solutions are loaded into different chambers separated by valves. When the valves are opened, a protein sample and a precipitant can be mixed. For example, a microfluidic system for FID method in 48 wells was created, allowing the testing of 48 crystallizing agents on one protein sample. Each well is composed of three pairs of compound reaction chambers. Each pair of compound reaction chambers are composed of two solution chambers, where a protein solution and a precipitant solution are loaded. In each well, the volume ratio of the pair of solution chambers creates three different mixing ratios — 4:1, 1:1, and 1:4. To realize one run of 144 experiments, 3 µL of protein solutions are sufficient while with the conventional method, 1 µL of protein solution is needed for one experiment. In addition to reducing the time and labor of the experiment, crystallization experiments on chip result in faster crystal growth along with higher hit rates than conventional methods (Fig. 6).[66]

In the droplet-based system, crystallization takes place inside droplets surrounded by an immiscible carrier fluid. To optimize crystallization conditions, hundreds of droplets containing different crystallization conditions are generated in a single experiment. Variation of crystallization conditions is accomplished by changing the flow rates of the streams of the aqueous reagent, protein, buffer, and precipitants. This system has been used in a microfluidic system for protein crystallization by microbatch and vapor diffusion methods (Fig. 7).[67] A composite PDMS/glass capillary microfluidic device has also been created. To avoid problems of uncontrolled evaporation of water during crystal growth, the PDMS section of the device is used to generate droplets and is connected to a glass capillary in which crystallization will be performed. To carry out the microbatch method, aqueous streams of the protein, additives, and precipitants are flown into a flowing hydrophobic medium of oil to create droplets, surrounded and carried by the oil flow. The composition of the droplets can be easily varied by varying the flow rates of the aqueous streams. The flow of oil carries the droplet into the glass capillary. When the capillary is filled, the flow is stopped and the capillary is detached, sealed, and

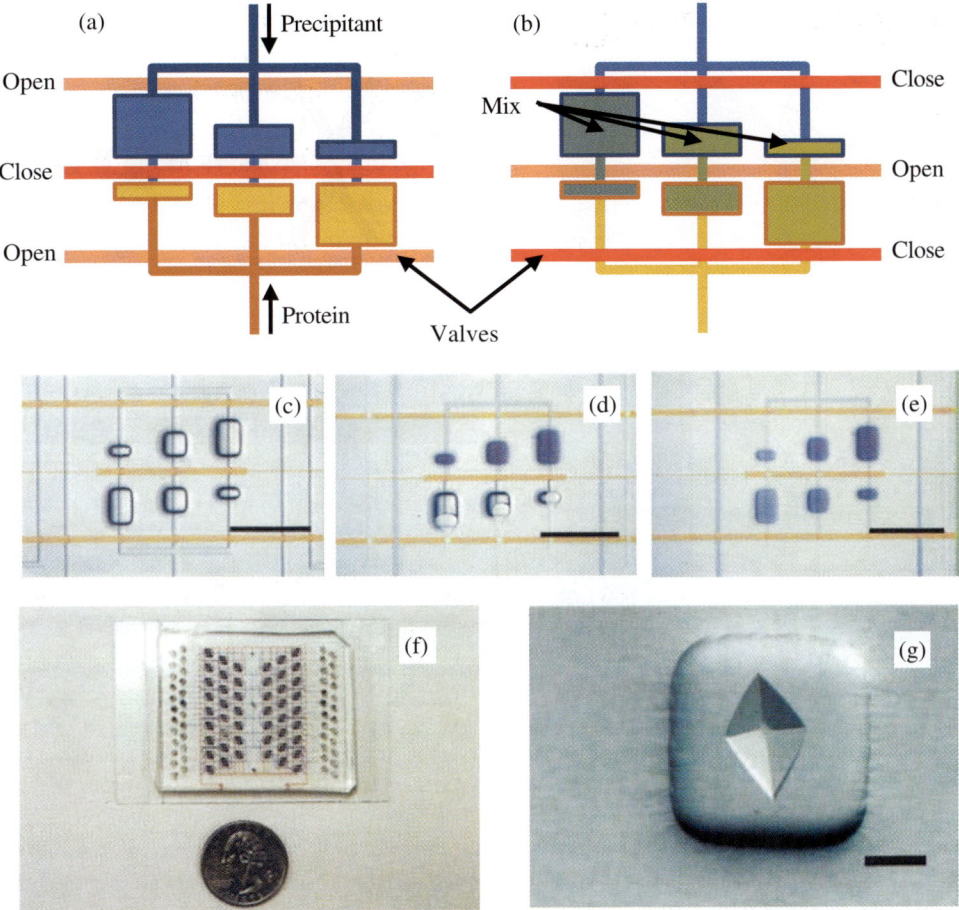

Fig. 6. Valve-based system for protein crystallization. Schematic representation of (a) the loading of precipitants and protein and (b) the process of free interface diffusion (FID) on-chip. (c)–(e) Three pairs of empty reaction microchambers (c) loaded of reagents (d) that will diffuse (e) (scale bar, 1 mm). (f) Protein crystallization chip with 144 parallel reaction chambers. (g) Crystals grown in one chamber. (Reproduced with permission from Ref. 66.)

incubated. Crystals formed in the droplet are monitored and analyzed by X-ray diffraction. To perform vapor diffusion in the same device, droplets of the crystallization mixture are alternated with droplets containing a high concentration of salt in the water-permeable oil, allowing water from the crystallization droplet to diffuse into the salt droplet. As the concentration increases in the droplet containing the protein, crystallization by vapor diffusion occurs.

In the SlipChip-based system, the solutions are loaded into two plates in close contact that can move relative to one another. By sliding the top plate relative to the bottom plate, the proteins and the precipitant are brought into

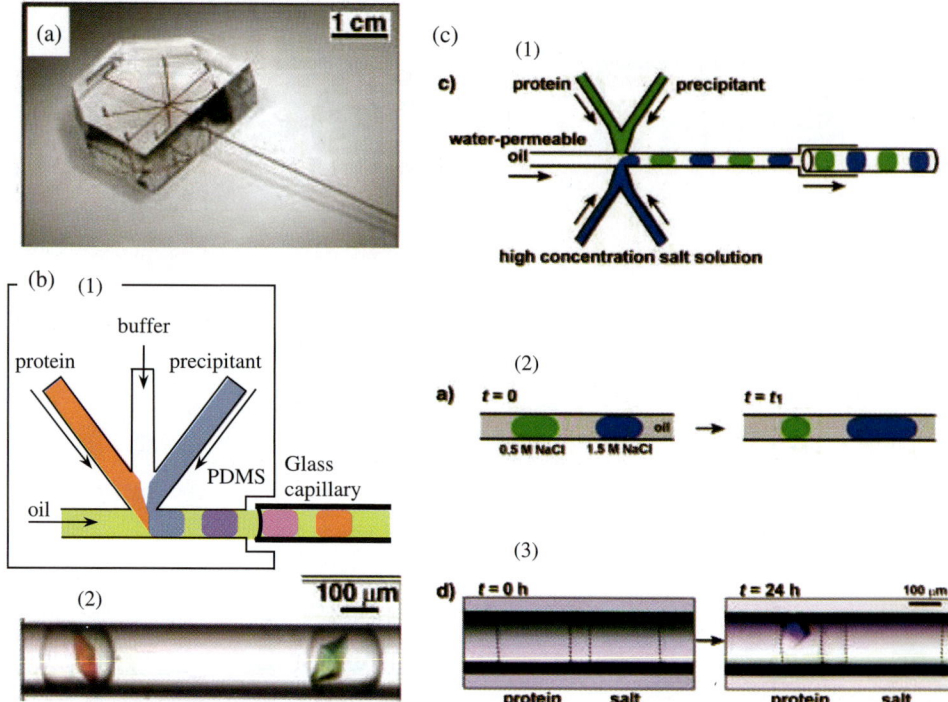

Fig. 7. Droplet-based system for protein crystallization. (a) Photograph of the device. (b) (1) Schematic representation of the microbatch process in the device. (2) Protein crystals obtained using this device. (c) (1) Schematic representation of the vapor diffusion process in the device. (2) Schematic illustration of the principle of vapor diffusion in a capillary. (3) Photographs that show a pair of droplets immediately after (left) and 24 h after (right) the setup. (Reproduced with permission from Ref. 67.)

contact, initializing the crystallization.[68] The two plates are made of glass, coated with photoresist. The photoresist is exposed to UV following the nanopattern of different photomasks and revealed. Glass is etched to create wells and channels. The SlipChip was configured to form 16 separate fluidic paths for the precipitants, each containing 10 wells, and a single fluidic path for the protein sample containing 160 wells. The SlipChip system can be used to perform protein crystallization by microbath and FID at a same time. For the microbath method analysis, the wells containing the precipitants are aligned with the wells containing the protein. For the FID method analysis, the microchannels that had formed the continuous fluidic path for the protein sample are slid to become the neck channel connecting the protein well to the precipitant well. Variations on the neck geometry vary the equilibration time between the two solutions. In addition, it is possible to obtain different mixing ratios by varying the well dimensions (Fig. 8).

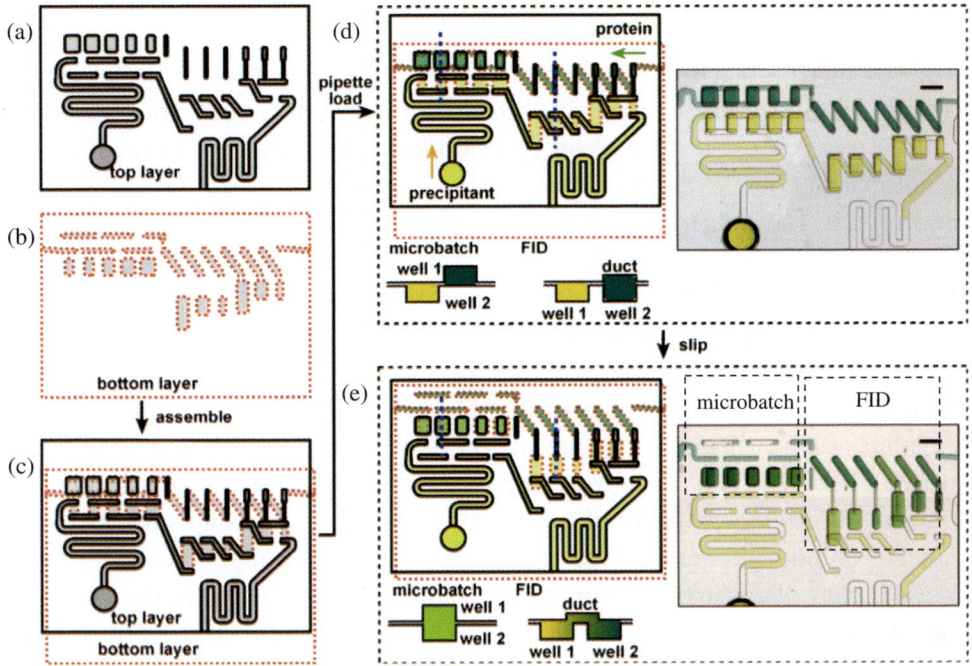

Fig. 8. SlipChip system for protein crystallization. (a)–(c) A schematic of the device showing the top plate containing wells for the protein and ducts for the protein and precipitant (a), the bottom plate containing ducts for the protein and wells for the precipitant and protein (b), and the two plates assembled (c). (d)–(e) The device after loading before "slipping" (d) and after (e). Left: A schematic of the device with solutions of protein (yellow) and precipitant (blue) filling wells. Right: photographs of the device with food dye. Bottom: A cross-sectional view (dotted blue lines in the schematic) showing how wells are connected in the microbatch (left) and in the FID experiments (right). (Reproduced with permission from Ref. 68.)

Identification of "hit" compounds

The next step in drug discovery is the identification of a "hit" compound, i.e., a small molecule that will interfere with the chosen target,[69] and eventually become a lead compound. For this step, high-throughput screening (HTS) is widely used.[70] Traditionally, HTS is performed by using 96-, 384-, or 1,536-well plates. While the volumes needed for the experiments are minimized, they are still reagent-consuming.[71] In addition, the difficulties in introducing nanoliter volumes in the microwells limited the development of smaller wells. Various microfluidic systems have been developed to screen drug candidate libraries on cells. These include high-throughput cell-based screening platforms and droplet microfluidic technology.

Microfluidic systems with high-density of cell microchambers for HST

High-throughput cell-based screening microfluidic platforms are of prime interest in the pharmaceutical industry for the rapid identification of "hit" compounds because of their short reaction times and smaller reagent volumes as compared to macroscale experiments. A practical microfluidic cell array requires the formation of discrete cell chambers. To address this challenge, microvalves can be used for the compartmentalization of the chambers. For example, a microfluidic array platform for high-throughput cell cytotoxicity screening has been developed (Fig. 9).[72] It is composed of microfluidic channels individually addressable in column and row directions (24 × 24). The channels for cell seeding are orthogonal to channels for chemical exposure. Six inlets (each connected to four channels) allow the inoculation of the cells and 12 inlets (each connected to two channels) allow the introduction of the solution. Within each channel intersection, there is a circular chamber that can be compartmentalized by integrated pneumatically actuated valves. Providing a cell culture environment is a significant challenge in the development of microfluidic devices. The local microenvironment of the cells should allow cell–cell interactions, communications, and exchanges as they are critical for cell growth and development.[73] One advantage of microfabricated systems is that they can present cells with these cues in a controllable and reproducible fashion that cannot easily be achieved by standard tissue culture.[74] In this device, to create a uniform distribution of seeded cells and an optimum cell density for development of the cell monolayer, each microchamber contains several U-shaped microcell sieves, allowing cell–cell interactions and the growth of healthy cell monolayers. The configuration of this device allows the culture of cells with adhesion and morphology patterns comparable to conventional 96-well plate cultures. As a proof-of-concept, toxicity assays were performed in the device. Five toxins at two concentrations (low and high) were inoculated in the cell chambers containing three different cell types. Cytotoxicity data comparison (microarrays *vs.* 96-well plate) exhibited a high degree of correlation. Hence, it was demonstrated that the repartition of the channels and the inlets of this system can allow the testing of up to 12 different chemicals and up to six different kinds of cells in a single experiment.

Droplet-based microfluidic devices for HTS

Droplet-based microfluidic approaches present various advantages that are useful for drug discovery, such as increased throughput for screening, reduced

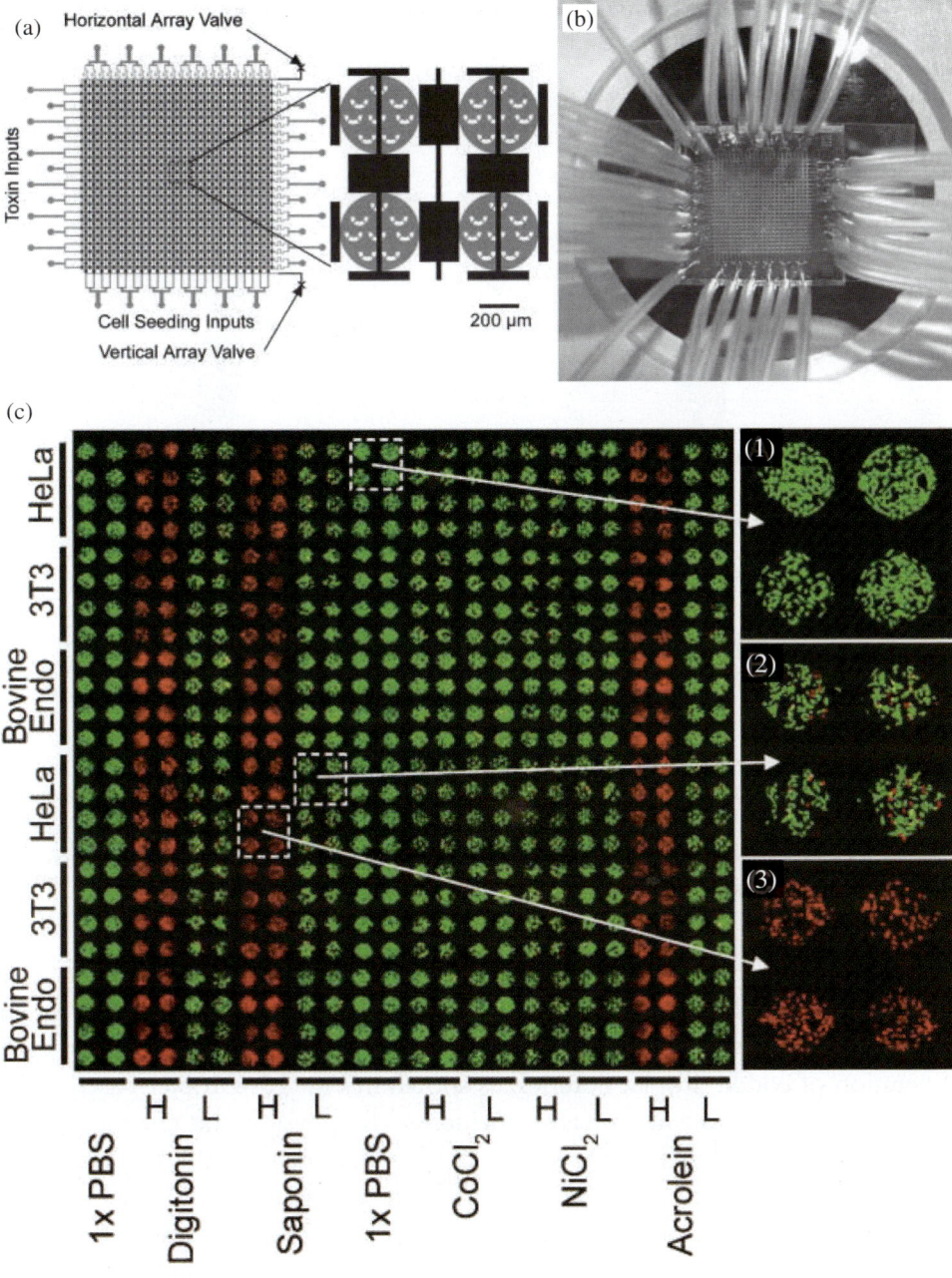

Fig. 9. Microfluidic system with cell culture chambers for drug screening. (a) Schematic of the 24 × 24 chamber microfluidic cytotoxicity array. Each chamber contains eight micro cell sieves for cell trapping. (b) Image of the microfluidic cytotoxicity array chip with fluid interconnects. (c) Composite image of the microfluidic cytotoxicity array chip after toxin incubation and live (green)/dead (red) staining. Insert shows control (1), cells incubated with low (L) concentration (2) or high (H) concentration (3) of toxin. (Reproduced with permission from Ref. 72.)

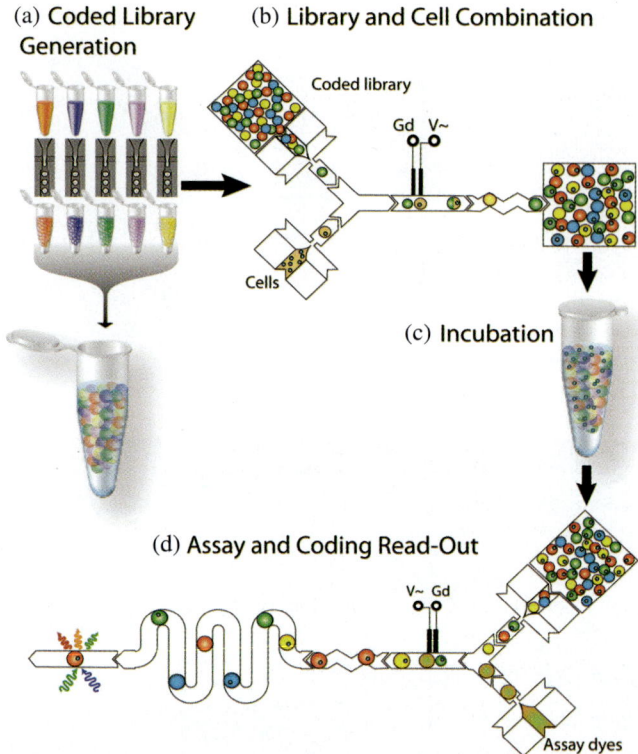

Fig. 10. Droplet-based HTS. The droplet screen has four steps: the coded library is generated (a) and each library member is merged with one cell-containing droplet (b). After incubation (c), merged droplets are re-injected into an assay chip to identify the compound effect on cells (d). (Reproduced with permission from Ref. 77.)

sample volumes, and single-cell analysis.[75,76] Recently, Brouzes et al.[77] have introduced a droplet-based microfluidic technology that enables HTS of single mammalian cells, combining an on-chip cell viability assay and the automated generation of coded droplet libraries, as shown in Fig. 10. Droplet-based technology presents various advantages, including the physical and chemical isolation of droplets, eliminating the risk of cross-contamination, the small amount of reagent (1 pL to 10 nL), the fast and efficient mixing of the reagents that occurs inside droplets, the ability to digitally manipulate droplets at a very high-throughput, the ability to incubate stable droplets off-chip and reintroduce them into the microfluidic environment for further processing and analysis, and the absence of moving parts. The small amount of cells needed in this technology makes it suitable for working on cells that have low availability, such as primary patient cells, taking one more step toward personalized medicine. For the realization of drug screening, three chip types were used — a library generation chip, a merge chip, and an assay chip. The screening of the chemical bank is realized in four steps: (1) a coded library is generated, where the drug library is created using the library generation chip into a droplet emulsion containing each optically labeled member; (2) the drug library droplets are combined with a cell containing droplet on the merge chip; (3) the combined droplets are incubated for cell treatment; and (4) the droplets are screened on the assay chip. The viability assay chip is composed of five parts that sequentially manipulate the droplets to perform viability tests on one single chip. First, a set of two nozzles introduces treated cell droplets and live/dead fluorescent dyes in the device. Then, one cell droplet and one dye droplet are merged by a fusion module that delivers an AC field. A mixing module facilitates the mixing of the cells with the viability dyes and a delay line incubates the droplet for 15 min, optimizing the incubation of the cells with the live/dead fluorescent dyes. Finally, a detection module collects the fluorescent signals excited with a laser slit.

Microdevices Useful in Lead Optimization Processes

Dose–response analysis and IC_{50} measurement

Nowadays, enzymes are the main target for drugs and represent half of the available marketed drugs.[3] The attraction of researchers toward enzymes as a drug target is due to the essentiality of their catalytic activity in physiology[78] and pathology, such as infection[79] or cancer,[80] and due to their susceptibility to be inhibited by small molecular weight molecules.[81] Hence, 50–75% of all new drug candidates tested by major pharmaceutical companies in the U.S. are focusing on enzymes as a primary target.[78] To evaluate and figure out the best drug candidate, the inhibitory potencies of all drug candidates have to be quantitatively characterized.[78] To compare the potential of different drug candidates, dose–response experiments are done wherein the half maximal inhibitory concentration (IC_{50}) of each candidate is used.[82–84] To realize such experiments, a wide range of gradient concentrations is necessary, as dose–response analysis must be done on a logarithmic scale.[85,86]

Conventionally, in IC_{50} experiments, the logarithmic scale concentration gradient is achieved by series of dilutions in test tubes, vials, or microplates.[87,88] However, series dilution is a multistep methodology that may generate errors at each step of the process. Each pipetting may affect the precision of the dilution ratio, and the addition of each imprecision may generate results that are not truly representative of the real drug potential.[89] Consequently, starting with incorrect data, drug candidates will inevitably have a higher failure rate in the later part of the drug discovery process when tested *in vivo*. A breakthrough

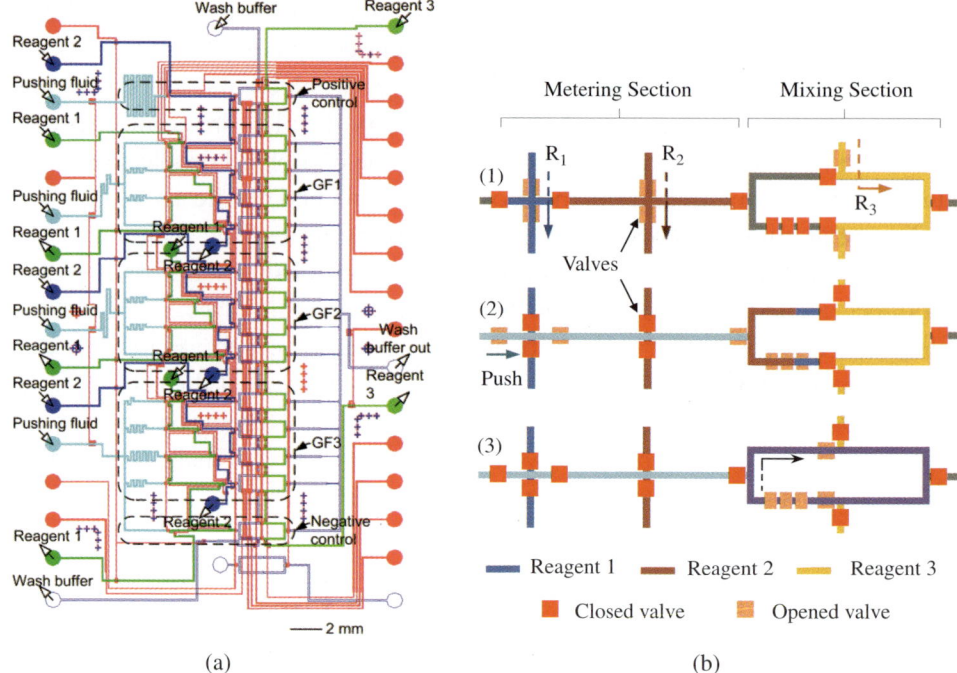

Fig. 11. Microfluidic system for dose–response analysis. (a) Schematic of the design. It consists of three gradient formers (GFs) and two control processors, positive and negative. The colored channels, green, blue, and fluorescent green, represent reagents 1, 2, and 3, respectively. Each GF contains four individual processors, each with different volume of reagent 1, 2, and 3. Respective different volumes of reagent 1 and 2 in each processor create specific concentrations. However, in each processor, the concentration of reagent 3 is constant. (b) Schematic and working of an individual processor for the titering (1), the conveying (2), and the mixing (3) processes. (Reproduced from Ref. 90.)

research on the realization of microfluidic system that is able to generate accurate logarithmic scale concentration gradient has been reported.[90] The chip for dose–response analysis was composed of two layers of channels — a fluidic layer, where reaction and measurement are done, and a control layer, which control the flow of the fluidic layer channels (Fig. 11). Channels of the control layer cross channels of the fluidic layer and by increasing the pressure in the control channels, the membrane separating the control and fluidic layers close the fluidic channels, acting as a valve. Valves are used to create a metering section controlling the mixing ratio of two reagents. By disposing valves along a channel at specific distances, a precise volume of liquid in the channel can be created. By controlling the disposition of three valves along a channel, it is possible to reproduce a mixing ratio between two reagents, controlling the volume of each reagent (Fig. 11(b)). The chip for dose–response analysis was composed of three gradient formers (GFs) and each GF was composed of four metering sections that can measure two reagents in the ratios 7.5:2.5, 5.0:5.0, 2.5:7.5, and 1.0:9.0. The chip also includes discrete processors for positive and negative controls. The three GFs were independent, allowing the introduction of three different reagent concentrations. Consequently, by introducing 10×, 100×, and 1,000× reagent concentrations in the three GFs in addition with 1,000× concentration in the positive control and 0 in the negative control, it is possible to create a wide range of concentration gradient: 0 (negative control), 1×, 2.5×, 5×, and 7.5× (for GF1), 10×, 25×, 50×, and 75× (for GF2), 100×, 250×, 500×, and 750× (for GF3), and 1,000× (positive control). After loading of the inhibitor and the buffer in the metering sections, the two solutions were pushed to a mixing section, where an enzyme has previously been loaded.

In this chip, the authors demonstrated the possibility of creating a logarithmic concentration gradient and the possibility to conduct dose–response experiments by determining the IC_{50} values of inhibitors of the matrix metalloproteinase MMP-9 enzyme from the corresponding plots of their logistic dose–response. A logarithmic concentration gradient from 0.063 nm to 250 nm (1:4,000) was created and inhibitors were incubated with the enzyme and its substrate (both at constant concentrations). From the dose–response plots for the three inhibitors, IC_{50} values could be determined. Those values could be compared with the values obtained from the conventional method.

ADME/Tox evaluation

A second area of drug discovery, in which enzymatic reactions are important, is the study of the metabolism of drug candidates and pharmacokinetics. Indeed, enzymes are responsible for the metabolic transformations of drug molecules, as for every xenobiotic, for their elimination from the systemic circulation.[91–93] For example, the cytochrome P450 family of hepatic enzymes is commonly involved in the oxidation of drugs.[94,95] Studies on the metabolism of drugs by the liver are part of the ADME/Tox parameters evaluated in the late stages of drug discovery.[96] Traditionally, drug metabolism is measured *in vivo* in various animal species, including human beings.[97,98] Alternatively, nowadays, it is common that researchers attempt to predict the metabolic transformation of drug candidates *in vitro* by studying the interaction of the molecule with the enzymes.[99] However, those experiments based on tissue or cell cultures or recombinant enzymes require labor-intensive work and milliliter-scale volumes of samples and reagents, which limits their use in the determination of kinetic parameters. Recently, Jambovane *et al*.[100] have developed an integrative microfluidic system to obtain the key parameters of

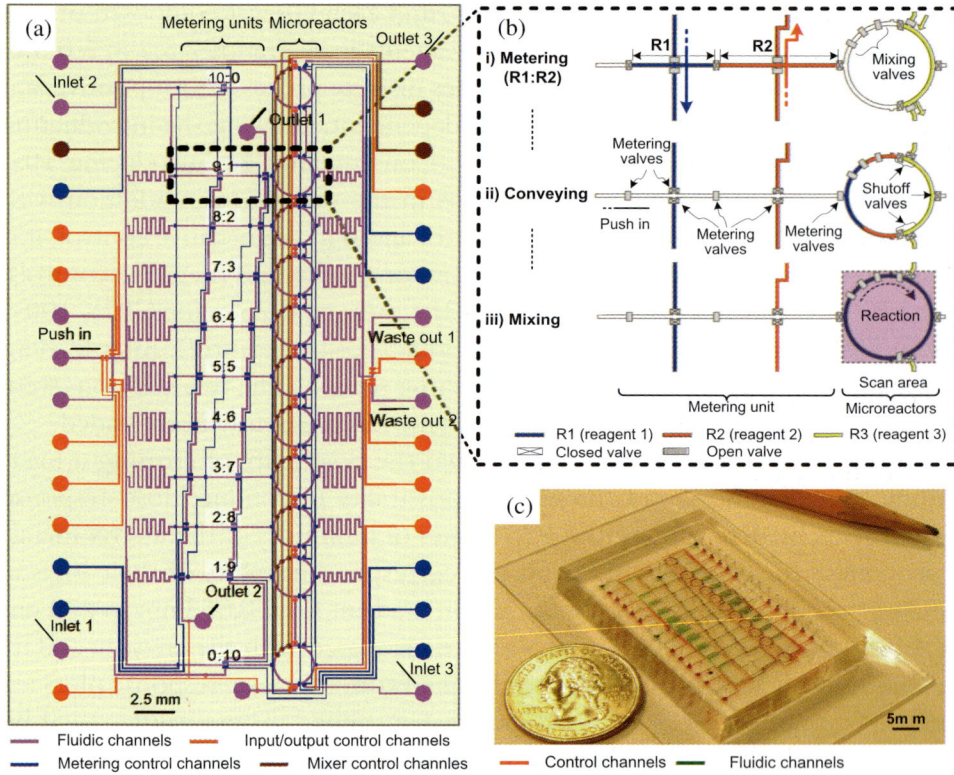

Fig. 12. Integrated microfluidic system for parallel processing. (a) A configuration of the microfluidic chip. (b) Schematic process flow of (i) titering, (ii) conveying, and (iii) mixing. (c) A photograph of the microfabricated microfluidic device. By using the chip, 11 different reaction conditions are tested at one time. (Reproduced from Ref. 100.)

enzyme kinetics, K_M and k_{cat}. The microfluidic chip was composed of 11 parallel processors, based on a previous chip design.[101] In this chip, analyses of 11 enzyme–substrate reactions were carried out at the same time by using less than 5 nL of reagents in each processor (Fig. 12). Each processor is composed of a metering section, as described previously, and a microreactor composed of a ring-shaped microfluidic system, where reagents are mixed and incubated. Each microreactor contains three mixing valves which can generate a fluid flow in the microreactor. The 11 processors allow the formation of a concentration gradient following the metering substrate. The buffer ratio of each processor was 10:0, 9:1, 8:2, 7:3, 6:4, 5:5, 4:6, 3:7, 2:8, 1:9, and 0:10. The chip was fabricated by using multi-step soft lithography. First, two different molds were created by a photolithography process to generate the fluidic channels and the control channels. The valves of the metering section and the mixing section were operated by external pressure controllers. The enzyme product was simultaneously detected in each parallel processor by optical detection inside a biochip reader (arrayWoRX®, Applied Precision). The data were collected and analyzed to estimate the values of the kinetic parameters K_M and k_{cat}. Those values could be compared with the values obtained in the conventional method.

Summary and Perspectives

Microfluidic technologies are powerful tools for drug discovery and are used at every level of the drug development process. In this chapter, we have described some of the systems and concepts that address a number of limitations imposed by conventional macroscale experimentations, including expensive methods, low throughput, large volume of reagents, and the long time needed. In the recent past, the miniaturization of electronic elements has revolutionized our world, decreased the size of components, and introduced computers at home, phones in the street, and so on. Today, in a similar fashion, with the miniaturization of pharmaceutical processes on a single chip, we are assisting in changes that are revolutionizing the pharmaceutical world, rendering drug discovery quicker, cheaper, and easier, aiming to increase the number of new drugs available in the future to cure common as well as orphan diseases. However, a number of challenges to achieve this idyllic vision in the future still persists. For example, microfluidic systems must remain simple to be used by researchers at the risk of being obsolete; they must have the capability to realize routine experiments for a long period of time, i.e., for chemical synthesis. Most importantly, mindsets need to be changed; researchers need to be convinced of the suitability of these systems by increasing the number of validation studies for the proposed processes. We are a now at the beginning of a new era in drug discovery, but a great deal of progress must be made in order to establish a successful entry in this age.

Acknowledgments

The authors would like to thank Austin Adamson, Chance Tara, and Kirn Cramer for their help during the preparation of the manuscript. They also thank the National Institutes for Health (NIH) for their support (funding EB008392).

References

1. International Human Genome Sequencing Consortium. Finishing the euchromatic sequence of the human genome. *Nature,* 2004, **431**(7011): p. 931–945.
2. Betz, U.A.K., Farquhar, R., Ziegelbauer, K., Genomics: Success or failure to deliver drug targets? *Curr Opin Chem Biol,* 2005, **9**(4): p. 387–391.
3. Hopkins, A.L., Groom, C.R., The druggable genome. *Nat Rev Drug Discovery,* 2002, **1**(9): p. 727–730.
4. Sleno, L., Emili, A., Proteomic methods for drug target discovery. *Curr Opin Chem Biol,* 2008, **12**(1): p. 46–54.
5. Page, M.J. *et al.*, Proteomics: A major new technology for the drug discovery process. *Drug Discovery Today,* 1999, **4**(2): p. 55–62.
6. Kubinyi, H., Drug research: Myths, hype and reality. *Nat Rev Drug Discovery,* 2003, **2**(8): p. 665–668.
7. Dobson, C.M., Chemical space and biology. *Nature,* 2004, **432**(7019): p. 824–828.
8. Gad, S.C., ed. *Drug Discoveryery Handbook,* Wiley-Interscience, Hoboken, New Jersey, 2005.
9. Cordell, G.A., Natural products in drug discovery. Creating a new vision. *Phytochem Rev,* 2002, **1**(3): p. 13.
10. DiMasi, J.A., Hansen, R.W., Grabowski, H.G., The price of innovation: New estimates of drug development costs. *J Health Econ,* 2003, **22**(2): p. 151–185.
11. Rydzewski, R.M., ed. *Real World Drug Discovery,* Elsevier, Amsterdam, 2008.
12. Service, R.F., Surviving the blockbuster syndrome. *Science,* 2004, **303**(5665): p. 1796–1799.
13. Chorghade, M.S., ed. *Drug Discovery and Development,* Vol. 1. Wiley-Interscience, 2006.
14. Knowles, J., Gromo, G., A guide to drug discovery: Target selection in drug discovery. *Nat Rev Drug Discovery,* 2003, **2**(1): p. 63–69.
15. Russ, A.P., Lampel, S., The druggable genome: An update. *Drug Discovery Today,* 2005, **10**(23–24): p. 1607–1610.
16. Spector, R., Vesell, E.S., The heart of drug discovery and development: Rational target selection. *Pharmacology,* 2006, **77**(2): p. 85–92.
17. Kell, D.B., Systems biology, metabolic modelling and metabolomics in drug discovery and development. *Drug Discovery Today,* 2006, **11**(23–24): p. 1085–1092.
18. Yan, Q., ed. *Methods in Molecular Biology,* Vol. 662. Springer Protocols, 2010.
19. Elowitz, M.B. *et al.*, Stochastic gene expression in a single cell. *Science,* 2002, **297**(5584): p. 1183–1186.
20. Rosenfeld, N. *et al.*, Gene regulation at the single-cell level. *Science,* 2005, **307**(5717): p. 1962–1965.
21. Blake, W.J. *et al.*, Noise in eukaryotic gene expression. *Nature,* 2003, **422**(6932): p. 633–637.
22. Cai, L., Friedman, N., Xie, X.S., Stochastic protein expression in individual cells at the single molecule level. *Nature,* 2006, **440**(7082): p. 358–362.
23. Thorsen, T., Maerkl, S.J., Quake, S.R., Microfluidic large-scale integration. *Science,* 2002, **298**(5593): p. 580–584.
24. Ottesen, E.A. *et al.*, Microfluidic digital PCR enables multigene analysis of individual environmental bacteria. *Science,* 2006, **314**(5804): p. 1464–1467.

25. Hong, J.W. et al., A nanoliter-scale nucleic acid processor with parallel architecture. *Nat Biotechnol*, 2004, **22**(4): p. 435–439.
26. Hellmich, W. et al., Single cell manipulation, analytics, and label-free protein detection in microfluidic devices for systems nanobiology. *Electrophoresis*, 2005, **26**(19): p. 3689–3696.
27. Taniguchi, Y. et al., Quantifying *E. coli* proteome and transcriptome with single-molecule sensitivity in single cells. *Science*, 2010, **329**(5991): p. 533–538.
28. Huang, J., Yamaji, H., Fukuda, H., Immobilization of *Escherichia coli* cells using porous support particles coated with cationic polymers. *J Biosci Bioeng*, 2007, **104**(2): p. 98–103.
29. Nolan, J.P., Sklar, L.A., The emergence of flow cytometry for sensitive, real-time measurements of molecular interactions. *Nat Biotechnol*, 1998, **16**(7): p. 633–638.
30. Schrum, D.P. et al., Microchip flow cytometry using electrokinetic focusing. *Anal Chem*, 1999, **71**(19): p. 4173–4177.
31. Preckel T, L.G., Chan, S., Wang, B., Dubrow, R., Buhlmann, C., Detection of cellular parameters using a microfluidic chip-based system. *J Assoc Lab Autom*, 2002, **7**(4): p. 5.
32. Chan, S.D. et al., Cytometric analysis of protein expression and apoptosis in human primary cells with a novel microfluidic chip-based system. *Cytometry A*, 2003, **55**(2): p. 119–125.
33. Lindon, J.C., Holmes, E., Nicholson, J.K., Metabonomics in pharmaceutical R & D. *FEBS J*, 2007, **274**(5): p. 1140–1151.
34. Morris, M., Watkins, S.M., Focused metabolomic profiling in the drug development process: Advances from lipid profiling. *Curr Opin Chem Biol*, 2005, **9**(4): p. 407–412.
35. Wishart, D.S., Applications of metabolomics in drug discovery and development. *Drugs R D*, 2008, **9**(5): p. 307–322.
36. Griffin, J.L., Metabonomics: NMR spectroscopy and pattern recognition analysis of body fluids and tissues for characterisation of xenobiotic toxicity and disease diagnosis. *Curr Opin Chem Biol*, 2003, **7**(5): p. 648–654.
37. Werner, E. et al., Mass spectrometry-based metabolomics: Accelerating the characterization of discriminating signals by combining statistical correlations and ultrahigh resolution. *Anal Chem*, 2008, **80**(13): p. 4918–4932.
38. Werner, E. et al., Mass spectrometry for the identification of the discriminating signals from metabolomics: Current status and future trends. *J Chromat B Anal Technol Biomed Life Sci*, 2008, **871**(2): p. 143–163.
39. Porter, S.E.G. et al., Analysis of four-way two-dimensional liquid chromatography-diode array data: Application to metabolomics. *Anal Chem*, 2006, **78**(15): p. 5559–5569.
40. Phelps, T.J., Palumbo, A.V., Beliaev, A.S., Metabolomics and microarrays for improved understanding of phenotypic characteristics controlled by both genomics and environmental constraints. *Curr Opin Biotechnol*, 2002, **13**(1): p. 20–24.
41. Kraly, J.R. et al., Review: Microfluidic applications in metabolomics and metabolic profiling. *Anal Chim Acta*, 2009, **653**(1): p. 23–35.
42. Lee, H. et al., Chip-NMR biosensor for detection and molecular analysis of cells. *Nat Med*, 2008, **14**(8): p. 869–874.
43. Liu, J.K. et al., Polymer microchips integrating solid-phase extraction and high-performance liquid chromatography using reversed-phase polymethacrylate monoliths. *Anal Chem*, 2009, **81**(7): p. 2545–2554.

44. Bai, H.Y. et al., Characterization and evaluation of two-dimensional microfluidic chip-HPLC coupled to tandem mass spectrometry for quantitative analysis of 7-aminoflunitrazepam in human urine. *Analyst*, 2010, **135**(10): p. 2737–2742.
45. Hall, D.G., Manku, S., Wang, F., Solution- and solid-phase strategies for the design, synthesis, and screening of libraries based on natural product templates: A comprehensive survey. *J Comb Chem*, 2001, **3**(2): p. 125–150.
46. Mitchell, M.C. et al., Microchip-based synthesis and total analysis systems (mu SYNTAS): Chemical microprocessing for generation and analysis of compound libraries. *J Chem Soc Perkin Trans*, 2001, **1**(5): p. 514–518.
47. Watts, P., Haswell, S.J., Microfluidic combinatorial chemistry. *Curr Opin Chem Biol*, 2003, **7**(3): p. 380–387.
48. Garcia-Egido, E., Wong, S.Y.F., Warrington, B.H., A Hantzsch synthesis of 2-aminothiazoles performed in a heated microreactor system. *Lab Chip*, 2002, **2**(1): p. 31–33.
49. Elderfield, R.C., ed. *Heterocyclic Compounds*, Wiley Interscience, New York, 1950.
50. Garcia-Egido, E. et al., Synthesis and analysis of combinatorial libraries performed in an automated micro reactor system. *Lab Chip*, 2003, **3**(2): p. 73–76.
51. Skelton, V. et al., The generation of concentration gradients using electroosmotic flow in micro reactors allowing stereoselective chemical synthesis. *Analyst*, 2001, **126**(1): p. 11–13.
52. Greenway, G.M. et al., The use of a novel microreactor for high throughput continuous flow organic synthesis. *Sens Actuators B*, 2000, **63**(3): p. 153–158.
53. Fernandez-Suarez, M., Wong, S.Y.F., Warrington, B.H., Synthesis of a three-member array of cycloadducts in a glass microchip under pressure driven flow. *Lab Chip*, 2002, **2**(3): p. 170–174.
54. Doku, G.N. et al., Electric field-induced mobilisation of multiphase solution systems based on the nitration of benzene in a micro reactor. *Analyst*, 2001, **126**(1): p. 14–20.
55. Sands, M. et al., The investigation of an equilibrium dependent reaction for the formation of enamines in a microchemical system. *Lab Chip*, 2001, **1**(1): p. 64–65.
56. Skelton, V. et al., The preparation of a series of nitrostilbene ester compounds using micro reactor technology. *Analyst*, 2001, **126**(1): p. 7–10.
57. De Mas, N. et al. Microchemical systems for direct fluorination of aromatic. In: *IMRET 5: Proc Fifth International Conference on Microreaction Technology*, Springer, Berlin, 2002.
58. Dexter, J.P., Parker, W., Parallel combinatorial chemical synthesis using single-layer poly(dimethylsiloxane) microfluidic devices. *Biomicrofluidics*, 2009, **3**(3): p.
59. Chayen, N.E., Saridakis, E., Protein crystallization: From purified protein to diffraction-quality crystal. *Nat Methods*, 2008, **5**(2): p. 147–153.
60. Wimberly, B.T. et al., Structure of the 30S ribosomal subunit. *Nature*, 2000, **407**(6802): p. 327–339.
61. Krengel, U. et al., Preliminary X-ray crystallographic analysis of the secreted chorismate mutase from *Mycobacterium tuberculosis*: A tricky crystallization problem solved. *Acta Crystallogr Sect F Struct Biol Cryst Commun*, 2006, **62**(Pt 5): p. 441–445.
62. Chayen, N.E. et al., An automated-system for microbatch protein crystallization and screening. *J Appl Crystallogr*, 1990, **23**: p. 297–302.
63. DeLucas, L.J. et al., Efficient protein crystallization. *J Struct Biol*, 2003, **142**(1): p. 188–206.

64. Wilson, L.J., AdcockDowney, L., Pusey, M.L., Monomer concentrations and dimerization constants in crystallizing lysozyme solutions by dialysis kinetics. *Biophys J*, 1996, **71**(4): p. 2123–2129.
65. Moreno, A., Saridakis, E., Chayen, N.E., Combination of oils and gels for enhancing the growth of protein crystals. *J Appl Crystallogr*, 2002, **35**: p. 140–142.
66. Hansen, C.L. et al., A robust and scalable microfluidic metering method that allows protein crystal growth by free interface diffusion. *Proc Natl Acad Sci U S A*, 2002, **99**(26): p. 16531–16536.
67. Zheng, B. et al., A droplet-based, composite PDMS/glass capillary microfluidic system for evaluating protein crystallization conditions by microbatch and vapor-diffusion methods with on-chip X-ray diffraction. *Angewandte Chemie Int Edn*, 2004, **43**(19): p. 2508–2511.
68. Li, L., Du, W., Ismagilov, R.F., Multiparameter screening on SlipChip used for nanoliter protein crystallization combining free interface diffusion and microbatch methods. *J Am Chem Soc*, 2010, **132**(1): p. 112–119.
69. Schnecke, V., Bostrom, J., Computational chemistry-driven decision making in lead generation. *Drug Discovery Today*, 2006, **11**(1–2): p. 43–50.
70. Sundberg, S.A., High-throughput and ultra-high-throughput screening: Solution- and cell-based approaches. *Curr Opin Biotechnol*, 2000, **11**(1): p. 47–53.
71. Mayr, L.M., Bojanic, D., Novel trends in high-throughput screening. *Curr Opin Pharmacol*, 2009, **9**(5): p. 580–588.
72. Wang, Z.H. et al., High-density microfluidic arrays for cell cytotoxicity analysis. *Lab Chip*, 2007, **7**(6): p. 740–745.
73. Bhatia, S.N. et al., Microfabrication of hepatocyte/fibroblast co-cultures: Role of homotypic cell interactions. *Biotechnol Prog*, 1998, **14**(3): p. 378–387.
74. El-Ali, J., Sorger, P.K., Jensen, K.F., Cells on chips. *Nature*, 2006, **442**(7101): p. 403–411.
75. Teh, S.Y. et al., Droplet microfluidics. *Lab Chip*, 2008, **8**(2): p. 198–220.
76. He, M. et al., Selective encapsulation of single cells and subcellular organelles into picoliter- and femtoliter-volume droplets. *Anal Chem*, 2005, **77**(6): p. 1539–1544.
77. Brouzes, E. et al., Droplet microfluidic technology for single-cell high-throughput screening. *Proc Natl Acad Sci U S A*, 2009, **106**(34): p. 14195–14200.
78. Copeland, R.A., ed. *Evaluation of Enzyme Inhibitors in Drug Discover*, Wiley-Interscience, Hoboken, New Jersey, 2005.
79. Prasad, V.R., Goff, S.P., Structure-function studies of HIV reverse transcriptase. *Ann N Y Acad Sci*, 1990, **616**: p. 11–21.
80. Edwards, D.R., Murphy, G., Cancer. Proteases — invasion and more. *Nature*, 1998, **394**(6693): p. 527–528.
81. Copeland, R., *Enzymes: A Practical Introduction to Structure, Mechanism, and Data Analysis*, Wiley-VCH, 2000.
82. Motulsky, H., Christopoulos, A., *Fitting Models to Biological Data using Linear and Nonlinear Regression: A Practical Guide to Curve Fitting*, Oxford University Press, USA, 2004.
83. Copeland, R.A., Mechanistic considerations in high-throughput screening. *Anal Biochem* 2003, **320**(1): p. 1–12.

84. Cortes, A. et al., Relationships between inhibition constants, inhibitor concentrations for 50% inhibition and types of inhibition: New ways of analysing data. *Biochem J*, 2001, **357**(Pt 1): p. 263.
85. Rozman, K. et al., A toxicologist's view of cancer risk assessment. *Drug Metab Rev*, 1996, **28**(1): p. 29–52.
86. Waddell, W.J., Thermodynamic basis for expressing dose logarithmically. *Toxicol Appl Pharmacol*, 2008, **228**(2): p. 156–157.
87. Seefeldt, S., Jensen, J., Fuerst, E., Log-logistic analysis of herbicide dose-response relationships. *Weed Technol*, 1995, **9**(2): p. 218–227.
88. Wang, M. et al., Curcumin induced HepG2 cell apoptosis-associated mitochondrial membrane potential and intracellular free Ca(2+) concentration. *Eur J Pharmacol*, 2011, **650**(1): p. 41–47.
89. Behringer, M.P., ed. *Techniques and Materials in Biology: Care and Use of Living Animals Plants and Microorganisms*, 2nd edn. Krieger Pub. Co., 1989.
90. Yun, J.Y., Jambovane, S., Kim, S.K., Cho, S.H., Duin, E.C., Hong, J.W., Log scale dose response of inhibitors on a chip. *Aral Chem*, 2011, **83**(16): p. 6148–6153.
91. Shargel, L., Yu, A.B.C., eds. *Applied Bipharmaceutics and Pharmacokinetics*, 4th edn. McGraw-Hill, New York, 1993.
92. DiPalma, J.R., DiGregorio, G.J., eds. *Basic Pharmacology in Medicine*, 3rd edn. McGraw-Hill, New York, 1990.
93. Hardman, J.G., Limbird, L.E., Molinoff, P.B., Ruddon, R.W., Gilman, A.G., eds. *Goodman and Gilman's The Pharmacological Basis of Therapeutics*, 9th edn. McGraw-Hill, New York, 1996.
94. Thummel, K.E., Wilkinson, G.R., In vitro and in vivo drug interactions involving human CYP3A. *Annu Rev Pharmacol Toxicol*, 1998, **38**: p. 389–430.
95. Kharasch, E.D. et al., Role of hepatic and intestinal cytochrome P450 3A and 2B6 in the metabolism, disposition, and miotic effects of methadone. *Clin Pharmacol Ther*, 2004, **76**(3): p. 250–269.
96. Tsaioun, K., Kates, S.A., eds. *ADMET for Medicinal Chemists: A Practical Guide*, Wiley Interscience, Hoboken, New Jersey, 2011.
97. Mei, H.Y., Czarnik, A.W., eds. *Integrated Drug Discovery Technologies*, Marcel Dekker, New York, 2005.
98. De Buck, S.S. et al., The prediction of drug metabolism, tissue distribution, and bioavailability of 50 structurally diverse compounds in rat using mechanism-based absorption, distribution, and metabolism prediction tools. *Drug Metab Dispos*, 2007, **35**(4): p. 649–659.
99. Copeland, R., *Evaluation of Enzyme Inhibitors in Drug Discovery: A Guide for Medicinal Chemists and Pharmacologists*, John Wiley & Sons, 2005.
100. Jambovane, S. et al., Determination of kinetic parameters, K_m and k_{cat}, with a single experiment on a chip. *Anal Chem*, 2009, **81**(9): p. 3239–3245.
101. Hong, J.W. et al., A nanoliter-scale nucleic acid processor with parallel architecture. *Nat Biotechnol*, 2004, **22**(4): p. 435–439.

Chapter 3

Microfluidic Glucose Sensors

Jithesh V. Veetil[*,†], *Sruthi Ravindranathan*[†], *Sha Jin*[†],
and Kaiming Ye[†,‡]

*Section of Cellular Biophotonics, National Institute of Health
5625 Fishers Lane, Bethesda, MD 20892, USA
†Biomedical Engineering Program, College of Engineering
203 Engineering Hall, University of Arkansas
Fayetteville, AR 72701, USA

Introduction

Effective healthcare management for diabetics includes the ability to monitor glucose levels in the body frequently or continuously. To this end, various glucose sensors have been developed that are based on different signal transduction mechanisms and recognition elements. These sensors can be categorized into two groups — i.e., (1) point sampling and (2) continuous glucose monitoring (CGM). Commercially available finger-prick glucometers and urine dipsticks exemplify the point sample method of glucose detection. CGM can be realized either invasively or with minimal invasiveness or noninvasively.[1] To date, there are only few CGM sensors available on the market. Most of them are electrochemical sensors that rely upon the enzyme, glucose oxidase (GOx). They can be only used in body for a few days due in part to the instability of GOx at body temperature.[1,2] Hence, there is tremendous interest in developing novel technologies for CGM. On the other hand, minimally or non-invasive CGM sensors can be coupled with an insulin pump

---

‡Corresponding author. Email: kye@uark.edu

to form a closed loop insulin control system that has the potential to prevent various complications associated with hypoglycemia or hyperglycemia.[3,4] As the whole spectrum of glucose sensors used for blood and interstitial fluids is beyond the scope of this chapter, here we focus on the microfluidic technology of and its application for glucose sensor development. More comprehensive discussions on glucose detection can be found elsewhere.[3,5–10]

Microfluidics is an emerging technology that can be used for designing multiplexed detection systems. Sensors developed based on microfluidic technologies can detect molecules in a tiny amount of solution. The reduction of sample volume would significantly accelerate analysis. Another advantage of microfluidic sensors over conventional ones is that they can be fabricated into a small size such as a chip.[11] For this reason, microfluidic sensors have recently gained enormous attention. In this chapter, we will describe several basic principles for developing microfluidic glucose sensors and their applications in point-of-care (POC) diagnostics.

Classification of Microfluidic Glucose Sensors Based on their Fabrication Methods

The basic idea of microfluidics is to control the flow and reaction of a small quantity of fluids in a miniaturized platform so that a system such as an analytic system can be placed on this platform, in a way similar to an electrical system miniaturized for integrated circuit (IC) industry. Due to its small size, many portable devices can be developed and used for POC, enabling cheaper, faster, and disposable tests in various clinical settings. Based on the design/physical nature of the substratum used in sensor fabrication, microfluidic sensors can be categorized into three different types: (1) lab-on-a chip, (2) lab-on-a paper, and (3) lab-on-a disk.

Lab-on-a-chip

This is the most common technology used for microfluidic sensor development. The basic technology for fabricating this type sensor is photolithography. Initially, most sensors were developed on silicon. Later, more materials, including glass and ceramics, have been explored for the development of lab-on-a-chip sensors. Recently, polydimethylsiloxane (PDMS), a cheap, rubber-like elastomer with good optical transparency and biocompatibility, has been widely used in fabricating these sensors. With the development of soft lithography, monolithic fabrication of all necessary microfluidic components within PDMS has become an attractive option.[12,13] Nevertheless, there are some limitations in PDMS-based lab-on-a-chip sensors. For example, the use of high temperature or harsh chemicals would lead to the deterioration of PDMS.

Lab-on-a-paper

This type of microfluidic sensor can also be referred as "lab-on-a-foil" or "lab-on-a-thin film." Due to demands for better optical characteristics, bio- or chemical compatibility, lower production costs, and faster prototyping, paper-based microfluidic sensors have been developed and used in various clinical settings such as home care and bedside.[14,15] In the past, conventional paper strip tests have gained great success in POC due to their simplicity, but they are normally not capable of performing multiplex and quantitative analysis.[16] These issues can, however, be overcome by using paper-based microfluidic devices. The ease of storage and portability adds convenience and reduces the cost of paper-based microfluidic devices. For example, Whitesides and colleagues have designed a paper-based electrochemical device that can detect heavy metal ions such as lead (Pb) in water at levels as low as one part per billion (ppb).[15] To fabricate these sensors, they printed three carbon and silver ink electrodes onto a piece of paper or a polyester film. The microfluidic channels created from polymers are patterned on the paper and form hydrophobic barriers to confine liquids inside the channels and over the electrodes. Apart from papers, plastic materials like thermoplastic, thermoset, and photosensitive polymers, and metals like aluminum substrates have all been explored for developing lab-on-a-paper microfluidic sensors.[17]

Lab-on-a-disk

This type of sensor can also be termed as "lab-on-a-CD" or centrifugal microfluidic devices. These sensors consist of a platform that is a rotating disk in a compact disk (CD) format, featuring fluidic channels and reaction chambers for chemical or biochemical analysis. These disposable disks, fabricated inexpensively like a standard CD, can be used with a reusable disk "player," which basically include a centrifuge. By spinning the disk, the centrifugal forces pump the samples and reagents from the inside to the outer perimeter of the disk. The flow-through can be "valved" by using simple hydrophobic barriers that block a stream of flow that is below a critical rotational frequency. Conversely, beyond that particular frequency, the barriers break and the liquid samples can pass by. In this design, metering of defined amounts of reagents can be achieved by microstructured chambers whose outlet is blocked by a hydrophobic valve and the remaining amount of liquid is drained into a waste chamber *via* an overflow channel. This way, liquid volumes from nano- to microliters can be defined. Extensive reviews on this type of sensor can be found elsewhere.[18,19]

Enzyme-based Microfluidic Glucose Sensors

Amperometric glucose sensors

Glucose sensors are generally composed of a recognition element that is interfaced to a signal transducer, correlating the concentration of the analyte to a measurable response. Enzymes such as GOx or glucose dehydrogenase can serve as a glucose recognition element for generating redox mediator species, resulting in electrochemical signals for detection. Enzymatic glucose sensors have been well developed and widely used for glucose detection. GOx catalyzes the oxidation of glucose, which generates hydrogen peroxide (H_2O_2), and the level of H_2O_2 released or the remaining O_2 associated with this reaction is then measured by a suitable electrode (Fig. 1). By amperometric detection, the output current generated in the aforementioned glucose oxidation is directly proportional to the concentration of glucose. Due to its simplicity, this mechanism has been widely used for developing microfluidic glucose sensors for CGM in body fluids.[20]

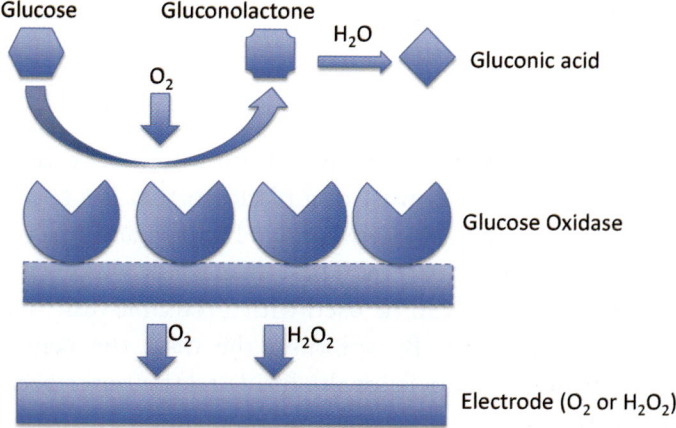

Fig. 1. Principle of enzymatic glucose detection. Glucose is oxidized into gluconolactone by the glucose oxidase enzyme (GOx) in the presence of O_2. The level of H_2O_2 released or the remaining O_2 associated with this reaction is then measured by a suitable electrode.

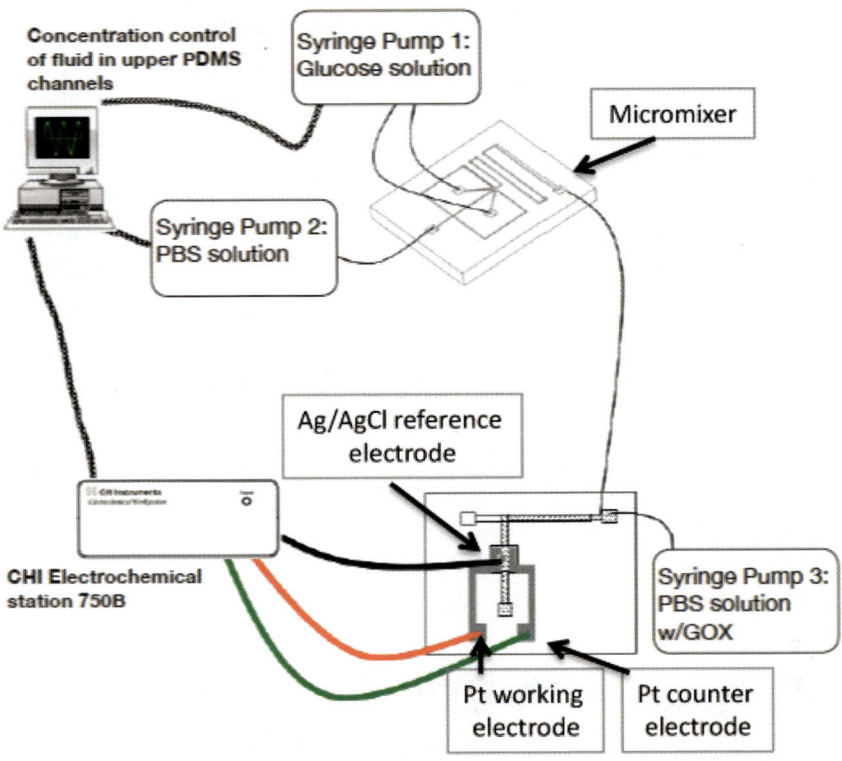

Fig. 2. Schematic diagram of the experimental setup for continuous glucose sensing within an on-chip microdialysis system. (Adapted from Ref. 21.)

Hsieh et al. developed a microdialysis system with platinum electrodes that are fabricated on clean glass slides with SU-8 2010 photoresist for fluidic channels.[21] Figure 2 illustrates the schematic diagram of their sensors. The semi-permeable membrane chosen for microdialysis in the sensor is the polycarbonate track-etch membrane treated by oxygen plasma for 90 s before applying to the fluidic channels. A reservoir channel is created to allow the adjustment of the glucose concentration in the reservoir for testing the sensitivity of the sensor. PDMS is used for fabricating the reservoir and an Ag/AgCl pellet electrode is inserted into the glucose sensing area. The device consists of a PDMS micromixer along with two syringe pumps, where one was filled with phosphate buffered solution (PBS) and the other with glucose solution. By varying the flow rates from the two syringe pumps, the concentration of glucose can be adjusted. A third syringe pump is used to produce a steady perfusion flow of GOx in PBS to the microfluidic channel. In the presence of GOx, H_2O_2 is generated as an end-product of glucose oxidation and is detected using a working electrode.

Enzyme-based optical microfluidic glucose sensors

Optical detection is another method widely employed in microfludic glucose sensors. In these sensors, an optical signal corresponding to light emission, reflection, or absorption due to the interaction of glucose with a specific reagent is captured, converted, and amplified into an electronic signal for detection. For example, colorimetric assays have been long used for the enzymatic detection of glucose.[22–24] Recently, an integrated centrifugal disk-type microfluidic device capable of detecting six different analytes, including glucose, from blood samples was reported (Fig. 3).[25] Specially designed microvalves on these disks could regulate the flow of a sample to various zones in the microfluidic disks. This device uses 350 µL of blood samples and can complete the assay within 22 min. Glucose is detected using hexokinase coupled with an endpoint colorimetric reagent whose saturated absorbance is at 340 nm.

The sensitivity and linear range of optical microfluidic glucose sensors can be further enhanced by connecting a chemiluminescence (CL) sensor to a peristaltic pump and detecting the signal using a luminescence analyzer.[26,27]

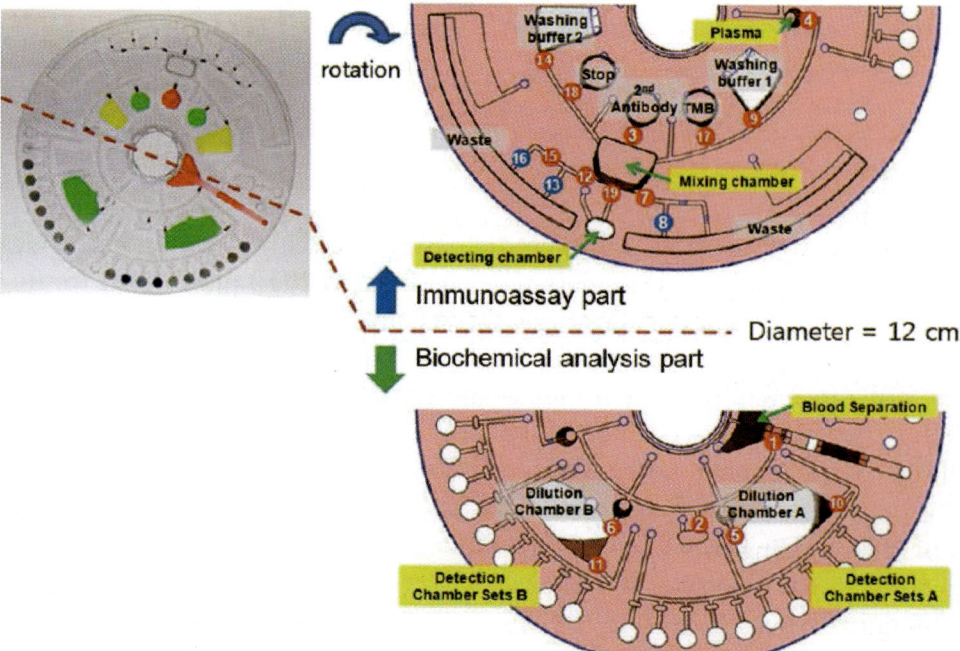

Fig. 3. Photograph of a lab-on-a-disk for multiple analytes. Detection wells are preloaded with lyophilized reagents. Other chambers for liquid-type reagent are shown with food dye solution for demonstration. The disk design shows the detailed microfluidic layout. The number indicates the order of the microvalve operation. (Adapted from Ref. 25.)

Schematic representation of a CL microfluidic sensor is shown in Fig. 4.[26] In this system, air serves as a carrier, which eliminates the possible formation of air bubbles as compared to when a solution is used as a carrier. Once formed, air bubbles will be difficult to remove from the inner walls of microfluidic channels. Air bubbles affect the stability and repeatability of detection considerably.

The fluidic movement is enabled by a negative pressure developed by a peristaltic pump. The CL sensor has two reservoirs (one is an enzyme reservoir and the other is a CL reagent reservoir) built on a glass plate. The two reservoirs are linked by a microchannel. The bottoms of both reservoirs are plugged with glass wool. GOx and analytical reagents, namely luminol and ferricyanide, are co-immobilized onto a controlled-pore glass (CPG) and anion exchange resin, respectively. The midpoint of the microchannel, which connects the enzyme reservoir to the CL reagent reservoir, is linked to a reactor cell (R) via another microchannel. The peristaltic pump is connected to the other end of this microchannel through channel C. The glass plate with the reservoirs and the microchannels is then bonded to a base plate with all except the reaction reservoir shaded by the black dope. Among several tested eluents, Na_3PO_4–NaOH exhibited the best performance in achieving a high CL intensity. For detection, the eluent can be injected into the CL reagent reservoir after the flow of a sample in the enzyme reservoir. The peristaltic pump will transfer and mix the H_2O_2 that is produced from the enzyme-catalyzed reaction of glucose with the CL reagents into the reactor cell, generating CL signals. These signals can be recorded using a computerized ultra-sensitive luminescence analyzer for glucose detection. This device has a linear glucose detection range from 1.1 to 110 mM. Its detection limit is 0.1 mM when applied to detect glucose in human serum.

Enzymatic glucose sensors with integrated sampling systems

For CGM, two approaches can be adopted. One is to implant a sensor directly in the human body, and the other is to connect a sensor to a sampling probe, which is in direct contact with the human body. The introduction of a sampling probe along with a sensor can minimize the issues related to the sensor's biocompatibility and the loss of sensor's activity due to biofouling.[28,29] Microdialysis, microperfusion, and iontophoresis are three sampling techniques that have been widely used for these purposes.[30,31] The principle of microdialysis sampling is quite simple. It is based on the transfer of molecules across a semi-permeable membrane. Only small molecules such as glucose can pass through the membrane and react with GOx to generate detectable signals. Thus, this sampling approach can significantly reduce

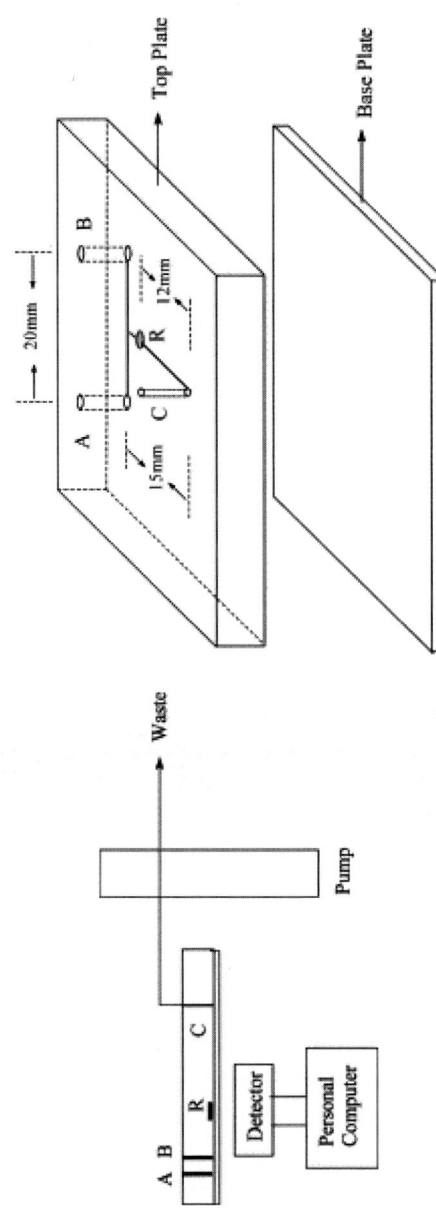

Fig. 4. Schematic diagram of a chemiluminescence microfluidic glucose sensor. A, enzyme reservoir, B, CL-immobilized reagents reservoir, R, reactor cell and C, channel. (Adapted from Ref. 26.)

interference of other molecules with the glucose detection, improving the reliability of the sensor.[33] However, one inherent problem associated with a microdialysis probe is the distance between the probe and sensor. The extended tubing which connects the probe with a sensor could result in signal loss or inconsistency between the measured and actual concentration in a sample. In addition, a long response time due to a high dead volume in the tubing is also problematic. To overcome these shortcomings, the Bio-MEMS (microelectromechanical system) device with an integrated microdialysis probe and a sensor array for glucose detection has been developed.[33,34] Figure 5 illustrates a Bio-MEMS integrated microdialysis glucose sensor.[35] In this sensor system, microdialysis probes are integrated onto a flow channel of a microfabricated electrochemical glucose sensor. The integration is performed using conventional printed circuit equipment that does not require any silicon wafer processing facilities. Dialysis probes are made of biocompatible polyacrylonitrile fibers with a molecular weight cut off value of 50 kDa. The microfabricated glucose sensor consists of conductive pads, a plug for a potentiostat, and a gold counter electrode. The sensor chip is

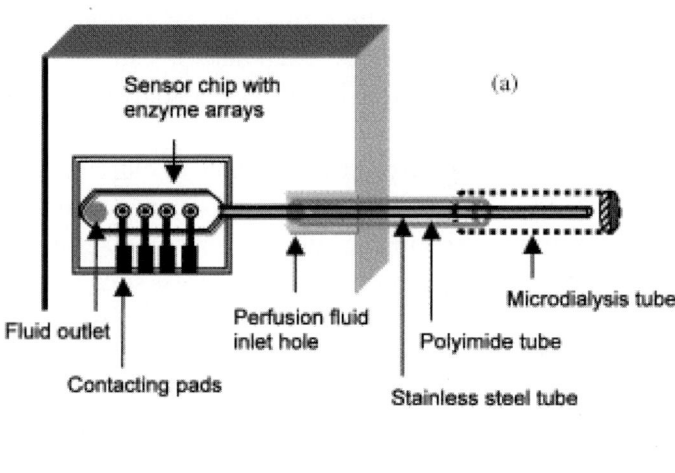

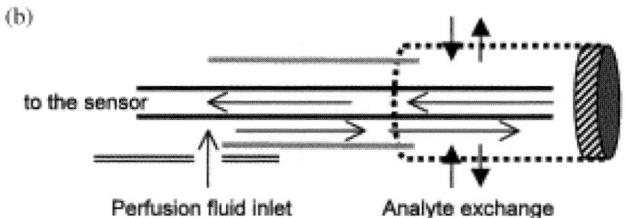

Fig. 5. Schematic diagram of a Bio-MEMS device with an integrated microdialysis probe and a sensor array for glucose detection. (a) A Bio-MEMs device and (b) a dialysis probe. (Adapted from Ref. 33.)

composed of a circular platinum working electrode and an Ag/AgCl reference electrode. GOx is immobilized onto the electrodes by gel entrapment in a photopatterned pHEMA (poly(2-hydroxyethyl methacrylate)) hydrogel membrane. This design reduces the dead volume to about 210 nL. The sensor has a linear glucose detection range from 0 to 30 mM with a sensitivity of 0.38 nA/mM glucose. Initial tests with this system in buffer or serum samples indicated that the sensor is stable and reliable. Experimental results suggested that the sensor has less than 3% variation in sensitivity during 16 days of continuous use in either a 5 mM glucose solution or normal horse serum. The device has an equilibration time of 5 min and a response time of ~2 min with a flow rate of 0.5 µL/min. The sensor can be further modified for continuous monitoring of both glucose and lactate by including two enzymatic reactions in a single device. Glucose can be detected through GOx-catalyzed reactions, whereas the lactate can be detected by lactate oxidase. Both enzymatic reactions generate H_2O_2 that is oxidized at a platinum working electrode, yielding currents for detection. It should be noted that the amperometric determination of multi-analytes are usually affected by the cross-talk of H_2O_2 from the glucose reaction bleeding through to the lactate-sensing area. Nevertheless, the cross-talk can be minimized in a Bio-MEMS device. In Bio-MEMS devices, each sensor can be physically separated, thereby restricting the bleeding of glucose oxidation products through the lactate-sensing area.

Nonspecific signals resulting from the presence of electroactive substances other than H_2O_2 in the sample that diffuses to the working electrodes is another problem in an amperometric glucose sensor system. A number of methods have been developed to eliminate or suppress such interferences. For instance, a membrane of 1,3-phenyldiamine by electropolymerization on the surface of a platinum working electrode was created. Using this strategy, it was found that interference from compounds such as ascorbic acid, 2-acetamidophenol, and 2,4-dihydroxybenzoic acid can be considerably suppressed.[33,34]

In the on-chip microdialysis system illustrated in Fig. 2, in-line sensing electrodes for determining electrical resistance of fluids that are flowed through microchannels have been employed.[21] In this approach, the driving force is the concentration difference between the perfusate and the component of interest, which is usually the interstitial fluid (ISF) in most of the glucose-sensing devices. A semi-permeable membrane, polycarbonate, has been chosen for the microdialysis. A reservoir channel is created to allow for the adjusting of the glucose concentration in the reservoir for testing the sensitivity of the system. The reservoir is fabricated using PDMS, and an Ag/

AgCl pellet electrode is inserted into the glucose-sensing area. Unlike other microfluidic enzymatic glucose sensors, the aforementioned system employs a perfusate containing GOx at a very high concentration of 1,000 units/mL in 5 mL PBS. The third syringe pump is used to produce a steady perfusion flow of GOx in PBS solution to the microfluidic channel. In the presence of GOx, H_2O_2 is produced as an end-product of the reaction. Its further oxidation can be detected by using a working electrode. Experimental results indicated that it takes 20 s for a perfusion flow to reach the sensing area at a flow rate of 0.5 µL/min.

The performance of an on-chip microdialysis sensor may deteriorate over time due to biofouling occurring on its surface. Hence, a number of strategies have been explored and developed to overcome biofouling and the associated decrease in the sensor's sensitivity. Piechotta and colleagues designed a MEMS silicon sensor for CGM based on the cavity principle, as shown in Fig. 6.[30] The sensor consists of a micromachined silicon chip with a cavity etched on it. The cavity ground is composed of a working electrode and a membrane of etched pores. GOx is immobilized onto a matrix of agarose gel and cross-linked by glutaraldehyde. The GOx-immobilized agarose gel acts as the active center of the sensor. When glucose is pumped into the microfluidic channel, it diffuses through the agarose gel and reacts with GOx in the cavity, which releases H_2O_2 that is further oxidized on the platinum electrode surface, giving rise to amperometric signals for detection. This design will allow for the controlling of the access of an analyte to an enzyme reservoir by varying the pore size of the membrane and agarose gel. Such adjustment could lead to increases in the sensor's linear range for both high and low glucose concentrations. When detecting higher glucose concentrations is needed, the number and size of the pores can be reduced to allow only a percentage of molecules

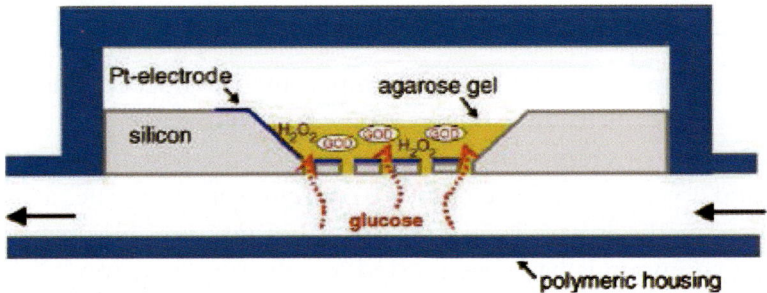

Fig. 6. Schematic representation of the function of a MEMS enzymatic glucose sensor. An analyte solution is pumped through the inlet canula in the flow through cell. While passing through the porous membrane, glucose diffuses through the pores into the cavity, where it is oxidized enzymatically. (Adapted from Ref. 30.)

to diffuse through the pores. In contrast, the size and the number of pores can be increased for detecting low concentrations of glucose. One of the major advances in this design is the inclusion of a second cavity, which acts as a "blank" (no enzyme, only agarose gel). Such a design allows for estimating the contribution of amperometric signals from interfering molecules present in the sample.

The micromachining techniques offer the advantage of high manufacturing precision of multiple cavities. The multiple cavities can be very well used for the parallel determination of multiple analytes such as glucose and lactate. This device has a sensitivity of 20 nA per mM glucose. However, the performance of these devices can be affected by a long response time due in part to the cathedral system and tubing employed for the sampling. At the range of 0.1 to 10 mM glucose concentration, the relative standard deviation varies between 29% and 8%. The flow rate of the sample in the system can be optimized at 0.5 µL/min.

One of the major focuses associated with microfluidic glucose-sensing devices is to develop an integrated system with real-time insulin administration based on the patient's health condition. Although multiple designs are proposed for microfluidic glucose detection chips, the reports on closed loop microfluidic glucose detection and insulin delivery are scarce.[11] A microfluidic chip comprising glucose-sensing electrodes, a flow sensor, and PDMS-based microfluidic structures such as micropumps, microvalves, and microchannels has been explored for real-time glucose sensing and automatic insulin injection.[11] This prototype had a detection range of 1.61 to 30 mM and a sensitivity of 0.99 nA/mM with sample consumption in the range of 30–100 µL.

Enzyme-free Glucose Detection Systems

Microfluidic glucose sensors on carbon nanotubes

The major advantage of GOx-based amperometric glucose sensors is their high chemical specificity and short response time. However, they are sensitive to the alteration of GOx activity and hence, enzyme-based sensors have low stability and a very limited shelf-life.[41] To circumvent the issue, a number of enzyme-free methods have been developed. For instance, boronic acid (BA)-containing compounds have been tested for glucose detection.[42,43] This type of sensor is based on a combination of affinity binding and its induced field effect transistor (FET) effect.[42] In the design proposed by Vlandas et al., the BA receptor was covalently coupled to the surface of trapped carbon nanotubes (CNTs). When glucose binds to BA, it changes the net charge on the

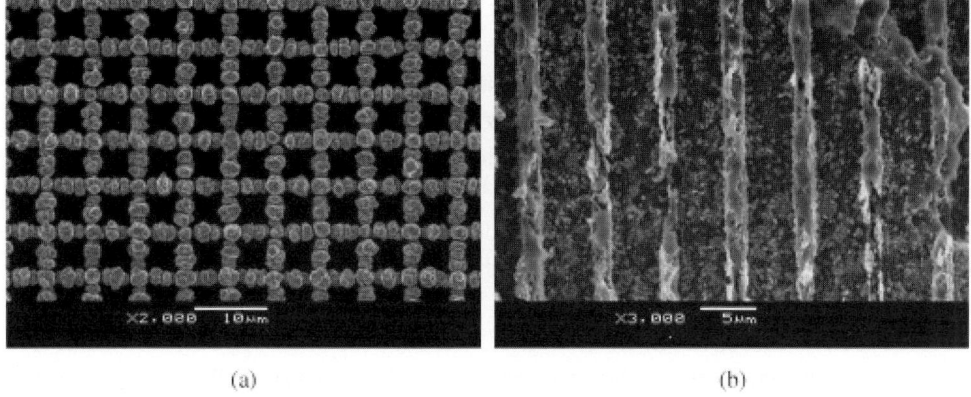

Fig. 7. SEM image of OPPy modified Pd/Si MCP electrode. (a) top view and (b) cross-section. (Adapted from Ref. 40.)

nanotubes, which can be detected through changes in the impedance of the tubes.[42]

Recently, an enzyme-free glucose sensor based on overoxidized polypyrrol (OPPy) on a silicon (Si) microchannel plate (MCP) with modified electrodepositing with palladium (Pd) nanoparticles has been reported, as depicted in Fig. 7.[40] In this design, a platinum (Pt) wire is used as a counter electrode, whereas a saturated calomel electrode is employed as a reference electrode. The microchannel design improves surface area by more than 50-fold compared to conventional planar electrodes. As mentioned earlier, it is possible that several compounds interfere with the sensor response by getting oxidized at the electrode surface. It was shown that the electrodeposited OPPy film considerably reduces such nonspecific signals compared to a bare Pd/Si MCP electrode. The prototype sensors retain about 92% of the initial signal after 30 days storage at room temperature. The sensor has a linear detection range up to 24 mM glucose, with a detection limit of 2.06 µM.

In general, non-enzymatic glucose sensors can undergo surface etching during electrochemical reactions or surface poisoning by chloride ions or intermediate species. This limits their linear and sensitivity detection ranges. Hence, attempts have been made to develop novel electrodes for enzyme-free glucose detection. For example, electrochemical glucose sensors have been designed using CNTs due to their increased electrochemical activity and reduced size.[35–37] In this direction, Yang *et al.* synthesized a novel nanocomposite of copper (Cu) nanocubes and multi-walled carbon nanotubes (MWCNTs). In their design, vertical arrays of MWCNTs are synthesized on tantalum (Ta) foils and connected to the surface of a glassy carbon (GC) electrode with conductive silver paint. The Cu nanocubes are electrochemically coated onto the MWCNT array by potentiostatic deposition. The embellishment of MWCNTs with Cu gives them a high aspect ratio and large surface area. The mechanism of detection by this sensor depends on the electrocatalytic oxidation of glucose by the metallic Cu via multiple steps on the MWCNT surface. The sensor developed based on this design has a sensitivity of 1,096 µA/mM/m^2 compared to the 367 µA/mM/cm^2 for the OPPy-coated Pd sensor.[40] The Cu–MWCNT glucose sensor has also been reported to have a low detection limit of 1 µM, with a response time of less than 1 s; however, its linear range is only up to 7.5 mM. Likewise, a Pd nanoparticle–single-walled carbon nanotube (Pd–SWCNT) hybrid nanostructure, explored for the non-enzymatic oxidation of glucose, yielded about 3 s response time, a linear range of 0.5–17 mM, with a lower detection limit of approximately 0.2 µM.[38]

In addition, the use of bare Pt electrodes for enzyme-free amperometric glucose detection has also been explored. The employment of bare Pt electrodes, however, is susceptible to poisoning by adsorbed intermediates. Thus, alloy electrocatalysts for use in enzyme-free amperometric glucose sensors have been investigated. It has been suggested that the Pt–Pb alloy provides better results since they are comparatively less susceptible to surface fouling than pure Pt electrodes.[44] These electrodes, however, still suffer from poisoning by chloride ions in spite of their better response. To overcome this problem, a microfluidic chip with nanoporous Pt and Pt/Pt oxide as a reference electrode has been developed.[45]

Microfluidic glucose sensors on paper and foil

As discussed earlier, one of the promising microfuidic glucose detection devices is designed using the principle of "lab-on-a-paper." This technology employs a paper as a fabrication substrate rather than a silicon wafer or a polymer-coated glass or plastic.[46] The system has been developed through the application of the photolithography technique and screen-printing technology to make microfluidic channels and electrodes.[47] Paper-based systems are relatively simple, inexpensive, disposable, and are also portable.[48] Figure 8 illustrates the design of an electrochemical lab-on-a-paper glucose detection system.[49] In this design, photolithography is used to pattern Whatman filter paper and the electrodes are screen-printed on it. Whitesides' group has developed a three-dimensional (3D) microfluidic paper analytical device (µPADs) by stacking alternating layers of paper and water-impermeable

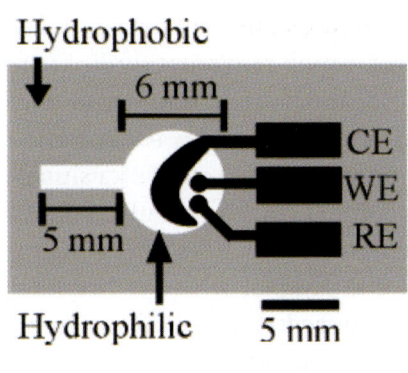 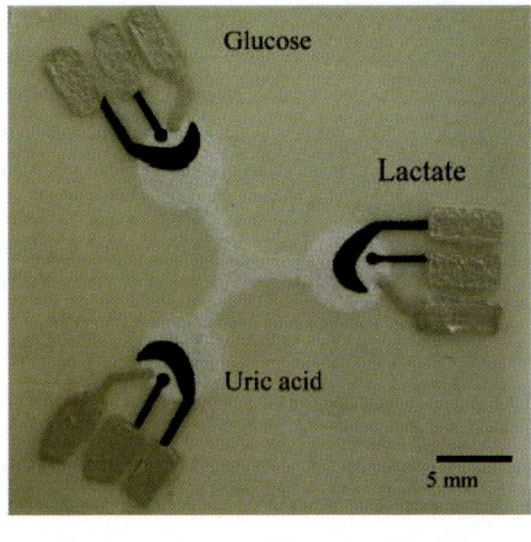

Fig. 8. (a) Schematic representation of a paper-based microfluidic device with electrochemical detection. RE: reference electrode, WE: working electrode, and CE: counter electrode. (b) Snapshot of multielectrode paper-based microfluidic device in which the hydrophilic area at the center wicks sample into three different electrodes immobilized with analyte-specific enzymes. The device size is about 16 cm². (Adapted from Ref. 49.)

double-sided adhesive tape both of which patterned in such a way that the flow of fluid is channeled within and between layers of paper.[50] In this design, the hydrophobic polymer patterned into the paper demarcates the channels through which the fluids move laterally. The layers of water-impermeable double-sided tape separate channels in neighboring layers of paper, and holes cut into 3D microfluidics chips are capable of shuttling fluids such as serum into discrete compartments (detection zones). This also offers an advantage in that one could take four different fluid samples and evenly distribute them into more than 1,000 separate detection zones. Apart from the cost advantage, the use of papers also aids in actively wicking fluids, moving them along separate channels, preventing the mixture and distribution of the liquid into analyzable quantities. Analyses on such fractionated liquid samples can assist in multiplex detection — for example, simultaneous tests for different pollutants in water samples and a suite of assays to detect disease antibodies in human blood or glucose levels in urine. The wicking property of the paper also eliminates the need for electric micropumps, which are required to move liquids through lab-on-chips made from glass or other polymeric substratum.

In another study, Noh and his colleagues integrated a "fluidic timer" on the µPAD design, which can track the endpoint of the time-based assay performed on paper-based microfluidic devices.[52,53] Basically, the timer consists of three components: (1) hydrophilic channels of paper that directly link the timer to the assay (the sample wicks through channels that lead simultaneously to the assay and the timer); (2) a metering region which is composed of paraffin wax in paper to control the flow rate of the fluid; and (3) a signaling component that provides an unambiguous signal to the end user so that the results of the assay can be read. The signaling component is dye, which is carried into the bottom layer of the timer by the sample, where it becomes visible. Overall, paper-based assays have demonstrated the capability of performing multiplexed assays on a piece of paper with small volumes of samples and can be realized simpler on-site analysis. However, the methods to generate patterned paper with photoresist or PDMS suffer from relatively complex fabrication processes. This is one of the major factors contributing to the higher cost of the materials. Hence, improvised methods for reducing the cost of paper-based microfluidic devices have been attempted. For example, a simple and low-cost production method to generate paper-based microfluidic devices with wax for portable bioassays has been reported by Lu and co-workers.[53] The wax patterning method on paper includes three different ways: (1) painting with a wax pen; (2) printing with an inkjet printer followed by painting with a wax pen; and (3) printing by a wax printer directly. The whole process is relatively simple and can be finished within 5–10 min without the use of a clean room, a UV lamp, and an organic solvent. The approach has reduced the cost for fabrication significantly.[53]

Another new system similar to paper-based microfluidic devices is the droplet-based glucose detection system. Droplet-based microfluidic systems use discretized microdroplets instead of the conventional continuous flow method for glucose detection. This type of system is comparatively easy to fabricate and reconfigure. The droplet-based microfluidic system consists of an electrowetting chip, which is a technique used to actuate the droplets.[54] A schematic diagram of a microfluidic electrowetting chip is shown in Fig. 9, wherein the LED placed on a perpendicular plane to the chip act as a light source. The intensity of the light absorbed is converted into a corresponding voltage by a photodiode, which is correlated to glucose concentration. In this design, the microdroplets of a sample and a reagent are mixed on the electrowetting chip where Trinder's reaction occurs (Figs. 9(b) and 9(c)). Glucose forms H_2O_2 in the presence of GOx, which in turn reacts with 4-amino antipyrine (4-AAP) and N-ethyl-N-sulfopropyl-m-toluidine in the presence of peroxide to form violet-colored quinoneimine. The concentration of glucose can then be

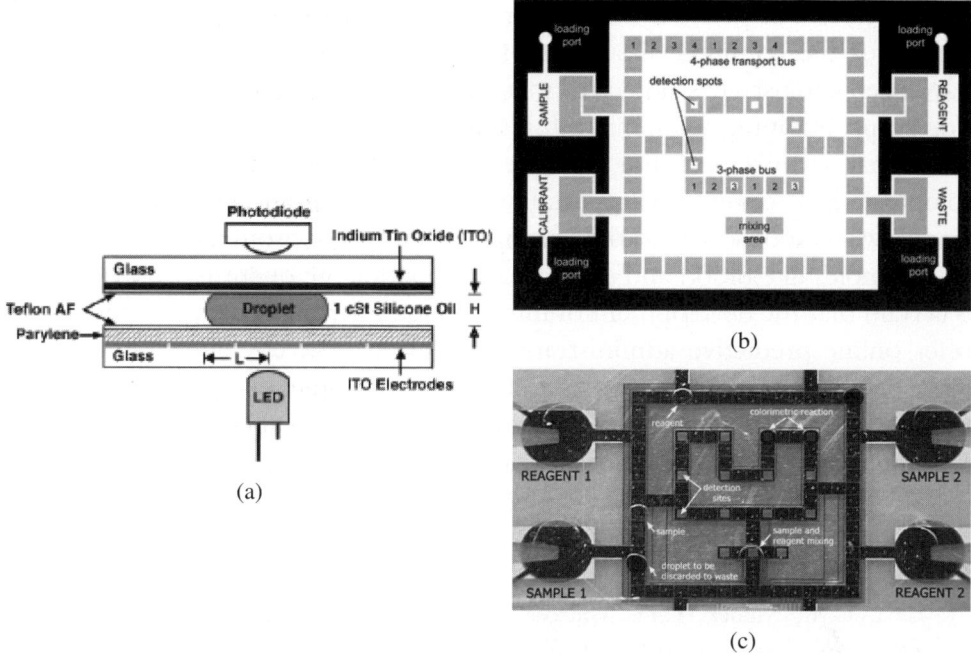

Fig. 9. (a) Vertical cross-section of the electrowetting set up along with the optical absorbance measurement instrument. In this set up, an LED (light source) and a photodiode (light to voltage converter) are assembled in a plane perpendicular to that of electrowetting chip. (b) Schematic of the lab-on-a-chip for the analysis of multiple analytes on a digital microfluidic platform. The lab-on-a-chip include reservoirs, droplet transport pathways (buses), and a waste area. The reservoir functions to store liquids, from which unit-sized droplets can be generated as well as to collect the waste. (c) A snapshot of the chip during the operation of the assay in the electrowetting microfluidic set up. The figure shows colorimetric reactions on two droplets, sample and reagent mixing, and another sample and reagent droplet being transported outer bus toward the mixing area, and a droplet being moved to the waste area — all occurring simultaneously. (Adapted from Ref. 54.)

calculated by a kinetic method wherein the concentration is proportional to the rate of change of absorbance.[54]

Challenges and Prospectives

In the field of diabetes technology, the "artificial pancreas" (the closed loop glucose detection and insulin delivery system) is the Holy Grail in this respect. The demand for such a system is huge, but is yet to be commercially realized, with the exception of a handheld device approved by the FDA as artificial pancreas outpatient trials began in May 2012. A number of microfluidic glucose sensors have been discussed above. In general, linearity, sensitivity, selectivity, and response time are four major parameters that need to be optimized for the commercial success of any glucose sensor. These parameters have been extensively studied for many commercially available glucose detection strips, but integrating microfluidic designs for glucose detection warrants additional optimization. For example, microfluidic devices that are intended to operate in a continuous/repeatable mode with biological samples generally face slower response due to the gradual adsorption of proteins on the surface, with samples like blood and serum containing a large quantity of proteins. It is certain that the development of miniaturized microfluidic CGM sensors can offer online predictive administration of insulin, thereby allowing a closed loop between glucose response and insulin dosing. Future generations of CGM sensors are likely to provide clinical benefits to diabetic patients including improved accuracy, ease of use, small size, and robustness.

References

1. Chia, C.W., Saudek, C.D., Glucose sensors: Toward closed loop insulin delivery. *Endocrinol Metab Clin North Am*, 2004, **33**(1): p. 175–195.
2. Wang, J., Electrochemical glucose biosensors. *Chem Rev*, 2008, **108**(2): p. 814–825.
3. Kondepati, V., Heise, H., Recent progress in analytical instrumentation for glycemic control in diabetic and critically ill patients. *Anal Bioanal Chem*, 2007, **388**(3): p. 545–563.
4. Klonoff, D.C., Continuous glucose monitoring. *Diabetes Care*, 2005, **28**(5): p. 1231–1239.
5. Koschinsky, T., Heinemann, L., Sensors for glucose monitoring: Technical and clinical aspects. *Diabetes Metab Res Rev*, 2001, **17**(2): p. 113–123.
6. Ellis, S.L., N.R., Gemperline, K., Garg, S.K., Use of continuous glucose monitoring in patients with type 1 diabetes. *Curr Diabetes Rev*, 2008, **4**(3): p. 207–217.
7. Renard, E., Implantable continuous glucose sensors. *Curr Diabetes Rev*, 2008, **4**(3): p. 169–174.
8. Heller, A., Feldman, B., Electrochemical glucose sensors and their applications in diabetes management. *Chem Rev*, 2008, **108**(7): p. 2482–2505.
9. Oliver, N.S., Toumazou, C., Cass, A.E., Johnston, D.G., Glucose sensors: A review of current and emerging technology. *Diabetes Med*, 2009, **26**(3): p. 197–210.
10. Renard, E., Implantable continuous glucose sensors. *Curr Diabetes Rev*, 2008, **4**(3): p. 169–174.
11. Huang, C.J., Chen, Y.H., Wang, C.H., Chou, T.C., Lee, G.B., Integrated microfluidic systems for automatic glucose sensing and insulin injection. *Sens Actuators B*, 2007, **122**(2): p. 461–468.
12. Mark, D., Haeberle, S., Roth, G., von Stetten, F., Zengerle, R., Microfluidic lab-on-a-chip platforms: Requirements, characteristics and applications. *Chem Soc Rev*, 2010, **39**(3): p. 1153–1182.
13. Sia, S.K., Whitesides, G.M., Microfluidic devices fabricated in poly(dimethylsiloxane) for biological studies. *Electrophoresis*, 2003, **24**(21): p. 3563–3576.

14. Mabey, D., Peeling, R.W., Ustianowski, A., Perkins, M.D., Diagnostics for the developing world. *Nat Rev Microbiol*, 2004, **2**(3): p. 231–240.
15. Nie, Z., Nijhuis, C.A., Gong, J., Chen, X., Kumachev, A., Martinez, A.W., Narovlyansky, M., Whitesides, G.M., Electrochemical sensing in paper-based microfluidic devices. *Lab Chip*, 2010,s **10**(4): p. 477–483.
16. Zhao, W., van der Berg, A., Lab on paper. *Lab Chip*, 2008, **8**(12): p. 1988–1991.
17. Focke, M., Kosse, D., Muller, C., Reinecke, H., Zengerle, R., von Stetten, F., Lab-on-a-Foil: Microfluidics on thin and flexible films. *Lab Chip*, 2010, **10**(11): p. 1365–1386.
18. Madou, M., Zoval, J., Jia, G., Kido, H., Kim, J., Kim, N., Lab on a CD. *Annu Rev Biomed Eng*, 2006, **8**: p. 601–628.
19. Gorkin, R., Park, J., Siegrist, J., Amasia, M., Lee, B.S., Park, J.M., Kim, J., Kim, H., Madou, M., Cho, Y.K., Centrifugal microfluidics for biomedical applications. *Lab Chip*, 2010, **10**(14): p. 1758–1773.
20. Moon, B.U., Koster, S., Wientjes, K.J., Kwapiszewski, R.M., Schoonen, A.J., Westerink, B.H., Verpoorte, E., An enzymatic microreactor based on chaotic micromixing for enhanced amperometric detection in a continuous glucose monitoring application. *Anal Chem*, 2010, **82**(16): p. 6756–6763.
21. Hsieh, Y.C., Zahn, J.D., On-chip microdialysis system with flow-through glucose sensing capabilities. *J Diabetes Sci Technol*, 2007, **1**(3): p. 375–383.
22. Cairns, A.J., Colorimetric microtiter plate assay of glucose and fructose by enzyme-linked formazan production: Applicability to the measurement of fructosyl transferase activity in higher plants. *Anal Biochem*, 1987, **167**(2): p. 270–278.
23. Kabasakalian, P., Kalliney, S., Westcott, A., Enzymatic blood glucose determination by colorimetry of N,N-diethylaniline-4-aminoantipyrine. *Clin Chem*, 1974, **20**(5): p. 606–607.
24. Wang, X-D., Chen, H-X., Zhou, T-Y., Lin, Z-J., Zeng, J-B., Xie, Z-X., Chen, X., Wong, K-Y., Chen, G-N., Wang, X-R., Optical colorimetric sensor strip for direct readout glucose measurement. *Biosens Bioelectron*, 2009, **24**(12): p. 3702–3705.
25. Lee, B.S, Lee, Y.U., Kim, H-S., Kim, T-H., Park, J., Lee, J-G., Kim, J., Kim, H., Lee, W.G., Cho, Y-K., Fully integrated lab-on-a-disc for simultaneous analysis of biochemistry and immunoassay from whole blood. *Lab Chip*, 2011, **11**(1): p. 70–78.
26. Lü, Y., Zhang, Z., Chen, F., Chemiluminescence microfluidic system sensor on a chip for determination of glucose in human serum with immobilized reagents. *Talanta*, 2003, **59**(3): p. 571–576.
27. Lee, S.H., Karim, M.M., ed. Recent chemiluminescence applications for glucose sensing. In *Glucose Sensing*, Vol. 11. Geddes, C.D., Lakowicz, J.R., eds. Springer, USA, 2006, p. 311–322.
28. Wisniewski, N., Moussy, F., Reichert, W.M., Characterization of implantable biosensor membrane biofouling. *Fresenius J Anal Chem*, 2000, **366**(6–7): p. 611–621.
29. Wisniewski, N., Reichert, M., Methods for reducing biosensor membrane biofouling. *Colloids Surf, B* 2000, **18**(3–4): p. 197–219.
30. Piechotta, G., Albers, J., Hintsche, R., Novel micromachined silicon sensor for continuous glucose monitoring. *Biosens Bioelectron*, 2005, **21**(5): p. 802–808.
31. Maran, A., Crepaldi, C., Tiengo, A., Grassi, G., Vitali, E., Pagano, G., Bistoni, S., Calabrese, G., Santeusanio, F., Leonetti, F. *et al.*, Continuous subcutaneous glucose monitoring in diabetic patients: A multicenter analysis. *Diabetes Care*, 2002, **25**(2): p. 347–352.

32. Clarke, W.L., Anderson, S., Farhy, L., Breton, M., Gonder-Frederick, L., Cox, D., Kovatchev, B., Evaluating the clinical accuracy of two continuous glucose sensors using continuous glucose error grid analysis. *Diabetes Care*, 2005, **28**(10): p. 2412–2417.
33. Petrou, P.S., Moser, I., Jobst, G., BioMEMS device with integrated microdialysis probe and biosensor array. *Biosens Bioelectron*, 2002, **17**(10): p. 859–865.
34. Petrou, P.S., Moser, I., Jobst, G., Microdevice with integrated dialysis probe and biosensor array for continuous multi-analyte monitoring. *Biosens Bioelectron*, 2003, **18**(5–6): p. 613–619.
35. Yang, J., Jiang, L.C., Zhang, W.D., Gunasekaran, S., A highly sensitive non-enzymatic glucose sensor based on a simple two-step electrodeposition of cupric oxide (CuO) nanoparticles onto multi-walled carbon nanotube arrays. *Talanta*, 2010, **82**(1): p. 25–33.
36. Yang, J., Zhang, W.D., Gunasekaran, S., An amperometric non-enzymatic glucose sensor by electrodepositing copper nanocubes onto vertically well-aligned multi-walled carbon nanotube arrays. *Biosens Bioelectron*, 2010, **26**(1): p. 279–284.
37. Jiang, L.C., Zhang, W.D., A highly sensitive nonenzymatic glucose sensor based on CuO nanoparticles-modified carbon nanotube electrode. *Biosens Bioelectron*, 2010, **25**(6): p. 1402–1407.
38. Meng, L., Jin, J., Yang, G., Lu, T., Zhang, H., Cai, C., Nonenzymatic electrochemical detection of glucose based on palladium-single-walled carbon nanotube hybrid nanostructures. *Anal Chem*, 2009, **81**(17): p. 7271–7280.
39. Kang, X., Mai, Z., Zou, X., Cai, P., Mo, J., A sensitive nonenzymatic glucose sensor in alkaline media with a copper nanocluster/multiwall carbon nanotube-modified glassy carbon electrode. *Anal Biochem*, 2007, **363**(1): p. 143–150.
40. Shi, J., Ci, P., Wang, F., Peng, H., Yang, P., Wang, L., Ge, S., Wang, Q., Chu, P.K., Nonenzymatic glucose sensor based on over-oxidized polypyrrole modified Pd/Si microchannel plate electrode. *Biosens Bioelectron*, 2011, **26**(5): p. 2579–2584.
41. Ueki, T., Aoki, S., Ishii, K., Imabayashi, S., Watanabe, M., Effect of a modification site on the electron-transfer reaction of glucose oxidase hybrids modified with phenothiazine via a poly(ethylene oxide) spacer. *Langmuir*, 2004, **20**(21): p. 9177–9183.
42. Vlandas, A., Kurkina, T., Ahmad, A., Kern, K., Balasubramanian, K., Enzyme-free sugar sensing in microfluidic channels with an affinity-based single-wall carbon nanotube sensor. *Anal Chem*, 2010, **82**(14): p. 6090–6097.
43. Tiwari, A., Terada, D., Yoshikawa, C., Kobayashi, H., An enzyme-free highly glucose-specific assay using self-assembled aminobenzene boronic acid upon polyelectrolytes electrospun nanofibers-mat. *Talanta*, 2010, **82**(5): p. 1725–1732.
44. Sun, Y., Buck, H., Mallouk, T.E., Combinatorial discovery of alloy electrocatalysts for amperometric glucose sensors. *Anal Chem*, 2001, **73**(7): p. 1599–1604.
45. Joo, S., Park, S., Chung, T.D., Kim, H.C., Integration of a nanoporous platinum thin film into a microfluidic system for non-enzymatic electrochemical glucose sensing. *Anal Sci*, 2007, **23**(3): p. 277–281.
46. Lu, Y., Lin, B., Qin, J., Patterned paper as a low-cost, flexible substrate for rapid prototyping of PDMS microdevices via "liquid molding". *Anal Chem*, 2011, **83**(5): p. 1830–1835.
47. Martinez, A.W., Phillips, S.T., Nie, Z., Cheng, C.M., Carrilho, E., Wiley, B.J., Whitesides, G.M., Programmable diagnostic devices made from paper and tape. *Lab Chip*, 2010, **10**(19): p. 2499–2504.

48. Martinez, A.W., Phillips, S.T., Whitesides, G.M., Carrilho, E., Diagnostics for the developing world: Microfluidic paper-based analytical devices. *Anal Chem,* 2010, **82**(1): p. 3–10.
49. Dungchai, W., Chailapakul, O., Henry, C.S., Electrochemical detection for paper-based microfluidics. *Anal Chem,* 2009, **81**(14): 5821–5826.
50. Martinez, A.W., Phillips, S.T., Whitesides, G.M., Three-dimensional microfluidic devices fabricated in layered paper and tape. *Proc Natl Acad Sci,* 2008, **105**(50): p. 19606–19611.
51. Noh, H., Phillips, S.T., Fluidic timers for time-dependent, point-of-care assays on paper. *Anal Chem,* 2010, **82**(19): p. 8071–8078.
52. Noh, H., Phillips, S.T., Metering the capillary-driven flow of fluids in paper-based microfluidic devices. *Anal Chem,* 2010, **82**(10): p. 4181–4187.
53. Lu, Y., Shi, W., Jiang, L., Qin, J., Lin, B., Rapid prototyping of paper-based microfluidics with wax for low-cost, portable bioassay. *Electrophoresis,* 2009, **30**(9): p. 1497–1500.
54. Srinivasan, V., Pamula, V.K., Fair, R.B., Droplet-based microfluidic lab-on-a-chip for glucose detection. *Anal Chim Acta,* 2004, **507**: p. 145–150.

Chapter 4

Applications of Microfabrication and Microfluidic Techniques in Mesenchymal Stem Cell Research

Abhijit Majumder[*,†,‡,§]*, Jyotsna Dhawan*[*,∥]*, Oren Levy*[†,‡,§]*, and Jeffrey M. Karp*[†,‡,§,††]

[*]*Institute for Stem Cell Biology and Regenerative Medicine (inStem)*
Bangalore 560065, India
[†]*Division of Biomedical Engineering, Department of Medicine*
Center for Regenerative Therapeutics
Brigham and Women's Hospital, Harvard Medical School
Cambridge, MA 02139, USA
[‡]*Harvard Stem Cell Institute, Harvard University*
Cambridge, MA 02138, USA
[§]*Harvard–MIT Division of Health Sciences and Technology*
Massachusetts Institute of Technology
Cambridge, MA 02139, USA
[∥]*Center for Cellular and Molecular Biology*
Council of Scientific and Industrial Research
Hyderabad 500007, India

Introduction

Mesenchymal stem cells and their importance in regenerative medicine

The great clinical promise attributed to mesenchymal stem cells (MSCs) in regenerative medicine stems from their colony-forming ability, multi-lineage

[††]Corresponding author. Email: jkarp@rics.bwh.harvard.edu

potential,[1–5] ability to promote angiogenesis, stabilize new blood vessels,[6–10] and to promote immunomodulatory factors that can downregulate inflammation.[11–15] Addressing these cells as "stem cells" proves to be rather controversial due to their limited differentiation capabilities and lack of clonal analysis and as a result, they are often referred to as non-hematopoietic bone marrow stromal cells (BMSCs) or multipotent mesenchymal stromal cells. According to the International Society for Cellular Therapy, MSCs must fulfill three criteria:[16] (1) adherance to tissue culture plastic; (2) expression of a typical set of surface markers such as CD73, CD90, and CD105, and a lack of lineage-specific markers such as CD34, CD14 or CD11b, CD79a or CD19, CD45, and HLA-DR; and (3) multi-lineage differentiation capacity. Under standard *in vitro* differentiation conditions, MSCs are able to differentiate into cells of the major mesodermal lineages, including osteoblasts, chondrocytes, adipocytes, and myoblasts, and arguably are also able to trans-differentiate into the neuronal lineage.[2,17–29] In addition to their multi-lineage differentiation capacity, MSCs are also relatively easy to isolate and culture *in vitro* due to their highly proliferative nature. Importantly, MSCs are completely devoid of any ethical issues associated with the use of embryonic stem cells (ESCs). In addition, MSCs were shown to have immunosuppressive and immunomodulatory functions upon transplantation.[14,30] As a result, both autologous and allogenic MSCs are considered clinically attractive for regenerative bone and muscle degenerative diseases and injuries, heart disease, graft versus host disease during bone marrow transplantation, inflammatory diseases, and cosmetic reconstructive therapies.[31–40] It is important to note that despite all these potential advantages, MSC transplantation has not yet become the standard of care for any therapeutic applications.

Challenges in MSC research and the role of microfabrication and microfluidics

To harvest the benefits of MSCs for regenerative medicine, it is important to expand them in culture and direct their phenotype with high efficiency and control. Critical challenges include: (1) achieving *ex vivo* expansion without compromising potency, as MSCs tend to lose their phenotype after six to eight passages, (2) maintaining cell survival following transplantation, and (3) efficiently directing the homing of MSCs to sites of injury or disease.[40–42] To address these issues, it should be useful to identify the factors that maintain the stem cell pool undifferentiated inside the body and efficiently employ them when required. It is now clear that in addition to soluble chemical signals, the extracellular tissue-specific microenvironment (the niche) serves an important role in determining cell behavior, including cell spreading, migration, proliferation, quiescence, survival, cytokine secretion, and differentiation.[43–48] The niche may impact the spatio-temporal availability of cytokines and growth factors and provide mechanical cues in the form of contact guidance, topology, stiffness, and ligand distribution.[44,49] To better understand the role of the niche, including favorable biomechanical conditions, it must be noted that tissue culture plastic is a poor replica of *in vivo* systems. First, a plastic dish is at least five to six orders of magnitude stiffer than any physiological niche.[50] Second, plastic dishes do not provide any three-dimensional (3D) cues, which are ubiquitous in all extracellular matrices (ECMs). Finally, standard plastic dishes do not offer any spatio-temporal control over cell size, cell shape, focal adhesion size/number, cell contact, nutrient supply, oxygen supply, and waste removal, which are all essential elements of any tissue system. In recent years, advances in biomaterials synthesis and 3D scaffold design have partially solved the problems of rigidity and dimension.[51–55] However, even a 3D scaffold fabricated from biomaterials of tissue-equivalent physicochemical properties may not provide the critical mechanical cues if the scaffold is made randomly without nanoscale precision. From this perspective, microfabrication and microfluidics provide important alternatives, which can accurately mimic important geometric features of an ECM-like topology, fluid flow, or stiffness distribution.[56–59] Moreover, microfabrication is also useful in decoupling different mechano-chemical signals, permitting separate investigation of each individual element. In summary, emerging microfabrication techniques are enabling the deconstruction and reconstruction of the niche to analyze and regulate cell behavior in greater detail.

Microfabrication: Cell-adhesive islands and micro-/nanotopography

The mechanical stimuli that critically influence MSC fate include ECM stiffness, cell shape and size, micro-/nanotopography, fluid shear force, and external vibration.[60–65] These physical/geometric factors are usually strongly coupled. For example, while cell shape and size can be two independent variables, they may also be tightly related to other physical/chemical stimuli such as material stiffness or ECM composition.[60,65,66] Therefore, it is important to decouple mechano-signals to fully understand their individual effects. Microcontact printing of cell-adhesive islands and substrates with micro/nano features are useful in this context.[67–69]

Micropatterning of cell-adhesive islands

Substrate patterning with cell-adhesive islands utilizes a variety of techniques such as microcontact printing, microfluidic patterning, inkjet printing, stencil-assisted patterning, photolithography, Dip pen lithography, e-beam lithography, and direct placing of the cells with laser tweezers.[70,71] The details of these methods are discussed elsewhere in this book. Other than laser-guided cell patterning, all other processes depend on preferential adhesion of the cells to specific materials. For example, in microcontact printing, cell-adhesive regions are often inked with ECM proteins such as fibronectin, laminin, collagen, or other charged molecules such as poly-L-lysine (PLL), while the rest of the substrate area is backfilled with cytophobic molecules such as PLL-grafted polyethylene glycol (PLL–PEG) (for a hydrophilic substrate) or Pluronics® (for a hydrophobic surface).[72] Falconnet et al.[71] have previously provided a thorough review on two-dimensional (2D) cell patterning techniques.

Effect of cell size

Patterning of the substrate with cell-adhesive islands is useful in restricting cell spreading and can be used to explore the effect of cell size on cellular physiological processes such as proliferation and differentiation. During the last four decades, numerous studies have demonstrated a close relation between cell spreading and different physiological activities.[73–78] The classical works of the Penman lab, using polyHEMA substrates, showed for the first time that the extent of cell spreading influences cell proliferation.[73,79–81] Spigelman and Ginty[74] showed that the fibronectin coating-mediated spreading of an adipogenic cell line (3T3-F442A) inhibits lipogenic differentiation. Although their work revealed the link between differentiation and cell size, cell spreading was essentially unrestricted. In 1997, it was shown that restricting endothelial cells to smaller sizes increased the level of apoptotic cell death.[77] In the case of MSCs that exhibit multiple lineage commitment options, the situation is more complex. Lineage commitment of human MSCs (hMSCs) depends on initial cell plating density;[2,78] at high plating density, there is a preference for adipogenic differentiation, whereas low plating density favors osteogenesis. Whether the plating density controls lineage commitment directly or controls the cell shape, which indirectly determines lineage, is not yet known. To probe this question, McBeath et al.[78] printed fibronectin islands of three different sizes (1,024, 2,025, and 10,000 μm^2) on a polydimethylsiloxane (PDMS) substrate (Fig. 1). Larger cell sizes favored osteogenic differentiation while the least spread, or rounded cells, favored adipogenic differentiation, suggesting that cell shape acts as an adipogenic–osteogenic switch in hMSC lineage commitment. The adhesion-dependent RhoA–ROCK pathway was identified as a crucial player in cell shape-dependant lineage commitment. In fact, directly altering RhoA–ROCK activity, using chemical inhibitors or genetic manipulation, proved to be an efficient substitute for soluble differentiation factors.[78] This work focused on cell spreading and not cell

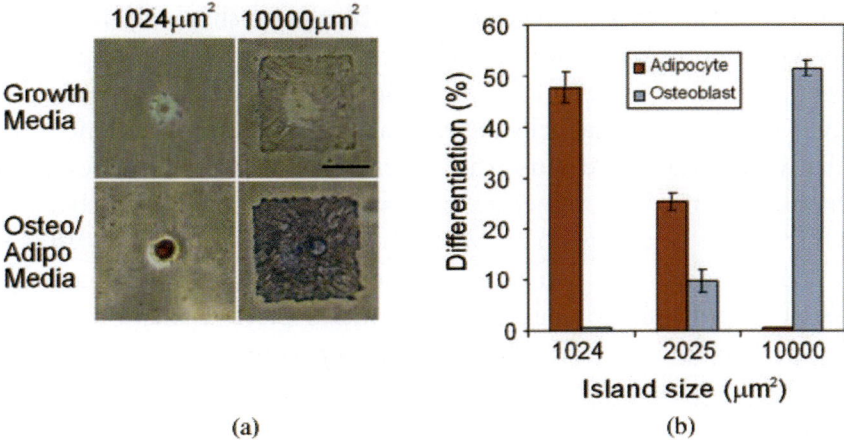

Fig. 1. hMSCs micropatterned on islands of different sizes after one week in growth media or mixed media show size-dependent differentiation fate. While a small cell size leads to adipogenesis, a large cell size enhances osteogenic differentiation. Red: lipid stain, blue: alkaline phosphate stain. (Reprinted from Ref. 78, Copyright 2004, with permission from Elsevier.)

shape. Cell and nuclear shapes also have a strong influence on MSC differentiation and gene expression, as discussed in the next section.

Effect of cell shape

The effect of cell shape can be decoupled from that of the cell size by using cell-adhesive islands of the same area but of different shapes. The shape, curvature, and aspect ratio of the adhesive island have been shown to have strong influence on MSC behavior. The most observable changes are in morphology, cytoskeletal arrangement, and focal adhesion distribution. The patterns that cause higher cytoskeletal tension induce the formation of more developed stress fibers and more prominent focal adhesions. For example, cells on a rectangular island, with high aspect ratio, show higher contractility and more assembled stress fibers compared to those on a square island of the same area. The differentiation of MSCs is strongly influenced by cell shape. When plated on microprinted parallel strips of fibronectin, the cells assume an elongated shape due to restriction in lateral spreading. Differentiation toward myocardial and neural lineages is favored in these elongated cells, while osteogenic differentiation is significantly suppressed.[82] The importance of geometry in MSC differentiation and lineage commitment was demonstrated by Luo *et al.*[83] and Kilian *et al.*[84] Luo *et al.*[83] printed hydrophobic alkanthiols on gold to generate cell-adhesive patterns of different geometric shapes, e.g., circle, triangle, trapezoid, etc. It was found that the shape of the island had a strong effect on hMSC adipogenic differentiation. For instance, comparing octagons and squares of similar areas revealed that the square shape induced a higher adipogenic differentiation rate than the octagon shape. Interestingly, this was observed even though each of the cell-adhesive islands contained multiple cells. As a result, individual cells did not take the shape of the island and only the cells at the edge could sense the pattern. Presumably, cells located at the border should sense only the edge or corner but not the whole shape of the island. Hence, the mechanism by which the shape of a pattern, which is much larger than a single cell, can influence cell behavior is not yet elucidated. A possible hypothesis is that cell–cell contact plays a major role in that process, perhaps by the transduction of mechano-sensing. It was previously reported that adipogenic differentiation increases with coordination number, a measure of average cell–cell contact per cell.[85]

Unlike Luo et al.,[83] Kilian and co-researchers[84] plated single hMSCs on islands of different geometric shapes so that cells could acquire the exact shape of the island. In that case, osteogenic differentiation increased with the increase in aspect ratio of the rectangular shape (the cell area was kept constant). Interestingly, for islands of pentagonal symmetry with different subcellular curvatures, subtle geometric differences brought a significant change in lineage specification, as shown in Fig. 2. This work highlighted the connection between MSC differentiation and myosin-generated contractility.

In addition to geometrically controlling differentiation through surface patterning, we previously examined the potential to control the ECM (bone nodule)

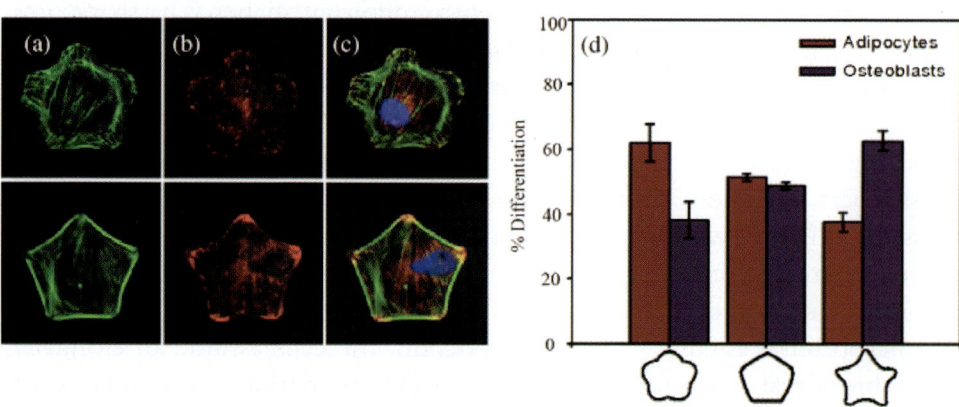

Fig. 2. (a)–(c) Immunofluorescence analysis of cells patterned on symmetric pentagonal islands with different curvatures (top and bottom panel) and the differences in cytoskeletal arrangement and focal adhesion assembly. Green: actin, red: vinculin, blue: nucleus. (d) Subtle geometric differences can bring about significant changes in lineage determination. Schematic shapes on horizontal axis shows the cell shape. (Reprinted from Ref. 84, Copyright 2010, National Academy of Sciences, USA.)

shape by studying the responses of SAOS-2 osteoblasts in a series of square, circular, and triangular geometries that could each accommodate multiple cells. The cells proliferated in the patterns to confluency within five days, expressed alkaline phosphatase, and produced a mineralized matrix in the center of each pattern as identified by von Kossa staining that approximated the shape of the patterns. This work demonstrated the potential to spatially control ECM production from cells via the presentation of specific patterns on the underlying surface.[86]

How cells sense shape and size remains an open question. One hypothesis suggests that altered cytoskeletal arrangements, caused by the adhesive islands, deforms the cell nucleus by physical strain, resulting in an alteration of gene transcription profiles.[87] Mechanical straining of the nucleus may change the nuclear pore size, perhaps leading to an altered influx of ions and/or regulatory molecules.[87–92] However, it must be noted that the nuclear pore is insensitive to molecules smaller than 30 kDa, which can pass through the nuclear membrane by passive diffusion. Whether the local compression of the nuclear membrane makes the pores perceptive to molecules smaller than 30 kDa is yet to be explored. Alternatively, mechanical stress applied on the nucleus may cause chromatin reorganization and decondensation. Recent findings suggest that the sub-nuclear architecture in the 3D space is just as important as the genetic code itself. Different levels of sub-nuclear design, such as chromatin arrangement, genome spatial organization, and compartmentalization of nuclear processes, may play key regulatory roles in multiple cellular processes.[93–98] Hence, it may be possible that a mechanical stress, applied on the nucleus via the cytoskeleton, regulates cell behavior by deforming the nucleus, ultimately altering the accessibility of different genes to the transcriptional machinery. Other possible mechanisms are stretch-activated ion channels and signaling via force-sensitive focal adhesion molecules, such as talin or integrin.[99–102] However, further research is required to explore how subtle changes in geometric signals, translated into slight contractility alterations, result in substantial shifts in gene expression and cellular processes.

To summarize, numerous studies exploring cell culture on micropatterns of different shapes and sizes have established that cell shape and size strongly influence the focal adhesion formation and maturation, contractility, proliferation, and differentiation of MSCs. However, there exist contradictory reports regarding the relative importance of cell shape and size. To add to the complexity, it seems that nuclear shape and volume also play an important role in many cellular processes.[103] It is clear that mechano-sensing via cell contractility is crucial in transmitting extracellular signals;[100,104–106] however, the precise molecular mechanism of converting geometric signals into altered physiological processes remains to be elucidated.

Effect of substrate topology on MSC fate

Importance of topography

Different cell types, even while residing in the same tissue, may experience different substrate topologies, ranging from macro- to nano-scale features.[107,108] These tissue-specific 3D features provide preferential cell–cell and cell–matrix contact and differential ligand presentations, causing changes in cell morphology, migration, and overall fate.[109,110]

Role of microfabrication

As topology is known to influence cell phenotype, it is important to fully understand the interaction between the substrate structure and cell behavior at different length scales. However, in a random structure such as a collagen scaffold, it is not possible to separately explore the features of different length scales and short- and long-range orders. Great progress in the field of microfabrication and nanotechnology has made it possible to create intricate surfaces with defined structures to explore the effects of feature size, shape, and order — separately and in conjunction. Using artificially created, structured substrates, it is now known that substrate structure controls cytoskeletal organization, cell polarization, proliferation, motility, and overall tissue morphogenesis.[107,109,115–120]

Artificially created topology: Structures and materials

Structures

Artificial topographical substrates can be broadly classified into two distinct categories. The first category includes structures that are mostly random or self-assembled, formed as a result of interaction between physical forces such as cohesion and surface tension, as in the case of self-assembled nanofibers (Fig. 3(a)). The second category includes substrate design that is strictly controlled and pre-defined. A good example is photolithographically designed microgrooves (Fig. 3(e)). Photolithographically patterned substrates may be classified into three categories — micropits, micropillars, and microgrooves (Figs. 3(c)–(e)).[121] In this discussion, we will use "micro" to define both micro- and nanoscale structures, unless otherwise specified.

Materials

In most of the work exploring the interaction of MSCs with surface topology, soft lithography and photolithography techniques have been employed to

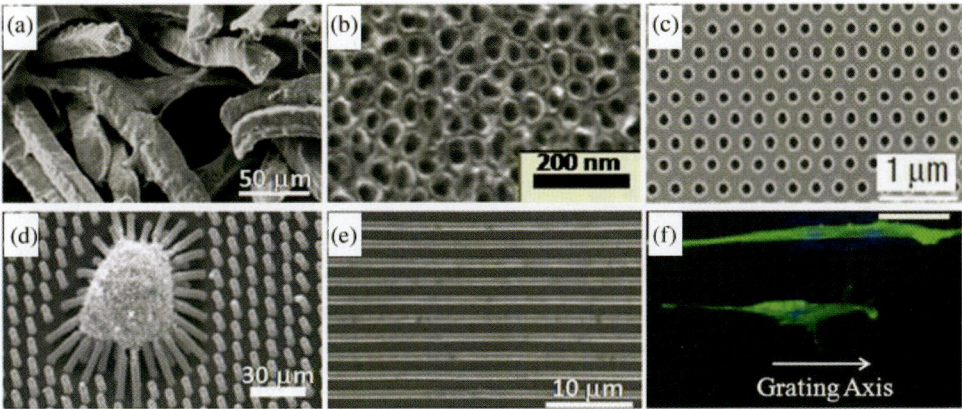

Fig. 3. (a)–(e) Microfabricated topological substrates used for MSC research. (a) Titanium fiber mesh (reprinted from Ref. 111. Copyright 2006, National Academy of Sciences, USA). (b) Carbon nanotube (reprinted from Ref. 112. Copyright 2009, National Academy of Sciences, USA). (c) Nanopits on PMMA (adapted with permission from Ref. 113. Copyright 2007, Macmillan Publishers, Ltd). (d) hMSC on PDMS micropillars (adapted with permission from Ref. 114. Copyright 2010, Macmillan Publishers Ltd). (e) PDMS microgrooves (reprinted from Ref. 28. Copyright 2008, Springer Netherlands). (f) hMSC morphology on PDMS microgrooves. Green: F-actin, blue: nucleus. Scale bar, 50 μm (reprinted with permission from Ref. 27. Copyright 2007, Elsevier).

mold PDMS to create substrates with microscale definition. Other than PDMS, hydrogels and other polymeric materials, such as poly-methyl methacrylate (PMMA), have also been used as substrates.[113,122–124] A great advantage of using topologically patterned soft materials (such as PDMS or hydrogel) is that their stiffness closely resembles that of the ECM. Therefore, by altering the topology as well as the mechanical and rheological properties of these materials, the *in vivo* environment can be better simulated. Plastics, ceramics, carbon, and silicon structures, which are mechanically rigid materials, are devoid of this benefit.

Micro-/nanogrooves

Effect of pattern size on MSC cytoskeleton, morphology, and alignment

Parallel microgrooves probably are the most studied topology in MSC research because of their simple geometry and ease of fabrication.[27,28,109,125–129] When MSCs are plated on surfaces with microgrooves, cells elongate and align themselves along the axis of the grooves within 2–4 h of plating (Fig. 3(f)).[27,28,125–130] Cells on microgrooves also show less prominent actin filaments. Upon the application of uni-axial cyclic strain, cells on smooth surfaces align perpendicular to the strain axis to reduce the stress on cytoskeleton.[125] Interestingly, for patterned substrates, the topological alignment dominates over the effect of strain. That is, cells continue to follow the contact guidance and do not change their alignment upon the application of strain.

Effect of pattern size on MSC differentiation

The first clear demonstration of pre-defined topology impact on MSC differentiation came from Kurpinski *et al.*,[125] where a micropatterned PDMS substrate subjected to cyclic uni-axial strain was used. DNA microarray, qRT-PCR, and Western blotting revealed that the strain caused genetic reprogramming and significant changes at both the RNA and protein levels. Furthermore, while strain promoted the phenotype of tension-bearing tissues (i.e., smooth muscle cells (SMCs)), it suppressed the expression of compression-bearing phenotypes (bone or cartilage). However, strain alone is not sufficient for terminal differentiation, suggesting the combinatorial need for additional factors (soluble cues). This study also revealed the differential sensitivity of MSC genes to the axis direction of applied strain relative to cell alignment. While some genes were not at all responsive to axis direction, a few showed small changes and only one gene (ARL7) showed dramatic change when the applied uni-axial strain axis was perpendicular to the direction of the cell alignment (i.e., the pattern direction). These results highlighted the intriguing complexity of mechano-transduction and demonstrated how micropatterning can potentially be used to decouple different mechanical cues.

The most striking effect of nanotopography on MSC differentiation is its ability to trans-differentiate the MSCs into the neuronal lineage even in the absence of any chemical cues.[27] It was shown that nanogratings (350-nm depth and 750-nm pitch) could upregulate the mature neuronal cell markers, microtubule-associated protein 2 (MAP2) and beta-tubulin III, with or without the presence of the neuronal induction factor retinoic acid (RA). In fact, MAP2 expression was higher in MSCs cultured on nanotopography *without* RA than that of MSCs cultured on planar PDMS *with* RA, demonstrating the dominant effect of mechanical cues over chemical cues in this particular case. Although trans-differentiation is a controversial area, another study showed that PDMS substrate with 1-µm–wide parallel grooves induces neuronal differentiation of umbilical cord blood (UCB)-derived MSCs more efficiently than a flat surface.[28] Together, these data (in addition to embryonic stem cell studies) demonstrate the effectiveness of parallel nanogrooves in the induction of neuronal differentiation.[131]

Effect of pattern size on proliferation

There is more contradiction than consensus regarding the effect of nano-/micrograting on cell proliferation rate. In Yim's study,[27] BrdU incorporation showed that MSCs on nanogratings are significantly less proliferative, supporting the findings of Thakar's group with bovine SMCs.[126] However, while Yim *et al.* showed that the hMSC proliferation rate was insensitive to patterns of width 10 μm or higher, Thakar *et al.* found that the SMC proliferation rate was significantly lower on a patterned substrate, even for a pattern of much larger dimensions (300 μm). In contrast, Kim *et al.*[28] showed that the proliferation rate of UCB-derived MSCs almost doubles on 1-μm–wide microgratings compared to flat PDMS. However, for 4-μm–wide patterns, no appreciable effect could be observed. Further research is required to explore whether the abovementioned contradictions can be attributed only to the difference in cell types.

Micro-/nanopits: Effect of disorder

Other than shape and size, short range order (or disorder) in the structure also plays an important role in directing cell phenotype.[113,132] For random structures, it is not possible to control the disorderliness and roughness. Two different structures with similar roughness parameters may have very different topological features in terms of orderliness and roughness distribution. To overcome this problem, Dalby *et al.* used electron beam lithography (EBL) to create nanopits with a precise degree of disorder and geometry on PMMA.[113] The PMMA substrate was embossed with 120-nm diameter and 100-nm deep nanopits in five different arrangements ranging from a perfect square and hexagonal array to a complete randomness covering two medium ranges of disorder. After 21 days of culturing hMSCs on these defined topologies *vs.* a control flat substrate, both the control and the random pit-cultured cells failed to express the bone cell markers osteopontin (OPN) and osteocalcin (OCN). However, cells cultured on substrates with disorderliness of 50 nm (length scale) expressed both OCN and OPN and an extended culture time of 28 days produced discrete bone nodules with mineralization. Osteo-specific microarray data and qPCR confirmed the osteogenic potential of the disordered nanopits, which was similar to that of the osteogenic induction factor, dexamethasone. The most important observation of this work is the higher osteogenic potential of a slightly disordered roughness compared to a surface with either perfectly ordered or completely random roughness.

A few other shapes: Effect of shape

The effect of shapes other than microgrooves on MSCs has not been studied in depth until recently. The structures that have been studied include round, square, and hexagonal pillars with varying sizes and lateral spacing.[122,123,133,134] The results from these studies confirm that hMSC morphology strongly depends on the pattern shape and size. However, effect of topology shape on cell differentiation and proliferation is not yet conclusive and demands further investigation. Shapes can also be generated through microcontact printing[84] and it may be useful to examine whether confined cell-adhesive islands and raised structures (i.e., pillared substrates) of the same shape and size have a similar effect on cell behavior.

Substrates with micropillars

Soft polymeric substrates with raised micropillars are often used in cell biology for traction force microscopy.[135–138] The pillars can deform individually under stress without effecting the configuration of the rest of the substrate, unlike a soft, smooth substrate, where the displacement of a single point is felt by all the other neighboring points. As a result, the shear force applied to each pillar can easily be calculated from its displacement.[139] Moreover, the rigidity or deformability of a pillar is a function of its length and diameter. Consequently, the stiffness of the substrate can be varied solely by changing the geometrical properties of the pillars without changing the bulk chemical properties of the pillar material. This approach has two major advantages: first, the chemical properties of the material are decoupled from its stiffness, and in addition, the force applied on each pillar can be calculated separately, thus making it possible to map the spatial distribution of the contractile force inside each individual cell. Exploiting the abovementioned advantages of micropillar substrates, Fu et al.[114] showed that the contractility, already at the very early phase of culture, has a lasting effect in determining hMSC fate. This study also showed that the osteogenic/adipogenic fate of hMSCs could be predicted with high accuracy from the early level of cytoskeletal tension.

Nanotubes and nanowires

In addition to polymers as substrates, novel materials are also being explored including single-walled and multi-walled carbon nanotubes (SWCNTs and MWCNTs), titanium oxide nanotubes and poly(ε-caprolactone) nanowires.[112,140,141] MSCs were shown to grow on CNTs without further treatment, although the spreading of MSCs has been reported to be higher on CNT-coated substrates compared to uncoated controls.[140] Interestingly, it has been shown that the resultant lineage of MSC differentiation can be altered solely by nanotube dimensions.[112] When hMSCs were cultured on vertically aligned TiO_2 nanotubes of different pore sizes for 14 days, 70–100-nm diameter nanotubes gave rise to highly elongated cells that expressed a high level of the osteogenic markers OCN, OPN, and alkaline phosphate, even in the absence of osteogenesis-inducing chemical factors. On the other hand, 30-nm tubes resulted in cells with rounded morphology and low expression level of osteogenic markers. The authors hypothesized that serum proteins (from the culture media) attach differently with nanotubes of varying diameters, leading to differential cell attachment in the initial culture period. As the diameter of the nanotubes increases, the cell adhesion on the tubes decreases, resulting in a low plating density. However, if the culture period is extended, this difference in cell density disappears due to cell proliferation. McBeath et al.[78] showed that lower plating density leads to cytoskeletal tension and osteogenic differentiation. It is possible that the same mechanism also applies here. However, for this hypothesis to be true, the impact of the initial culture phase must be lasting and fate-determining, similar to recent study.[114]

To summarize, cytoskeletal tension, an indicator of cell shape and size, plays an important role in determining MSC lineage fate. It appears that increased contact area with a substrate and high cytoskeletal tension leads to osteogenic differentiation, whereas lower tensions favor myogenic, adipogenic, or neurogenic lineages. However, contradictory results showed that the relationship between cell morphology and lineage fate is much more complex. Single parameters such as contact area or aspect ratio are not sufficient to predict cell fate. Instead, an array of parameters, obtained from high content imaging of single cells, is required to predict cell fate with reasonable accuracy.[142] While microcontact printing and microfabricated topology have contributed toward this goal, we are still far from identifying and assessing the relative importance of different mechanical and geometric parameters in directing MSC fate.

Application of Microfluidics

Importance of flow in cell biology

In the previous two sections, we have discussed how mechanical cues can be incorporated into *in vitro* culture systems. However, a key parameter is still missing — time. Within tissues, dynamics are critical. Specifically, diffusion and convention of fluid flow are ubiquitous. As a result, nutrients, chemokines, are and cytokines are supplied and metabolic wastes are removed in a time- and space-dependent manner. Due to fluid flow, a cell or a tissue may also experience pressure or a temperature gradient. Moreover, flow applies shear force, which may cause "stretching" at different length scales, from protein folding to expanding the dimensions of a tissue. To recapitulate *in vivo* conditions, fluid flow should be considered within *in vitro* models and may be achieved through the use of microfluidic technology.[143]

Advantages of microfluidics

The first and most prominent advantage of microfluidics is its ability to handle small volumes of fluid. As a result, precious cells and reagents can be utilized judiciously. Also, multiple assays can be run simultaneously by designing intelligent channel circuitry to save time and human effort. Another important feature of microfluidic systems is that the fluid flow is typically laminar in a microchannel, unless specific modifications are made. As a result, mixing takes place only by diffusion, and no chaotic or turbulent mixing is observed.[144] This unique feature of a microfluidic system makes it possible to flow fluids with different properties (e.g., temperature) side by side in the same channel without mixing.[145] The laminar nature of flow permits the introduction and withdrawal of individual or combinatorial signals in a systematic and well-defined manner. However, if mixing is required, it can be achieved using specific modifications.[146,147] Most of the microfluidic systems are made of PDMS, a transparent polymer. Therefore, imaging is possible without disturbing the cells and without compromising the sterility of the cultures. Finally, with progress in microfabrication and microelectronics, it is possible to integrate a microfluidic system with computer control to modulate the flow circuitry without manual intervention.[148]

Problems of microfluidics

Despite the abovementioned advantages, microfluidic systems have not yet been fully adopted by the biology community.[59,149] This stems from several pitfalls of current microfluidic systems. First, PDMS, the material mostly used for fabricating microfluidic devices, is hydrophobic and non-adhesive to cells. As a result, it requires special treatment such as plasma oxidation and coatings before it can be used for cell culture. In addition, maintaining pH levels and constant nutrient and oxygen supply, and removing metabolic waste and cell debris require elaborate instrumentation, including pumps, microvalves, and gas–liquid mixers.[149] The effort required to assemble and maintain the system under sterile conditions may become impractical. Moreover, many of the biological laboratories may not have access to microfabrication facilities and microfluidic platforms. Consequently, microfluidic systems are not yet fully employed in stem cell research in general, in particular MSC research.[59]

Microfluidics in MSC research

The first successful culture of hESCs using a microfluidic platform was demonstrated by Abhayankar et al.[150] and the first differentiation-on-a-chip was demonstrated by Tourovskaia et al.,[151] who used the microfluidic platform to differentiate skeletal muscle cells, C2C12. Utilizing laminar flow, they showed that cells can be tagged locally, without disturbing neighboring cells in a population.[152] These data raise the attractive possibility of treating cells with different chemicals, inhibitors, or differentiation factors side by side in the same culture, thus keeping all other parameters unchanged. Valero et al.[153] used a microfluidic system to electroporate single C2C12 cells and hMSCs with vector DNA encoding GFP. A microfluidic device, etched on silicon and equipped with a platinum electrode, was used to trap nine cells at a time and electroporate them with vector DNA, encoding a GFP–ERK1 fusion protein. The transfection efficiency obtained was as high as 75%. In principle, it should be possible to develop a higher throughput device using the same design principle to handle a larger number of cells in parallel and to conduct high quality research on a single-cell basis.

Microfluidic chips have also been used to demonstrate the maintenance and differentiation of MSCs into hepatogenic, osteogenic, and adipogenic lineages.[148,154–158] However, the application of microfluidics in MSC research is still in its infancy. One of the questions that could be successfully addressed by microfluidic systems is the effect of shear stress on MSC differentiation. While it has been shown by using fiber mesh or bioreactors that shear stress enhances osteogenic differentiation, flow rate and shear stress can be separately and more precisely controlled in a microfluidic device than in a random mesh of fibers.[111,159,160] In a single chip, the effect of multiple combinations of differentiation factors and other mechanical signals, including shear stress, can be studied.

Conclusions and Outlook

Significant progress in the fields of microfabrication and microfluidics has enabled biologists to address previously unanswerable questions. In particular, mechano-sensing and the effects of mechano-signals on cell behavior can be explored on a single-cell, and even on the organelle or cytoskeletal level, using microfabricated platforms. The microscale approach is attractive to reduce the usage of expensive reagents and cells. As MSCs lose their phenotype and differentiation potential after six to eight passages, assays that utilize fewer cells may reduce the need for primary isolation (e.g., from the bone marrow). Other than these practical benefits, microfabricated platforms provide deeper insight into the niche, in terms of providing mechanical and geometric cues. Future research efforts will likely include a reductionist's approach, to reduce complexity of standard assays to enable the study of individual variables and their synergistic effects to improve our understanding of the niche.[157,158] The other path is to study the commonalities among different mechano-signals in terms of their effects on cellular molecular pathways and physiological processes. For example, a large cell-adhesive pattern, a stiff substrate, vibration, semi-disorder, substrate strain, and shear stress all appear to increase contractility and enhance osteogenic differentiation. It is not yet fully known how a cell perceives all of these signals. Advances in microfabrication and microfluidic technology and their integration with biological systems should enable the design of scaffolds that mimic all the essential characteristics of a niche.

Suggested Reviews:

Stem cells and MSCs

1. "Prospects for stem cell-based therapy" by Daley and Scadden, 2008 (Ref. 40).
2. "Repair of tissues by adult stem/progenitor cells (MSCs): Controversies, Myths and Changing paradigms" by Prockop, 2009 (Ref. 42).
3. "Mesenchymal stem cell homing: The devil is in the details" by Karp and Teo, 2009 (Ref. 41).
4. Ankrum, J. and Karp, J.M., Mesenchymal stem cell therapy: Two steps forward, one step back. *Trends Mol Med* 2010, **16**(5): p. 203–209.

Microscale techniques and biology

1. Voldman, J., Gray, M.L., Schmidt, M.A., Microfabrication in biology and medicine. *Annu Rev Biomed Eng* 1999, **1**: p. 401–425.
2. Folch, A., Toner, M., Microengineering of cellular interactions. *Annu Rev Biomed Eng* 2000, **2**: p. 227–256.
3. "Microscale technologies for tissue engineering and biology" by Khademhosseini *et al.*, 2006 (Ref. 57).

4. Ferreira, L., Karp, J.M., Nobre, L., Langer, R., New opportunities: The use of nanotechnologies to manipulate and track stem cells. *Cell Stem Cell* 2008, **3**: p. 136–146.
5. "Biomaterials for stem cell differentiation" by Dawson *et al.*, 2008 (Ref. 53).
6. "Mimicking stem cell niches to increase stem cell expansion" by Dellatore *et al.*, 2008 (Ref. 48).
7. "Designing materials to direct stem-cell fate" by Lutolf *et al.*, 2009 (Ref. 55).
8. "High-throughput methods to define complex stem cell niches" by Kobel and Lutolf, 2010 (Ref. 59).
9. Fisher, O.Z., Khademhosseini, A., Langer, R., Peppas, N.A., Bioinspired materials for controlling stem cell fate. *Acc Chem Res* 2010, **43**: p. 419–428.
10. Park, J.Y., Takayama, S., Lee, S.H., Regulating microenvironmental stimuli for stem cells and cancer cells using microsystems. *Integr Biol* 2010, **2**: p. 229–240.

Mechanotransduction

1. Assoian, R.K. and E.A. Klein, Growth control by intracellular tension and extracellular stiffness. *Cell* 2008. **18**: p. 347–352.
2. "Mechanotransduction at a distance: mechanically coupling the extracellular matrix with the neucleus" by Wang *et al.*, 2009 (Ref. 87).
3. Janmey, P. A., Winer, J.P., Murray, M.E., Wen, Q., The hard life of soft cells. *Cell Motil Cytoskeleton* 2009, **66**: p. 597–605.
4. "Growth factors, matrices, and forces combine and control stem cells" by Discher *et al.*, 2009 (Ref. 63).
5. "Control of stem cell fate by physical interaction with the extracellular matrix" by Guilak *et al.*, 2009 (Ref. 62).
6. Titushkin, I., Sun, S., Shin, J., Cho, M., Physicochemical control of adult stem cell diferentiation: Shedding light on potential molecular mechanism. *J Biomed Biotechnol* p. 1–14.
7. Potier, E., Noailly, J., Ito, K., Directing bone marrow-derived stromal cell function with mechanics. *J Biomech* 2010, **43**: p. 807–817.

Micropatterning

1. "Surface engineering approaches to micropattern surfaces for cell-based assays" by Falconnet *et al.*, 2006 (Ref. 72).
2. "Stamps, inks and substrates: polymers in microcontact printing" by Kaufmann and Ravoo, 2010 (Ref. 71).

Effect of topology

1. "Engineering substrate topography at the micro- and nanoscale to control cell function" by Bettinger *et al.*, 2009 (Ref. 121).
2. Ravichandran, R., Liao, S., Ng, C.C., Chan, C.K., Raghunath, M., Effects of nanotopography on stem cell phenotypes. *World J Stem Cells* 2009, **1**: p. 55–66.
3. "Substrate topography shapes cell function" by Kulangara and Leong, 2009 (Ref. 120).
4. McNamara, L.E., McMurray, R.J., Biggs, M.J.P., Kantawong, F., Oreffo, R.O.C., Dalby, M.J., Nanotopographical control of stem cell differentiation. *J Tissue Eng* p. 1–13.

Microfluidics

1. "Stem cells in microfluidics" by van Noort *et al.*, 2009 (Ref. 149).
2. "High-throughput methods to define complex stem cell niche" by Kobel and Lutolf, 2010 (Ref. 59).

Acknowledgement

This work was supported by the National Institute of Health grant HL097172, HL095722 and DE019191 and by the American Heart Association grant #0970178N to JMK.

References

1. Prockop, D.J., Marrow stromal cells as stem cells for nonhematopoietic tissues. *Science* 1997, **276**: p. 71–74.
2. Pittenger, M.F., Mackay, A.M., Beck, S.C., Jaiswal, R.K., Douglas, R., Mosca, J.D., Moorman, M.A., Simonetti, D.W., Craig, S., Marshak, D.R., Multilineage potential of adult human mesenchymal stem cells. *Science* 1999, **284**: p. 143–147.
3. Colter, D.C., Sekiya, I., Prockop, D.J., Identification of a subpopulation of rapidly self-renewing and multipotential adult stem cells in colonies of human marrow stromal cells. *Proc Natl Acad Sci*, 2001, **98**(14): p. 7841–7845.
4. Li, W.J., Tuli, R., Huang, X., Laquerriere, P., Tuan, R.S., Multilineage differentiation of human mesenchymal stem cells in a three-dimensional nanofibrous scaffold. *Biomaterials* 2005, **26**: p. 5158–5166.
5. Nagai, A., Kim, W.K., Lee, H.J., Jeong, H.S., Kim, K.S., Hong S.H., Park I.H., Kim, S.U., Multilineage potential of stable human mesenchymal stem cell line derived from fetal marrow. *PloS One* 2007, **2**: p. e1272.
6. Al-Khaldi, A., Eliopoulos, N., Martineau, D., Lejeune, L., Lachapelle, K., Galipeau, J., Postnatal bone marrow stromal cells elicit a potent VEGF-dependent neoangiogenic response *in vivo*. *Gene Ther* 2003, **10**(8): p. 621–629.

7. Gruber, R., Kandler, B., Holzmann, P., Vogele-Kadletz, M., Losert, U., Fischer, M.B., Watzek, G., Bone marrow stromal cells can provide a local environment that favors migration and formation of tubular structures of endothelial cells. *Tissue Eng* 2005, **11**: p. 896–903.
8. Wu, Y., Chen, L., Scott, P.G., Tredget, E.E., Mesenchymal stem cells enhance wound healing through differentiation and angiogenesis. *Stem Cells* 2007. **25**(10): p. 2648–2659.
9. Shyu, K.G., Wang, B.W., Hung, H.F., Chang, C.C., Shih, D.T., Mesenchymal stem cells are superior to angiogenic growth factor genes for improving myocardial performance in the mouse model of acute myocardial infarction. *J Biomed Sci* 2006, **13**: p. 47–58.
10. Kasper, G., Dankert, N., Tuischer, J., Hoeft, M., Gaber, T., Glaeser, J.D., Zander, D., Tschirschmann, M., Thompson, M., Matziolis, G., Duda, G.N., Mesenchymal stem cells regulate angiogenesis according to their mechanical environment. *Stem Cells* 2007, **25**: p. 903–910.
11. Krampera, M., Glennie, S., Dyson, J., Scott, D., Laylor, R., Simpson, E., Dazzi, F., Bone marrow mesenchymal stem cells inhibit the response of naive and memory antigen-specific T cells to their cognate peptide. *Blood* 2003, **101**: p. 3722–3729.
12. Aggarwal, S., Pittenger, M.F., Human mesenchymal stem cells modulate allogeneic immune cell responses. *Blood* 2005, **105**(4): p. 1815–1822.
13. Beyth, S., Borovsky, Z., Mevorach, D., Liebergall, M., Gazit, Z., Aslan, H., Galun, E., Rachmilewitz, J., Human mesenchymal stem cells alter antigen-presenting cell maturation and induce T-cell unresponsiveness. *Blood* 2005, **105**(5): p. 2214–2219.
14. Nauta, A.J., Fibbe, W.E., Immunomodulatory properties of mesenchymal stromal cells. *Blood* 2007, **110**: p. 3499–3506.
15. Zheng, Z.H., Li, X.Y., Ding, J., Jia, J.F., Zhu, P., Allogeneic mesenchymal stem cell and mesenchymal stem cell-differentiated chondrocyte suppress the responses of type II collagen-reactive T cells in rheumatoid arthritis. *Rheumatol* 2008, **47**: p. 22–30.
16. Dominici, M., Le Blanc, K., Mueller, I., Slaper-Cortenbach, I., Marini, F., Krause, D., Deans, R., Keating, A., Prockop, D.J., Horwitz, E., Minimal criteria for defining multipotent mesenchymal stromal cells. The international society for cellular therapy position statement. *Cytotherapy* 2006, **8**: p. 315–317.
17. Friedenstein, A.J., Chilakhjan, R.K., Lalykina, K.S., The developement of fibroblast colonies in monolayer cultures of gunea-pig bone marrow and spleen cells. *Cell Tissue Kinet*, 1970, **3**: p. 393–403.
18. Jaiswal, N., Haynesworth, S.E., Caplan, A.I., Bruder, S.P., Osteogenic differentiation of purified culture-expanded uman mesenchymal stem cells *in vitro*. *J Cell Biochem* 1997, **64**: p. 295–312.
19. Jhonstone, B., Hering, T.M., Caplan, A.I., Goldberg, V.M., Yoo, J.U., *In vitro* chondrogenesis of bone marrow-derived mesenchymal progentor cells. *Exp Cell Res* 1998, **238**: p. 265–272.
20. Krampera, M., Pizzolo, G., Aprili, G., Franchini, M., Mesenchymal stem cells for bone, cartilage, tendon and skeletal muscle repair. *Bone* 2006, **39**: p. 678–683.
21. Chamberlain, G., Fox, J., Ashton, B., Middleton, J., Concise review: Mesenchymal stem cell: Their phenotype, differentiation capacity, immunological feautures and potential for homing. *Stem Cells* 2007, **25**: p. 2739–2749.
22. Woodbury, D., Schwarz, E.J., Prockop, D.J., Black, I.B., Adult rat and human bone marrow stromal cells differentiate into neurons. *J Neurosci Res* 2000, **61**: p. 364–370.

23. Sanchez-Ramos, J., Song, S., Cardozo-Pelaez, F., Hazzi, C., Stedeford, T., Willing, A., Freeman, T.B., Sapota, S., Janssen, W., Patel, N., Cooper, D.R., Sanberg, P.R., Adut bone marrow stromal cells differentiate into neuronal cells *in vitro*. *Exp Neurol* 2000, **164**: p. 247–256.
24. Deng, W., Obrocka, M., Fischer, I., Prockop, D.J., *In vitro* differentiation of human marrow stromal cells into early progenitors of neural cells by conditions that increase intracellular cyclic AMP. *Biochem Bhiophys Res Commun* 2001, **282**: p. 148–152.
25. Qian, L., Saltzman, W.M., Improving the expansion and differentiation of mesenchymal stem cells through culture surface modification. *Biomaterials* 2004, **25**: p. 1331–1337.
26. Wiselet-Gendebien, S., Hans, P., Leprine, P., Rigo, J.M., Moonen, G., Rogister, B., Plasticity of cultured mesenchymal stem cells; switch from nestin positive to neuron-like phenotype. *Stem Cells* 2005, **23**: p. 392–402.
27. Yim, E.K.F., Pang, S.W., Leong, K.W., Synthetic nanostructures inducing differentiation of human mesenchymal stem cells into neuronal lineage. *Exp Cell Res* 2007, **313**: p. 1820–1829.
28. Kim, S., Lee, J.K., Kim, J.W., Jung, J.W., Seo, K., Park, S.B., Roh, K.H., Lee, S.R., Hong, Y.H., Kim, S.J., Lee, Y.S., Kim, S.J., Kang, K.S., Surface modification of polydimehylsiloxane (PDMS) induced roliferation and neural-like cells differentiation of umbilical cord blood-derived mesenchymal stem cells. *J Mat Sci Mat Med* 2008, **19**: p. 2953–2962.
29. Prabhakaran, M.P., Venugopal, J.R., Ramakrishna, S., Mesenchymal stem cell differentiation to neuronal cells on electrospun nanofibrous substrates for nerve tissue engineering. *Biomaterials* 2009, **30**: p. 4996–5003.
30. Maitra, B., Szekely, E., Gjini, K., Laughlin, M.J., Dennis, J., Haynesworth, S.E., Koc, O.N., Human mesenchymal stem cells support unrelated donor hematopoietic stem cells and suppress T-cell activation. *Bone Marrow Transplant* 2004, **33**: p. 597–604.
31. Kadiyala, S., Jaiswal, N., Bruder, S., Culture-expanded bone marrow-derived mesenchymal stem cell can regenerate a critical sized segmental bone defect. *Tissue Eng* 1997, **3**: p. 173–185.
32. Horwitz, E.M., Prockop, D.J., Gordon, P.L., Koo, W.W., Fitzpatrick, L.A., Neel, M.D., McCarville, M.E., Orchard, P.J., Pyeritz, R.E., Brenner, M.K., Clinical response to bone marrow transplantation in children with severe osteogenesis imperfecta. *Blood* 2001, **97**: p. 1227–1231.
33. Quarto, R., Mastrogiacomo, M., Cancedda, R., Kutepov, S.M., Mukhachev, V., Lavroukov, A., Kon, E., Marcacci, M., Repair of large bone defects with the use of autologous bone marrow stromal cells. *N Engl J Med* 2001, **344**: p. 385–386.
34. Vacanti, C.A., Bonassar, L.J., Vacanti, M.P., Shufflebarger, J., Replacement of an avulsed phalanx with tissue-engineered bone. *N Engl J Med* 2001, **344**: p. 1511–1514.
35. Wakitani, S., Imoto, K., Yamamoto, T., Saito, M., Murata, N., Yoneda, M., Human autologous culture expanded bone marrow mesenchymal cell transplantation for repair of cartilage defects in osteoarthritic knees. *Osteoarthritis Cartilage* 2002, **10**: p. 199–206.
36. Chen, S.L., Fang, W.W., Qian, J., Ye, F., Liu, Y.H., Shan, S.J., Zhang, J.J., Lin, S., Liao, L.M., Zhao, R.C., Improvement of cardiac function after transplantation of autologous bone marrow mesenchymal stem cells in patients with acute myocardial infraction. *Chin Med J* 2004, **117**: p. 1443–1448.
37. Guo, X., Wang, C., Zhang, Y., Xia, R., Hu, M., Duan, C., Zhao, Q., Dong, L., Lu, J., Qing Song, Y., Repair of large articular cartilage defects with implants of autologous

mesenchymal stem cells seeded into beta-tricalcium phosphate in a sheep model. *Tissue Eng* 2004. **10**: p. 1818–1829.
38. Dai, W., Hale, S.L., Martin, B.J., Kuang, J.Q., Dow, J.S., Wold, L.E., Kloner, R.A., Allogenic mesenchymal stem cell transplantaion in postinfracted rat myocardium:short- and long-term effects. *Circulation* 2005, **112**: p. 214–223.
39. Conrad, C., Huss, R., Adult stem cell lines in regenrative medicine and reconstructive surgery. *J Surg Res,* 2005, **124**: p. 201–208.
40. Daley, G.Q., Scadden, D.T., Prospects for stem cell-based therapy. *Cell* 2008, **132**: p. 544–548.
41. Karp, J.M., Teo, G.S.L., Mesenchymal stem cell homing: The devil is in the details. *Cell Stem Cell* 2009, **4**: p. 206–216.
42. Prockop, D.J., Repair of tissues by adult stem/progenitor cells (MSCs): Controversies, myths and changing paradigms. *Mol Ther* 2009, **17**: p. 939–946.
43. Discher, D.E., Janmey, P., Wang, Y., Tissue cells feel and respond to the stiffness of their substrate. *Science* 2005, **310**: p. 1139–1143.
44. Scadden, D.T., The stem cell niche as an entity of action. *Nature* 2006, **441**: p. 1075–1079.
45. Zajac, A.L., Discher, E.D., Cell differentiation through tissue elasticity-coupled, myosi-driven remodeling. *Curr Opin Cell Biol* **20**: 609–615.
46. Wozniak, M.A., Chen, C.S., Mechanotransduction in development: A growing role for contractlity. *Nat Rev* 2009, **10**: p. 34–43.
47. Gardel, M., Schwarz, U., Cell-substrate interactions. *J Phys Condens Matter* 2010, **22**: p. 1–3.
48. Dellatore, S.M., Garcia, A.S., Miller, W.M., Mimicking stem cell niches to increase stem cell expansion. *Curr Opin Cell Biol* 2008, **19**: p. 534–540.
49. Morrison, S.J., Spradling, A.C., Stem cells and niches: Mechanisms that promote stem cell maintenance throughout life. *Cell* 2008, **132**: p. 598–611.
50. Gilbert, P.M., Havenstrite, K.L., Magnusson, K.E., Sacco, A., Leonardi, N.A., Kraft, P., Nguyen, N.K., Thrun, S., Lutolf, M.P., Blau, H.M., Substrate elasticity regulates skeletal muscle stem cell self-renewal in culture. *Science* 2010, **329**: p. 1078–1081.
51. Lutolf, M.P., Hubbel, J., Synthetic biomaterials as instructive extracellular micro-environments for morphogenesis in tissue engineering. *Nat Biotechnol* 2005, **23**: p. 47–55.
52. Saha, K., Pollock, J.F., Schaffer, D.V., Healy, K.E., Designing synthetic materials to control stem cell phenotype. *Curr Opin Chem Biol* 2007, **11**: p. 381–387.
53. Dawson, E., Mapili, G., Erickson, K., Taqvi, S., Roy, K., Biomaterials for stem cell differentiation. *Adv Drug Delivery Rev* 2008, **60**: p. 215–228.
54. Dutta, R.C., Dutta, A.K., Cell-interactive 3D-scaffold; advances and applications. *Biotechnol Adv* 2009, **27**: p. 334–339.
55. Lutolf, M.P., Gilbert, P.M., Blau, H.M., Designing materials to direct stem-cell fate. *Nature* 2009, **462**: p. 433–441.
56. Karp, J.M. Microscale approaches for bone tissue engineering. In: Khademhosseini, A., Borenstein, J. Toner, M., Takayama, S., eds., Artech House Publishing, Boston, 2008, p. 417–430.
57. Khademhosseini, A., Langer, R., Borenstein, J., Vacanti, J.P., Microscale technologies for tissue engineering and biology. *Proc Natl Acad Sci* 2006, **103**: p. 2480–2487.
58. Karp, J.M., Mahdavi, A., Hong, S., Khademhosseini, A., Langer, R., Microscale approaches for bone tissue engineering. *Micro and Nanoengineering of the Cellular*

Microenvironment: Applications and Technologies. Artech House Publishing, Boston, 417–430, 2008.
59. Kobel, S., Lutolf, M.P., High-throughput methods to define complex stem cell niches. *BioTechniques* 2010, **48**: p. IX–XXI.
60. Engler, A.J., Sen, S., Sweeney, H.L., Discher, D.E., Matrix elasticity directs stem cell lineage specification. *Cell* 2006, **126**: p. 677–689.
61. Even-Ram, S., Artym, V.Y., Yamada, K.M., Matrix control of stem cell fate. *Cell* 2006, **126**: p. 645–647.
62. Guilak, F., Cohen, D.M., Estes, B.T., Gimble, J.M., Liedtke, W., Control of stem cell fate by physical interactions with extracelular matrix. *Cell Stem Cell* 2009, **5**: p. 17–26.
63. Discher, D.E., Mooney, D.J., Zandstra, P.W., Growth factors, matrices, and forces combine and control stem cells. *Science* 2009, **324**: p. 1673–1677.
64. Luu, Y.K., Capilla, E., Rosen, C.J., Gilsanz, V., Pessin, J.E., Judex, S., Rubin, C.T., Mechanical stimulation of mesenchymal stem cell proliferation and differentiation promotes osteogenesis hile reventing dietary-induced obesity. *J Bone Miner Res* 2009, **24**: p. 50–61.
65. Yeung, T., Georges, P.C., Flanagan, L.A., Marg, B., Ortiz, M., Funaki, M., Zahir, N., Ming, W., Weaver, V., Janmey, P.A., Effects of substarte stiffness on cell morphology, cytoskeletal structure and adhesion. *Cell Motil Cytoskeleton* 2005, **60**: p. 24–34.
66. Rowland, A.S., George, P.A., Cooper-White, J.J., Directing osteogenic and myogenic differentiation of MSCs: Interplay of stiffness and adhesive ligand presentation. *Am J Physiol Cell Physiol* 2008, **295**: p. 1037–1044.
67. Kane, R.S., Takayama, S., Ostuni, E., Ingber, D.E., Whitesides, G.M., Patterning proteins and cells using soft lithography. *Biomaterials* 1999, **20**: p. 2363–2370.
68. Singhvi, R., Kumar, A., Lopez, G.P., Stephanopoulos, G.N., Wang, D.I., Whitesides, G.M., Engineering cell shape and function. *Science* 1994, **264**: p. 696–698.
69. Xia, Y.N., Whitesides, G.M., Soft lithography. *Annu Rev Mater Sci* 1998, **28**: p. 153–184.
70. Whitesides, G.M., Ostuni, E., Takayama, S., Jiang, X., Ingber, D.E., Soft lithography in biology and biochemistry. *Annu Rev Biomed Eng* 2001, **3**: p. 335–373.
71. Falconnet, D., Csucs, G., Grandin, H.M., Textor, M., Surface engineering approaches to micropattern surfaces for cell-based assays. *Biomaterials* 2006, **27**: 3044–3063.
72. Kaufmann, T., Ravoo, B.J., Stamps, inks and substrates: Polymers in microcontact printing. *Polym Chem* 2010, **1**: p. 371–387.
73. Benecke, B.J., Ben-Ze'ev, A., Penman, S., The control of mRNA production, translation and turnover in suspended and reattached anchorage-dependent fibroblasts. *Cell* 1978, **14**(4): p. 931–939.
74. Spiegelman, B.M., Ginty, C.A., Fibronectin modulation ofcell shape and lipogenic gene expression in 3T3-adipocytes. *Cell* 1983, **35**: p. 657–666.
75. Watt, F.M., Jordan, P.W., O'Neill, C.H., Cell shape controls terminal differentiation of human epidermal keratinocytes. *Proc Natl Acad Sci* 1988, **85**: p. 5576–5580.
76. Rosekelley, C.D., Despre, P.Y., Bissell, M.J., Extracellular matrix-dependant tissue-specific gene expression in mammary epithelial cells requires both physical and biochemical transduction. *Proc Natl Acad Sci* 1994, **91**: p. 12378–12382.
77. Chen, C.S., Mrksich, M., Huang, S., Whitesides, G.M., Ingber, D.E., Geometric control of cell life and death. *Science* 1997, **276**: p. 1425–1428.

78. McBeath, R., Pirone, D.M., Nelson, C.M., Bhadriraju, K., Chen, C.S., Cell shape, cytoskeletal tension, and RhoA regulate stem cell lineage commitment. *Dev Cell* 2004, **6**: p. 483–495.
79. Farmer, S.R., Ben-Ze'av, A., Benecke, B.J., Penman, S., Altered translatability of messenger RNA from suspended anchorage-dependent fibroblasts: Reversal upon cell attachment to a surface. *Cell* 1978, **15**: p. 627–637.
80. Ben-Ze'ev, A., Farmer, S.R., Penman, S., Protein synthesis requires cell-surface contact while nuclear events respond to cell shape in anchorage-dependent fibroblasts. *Cell* 1980, **21**(2): p. 365–372.
81. Wittelsberger, S.C., Kleene, K., Penman, S., Progressive loss of shape-responsive metabolic controls in cells with increasingly transformed phenotype. *Cell* 1981, **24**: p. 859–866.
82. Tay, C.Y., Yu, H., Pal, M., Leong, W.S., Tan, N.S., Ng, K.W., Leong, D.T., Tan, L.P., Micropatterned matrix directs differentiation of human esenchymal stem cells towards myocardial lineage. *Exp Cell Res* 2010, **316**: p. 1159–1168.
83. Luo, W., Jones, S.R., Yousaf, M.N., Geometric control of stem cell differentiation rate on surfaces. *Langmuir* 2008, **24**: p. 12129–12133.
84. Kilian, K.A., Bugarija, B., Lahn, B.T., Mrksich, M., Geometric cues for directing the differentiation of mesnchymal stem cells. *Proc Natl Acad Sci* 2010, **107**: p. 4872–4877.
85. Tang, J., Peng, R., Ding, J., The regulation of stem cell differentiation by cell-cell contact on micropatterned material surfaces. *Biomaterials* 2010, **31**: p. 2470–2476.
86. Karp, J.M., Yeo, Y., Geng, W. Cannizaro, C., Yan, K., Kohane, D.S., Vunjak-Novakovic, G., Langer, R.S., Radisic, S., A photolithographic method to create cellular micropatterns. *Biomaterials* 2006, **27**: p. 4755–4764.
87. Wang, N., Tytell, J.D., Ingber, D.E., Mechanotransduction at a distance: Mechanically coupling the extracellular matrix with the nucleus. *Nat Rev Mol Cell Biol* 2009, **10**: p. 75–82.
88. Jiang, L.W., Schindler, M., Nuclear transport in 3T3 fibroblasts: Effects of growth factors, transformation, and cell shape. *J Cell Biol* 1988, **106**: p. 13–19.
89. Feldherr, C.M., Akin, D., The permeability of the nuclear envelope in dividing and non-dividing cell cultures. *J Cell Biol* 1990, **111**: p. 1–8.
90. Thomas, C.H., Collier, J.H., Sfeir, C.S., Healy, K.E., Engineering gene expression and protien synthesis by modulation of nuclear shape. *Proc Natl Acad Sci* 2002, **99**: p. 1972–1977.
91. Taylor, W., Gokay, K.E., Capaccio, C., Davis, E., Glucksberg, M., Dean, D.A., The effects of cyclic stretch on gene transfer in alveolar epithelial cells. *Mol Ther J Am Soc Gene Ther* 2003, **7**: p. 542–549.
92. Lammerding, J., Schulze, P.C., Takahashi, T., Kozlov, S., Sullivan, T., Kamm, R.D., Stewart, C.L., Lee, R.T., Lamin A/C deficiency causes defective nuclear mechanics and mechanotransduction. *J Clin Invest* 2004, **113**: p. 370–378.
93. Demecret, C., Vassetzky, Y., Mechali, M., Chromatin remodelling and DNA replication: From nucleosome to loop domains. *Oncogene* 2001, **20**: p. 3086–3093.
94. Misteli, T., Spatial positioning; a new dimension in genome function. *Cell* 2004, **119**: p. 153–156.
95. Chuang, C.H., Belmont, A.S., Close encounters between active genes in the nucleus. *Genome Biol* 2005, **6**: p. 237.1–237.5.

96. Misteli, T., Beyond the sequence: Cellular organization of genome function. *Cell* 2007, **128**(4): p. 787–800.
97. Mazumder, A., Roopa, T., Basu, A., Mahadevan, L., Shivashankar, G.V., Dynamics of chromatin decondensation reveals the structural integrity of a mechanically prestressed nucleus. *Biophys J* 2008, **95**: p. 3028–3035.
98. Hu, Y., Kireev, I., Plutz, M., Ashourian, N., Belmont, A.S., Large-scale chromatin structure of inducible genes: Transcription on a condensed, linear template. *J Cell Biol* 2009, **185**: p. 87–100.
99. Geiger, B., Spatz, J.P., Bershadsky, A.D., Environmental sensing through focal adhesions. *Nat Rev Mol Cell Biol* 2009, **10**: p. 21–33.
100. Zaidel-Bar, R., Geiger, B., The switchable integrin adhesome. *J Cell Sci* 2010, **123**: p. 1385–1388.
101. Moore, S.W., Roca-Cusachs, P., Sheetz, M.P., Stretchy proteins on stretchy substrates: The important elements of integrin-mediated rigidity sensing. *Dev Cell* 2010, **19**(2): p. 194–206.
102. Schwartz, M.A., Integrins and extracellular matrix in mechanotransduction. *Cold Spring Harbor Perspect Biol* 2010, **2**: p. a005066.
103. Roca-Cusachs, P., Alcaraz, J., Snyer, R., Samitier, J., Farre, R., Navajas, D., Micropatterning of single endothelial cell shape reveals a tight coupling between nuclear volume in G1 and proliferation. *Biophys J* 2008, **94**: p. 4984–4995.
104. Cai, S., Pestic-Dragovich, L., O'Donnel, M.E., Wang, N., Ingber, D., Elson, E., De Lanerolle, P., Regulation of cytoskeletal mechanics, and cell growth by myosin light chain phosphorylation. *Am J Physiol Cell Physiol* 1998, **44**: C1349–C1356.
105. Dhawan, J., Helfman, D.M., Modulation of acto-myosin contractility in skeletal muscle myoblasts uncouples growth arrest from differentiation. *J Cell Sci* 2004, **117**: p. 3735–3748.
106. Mammoto, A., Huang, S., Moore, K., Oh, P., Ingber, D.E., Role of RhoA, mDia, and ROCK in cell shape-dependant control of the Skp2-p27(kip 1) pathway and G(1)/S transition. *J Biol Chem* 2004, **279**: p. 26323–26330.
107. Clark, P., Conolly, P., Curtis, A.S.G., Dow, J.A., Wilkinson, C.D., Topographical control of cell behavior. I. Simple step cues. *Development* 1987, **99**: p. 439–448.
108. Curtis, A. and Riechle, M., Tissue engineering in biophysical background. *Phys Med Biol* 2001, **46**: p. 47–65.
109. Curtis, A.S., Varde, M., Control of cell behavior: Topological factors. *J Nat Cancer Ins* 1964, **33**: p. 15–26.
110. Harrison, R.G., On the stereotropism of embryonic cells. *Science* 1911, **34**: p. 279–281.
111. Datta, N., Pham, Q.P., Sharma, U., Sikavitsas, V.I., Jansen, J.A., Mikos, A.G., In vitro generated extracellular matrix and fluid shear stress synergistically enhance 3D osteoblastic differentiation. *Proc Natl Acad Sci* 2006, **103**: p. 2488–2493.
112. Oh, S., Brammer, K.S., Li, Y.S.J., Teng, D., Engler, A.J., Chien, S., Jin, S., Stem cell fate dictated solely by altered nanotube dimension. *Proc Natl Acad Sci* 2009, **106**: p. 2130–2135.
113. Dalby, M.J., Gadegaard, N., Tare, R., Andar, A., Riehle, M.O., Harzyk, P., Wilkinson, C.D.W., Oreffo, R.O.C., The control of human mesenchymal cell differentiation using nanoscale symmetry and disorder. *Nat Mater* 2007, **6**: p. 997–1003.
114. Fu, J., Wang, Y.K., Yang, M.Y., Desai, R.A., Yu, X., Liu, Z., Chen, C.S., Mechanical regulation of cell function with geometrically modulated elastomeric substrates. *Nat Methods* 2010, **7**: p. 733–736.

115. Dalby, M.J., Riehle, M.O., Johnstone, H.J.H., Affrossman, S., Curtis, A.S.G., Polymer-demixed nanotopography: Control of fibroblast spreading and proliferation. *Tissue Eng* 2002, **8**: p. 1099–1108.
116. Gallagher, J.O., McGhee, K.F., Wilkinson, C.D.W., Riehle, M.O., Interaction of animal cells with ordered nanotopography. *IEEE Trans Nanobiosci* 2002, **1**: p. 24–28.
117. Berry, C.C., Campbell, G., Spadiccino, A., Robertson, M., Curtis, A.S.G., The influence of microscale topography on fibroblast attachment and motility. *Biomaterials* 2004, **25**: p. 5781–5788.
118. Powell, H.M., Kniss, D.A., Lannutti, J.J., Nanotopographic control of cytoskeletal organization. *Langmuir* 2006, **22**: p. 5087–5094.
119. Biggs, M.J.P., Richards, R.G., Gadedgaard, N., Wilkinson, C.D.W., Dalby, M.J., The effects of nanoscale pits on primary human oseoblast adhesion formation and cellular spreading. *J Mater Sci Mater Med* 2007, **18**: p. 399–404.
120. Kulangara, K., Leong, K.W., Substrate topography shapes cell function. *Soft Matter* 2009, **5**: p. 4072–4076.
121. Bettinger, C.J., Langer, R., Borenstein, J.T., Enginnering substrate topography at the micro- and nanoscale to control cell function. *Angew Chem Int Ed* 2009, **48**: p. 5406–5415.
122. Engel, E., Martinez, E., Mills, C.A., Funes, M., Planell, J.A., Samitier, J., Mesenchymal stem cell differentiation on microstructured poly(methyl methacrylate) substartes. *Ann Anat* 2009, **191**: p. 136–144.
123. Poellmann, M.J., Harrell, P.A., King, W.P., Wagoner Johnson, A.J., Geometric microenvironment directs cell morphology on topographically patterned hydrogel substrates. *Acta Biomater* 2010, **6**: p. 3514–3523.
124. Dalby, M.J., Andar, A., Nag, A., Affrossman S., Tare, R., McFarlane, S., Oreffo, R.O.C., Genomic expression of mesenchymal stem cells to altered nanoscale topographies. *J R Soc* 2008, **5**: p. 1055–1065.
125. Kurpinski, K., Chu, J., Hashi, C., Li, S., Anisotropic mechanosensing by mesenchymal stem cells. *Proc Natl Acad Sci* 2006, **103**: p. 16095–16100.
126. Thakar, R.G., Ho, F., Huang, N.F., Liepmann, D., Li, S., Regulation of vascular smooth muscle cells by micropatterning. *Biochem Bhiophys Res Commun* 2003, **307**: p. 883–890.
127. Zahor, D., Radko, A., Vago, R., Gheber, L.A., Organization of mesenchymal stem cell is controlled by micropatterned silicon substrates. *Mater Sci Eng C* 2007, **27**: p. 117–121.
128. Biggs, M.J.P., Richards, R.G., McFarlane, S., Wilkinson, C.D.W., Oreffo, R.O.C., Dalby, M.J., Adhesion formation of primary human osteoblast and the functional response of mesenchymal stem cells to 330 nm deep microgrooves. *J R Soc, Interface* 2008, **5**: p. 1231–1242.
129. Biggs, M.J.P., Richards, R.G., Gadedgaard, N., Wilkinson, C.D.W., Oreffo, R.O.C., Dalby, M.J., The use of nanoscale topography to modulate the dynamics of adhesion formation in primary osteoblasts and ERK/MAPK signalling in STRO-1+ enriched skeletal stem cells. *Biomaterials* 2009, **30**: p. 5094–5103.
130. Fujita, S., Ohshima, M., Iwata, H., Time-lapse observation of cell alignment on nanogrooved patterns. *J R Soc Interface* 2009, **6**: p. S269–S277.
131. Lee, M.R., Kwon, K.W., Jung, H., Kim, H.N., Suh, K.H., Kim, K., Kim, K.S., Direct differentiation of human embryonic stem cells into selective neurons on nanoscale ridge/groove pattern arrays. *Biomaterials* 2010, **31**: p. 4360–4366.

132. Gadegaard, N., Dalby, M.J., Riehle, M.O., Wilkinson, C.D.W., Optimizing substrate disorder for bone tissue engineering of mesenchymal stem cells. *J Vac Sci Technol B* 2008, **26**: p. 2554–2557.
133. Davidson, P.M., Ozcelik, H., Hasirci, V., Reiter, G., Anselme, K., Microstructured surfaces cause severe but non-detrimental deformation of the cell nucleus. *Adv Mater* 2009, **21**: p. 3586–3590.
134. Wan, L.Q., Kang, S.M., Eng, G., Grayson, W.L., Lu, X.L., Huo, B., Gimble, J., Guo, X.E., Mow, V.C., Vunjak-Novakovic, G., Geometric control of human stem cell morphology and differentiation. *Integr Biol* **2**: p. 346–353.
135. Balaban, N.Q., Scwarz, U.S., Riveline, D., Goichberg, P., Tzur, G., Sabanay, I., Mahalu, D., Safran, S., Bershadsky, A., Addadi, L., Geiger, B., Force and focal adhesion assembly: A close relationship studied using elastic micropatterned substrate. *Nat Cell Biol* 2001, **3**: p. 462–472.
136. Munevar, S., Wang, Y., Dembo, M, Traction force microscopy of migrating normal H-ras transformed 3T3 fibroblasts. *Biophys J* 2001, **80**: p. 1744–1757.
137. Tan, J.L., Tien, J., Chen, C.S., Microcontact printing of proteins on mixed self-assembled monolayers. *Langmuir* 2002, **18**: p. 519–523.
138. Tan, J.L., Tien, J., Pirone, D.M., Gray, D.S., Bhadriraju, K., Chen, C.S., Cells lying on a bed of microneedles: An approach to isolate mechanical force. *Proc Natl Acad Sci* 2003, **100**: p. 1484–1489.
139. McGarry, J.P., Fu, J., Yang, M.T., Chen, C.S., McMeeking, R.M., Evans, A.G., Deshpande, V.S., Simulation of the contractile response of cells on an array of micro-posts. *Phil Trans R Soc A* 2009, **367**: p. 3477–3497.
140. Park, S.Y., Park, S.Y., Namgung, S., Kim, B., Im, J., Kim, J.Y., Sun, K., Lee, K.B., Nam, J., Park, Y., Hong, S., Carbon nanotube monolayer paterns for directed growth of mesenchymal stem cells. *Adv Mater* 2007, **19**: p. 2530–2534.
141. Porter, J.R., Henson, A., Popat, K.C., Biodegradable poly(ε-caprolactone) nanowires for bone tissue engineering applications. *Biomaterials* 2009. **30**: 780–788.
142. Treiser, M.D., Yang, E.H., Gordonov, S., Cohen, D.M., Androulakis, I.P., Kohn, J., Chen, C.S., Moghe, P.V., Cytoskeleton-based forecasting of stem cell lineage fates. *Proc Natl Acad Sci* 2010, **107**: p. 610–615.
143. Meyvantsson, I., Beebe, D.J., Cell culture models in microfluidic systems. *Annu Rev Anal Chem* 2008, **1**: p. 423–449.
144. Whitesides, G.M., The origins and the future of microfluidics. *Nature* 2006, **442**: p. 368–373.
145. Lucchetta, E.M., Lee, J.H., Fu, L.A., Patel, N.H., Ismagilov, R.F., Dynamics of drosophila embryonic patterning network perturbed in space and time using microfluidics. *Nature* 2005, **434**: p. 1134–1138.
146. Verma, M.K., Majumder, A., Ghatak, A., Embedded template-assisted fabrication of complex microchannels in PDMS and design of a microfluidic adhesive. *Langmuir* 2006, **22**: p. 10291–10295.
147. Verma, M.K., Ganneboyina, S.R., Rakshit, V.R., Ghatak, A., Three-dimensional multihelical microfluidic mixers for rapid mixing of liquids. *Langmuir* 2008, **24**: p. 2248–2251.
148 Gomez-Sjoberg, R., Layrat, A.A., Dirone, D.M., Chen, C.S., Quake, S.R., Versatile, fully automated, microfluidic cell culture system. *Anal Chem* 2007, **79**: p. 8557–8563.
149. van Noort, D., Ong, S.M., Zhang, C., Zhang, S., Arooz, T., Yu, H., Stem cells in microfluidics. *Biotechnol Prog* 2009, **25**: p. 52–60.

150. Abhayankar, V.V., Bittner, G.N., Causey, J.A., Kamp, T.J., Beebe, D.J., Human embryonic stem cell culture in microfluidic channels. *7th Int Conf Miniaturized Chem Biochem Anal Syst* 2003, California, USA.
151. Tourovskaia, A., Figueroa-Masot, X., Folch, A., Differentiation-on-a-chip: A microfluidic platform for long-term cell culture studies. *Lab Chip* 2005, **5**: p. 14–19.
152. Tourovskaia, A., Figueroa-Masot, X., Folch, A., Long-term microfluidic culture of myotube microarrays for high-throughput focal stimulation. *Nat Protoc* 2006, **1**: p. 1092–1104.
153. Valero, A., Post, J.N., van Nieuwkasteele, J.W., ter Barak, P.M., Kruijer, W., van den Berg, A, Gene transfer and proten dynamics in stem cells using cell electroporation in a microfluidic device. *Lab Chip* 2008, **8**: p. 62–67.
154. Ju, X., Gao, N., Shi, Q., Hou, H., Hepatogenic differentiation of mesenchymal stem cells using microfluidic chips. *Biotechnol J* 2008, **3**: p. 383–391.
155. Ni, X.F., Crozatier, C., Sensebe, L., Langonne, A., Wang, L., Fan, Y., He, P.G., Chen, Y., On-chip differentiation of human mesenchymal stem cells into adipocytes. *Microelectron Eng* 2008, **85**: p. 1330–1333.
156. Zhang, F., Sensebe, L., Zhou, Y.L., Lin, C.J., Chen, Y., Osteogenic differentiation of human mesenchymal stem cells on chip: A comparison between two nutrient feeding methods. *Microelectron Eng* 2009, **86**: p. 1459–1461.
157 Zhang, Y. Gazit, Z., Pelled, G., Gazit, D., Vunjak-Novakovic, G., Patterning osteogenesis by inducible gene expression in microfluidic system. *Integr Bio*, 2011, **3**: p. 39–47.
158. Tenstad, E., Tourovskaia, A., Folch, A., Myklebost, O., Rian, E., Extensive adipogenic and osteogenic differentiation of patterned human mesenchymal stem cells in a microfluidic device. *Lab Chip* 2010, **10**: p. 1401–1409.
159. Zhao, F., Chella, R., Ma, T., Effects of shear stress on 3D human mesenchymal stem cell construct development in a perfusion bioreactor system: Experiments and hydrodynamic modeling. *Biotechnol Bioeng* 2007, **96**: p. 584–595.
160. Glossop, J.R., Cartmell, S.H., Effect of fluid flow-induced shear stress on human mesenchymal stem cell: Differential gene expression of IL1B and MAP3K8 in MAPK signaling. *Gene Expression Patterns* 2008, **9**: p. 381–388.

Chapter 5

Patient-Specific Modeling of Low-Density Lipoprotein Transport in Coronary Arteries

Ufuk Olgac

Department of Mechanical Engineering
Koc University, Rumeli Feneri Yolu
34450 Sariyer, Istanbul, Turkey
uolgac@ku.edu.tr

Introduction

Atherosclerosis, a progressive disease characterized by the accumulation of lipids in the arterial walls, is the primary cause of heart disease and stroke.[1] Locally elevated concentrations of low-density lipoprotein (LDL) are considered to be the initiator of atherosclerotic plaque formation.[1–3] Therefore, the transport of LDL into arterial walls has been the subject of various experimental[3–8] and computational investigations.[9–17]

Computational LDL transport models, as recently reviewed by Khakpour et al.,[18] are categorized into three types — wall-free models, homogenous wall models, and multi-layer wall models. In the simple wall-free models, fluid dynamics and LDL transport are calculated solely in the artery lumen. A constant filtration velocity and LDL flux defined by an overall mass transfer coefficient are generally applied on the lumen-side artery surface. The main drawback of wall-free models is that they do not provide any information on the transmural flow and solute dynamics in the arterial wall.[13,14] Homogeneous wall models account for both the artery lumen and arterial wall transport and treat the arterial wall as a single-layer homogeneous porous medium. Stangeby et al.[11] and Sun et al.[12] used homogeneous wall models employing an overall mass transfer coefficient and the Kedem–Katchalsky equations to account for

LDL transport through the endothelium. Recently, a local wall shear stress (WSS)-dependent three-pore model based on the homogeneous wall approach representing normal junctions, leaky junctions, and the vesicular pathway, has been proposed.[10] This three-pore model proved to be able to better represent the dependence of blood plasma and LDL transport through the endothelium on local blood flow characteristics than the single pathway approaches of Stangeby *et al.*[11] and Sun *et al.*[12] Finally, multi-layer wall models distinguish the intima from the media in the arterial wall.[15–17,19] They represent transmural flow and LDL dynamics better than homogeneous wall models, but are more complex.

In this chapter, a homogeneous wall model is presented, which calculates the transport of LDL from the artery lumen into the arterial wall in a three-dimensional (3D) left coronary artery with and without calcified plaque at normal and hypertensive blood pressures. To this end, the anatomy data of a patient with coronary artery disease are acquired using computed tomography (CT). An approximation of the patient's artery geometry in its healthy state is obtained through the segmentation of the vessel lumen including the detected calcified plaque. The calcified plaque is subsequently removed from the segmentation mask to yield the patient's artery geometry in its diseased state. The obtained healthy and diseased models of the artery are meshed to be used in the reconstruction of the arterial blood flow field using computational fluid dynamics (CFD), which is a prerequisite for the calculation of the spatial shear stress distribution at the endothelium. Intraluminal pressures of 70 and 120 mmHg are used in the CFD calculations to mimic normal and hypertensive conditions. The artery lumen surface meshes of the two models are uniformly extruded in the local surface normal vector direction to obtain arterial walls of constant thickness. The interface between the artery lumen and the arterial wall, i.e., the endothelium, is represented by a three-pore model, taking into account LDL transport through normal junctions, leaky junctions, and the vesicular pathway. In this three-pore model, the local fraction of leaky junctions is calculated based on the local WSS field. LDL is either transported actively across the endothelium via vesicles or passively through leaky junctions. However, leaky junctions are the main pathway, accounting for over 90% of the overall LDL transport into the arterial wall.[7] Using the pore theory, the local blood plasma and LDL fluxes into the arterial wall are determined based on the local fraction of leaky junctions, as described in detail in Ref. 10. These locally calculated fluxes are then used as boundary conditions in CFD calculations to reconstruct the flow and LDL concentration field in the arterial wall.

It is hypothesized that it may be possible to predict potential sites of plaque formation through patient-specific simulations of LDL accumulation.

Previous studies on the prediction of atherosclerotic plaque locations rely on WSS and from there, hemodynamic parameters were derived as their indicators.[20-24] In comparison, the presented model allows for the direct identification of LDL accumulation sites, which are generally accepted as prerequisites for plaque formation. The model further enables the effect of hypertension on LDL accumulation to be taken into account.

Nevertheless, 3D computational models that deal with LDL transport are scarce. Most of the published models include LDL transport in the artery lumen only and do not treat LDL passage through the endothelium and into the arterial wall.[13,14] One recent study by Koshiba *et al.*[9] included LDL transport in both the artery lumen and arterial wall of a curved 3D artery, but it only employed a constant permeability for the endothelium with a single pathway approach. The novelty of the herein presented model lies in the modeling of LDL transport in a 3D patient-specific coronary artery, in which the endothelium is represented by a three-pore model that enables the inclusion of the effects of local WSS on various pathways of blood plasma and LDL flux.

Methods

Acquisition and processing of anatomy data

A 74-year-old female patient with atypical anginal pain was examined on a dual-source CT system (Somatom Definition, Siemens Medical Solutions, Forchheim, Germany). The study protocol was approved by the local ethics committee who waived the written informed consent requirement. An initial non-contrast enhanced scan was performed for calcium scoring. Thereafter, the patient was injected with a contrast agent in a right antecubital vein. The contrast agent application was controlled by bolus-tracking in the ascending aorta. The scanning parameters followed a standard protocol:[25] detector collimation 2 mm × 32 mm × 0.6 mm, slice collimation 2 mm × 64 mm × 0.6 mm by means of a z-flying focal spot, gantry rotation time of 330 ms, pitch of 0.2, tube current time product of 330 mAs per rotation, and tube potential of 120 kV. Both non-enhanced and contrast-enhanced CT scans were performed from the level of the tracheal bifurcation to the diaphragm. The non-enhanced CT scan was reconstructed with a B35f kernel at 70% of the R–R interval using 3.0-mm non-overlapping slices. The contrast-enhanced CT angiography data set was reconstructed during mid-diastole at 70% of the R–R interval using a slice thickness of 0.75 mm (in plane resolution 0.256 mm at 512 × 512 voxels), a reconstruction increment of 0.5 mm, and using the soft-tissue convolution kernel B26f.

Following CT data acquisition, the coronary arteries were automatically segmented with a progressive region growing technique.[26] The resulting binary segmentation mask enclosed the vessel lumen as well as potential calcified plaques. The segmentation mask was converted into a surface mesh using a marching cube algorithm.[27] This mesh, which surrounded the vessel lumen and calcified plaques, was regarded as a representation of the lumen of the artery in its healthy state. A calcified plaque was detected in the proximal left anterior descending artery (segment six according to the model of the American Heart Association[28]). The plaque was manually segmented by applying a region growing technique; starting from a seed point within the plaque, all connected voxels with intensity values greater than a threshold value were regarded as representations of plaque. This threshold was determined to be 516 Hounsfield units (HU) by the visual inspection of the segmentation results superimposed on the original data set (Fig. 1). All voxels belonging to the plaque were then removed from the segmentation mask and

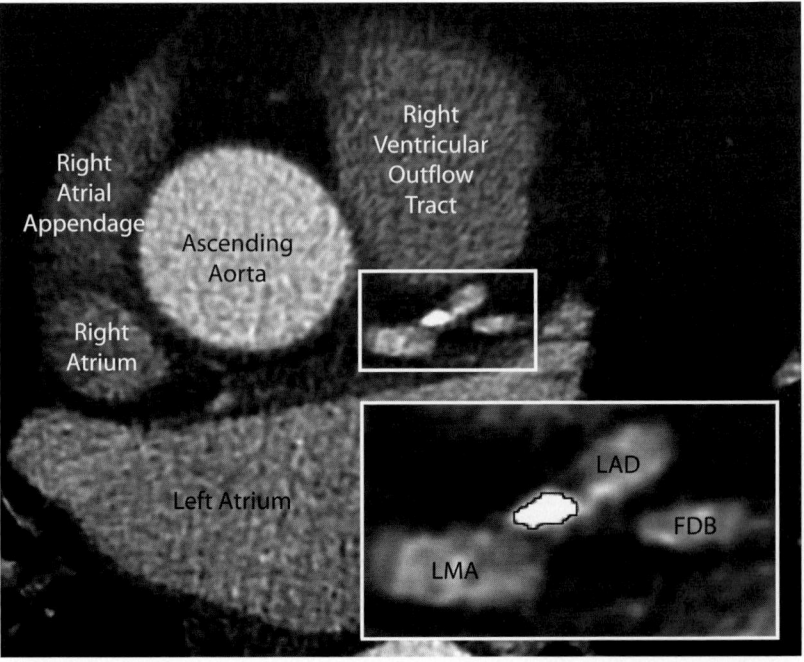

Fig. 1. Single, transverse slice at the level of the calcified plaque from the original CT coronary angiography data set. The calcified plaque is shown in the enlarged view (inset) with its superimposed segmentation visible as a black enveloping line. LMA, LAD, and FDB stand for left main coronary artery, left anterior descending artery, and first diagonal branch of the left anterior descending artery, respectively.[29]

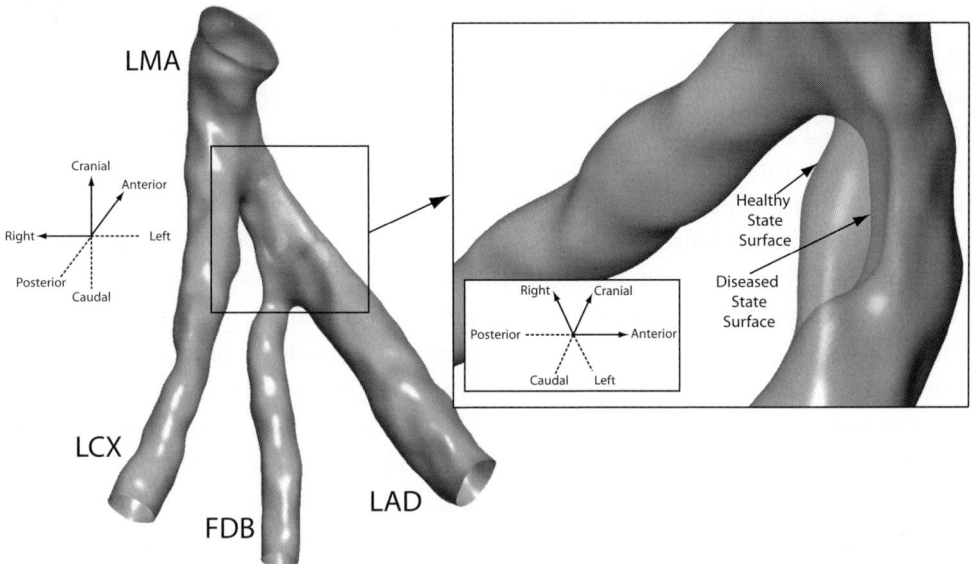

Fig. 2. Superposition of the surfaces of the patient's coronary arteries in their healthy and diseased states. The difference between the two states, i.e., the calcified plaque, is shown in the inset (note rotated coordinate system). LMA, LCX, LAD, and FDB stand for left main coronary artery, left circumflex artery, left anterior descending artery, and first diagonal branch of the left anterior descending artery, respectively.[29]

a surface mesh enclosing the artery lumen without plaque was constructed. This mesh represented the lumen of the artery in its diseased state. Both meshes, representing the healthy and diseased states, respectively, were converted to NURBS (non-uniform rational B-spline) surfaces. This approach ensures adequate smoothing of the lumen surface, which is essential for the subsequent generation of a high-quality computational grid. Figure 2 shows the NURBS representation of the artery lumen in its healthy and the diseased states.

Reconstruction of the flow field in the artery lumen

Under the assumption of constant blood density with Newtonian rheology, the fluid dynamics in the artery lumen are governed by the incompressible Navier–Stokes and continuity equations:

$$\partial_t \mathbf{u}_l + (\mathbf{u}_l \cdot \nabla)\mathbf{u}_l - \mu \Delta \mathbf{u}_l + \frac{1}{\rho}\nabla p_l = 0 \text{ and} \qquad (1)$$

$$\Delta \cdot \mathbf{u}_l = 0, \qquad (2)$$

where ∇ and Δ denote the gradient and the Laplacian operator, respectively, u_1 is the velocity vector field in the artery lumen, pi is the blood pressure in the lumen, and ρ and μ are the density and viscosity of blood, respectively, taken as $\rho = 1{,}050$ kg/m^3 and $\mu = 0.0035$ Pa·s.[30]

Based on the NURBS representation of the artery lumen, four sets of non-uniform unstructured computational volume grids were constructed for both the healthy and diseased states. The respective grids consisted of approximately 376,000, 686,000, 976,000, and 1,382,000 tetrahedral and prismatic volumes. A grid independence study showed that the grid with 976,000 volumes was sufficient for the reconstruction of the flow field in the artery lumen. Equations (1) and (2) were solved iteratively on this grid with the finite-volume CFD code ANSYS CFX 11.0 (ANSYS, Inc., Canonsburg, PA, USA) using an Algebraic Multigrid scheme[31] and a pressure–velocity coupling adapted from Ref. 32. A second-order backward Euler scheme was used for the discretization of the transient term, i.e., the first term on the left hand side of Eq. (1). Time-step independence was satisfied with a time-step size of 1/80 of the patient's cardiac cycle. Period independence was reached at the second pulsatile cycle. As inlet boundary conditions, the pulsatile blood inflow profile into the left main coronary artery (LMA) given in Ref. 33 was specified. Constant pressure boundary conditions of 70 and 120 mmHg[8,10] for the normal and hypertensive cases, respectively, were used at the three inferior ends of the domain and a no-slip boundary condition was specified at the rigid wall.

Next to these transient flow calculations, steady state flow conditions in the artery lumen were also determined by solving Eqs. (1) and (2) without the transient term in Eq. (1). As the inlet boundary condition, a volumetric inflow rate of 57 mL/min, i.e., the time-averaged blood inflow rate based on the pulsatile inflow profile,[33] was specified at the superior end of the artery lumen. The remaining boundary conditions were the same as in the transient calculations.

Determination of blood plasma and LDL fluxes through the endothelium

The employed three-pore model, which describes the blood plasma and LDL transport through the endothelium, takes into account transport through normal junctions, leaky junctions, and the vesicular pathway.[10] The normal endothelial junctions were modeled as cylindrical pores on the junction strands at the endothelial clefts, and the leaky junctions were modeled as ring-shaped pores surrounding the leaky cells. Blood plasma flux occurs through both normal and leaky junctions. Filtering through the normal junctions, blood plasma can only carry solutes with a radius smaller than 2 nm.[34] LDL molecules with their radius of 11 nm are not able to pass through normal junctions, but can enter the arterial wall through leaky junctions that surround mitotic and dying cells.[5,35] While LDL is also actively transported across the endothelium via vesicles, leaky junctions are the main pathway, accounting for over 90% of the overall LDL transport into the arterial wall.[7]

Atherosclerosis occurs primarily in regions of the arterial tree where recirculation zones or other flow disturbances causing low steady or highly oscillatory WSS are present.[4] Endothelial cells respond to mechanical stimuli such as shear stress and pressure through intracellular signaling pathways.[4,36] Endothelial cells exposed to low steady or highly oscillatory shear stress are roundly shaped and show a high proliferation rate, leading to the formation of leaky junctions. When exposed to high steady shear stress, endothelial cells are elongated in the flow direction with a lower rate of mitosis and hence, show a smaller number of leaky junctions compared to the ones exposed to low steady or highly oscillatory shear stress.[4,6,37–39] As a result, endothelial shear stress influences LDL transport into the arterial wall by promoting or demoting the formation of leaky junctions, i.e., the primary LDL transport pathway.

The local fraction of leaky junctions, ϕ, which is defined as the ratio of the area of leaky cells to the area of all cells,[40] is calculated as a function of local WSS. The WSS values are obtained using the CFD model detailed in the previous section. The following correlations derived in Ref. 10 and obtained from Refs. 4–6 and 37–39 are used to this end:

$$SI = 0.380 e^{-0.790\,WSS} + 0.225 e^{-0.043\,WSS}, \tag{3}$$

$$\#MC = 0.003797 e^{14.75\,SI}, \tag{4}$$

$$\#LC = 0.307 + 0.805 \cdot (\#MC), \quad \text{and} \tag{5}$$

$$\phi = \frac{\#LC \times \pi R_{cell}^2}{\text{unit area}}, \tag{6}$$

where SI, #MC, and #LC are the shape index of cells (equal to one for a circular cell and approaches zero for a highly elongated cell[6,37]), number of mitotic cells, and number of leaky cells, respectively, R_{cell} and unit area are the radius of a single endothelial cell (15 μm[40]) and a unit area of 0.64 mm,[2,4] respectively, and WSS is the wall shear stress in Pa.

In order to determine the blood plasma and LDL flux through the endothelium, the transport properties of each pathway have to be established.

The transport properties of the leaky junction pathway, i.e., hydraulic conductivity, $L_{p,lj}$, diffusive permeability, P_{lj}, and solvent drag reflection coefficient, $\sigma_{f,lj}$, are calculated with the pore theory[41,42] using the local fraction of leaky junctions obtained with Eqs. (3)–(6). The hydraulic conductivity of the normal junctions is adopted from Ref. 10 as $L_{p,nj} = 1.16 \times 10^{-9}$ ms^{-1} mmHg^{-1} and the diffusive permeability of normal junctions to LDL, P_{nj}, and the solvent drag reflection coefficient of the normal junctions, $\sigma_{f,nj}$, are set to zero and one, respectively, since there is no LDL passage through normal junctions.

The fluid dynamics through the endothelium and in the arterial wall are represented by an electrical analogy in which fluid flow is considered to be analogous to current in an electric circuit. The driving pressure is then analogous to the potential difference. Denoting the total flow resistance formed by the endothelial resistance, R_{end}, and wall resistance, R_{wall}, as $R_T = R_{end} + R_{wall}$, the filtration velocity can be expressed as

$$J_v = \frac{p_l^{end} - p_{adv}}{R_T}, \qquad (7)$$

where p_l^{end} and p_{adv} are the lumen side pressure at the endothelium and the pressure at the media–adventitia interface, respectively. $R_{wall} = \mu_p t_a / \kappa_p$ is the flow resistance in the arterial wall derived from Darcy's law given in the next section and μ_p, t_a, and κ_p are the viscosity of blood plasma, thickness of the arterial wall, and the Darcy permeability of the arterial wall, respectively. R_{end} is the combined flow resistance of the normal and leaky junctions — two parallel flow pathways — with the respective flow resistances of $R_{nj} = 1/L_{p,nj}$ and $R_{lnj} = 1/L_{p,lj}$, given as

$$\frac{1}{R_{end}} = \frac{1}{R_{nj}} + \frac{1}{R_{lj}}. \qquad (8)$$

LDL flux, J_s, is given as the product of the total apparent permeability of the endothelium, P_{app}, and the lumen side concentration of LDL at the endothelium, c_l^{end}:

$$J_s = P_{app} \, c_l^{end}. \qquad (9)$$

It was shown in Ref. 10 that the maximum spatial variation of LDL concentration in the artery lumen can be expected to be very small (under 1.0% in that particular geometry). It has been shown by others[13,16,17] that the concentration polarization — the increase of solute concentration on the endothelial surface — is promoted with increasing intraluminal pressure and consequently, increasing filtration velocity. For an intraluminal pressure of 70 mmHg, i.e., one of the pressures used in this study, Ai and Vafai[16] and Yang and Vafai[17] calculated the increase in the surface concentrations of LDL to be under 2% and 3%, respectively. For 120 mmHg, these are 5% and 6%, respectively. Therefore, in the current work, the lumen-side concentration of LDL is assumed to be constant and is assigned the bulk fluid LDL concentration, C_0, of 1.0×10^{-3} mol/m^3 in accordance with the typical human blood LDL concentration level.[43] As LDL flux occurs via both the vesicular pathway and through leaky junctions, the total apparent permeability, P_{app}, is the sum of the vesicular permeability, P_v, taken as 1.92×10^{-11} m/s,[10] and the apparent permeability of the leaky junction pathway, $P_{app,lj}$, which is calculated locally based on the diffusive permeability of the leaky junction pathway, P_{lj}, and the blood plasma flux through the leaky junction pathway, $J_{v,lj}$.

Reconstruction of flow and LDL concentration fields in the arterial wall

The arterial wall is regarded as a single-layer porous medium with a constant thickness of 0.34 mm, the combined thickness of the intima and media of the human left anterior descending coronary artery (LAD).[44] The arterial wall is further assumed to be stationary and rigid. The transmural velocity vector field in the arterial wall, u_w, is calculated with Darcy's law,

$$\mathbf{u_w} = \frac{\kappa_p}{\mu_p} \nabla p_w \text{ and} \tag{10}$$

$$\nabla \cdot \mathbf{u_w} = 0, \tag{11}$$

where p_w is the pressure in the wall and $\mu_p = 0.001$ Pa·s is the blood plasma viscosity.[30] The Darcy permeability of the arterial wall is adopted from Ref. 10 as $\kappa_p = 1.2 \times 10{-}18$ m^2. The lumen side pressure at the endothelium, p_l^{end}, as obtained with the CFD model of the artery lumen, is used in Eq. (7) to calculate the local blood plasma flux at the endothelium, J_v, which is then applied as the inlet boundary condition for the arterial wall domain. A constant pressure boundary condition of $p_{adv} = 17.5$ mmHg[10] and $p_{adv} = 30$ mmHg[16,17] are applied at the media–adventitia interface for the intraluminal pressures of 70 and 120 mmHg, respectively. Zero flux boundary conditions were used at the longitudinal ends of the wall.

LDL transport in the arterial wall is governed by the convection–diffusion reaction equation

$$K_{lag}\mathbf{u_w} \cdot \nabla c_w - D_w \Delta c_w + r_w c_w = 0, \qquad (12)$$

where c_w is the LDL concentration in the wall, $K_{lag} = 0.1486$,[15] $D_w = 8.0 \times 10^{-13}$ m²/s[19] and $r_w = 3.0 \times 10^{-4}$ s⁻¹[10] are the LDL lag coefficient, diffusion coefficient of LDL, and LDL degradation rate, respectively. The LDL flux, J_s, locally calculated with Eq. (9), is applied at the endothelium, and a constant LDL concentration value of $c_{adv}/C_0 = 0.005$ and $c_{adv}/C_0 = 0.015$ are imposed at the media–adventitia interface for the intraluminal pressures of 70 and 120 mmHg, respectively.[8] At the longitudinal ends of the wall, isolation boundary conditions were used.

Representations of the arterial walls were obtained by uniformly extruding the artery lumen surface grids in the direction of the local outer surface normal vectors, thereby generating non-uniform unstructured volume grids consisting of approximately 468,190, 936,380, and 1,404,570 prismatic elements. A grid independence study showed that the grid with 936,380 elements was sufficient for the reconstruction of the flow and concentration fields in the arterial wall with the finite-element CFD code Comsol Multiphysics, Version 3.3 (COMSOL AB, Stockholm, Sweden) using a conjugate gradients iterative solver[45] in conjunction with an Algebraic Multigrid preconditioner.

Results

Endothelial WSS, together with intraluminal pressure, governs LDL flux into the arterial wall. The characteristic time of endothelial cell response to shear stress in terms of proliferation and death — both phenomena that lead to leaky junction formation — is much longer than that of WSS oscillation (order of hours[6,37,46] compared to less than a second). For this reason, WSS values are generally time averaged before they are used to calculate their effect on leaky junction formation. Time averaging of WSS yields average WSS (AWSS). Alternatively, instead of time averaging transient WSS values, steady WSS (SWSS) can be derived from steady-state flow calculations. Comparing Figs. 3(b) and 3(c), which show SWSS and AWSS distribution in the treated domains, respectively, little difference between the two can be seen. Similar good comparisons between SWSS and AWSS were found by Joshi et al.[47] and Myers et al.[48] for right coronary arteries and Ramaswamy et al.[49] for left coronary arteries. Since most of the experimental data that were used to correlate

WSS to cell shape index[6,37,38] are based on steady flow experiments and since the differences between SWSS and AWSS are negligible, SWSS is used as the input to the correlations to obtain the fraction of leaky junctions with respect to shear stress given by Eqs. (3)–(6).

The blood flow characteristics in the lumen of the artery in its "healthy" state (by regarding the calcified plaque as part of the artery lumen) and its diseased state differ considerably. Past the bifurcation of the LMA into the left circumflex artery (LCX) and the LAD, velocity magnitudes of up to 0.22 m/s are reached in the healthy state as the streamlines in Fig. 3(a) demonstrate. This velocity is increased to 0.26 m/s in the diseased state due to the decreased cross-sectional area of the LAD caused by the presence of atherosclerotic plaque. However, the increase in velocity is not sufficient to maintain the healthy state blood flow rate through the LAD. In the healthy state, 65.8% of the blood flows through the LAD and 9.1% through the first diagonal branch (FDB) and in the diseased state, these figures are 56.2% and 7.6%, respectively, showing that plaque formation in the proximal segment of the LAD reduces the blood supply to the distal segment of the LAD and the FDB.

Considering the SWSS magnitudes at the endothelium in the healthy state in Fig. 3(b), a small region of high shear stress is distinguished at the bifurcation where the LMA branches into the LAD and the LCX, followed by a low shear stress region in the proximal section of the LCX and a low shear stress region in the proximal LAD, coinciding with the location of the plaque in the diseased model. These two low shear stress regions (SWSS < 0.2 Pa) are a direct result of the characteristics of the flow field at this bifurcation; consulting again the streamlines in the healthy state in Fig. 3(a), the flow through the LMA accelerates over the bifurcation, thereby forming the small region of high shear stress and separating off the wall, generating the two low shear stress regions distal to the bifurcation. In the diseased model, the flow through the LMA accelerates over the plaque, forming a high shear stress region at the proximal end of the plaque. The low shear stress region in the proximal section of the LCX remains approximately at the same location, yet covers a smaller area compared to the healthy state, whereas the second low shear stress region in the proximal LAD shifts downstream of the plaque. The increase in the LCX relative blood flow rate from 25.1% in the healthy state to 36.2% in the diseased state causes the low shear stress region to shrink in size. In Fig. 3(a), the streamlines in the diseased state show that the flow over the plaque separates from the wall due to the larger cross-sectional area at the distal part of the plaque, leading to the formation of a recirculation zone that is responsible for the low shear stress region downstream of the plaque.

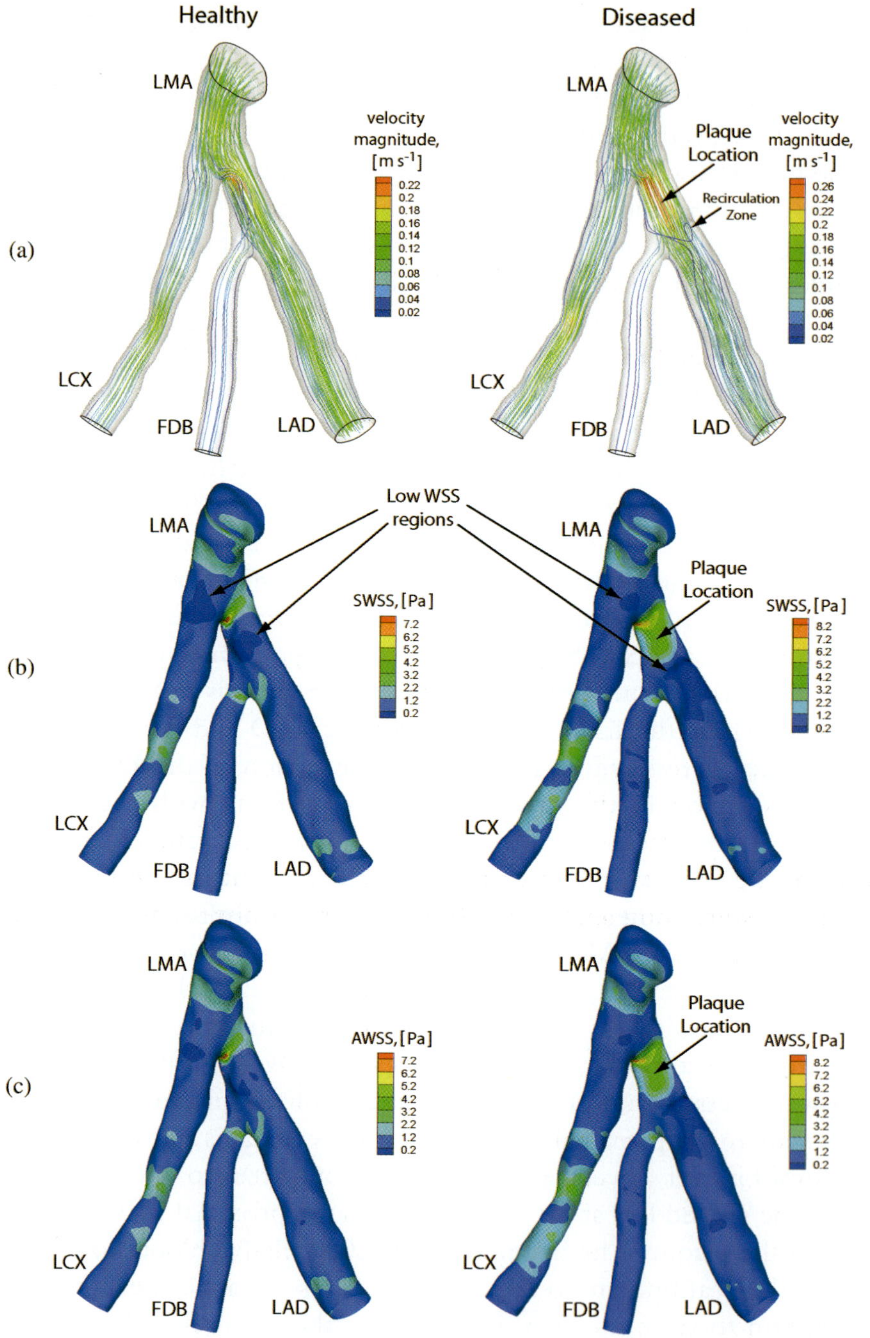

Fig. 3. Comparison of the models representing the healthy and plaque-free state of the patient's coronary arteries (left) and the diseased state (right). (a) Streamlines for the steady case colored based on velocity magnitude, (b) steady wall shear stress (SWSS) magnitude at the endothelium, and (c) average wall shear stress (AWSS) magnitude at the endothelium for the pulsatile case.[29]

Figures 4(a) and 4(b) juxtapose the total filtration velocities through the endothelium and the filtration velocities through the leaky junction pathway in the healthy and diseased states, respectively, at the normal intraluminal pressure of 70 mmHg. In the healthy model, increased filtration velocities are observed at the abovementioned two low shear stress regions. The low shear stresses at these regions lead to an elevated number of leaky junctions, which in turn leads to an increased filtration of blood plasma through the leaky junction pathway into the arterial wall. The magnitude of the filtration velocity through the leaky junction pathway increases up to 30-fold in the low shear stress regions compared to the region with the highest shear stress at the bifurcation where the LMA branches into the LAD and the LCX. In the diseased model, distal to the plaque within the recirculation zone, the filtration velocities are also increased. The magnitudes of the total filtration velocities are in the range of 1.5–2.05×10^{-8} m/s in both the healthy and the diseased states, which is in agreement with the literature value of 1.78×10^{-8} m/s for the intraluminal pressure of 70 mmHg.[8] In Fig. 4(c), the wall-side LDL concentrations at the endothelium are compared. In the healthy model, two high LDL concentration regions (normalized LDL concentration >0.2) are observed downstream of the bifurcation of the LMA into the LCX and the LAD. These regions correspond to those with high filtration velocity through the leaky junction pathway. This increase in blood plasma flux through the leaky junctions increases the amount of LDL convected into the arterial wall. Of these two regions, the one in the proximal section of the LCX mostly preserves its location, but diminishes in size in the diseased state compared to the healthy state. In the second region at the proximal section of the LAD, the plaque observed in the diseased state is formed and the high LDL concentration region is shifted immediately distal to the plaque.

Figures 5(a) and 5(b) show the results for the total filtration velocity through the endothelium and the normalized wall-side LDL concentration at the endothelium, respectively, for the hypertensive case with intraluminal pressure of 120 mmHg in the healthy state of the artery. With this increased blood pressure, the magnitude of the total filtration velocity increases to 3.3–3.9×10^{-8} m/s. This range of values is in agreement with the values of 3.05×10^{-8} m/s and 3.95×10^{-8} m/s from the computational studies of Ai and Vafai[16] and Yang and Vafai,[17] respectively, and slightly higher than the experimental value of 2.55×10^{-8} m/s reported by Meyer et al.[8] The discrepancy with the experimental value is most likely related to the fact that increasing intraluminal pressure compresses the arterial wall and thereby decreases its permeability.[50]

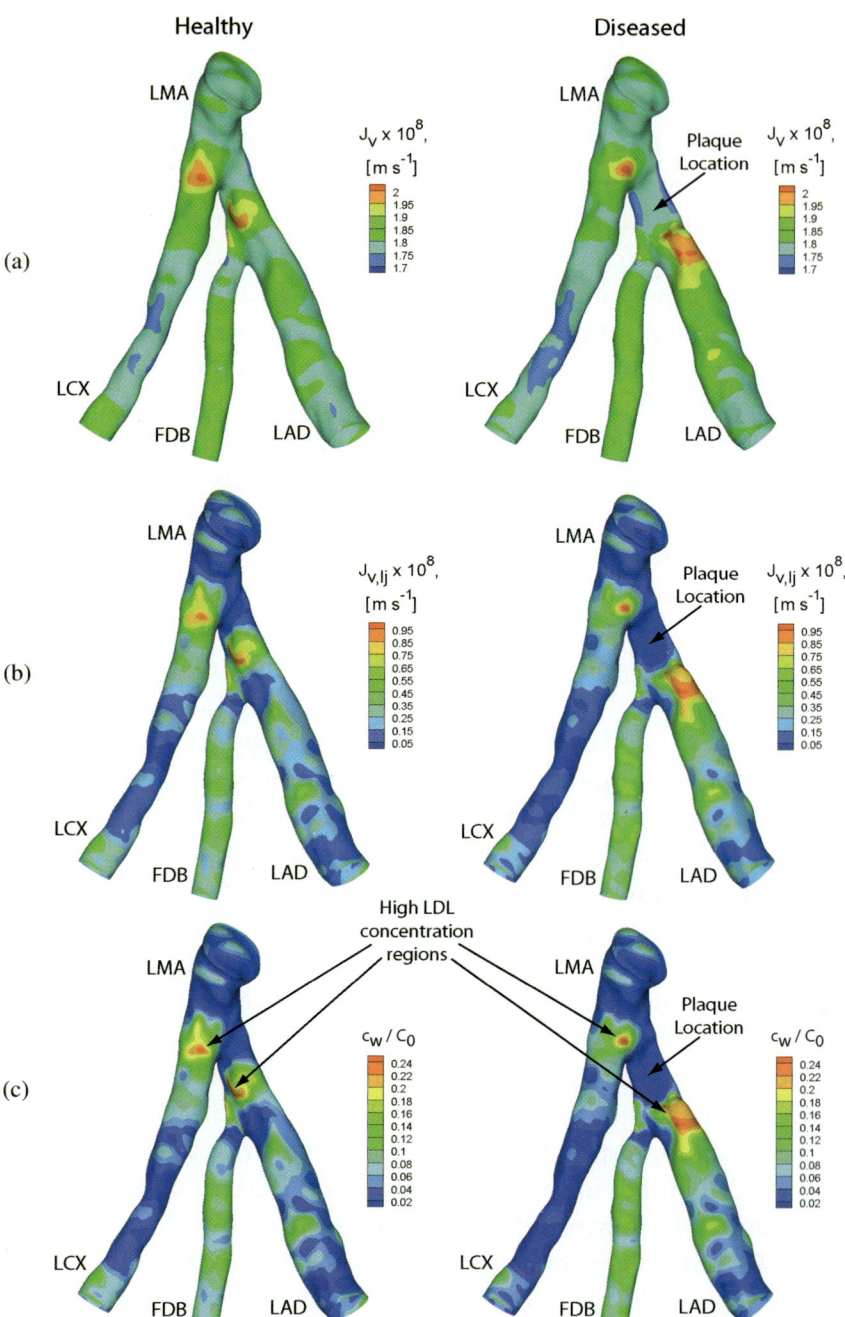

Fig. 4. Comparison of the models representing the healthy and plaque-free state of the patient's coronary arteries (left) and the diseased state (right) for the normal intraluminal pressure of 70 mmHg. (a) Total filtration velocity through the endothelium, (b) filtration velocity through the leaky junction pathway, and (c) wall-side LDL concentration at the endothelium normalized by the bulk fluid concentration C_0.[29]

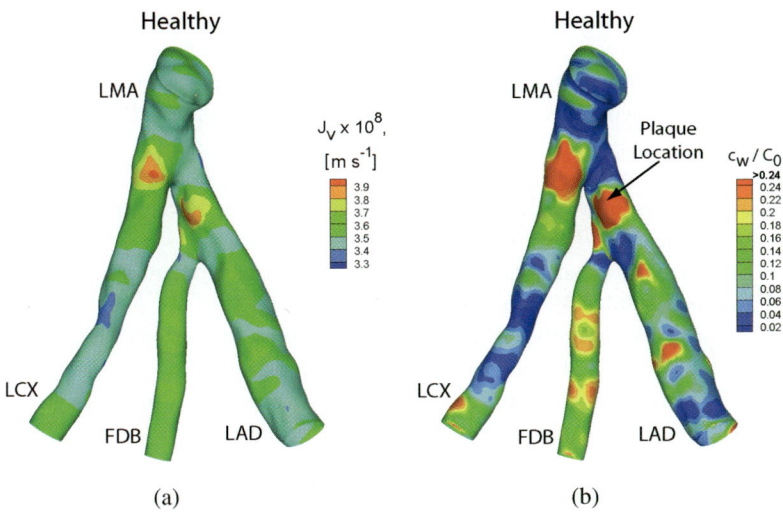

Fig. 5. The healthy and plaque-free state of the patient's coronary arteries for the hypertensive case with 120 mmHg intraluminal pressure. (a) Total filtration velocity through the endothelium and (b) wall-side LDL concentration at the endothelium normalized by the bulk fluid concentration C_0.[29]

If normalized LDL concentrations higher than 0.2 are assumed to be indicative of potential atherosclerotic lesion-prone sites, then Fig. 5(b) demonstrates that the number of such locations increases in the hypertensive case compared to the situation with normal intraluminal pressure shown in Fig. 4(c). In addition, the original two critical atherosclerotic lesion-prone sites are much larger in the hypertensive case.

Summary

A 3D, patient-specific model of LDL transport from blood into the arterial wall has been presented. The novelty of this work lies in the representation of the arterial endothelium as a three-pore model with shear stress-dependent transport properties coupled with a patient-specific reconstruction of the luminal blood flow field. This approach, as opposed to prior single pathway models with constant transport properties, allows for a patient-specific localization of arterial wall regions with high LDL concentrations. The herein presented model also quantifies the levels of LDL concentration in the arterial wall for various intraluminal pressures. The application of the model to the healthy and the diseased states of the arteries confirms the relationship between high LDL concentration sites and actual plaque locations, supporting the initial hypothesis that the potential sites of plaque formation may be predicted through patient-specific, shear-dependent simulations of LDL accumulation.

References

1. Lusis, A.J., Atherosclerosis. *Nature*, 2000, **407**(6801): p. 233–241.
2. Ross, R., Atherosclerosis: A defense mechanism gone awry. *Am J Pathol*, 1993, **143**(4): p. 987–1002.
3. Schwenke, D.C., Carew, T.E., Initiation of atherosclerotic lesions in cholesterol-fed rabbits focal increases in arterial LDL concentration precede development of fatty streak lesions. *Arteriosclerosis*, 1989, **9**(6): p. 895–907.
4. Chien, S., Molecular and mechanical bases of focal lipid accumulation in arterial wall. *Prog Biophy Mol Biol*, 2003, **83**(2): p. 131–151.
5. Lin, S.J. et al., Transendothelial transport of low-density lipoprotein in association with cell mitosis in rat aorta. *Arteriosclerosis*, 1989, **9**(2): p. 230–236.
6. Sakamoto, N., Ohashi, T., Sato, M., Effect of shear stress on permeability of vascular endothelial monolayer cocultured with smooth muscle cells. *JSME Int J Ser C*, 2004, **47**(4): p. 992–999.
7. Cancel, L.M., Fitting, A., Tarbell, J.M., In vitro study of LDL transport under pressurized (convective) conditions. *Am J Physiol Heart Circ Physiol*, 2007, **293**(1): p. H126–H132.
8. Meyer, G., Merval, R., Tedgui, A., Effects of pressure-induced stretch and convection on low-density lipoprotein and albumin uptake in the rabbit aortic wall. *Circ Res*, 1996, **79**(3): p. 532–540.
9. Koshiba, N. et al., Multiphysics simulation of blood flow and LDL transport in a porohyperelastic arterial wall model. *J Biomech Eng*, 2007, **129**(3): p. 374–385.
10. Olgac, U., Kurtcuoglu, V., Poulikakos, D., Computational modeling of coupled blood-wall mass transport of LDL: Effects of local wall shear stress. *Am J Physiol Heart Circ Physiol*, 2008, **294**(2): p. H909–H919.
11. Stangeby, D.K., Ethier, C.R., Computational analysis of coupled blood-wall arterial LDL transport. *J Biomech Eng-Trans ASME*, 2002, **124**(1): p. 1–8.
12. Sun, N. et al., Fluid-wall modelling of mass transfer in an axisymmetric stenosis: Effects of shear-dependent transport properties. *Ann Biomed Eng*, 2006, **34**(7): p. 1119–1128.
13. Wada, S., Karino, T., Theoretical prediction of low-density lipoproteins concentration at the luminal surface of an artery with a multiple bend. *Ann Biomed Eng*, 2002, **30**(6): p. 778–791.
14. Wada, S., Koujiya, M., Karino, T., Theoretical study of the effect of local flow disturbances on the concentration of low-density lipoproteins at the luminal surface of end-to-end anastomosed vessels. *Med Biol Eng Comput*, 2002, **40**(5): p. 576–587.
15. Sun, N. et al., Effects of transmural pressure and wall shear stress on LDL accumulation in the arterial wall: A numerical study using a multi-layered model. *Am J Physiol Heart Circ Physiol*, 2007.
16. Ai, L., Vafai, K., A coupling model for macromolecule transport in a stenosed arterial wall. *Int J Heat Mass Transfer*, 2006, **49**(9–10): p. 1568–1591.
17. Yang, N., Vafai, K., Modeling of low-density lipoprotein (LDL) transport in the artery-effects of hypertension. *Int J Heat Mass Transfer*, 2006, **49**(5–6): p. 850–867.
18. Khakpour, M., Vafai, K., Critical assessment of arterial transport models. *Int J Heat Mass Transfer*, 2008, **51**(3–4): p. 807–822.
19. Prosi, M. et al., Mathematical and numerical models for transfer of low-density lipoproteins through the arterial walls: A new methodology for the model set up with applications to the study of disturbed lumenal flow. *J Biomech*, 2005, **38**(4): p. 903–917.

20. Chatzizisis, Y.S. et al., Prediction of the localization of high-risk coronary atherosclerotic plaques on the basis of low endothelial shear stress: An intravascular ultrasound and histopathology natural history study. *Circulation*, 2008, **117**(8): p. 993–1002.
21. Feldman, C.L. et al., Determination of *in vivo* velocity and endothelial shear stress patterns with phasic flow in human coronary arteries: A methodology to predict progression of coronary atherosclerosis. *Am Heart J*, 2002, **143**(6): p. 931–939.
22. Zhang, J.M. et al., Numerical investigation and identification of susceptible sites of atherosclerotic lesion formation in a complete coronary artery bypass model. *Med Biol Eng Comput*, 2008.
23. Huo, Y., Wischgoll T., Kassab, G.S., Flow patterns in three-dimensional porcine epicardial coronary arterial tree. *Am J Physiol Heart Circ Physiol*, 2007, **293**(5): p. H2959–H2970.
24. Lei, M., Kleinstreuer, C., Truskey, G.A., Numerical investigation and prediction of atherogenic sites in branching arteries. *J Biomech Eng-Trans ASME*, 1995, **117**(3): p. 350–357.
25. Alkadhi, H. et al., Dual-source computed tomography coronary angiography: Influence of obesity, calcium load, and heart rate on diagnostic accuracy. *Eur Heart J*, 2008, **29**(6): p. 766–76.
26. Bock, S. et al., Robust vessel segmentation — art. no. 691539. In: *Medical Imaging 2008 Conference*, Spie-Int Soc Optical Engineering, San Diego, CA, 2008.
27. William, E.L., Harvey, E.C., Marching cubes: A high resolution 3D surface construction algorithm. *SIGGRAPH Comput Graph*, 1987, **21**(4): p. 163–169.
28. Austen, W.G. et al., A reporting system on patients evaluated for coronary artery disease. Report of the Ad Hoc Committee for Grading of Coronary Artery Disease, Council on Cardiovascular Surgery, American Heart Association. *Circulation*, 1975, **51**(4): p. 5–40.
29. Olgac, U. et al., Patient-specific three-dimensional simulation of LDL accumulation in a human left coronary artery in its healthy and atherosclerotic states. *Am J Physiol Heart Circ Physiol*, 2009, **296**(6): p. H1969–H1982.
30. Milnor, W.R., *Hemodynamics*, 2nd edn. Williams & Wilkins, Baltimore, 1989, p. 419.
31. Hutchinson, B.R., Raithby, G.D., A multigrid method based on the additive correction strategy. *Numer Heat Transfer*, 1986, **9**(5): p. 511–537.
32. Majumdar, S., Role of underrelaxation in momentum interpolation for calculation of flow with nonstaggered grids. *Numer Heat Transfer*, 1988, **13**(1): p. 125–132.
33. Berne, R.M., Levy, M.N., *Cardiovascular Physiology*, 5th edn. Mosby, St Louis, 1986.
34. Ogunrinade, O., Kameya, G.T., Truskey, G.A., Effect of fluid shear stress on the permeability of the arterial endothelium. *Ann Biomed Eng*, 2002, **30**(4): p. 430–446.
35. Lin, S.J., Jan, K.M., Chien, S., Role of dying endothelial cells in transendothelial macromolecular transport. *Arteriosclerosis*, 1990, **10**(5): p. 703–709.
36. Chien, S., Effects of disturbed flow on endothelial cells. *Ann Biomed Eng*, 2008, **36**(4): p. 554–562.
37. Chiu, J.J. et al., Effects of disturbed flow on endothelial cells. *J Biomech Eng-Trans ASME*, 1998, **120**(1): p. 2–8.
38. Levesque, M.J. et al., Correlation of endothelial-cell shape and wall shear-stress in a stenosed dog aorta. *Arteriosclerosis*, 1986, **6**(2): p. 220–229.
39. Suciu, A., Effects of External Forces on Endothelial Cells. EPFL, Switzerland, 1997.
40. Huang, Y. et al., A fiber-matrix model for the growth of macromolecular leakage spots in the arterial intima. *J Biomech Eng-Trans ASME*, 1994, **116**(4): p. 430–445.

41. Curry, F.E., Hydrodynamic description of osmotic reflection coefficient with application to pore theory of transcapillary exchange. *Microvasc Res,* 1974, **8**(2): p. 236–252.
42. Curry, F.E., Mechanics and thermodynamics of transcapillary exchange, In: Renkin, E.M., Charles Michel, C., eds., *Handbook of Physiology, The Cardiovascular System, Microcirculation,* Sec. 2, Vol. 4, Part. 1, American Physiological Society, Bethesda, 1984, p. 309–374.
43. Goldstein, J.L., Brown, M.S., The low-density lipoprotein pathway and its relation to atherosclerosis. *Annu Rev Biochem* 1977, **46**: p. 897–930.
44. Gradus-Pizlo, I. et al., Left anterior descending coronary artery wall thickness measured by high-frequency transthoracic and epicardial echocardiography includes adventitia. *Am J Cardiol,* 2003, **91**(1): p. 27–32.
45. Hestenes, M.R., Stiefel, E., Methods of conjugate gradients for solving linear systems. *J Res Nat Bur Stand,* 1952, **49**(6): p. 409–436.
46. Kadohama, T. et al., Effects of different types of fluid shear stress on endothelial cell proliferation and survival. *J Cell Physiol,* 2007, **212**(1): p. 244–251.
47. Joshi, A.K. et al., Intimal thickness is not associated with wall shear stress patterns in the human right coronary artery. *Arterioscler Thromb Vasc Biol,* 2004, **24**(12): p. 2408–2413.
48. Myers, J.G. et al., Factors influencing blood flow patterns in the human right coronary artery. *Ann Biomed Eng,* 2001, **29**(2): p. 109–120.
49. Ramaswamy, S.D. et al., Fluid dynamic analysis in a human left anterior descending coronary artery with arterial motion. *Ann Biomed Eng,* 2004. **32**(12): p. 1628–1641.
50. Huang, Y.Q. et al., A fiber matrix model for the filtration through fenestral pores in a compressible arterial intima. *Am J Physiol-Heart Circ Physiol,* 1997, **41**(4): p. H2023–H2039.

Chapter 6

Point-of-Care Microdevices for Global Health Diagnostics of Infectious Diseases

*Sau Yin Chin, Tassaneewan Laksanasopin,
Curtis D. Chin, and Samuel K. Sia**

*Department of Biomedical Engineering
Columbia University, 351 Engineering Terrace
1210 Amsterdam Ave, New York City
NY 10027, USA*

Introduction

Infectious diseases are the second leading cause of death worldwide (behind only cardiovascular diseases), accounting for ~16% of deaths globally and ~40% of deaths in low-income countries.[1] They account for almost a third of healthy life years lost to premature death and disability in sub-Saharan Africa (Figs. 1 and 2). Among the most vulnerable population groups are young children and women of reproductive age, as well as those living in remote rural areas and in areas of conflict, who have poor access to clinics.[2]

Infectious diseases also decrease the economic productivity and growth in low-income countries due to death and disability. According to the World Health Organization (WHO), the economic loss due to deaths caused by HIV/AIDS alone in 1999 was estimated to be an astounding 12% of the gross national product in sub-Saharan African countries. This value is projected to increase to 20% or more in the next decade or so.[3] The global burden of tuberculosis (TB) can also be estimated, given that 8.4 million people contract TB yearly.[4] Assuming a 30% decline in productivity due to the disease and annual deaths of two million, the annual economic loss is about US$12 billion.[5]

---

*Corresponding author. Email: ss2735@columbia.edu

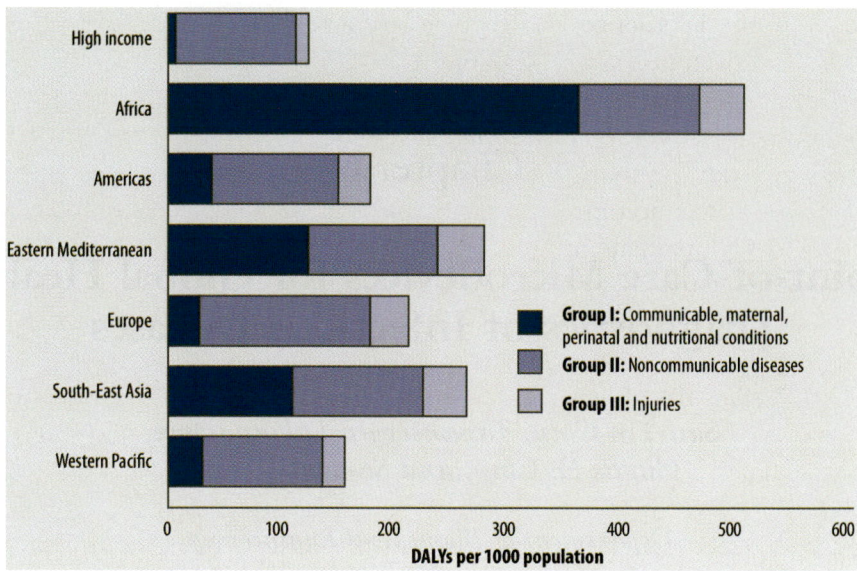

Fig. 1. Distribution of disease burden (in disability-adjusted life years, or DALYs, per 1,000 people), according to region (high- vs. low- and middle-income countries) and cause.[1]

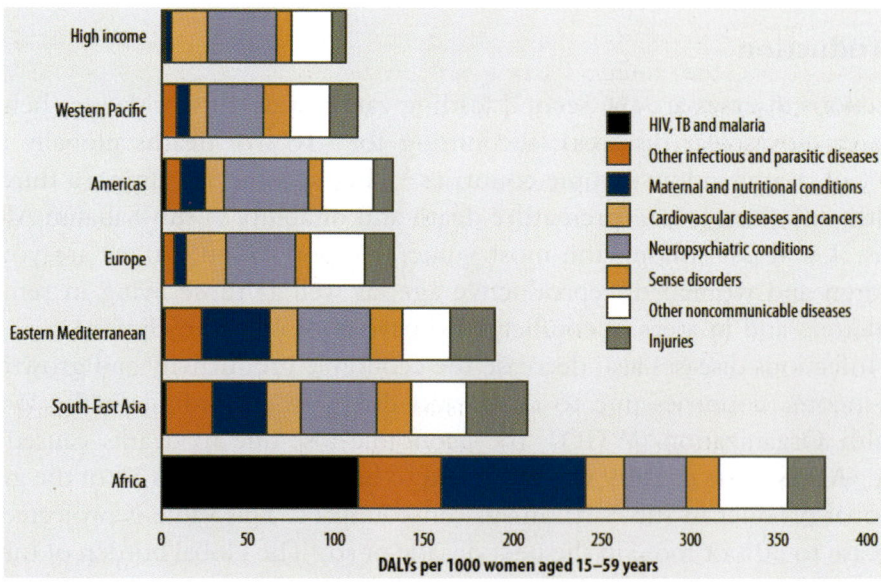

Fig. 2. Major causes of disease burden in women aged 15–59 years in high-, middle-, and low-income countries, WHO, 2004.[1]

Innovations in science and technology can reduce the health disparity between the developed and developing worlds.[6,7] In particular, new technologies are needed to provide healthcare beyond the centralized laboratory and into health clinics at the community level (defined here as "point-of-care" settings). A study by the Bill and Melinda Gates Foundation and the NIH identified new technologies for diagnostics, particularly those "that allow assessment of individuals for multiple conditions or pathogens at point-of-care," as one of the "Grand Challenges for Global Health."[8] This chapter mainly focuses on the use of microfluidic technology to improve global health by bringing infectious disease diagnostics to the point-of-care.

Microfluidics-based devices are being developed to detect a wide range of analytes from a variety of biological fluid matrices. These devices are attractive because they have the potential to miniaturize tests on portable devices at a fraction of the cost and time needed by complex benchtop tests, while maintaining or exceeding conventional analytical performance. The classes of analytes detected include proteins, nucleic acids, cells, and small molecules.[9]

Proteins: Immunoassays

A number of clinical specimens (e.g., whole blood, sera, saliva) contain free-floating proteins that are used for the diagnosis and monitoring of infectious diseases.[10] Point-of-care tests that are commercially available include diagnostics kits for bacterial infections (e.g., *Streptococcus A* and *B*, *Chlamydia trachomatis*, *Treponema pallidum*), viral infections (e.g., HIV, influenza), and parasitic infections (e.g., malaria).[9,11,12]

One of the most common assays that are adapted for point-of-care microfluidics-based diagnostics based on the detection of proteins is immunoassay, which makes use of the binding interactions between antigens and antibodies to detect protein markers from either pathogen or host immune responses. Analytes for immunoassays span the entire range of pathogen types, from viruses (e.g., anti-HIV antibodies and p24 antigen for HIV) to bacteria (e.g., anti-treponemal antibodies for syphilis and early secretory antigenic target 6 for TB) and parasites (e.g., histidine-rich protein 2 for malaria). (Immunoassays are also commonly used for the detection of non-communicable diseases such as prostate cancer via the measurement of levels of prostate-specific antigen, as in a system by Claros Diagnostics, or heart disease via the measurement of levels of B-type natriuretic peptide, as in the Triage BNP Test by Biosite.[13])

Heterogeneous immunoassays are a popular immunoassay format, where the capture of analytes occurs at protein-modified surfaces. The reference standard of many heterogeneous immunoassays is the enzyme-linked immunosorbent assay (ELISA), which uses multiwell plates and can be automated for high-throughput processing at well-equipped central laboratories. Most ELISAs use colorimetry- or chemiluminescence-based detection and have detection limits typically in the pM range due to enzyme-mediated signal amplification and serial washing. Unfortunately, the traditional ELISA format, which requires expensive and bulky instrumentation (for liquid handling and signal detection) and trained workers, is ill-suited for point-of-care testing in low-resource settings.

Microfluidics is an attractive technology for point-of-care immunoassays for global health. Heterogeneous immunoassays, because they involve capture of analytes at surfaces, are well suited to exploit the large surface-to-volume ratios encountered in microfluidics. Faster analysis times can be achieved because of the replenishment of analytes and detection reagents in the boundary layer above the surface in standard well plate formats.[14] Below, we highlight recent advances in microfluidic immunoassays that have the potential to be used in resource-poor settings.

Fluid control

Valves are an essential component of fluid control on microdevices. Pneumatic deflection of elastomers, such as polydimethylsiloxane (PDMS), in multilayer systems has been demonstrated to be robust, leak-free, and capable of facilitating highly multiplexed systems.[15] They have been utilized in the development of bead-based microfluidic ELISA systems.[16] However, these micromechanical valves require high pressures, often supplied by gas tanks. Active microvalves that are based on elastomeric deflections and better suited at point-of-care settings include hydraulic (liquid) deflection based on magnetic actuation with solenoids[17] and torque-actuated deflection using small machine screws.[18]

Passive approaches, which leverage differences in fluid behavior from varying microchannel geometries in capillary systems, are attractive because they do not require external power and moving off-chip parts. Such control mechanisms include delay valves, which merge smaller channels into larger channels to allow for the smooth collection of incoming fluid streams at different flow rates; stop valves, which reduce the width of a microfluidic path using a restriction and enlarge it abruptly to reduce the capillary pressure of a liquid front to zero; and trigger valves, which are the assembly of multiple stop valves preventing further fluid flow into a common outlet until the arrival of all inlet streams.[19] Some of these microfluidic control elements have been integrated on a chip for detection of C-reactive protein.[20] In addition, check valves have been implemented on a microfluidic device for multistep ELISA detecting botulinum neurotoxin.[21] Valveless delivery of reagents in microfluidic systems has also been demonstrated for detecting anti-HIV antibodies.[22]

Fluid actuation and delivery

Movement of fluids by capillary forces is reliable and does not require external power or moving parts. Miniaturized immunoassays based on capillary forces have been used for detecting cardiac markers[23] and luteinizing hormone.[24] Paper-based microfluidic systems also leverage capillary flow and have been used to detect anti-HIV antibodies.[25] Despite the need for external power, electrophoretic immunoassays in capillaries can be utilized if the power requirement is low (allowing for battery operation) and external instrumentation integrated in a single, easy-to-use device; such has been demonstrated in a promising proof-of-concept device for integrated, rapid point-of-care testing of biotoxins ricin, Shiga toxin I, and Staphylococcal enterotoxin B,[26] as well as in a rapid bioassay for endogenous matrix metalloproteinase-8 in saliva.[27] Pneumatics-based actuation of fluids can be suited for point-of-care settings, for example in the manual operation of on-card bellows in the commercially available ABO blood-typing chip by Micronics. A hand pump can also be used for pneumatic fluid actuation in microfluidic immunoassays.[22,28] Injection-molded centrifugal-based platforms (CDs) rely on spin frequency to drive fluid movement, and movements are gated by capillary or hydrophobic valves; these can be suited for point-of-care testing in resource-limited settings, where, in one example, a centrifugal bead-based immunoassay was developed for the detection of antigen and antibody to the hepatitis B virus.[29]

Signal detection

The reduction of silver ions on gold nanoparticles is an attractive method of signal amplification and detection because the signals can be developed under continuous flow and can be read using low-cost optics; silver–gold amplification has been used to detect anti-HIV antibodies.[28] Our group has combined this concept with plug-based reagent delivery[22] in cassettes made of injection-molded plastic (from Claros Diagnostics) for the simultaneous detection of HIV and syphilis antibodies (Fig. 3). In a preclinical evaluation at several clinical sites in Rwanda, we showed sensitivity and specificity rivaling those of reference benchtop assays on hundreds of patient samples, with a time-to-result of no more than 20 min.[30] We also demonstrated excellent performance of HIV diagnosis using only 1 μL of unprocessed whole blood. The social impact of this test is potentially very high when used in remote settings (e.g., clinics providing antenatal care), since treatments for HIV and syphilis are affordable and effective for minimizing disease transmission from mother to child. Surface plasmon resonance is another suitable detection method, given the appropriate signal amplification schemes; Sensata

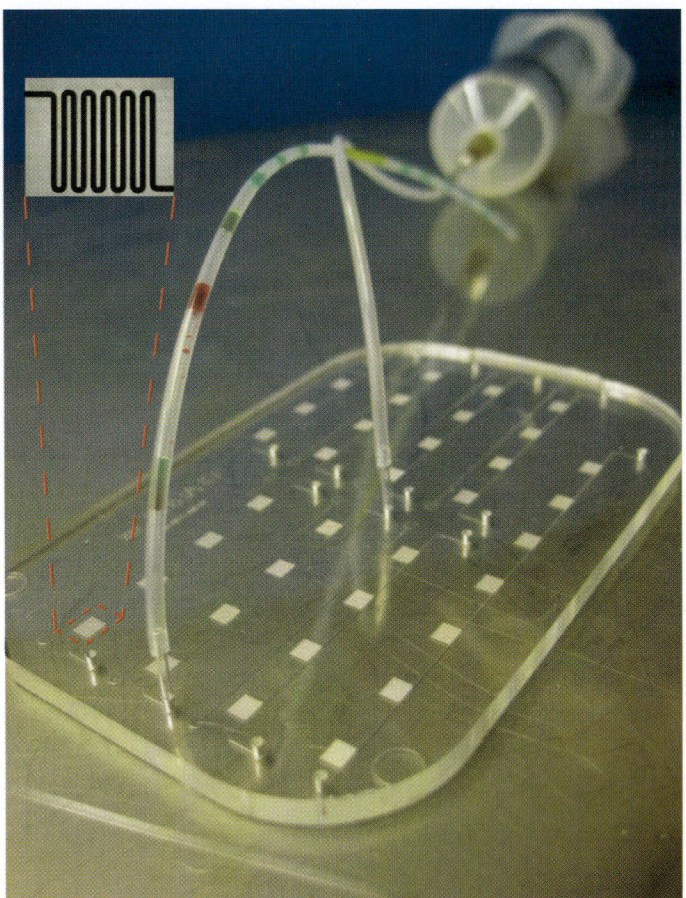

Fig. 3. Plastic cassette and fluidic set-up for simultaneous detection of HIV and syphilis. Injection-molded plastic cassettes are fitted with tubings pre-loaded with reagent plugs and a syringe providing negative pressure. Inset shows a detailed image of the microfluidic design of a single detection zone with silver development.

Technologies has previously developed a low-cost, disposable device that can detect heterogeneous antigen/antibody binding with changes in the refractive index measured at the gold surface.

Nucleic Acid-based Testing

Clinical diagnoses can be made based on the analysis of DNA or RNA sequences. Nucleic acid detection and analysis can identify the type of infection or pathogen and multistage diseases. It can be used in prenatal diagnosis of inherited disorders, clinical disease diagnosis (genetic disease, infection, disease staging, drug resistance mutation, and pathogen presence/abundance), and forensic investigations. Nucleic acid testing offers detection that is highly sensitive (due to sample amplification) and specific (due to base pairing of complementary nucleotides).

In order to perform nucleic acid testing on a microfluidic chip, the main functionalities that have to be integrated on-chip include sample preparation, nucleic acid amplification, and the detection of the amplified product. While these steps can be easily integrated together and performed in a general laboratory, the miniaturization of such disparate processes onto a single microfluidic chip is still a topic of intense research.[31-33] The design and integration of microchips for nucleic acid detection have been developed for three main processing steps: sample preparation, signal amplification, and product detection. A fully integrated system could avoid contamination, reduce worker steps, and deliver rapid results. However, most of the devices that have been reported are single-function and require separate modules for detection and analysis due to the challenges involved in combining and miniaturizing these functionalities on-chip, which require much effort and time.[33] A comparison of conventional approaches into microfluidic approaches is summarized in Table 1 (adapted from Ref. 34).

Sample preparation

After biological samples are collected, cell isolation and lysis followed by nucleic acid extraction, purification, and pre-concentration may be performed.[35] This "sample processing" or "sample preparation" step has been less developed than other assay steps because of its intrinsic complexity. Moreover, contamination, inhibitors for subsequent amplification steps, and nucleic acid degradation are also critical and influence diagnostic testing as these factors impede the quantitative assessment of the analyte in question, leading to misinterpretation of results.[32,34] Therefore, sample preparation tends to be performed off-chip (using laboratory equipment such as the centrifuge), while amplification and detection can be accomplished in microfluidic systems.[36] Nonetheless, efforts have been made to integrate cell capture, cell lysis, mRNA purification, cDNA synthesis, and cDNA purification into a single microfluidic device.[37] The integration of sample pre-treatment with analysis could lead to improvements in sensitivity (as less sample is lost in between steps) and convenience.[9] One of the first microfluidics-based DNA purification procedures has been demonstrated using silicon microchips, taking advantage of high aspect ratio features to increase the capture surface area and hence, the loading capacity.[38] Other work has focused on single-cell mRNA extraction and analysis via cDNA synthesis on a microfluidic chip.[37]

Table 1. A comparison of conventional microbiological approaches with microfluidic approaches for nucleic acid detection.

Nucleic Acid Detection Steps	Conventional Techniques	Microfluidic Technology
Microbiological cell isolation	— Magnetic capture — Dielectrophoresis — Mechanical filtration	— Micromixers — Functionalized magnetic beads — Dielectrophoresis chips — Microfilters
Cell lysis	— Chemical lysis — Mechanical lysis — Thermal lysis — Electroporation	— Micromixers — Minisonicators — Microfilters — Microelectrode-based electrical lyses
Nucleic acid extraction and purification	— Silica-based resins — Glass matrix — Capture membranes	— On-chip channels of cellulose, silica beads, and silica resins — Silicon dioxide pillar arrays — Photoactivated PC surfaces
Nucleic acid amplification	Polymerase chain reaction	— Microfabricated reaction chambers — Flow-through PCR chips — Resistive heaters — Convection-driven PCR chips
Product detection	— Optical — Electrochemical — Mechanical	On-chip detection using fluorescence, SPR, SERS, nanoparticles, electrochemical methods, QCM, and microcantilevers

Source: Adapted from Ref. 34.

Nucleic acid amplification

Since the amount of nucleic acid acquired from either the preparation step or from the raw sample is usually low for immediate identification and quantification,[33] a method of amplification is needed to obtain a sufficient nucleic acid detection signal. The most common technique is the polymerase chain reaction (PCR), for which miniaturization promotes the ability to reduce the reagent consumption, reduce the cycle time, and automate the process.[9] The miniaturization of PCR has many advantages, such as decreased cost of fabrication and operation, decreased reaction time for DNA amplification, reduced cross-talk of the PCR reaction, and the ability to perform large numbers of parallel amplification analyses on a single PCR microfluidic chip. Also, microfluidics allows for increased portability and integration of the PCR device.

One of the first silicon-based stationary PCR chips was described several years after the introduction of PCR itself.[39] Since then, many research groups have begun to develop microchip-based PCR devices. Most of these devices are based on silicon and glass, but more recently, polymer materials such as PDMS,[40] poly(methyl methacrylate) (PMMA), polycarbonate, SU-8, polyimide, poly(cyclic olefin), and epoxy are being used.

The three main design concepts for PCR microfluidics are chamber stationary PCR, flow-through PCR, and thermal convection-driven PCR.[39] For the stationary design, a droplet-based microfluidic PCR, which is performed similarly to conventional PCR methods, has been demonstrated.[41] In this design, the PCR fluid is kept stationary while the temperature of the reaction chamber is cycled between three different temperatures (melting, annealing, and extension phases). The authors demonstrated the use of a microfluidic device to perform PCR on aqueous-in-oil droplets with volumes in the nanoliter range for potential use at the point-of-care. Flow-through PCR microfluidics allow for more flexibility in changing reaction rates and times due to a time–space conversion concept. For example, IMM (Institut für Mikrotechnik Mainz GmbH, Mainz, Germany) is developing a micropump-based system that uses a ferrofluidic actuator and magnets to move magnetic fluids throughout the microchip to react with specific reagents and access different temperature zones.[42] Therefore, the duration of each cycle can be controlled by the fluid velocity. This design allows rapid heat transfer and thermal cycling and the run–time for such assays is in the order of minutes. HandyLab (HandyLab, Inc., Michigan, USA; HandyLab is now part of Becton Dickinson) has also developed a disposable microfluidic chip that implements heat and pressure gradients to move microliter-sized plugs via valves and gates through different temperature zones within the chip.[42,43] Flow-through designs also decrease the possibility of cross-contamination between samples and allow for the incorporation of many other functions, which is appealing as it leads toward the development of a micro total analysis system. Thermal convection-driven PCR microfluidics involves driving the reaction solutions through two temperature zones. This concept is similar to the flow-through design, except that the driving force of the sample is buoyancy. In this manner, a temperature gradient is generated across a reaction vessel, which drives thermal convection to circulate the solution between the hot and cold regions for PCR.[44] As PCR is a temperature-controlled, enzyme-catalyzed biochemical reaction system, the method in which the different temperature zones are generated and maintained is crucial to the design of PCR microfluidics. Various heating methods have been employed and they can be broadly categorized into contact and non-contact heating methods. Contact heating has been implemented in PCR microfluidics using integrated thin-film platinum resistors as both the heating and sensing elements on these chips[45] and non-contact methods include hot-air cycling.[46] Both methods have low power consumption and are amenable for use at the point-of-care.

Isothermal techniques have been developed to perform nucleic acid amplification without thermal cycling.[47,48] This feature removes the need of using different temperatures by making use of enzymes to perform the amplification at a single temperature. For example, Biohelix (Beverly, MA, USA) has developed a method for performing isothermal amplification called helicase-dependent amplification that is amenable for point-of-care use.[49]

Product detection

There are many methods to detect nucleic acids, of which one of the primary methods is quantifying nucleic acids by ultraviolet light (260 nm) absorption. Nonetheless, fluorescence-based techniques remain the most commonly employed due to its high level of sensitivity and low background noise.[34] Fluorescent dyes can either bind to non-specific locations (general interactions) or specific locations of molecule depending on the application. Cepheid, Inc., for example, has developed a real-time fluorescent PCR detection (fluorescently labeled probes) requiring instruments that may be used in some but not all point-of-care settings. Cepheid's GeneXpert test platform (Fig. 4) has been tested for clinical trials in four developing countries and showed promising results for detecting TB.[50] However, this system requires an uninterrupted and stable electrical power supply and annual validation of the system, and generates considerably more waste than microscopy techniques,[51] which may lead to problems of waste management in resource-limited settings.

At the point-of-care, electrochemical methods may also be suitable due to their compatibility with low-cost and portable analyzers.[33] For example, Nanosphere is building a scanner-based detector to detect DNA via nanoparticle probes.[52] Gold nanoparticles functionalized with oligonucleotides are used as probes for DNA sequences complementary to the sequences of those oligonucleotides (Fig. 5). The visualization of the gold nanoparticle and hence DNA content, is performed using a signal amplification method in which silver is reduced at the surface of the gold, and a scanner is used to measure the amount of light scattered.

Another commercial effort toward a point-of-care nucleic acid device includes Spartan RX CYP2C19 by Spartan Bioscience, which is the first point-of-care genetic testing system to receive the CE marking.[53]

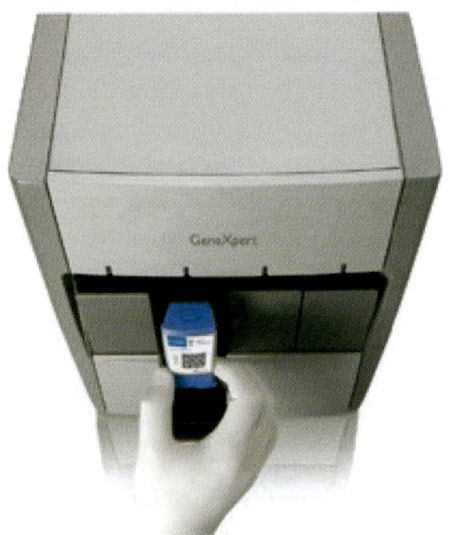
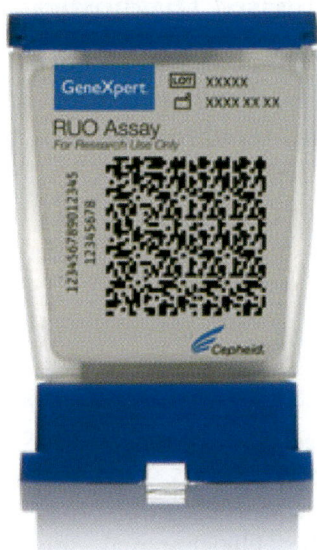

Fig. 4. Cepheid's GeneXpert test platform (left) integrates sample processing and PCR in a disposable plastic cartridge (right) containing reagents for cell lysis, nucleic acid extraction, amplification, and amplicon detection. This system has been used to detect drug-resistant TB cases by amplifying specific sequence of the *rpoB* gene of *Mycobacterium tuberculosis* and probing with molecular beacons for mutations conferring rifampicin resistance.[50]

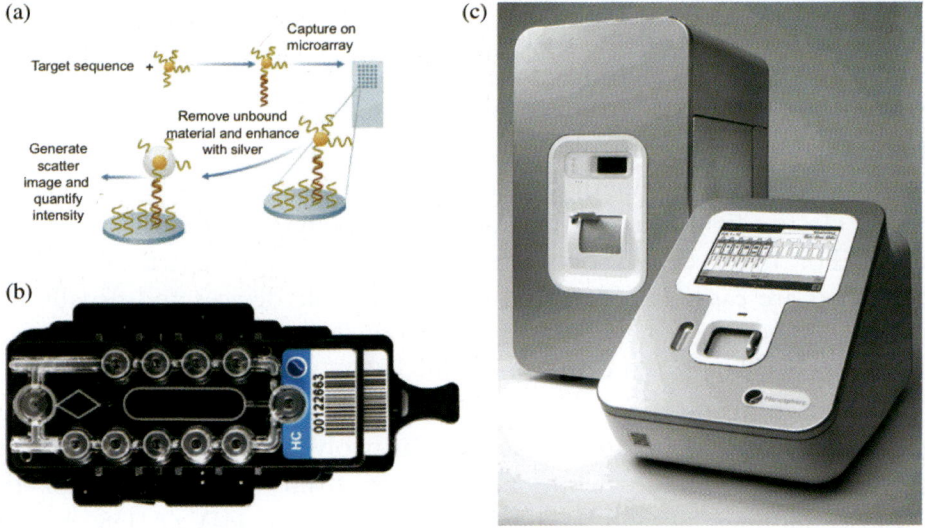

Fig. 5. Nanosphere Verigene system and capture schematic. (a) Capture and detection of target sequence in the Nanosphere system, which uses gold nanoparticles functionalized with oligonucleotides. Following silver development, the amount of light scattered is measured using a scanner-based device. (b) Picture of a Nanosphere cartridge with loaded reagent wells. (c) Picture of Verigene readers with touchscreen control panels.

Cell-based Diagnostics

Cell counting is an important component in hematological analysis as it often proves useful in infectious disease diagnosis and monitoring. Full blood counts (erythrocytes and white blood cells) can be useful in the monitoring of diseases such as anemia and HIV/AIDS. Various works have been done to perform cell counting on microfluidic chips, particularly $CD4^+$ lymphocyte counting for HIV/AIDS monitoring,[54–56] in order to substitute the current gold standard.

The conventional method for cell counting is via flow cytometry, which is expensive and requires ground electricity and trained personnel. Techniques for cell sorting can be broadly categorized into size-, density- and affinity-based techniques (using chemical, electrical, or magnetic means).[57] Size-based methods do not use any surface markers or proteins and are based solely on the concept that the size of the cells of interest are larger than that of the cells that make up the bulk of the sample (e.g., erythrocytes). For example, there have been efforts to develop a method to isolate larger neuroblastomas and mononuclear cells are separated from the smaller erythrocytes present in the sample.[58] The inherent motility of sperm cells has also been used to separate motile and immotile sperm cells in laminar flow within a microfluidic chip.[59] Other novel size-based separation methods force a stream of cells through a series of channels by hydrodynamic flow switching, electrokinetic flow switching, dielectrophoresis, electrowetting-assisted flow switching, or valves,[9] which provide predictable and reproducible cell movement. Affinity-based methods include the specific capture of cells using antibodies such as with magnetic beads coated with antibodies, which is a common laboratory protocol. Another affinity-based method makes use of dielectrophoresis. This method makes use of the differential response of cells to electric fields due to their density and physiological and metabolic states. (For more details on dielectrophoresis, see Chapter 14.) These approaches are easily implemented on microfluidic devices and usually followed by downstream analysis such as fluorescence microscopy or proteomics/genomics analysis following the lysing of the captured cells.[60]

Cell-based diagnostics at the point-of-care has been increasingly important in hematology laboratory practices. It involves the numeration of specific cell types, typically white blood cells, from whole blood for the diagnosis, monitoring, and staging of diseases such as HIV/AIDS. A common use of microfluidics for cellular analysis is to replace traditional, resource-intensive technologies such as those used for cytometry and cellular biosensors. Such microfluidic methods may involve cell (or microparticle) separation based on size[53] and are based on methods such as filtration,[61] laminar flow,[62] dielectrophoretic force,[63] optical gradient force,[64] magnetic force,[65] and acoustic force.[66] Other methods are based on antibody-specific capture of cells.[67]

Hematology practices at the point-of-care have been increasingly important, especially in obtaining white cell counts. The HemoCue WBC system (HemoCue AB, Sweden) measures total white blood cells in just 10 µL of sample (Fig. 6).[54] The point-of-care analyzer consists of a photomicroscope, microcuvette holder, and an LCD display. It also runs on batteries and has low power consumption. The Chempaq analyzer uses differential impedance to detect different types of cells to return a white blood cell count or full blood counts, while photometry measurements are used for hemoglobin assessments (Fig. 6).[68] All reagents are self-contained in this system and results are available in 3 min.

Due to the pressing need for a point-of-care CD4 T lymphocyte enumeration method for the management of HIV/AIDS epidemic, especially in underserved regions, much work has been done to develop a microfluidic approach. A lysate impedance measurement method has been proposed, where the specific cells of interest are captured and lysed and the resulting change in impedance is measured using electrodes integrated on-chip.[69] Another method uses semiconductor quantum dots to label captured CD4 cells, minimizing the optical requirements needed to image and obtain the CD4 count. An automated software is then used to analyze the fluorescent images.[70] An interesting use of microfluidics to create point-of-care devices for cell-based assays is the integration of commercially available fiber optics with the microchips so as to miniaturize the whole system by reducing any required optical and mechanical hardware.[71]

Future Applications

Companion diagnostics

In global health, the use of companion diagnostics or theranostics — therapy-specific diagnostics — could help optimize therapeutic interventions for bacterial infections;[72–74] these infections are commonly treated empirically with broad-spectrum antimicrobials because clinical microbiology procedures require days before delivering results. With rapid detection and identification of microbial pathogens and strain typing, specific antimicrobials for treatment can be managed and selected, thereby yielding lower toxicity and healthcare costs and minimizing the buildup of microorganism resistance.[72] The effect of genotype on drug efficacy and metabolism will be a crucial consideration for

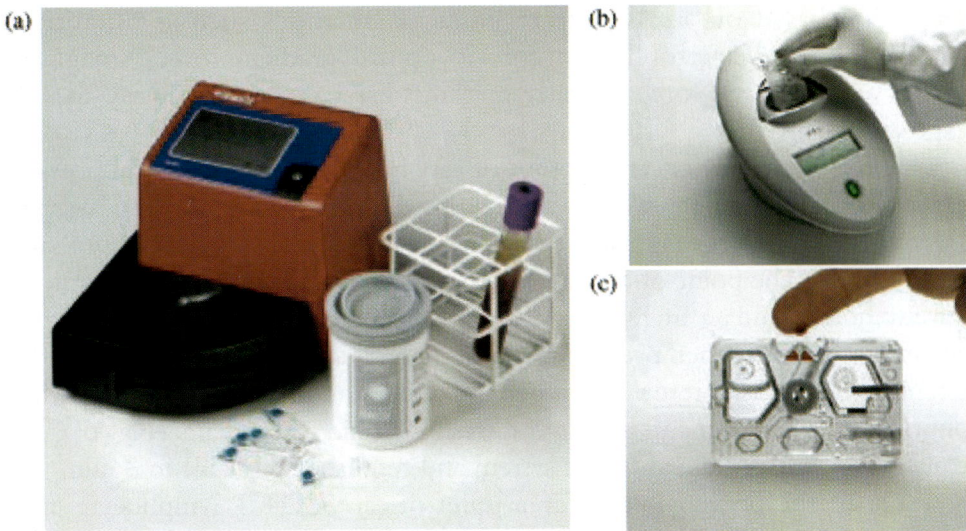

Fig. 6. HemoCue WBC System and ChemPaq. (a) Picture of the HemoCue WBC System for total white blood cell count. (b) Picture of the Chempaq XBC multiplatform analyzer. (c) Picture of the PAQ cartridge to be used with the Chempaq analyzer. The PAQ cartridge contains all reagents required for diagnostic test.

theranostics. In addition, microarrays, which have been approved for pharmacogenomic profiling of drug metabolism, represent a step toward multiplexed genetic analysis of pathogens and host genomic expression and polymorphism profiling for managing patient treatments.

Proteomics

Proteomics is the large-scale analysis of gene products and their interactions both within and outside the cellular space.[75,76] For diagnostics, the detection of protein markers from different groups (e.g., antigens, antibodies, cytokines, membrane-bound and intracellular targets) and from different sources (e.g., tissue sections, tumor cells, blood, saliva, urine) is sometimes important in fully characterizing disease states.[76,77] Epidemiological surveillance and disease management can potentially benefit from the proteomic-wide classification of pathogen strains and subtypes, which can display extensive heterogeneity within species.[78,79] Protein microarrays allow the parallel analysis of hundreds to thousands of proteins at small sample volumes. In theory, they can merge with microfluidics[80] to yield products geared for point-of-care settings: (1) Biosite's (San Diego, CA, USA) Triage protein array processes 100 different proteins on one array with submicroliter sample volumes. Sample handling of whole blood is passive (passive separations and lysis), and fluids flow in capillary structures on injection-molded plastic; (2) Raybiotech's (Norcross, GA, USA) cytokine kit detects expression levels of 126 cytokines from serum samples, supernatants, and cell lysate samples, in which the kit uses 100 µL of sample and employs chemiluminescence detection; and (3) Randox Laboratories (Crumlin, UK) has a protein biochip two-dimensional microarray technology, with each spot containing a unique immobilized protein on a 9-mm^2 solid substrate. This system also uses chemiluminescence-based detection.

Summary and Outlook

After much research in academic, government, nonprofit, and commercial settings, microfluidic methods for infectious disease diagnostics are finally being realized. As microfluidic methods for individual steps are increasingly proven, a critical challenge is to integrate these steps into a coherent integrated device. While some of these integrated products are coming into the market for diagnosing infectious diseases in Europe and the U.S., they must overcome additional regulatory and socioeconomic factors before they can be deployed in the developing world. Bridging the gap between demonstration of proof-of-concept and construction of a functional integrated device remains a critical challenge for researchers.

References

1. Kost, G.J., *Principles and Practice of Point-of-Care Testing*. Lippincott Williams & Wilkins, 2002.
2. WHO, UNICEF, UNAIDS, *Towards Universal Access: Scaling Up Priority Interventions in the Health Sector: Progress Report 2009*. World Health Organization, Geneva, 2009.
3. *Removing Obstacles to Healthy Development*. World Health Organization, Geneva, 1999.
4. *The Burden of Tuberculosis: Economic Burden*. World Health Organization: Geneva, 2008.
5. Fonkwo, P.N., Pricing infectious disease — the economic and health implications of infectious diseases. *EMBO Rep*, 2008, **9**: p. S13–S17.
6. *Human Development Report 2001: Making New Technologies Work for Human Development*. United Nations Development Programme (UNDP), New York, 2001.
7. *Genomics and World Health: Report of the Advisory Committee on Health Research*. World Health Organization, Geneva, 2002.
8. Varmus, H. *et al.*, Public health. Grand challenges in global health. *Science*, 2003, **302**(5644): p. 398–399.
9. Chin, C.D., Linder, V., Sia, S.K., Lab-on-a-chip devices for global health: Past studies and future opportunities. *Lab Chip*, 2007, **7**(1): p. 41–57.
10. Hall, D.A., Ptacek J., Snyder, M., Protein microarray technology. *Mech Ageing Dev*, 2007, **128**(1): p. 161–167.

11. Madhivanan, P. et al., Simple and inexpensive point-of-care tests improve diagnosis of vaginal infections in resource constrained settings. *Trop Med Int Health*, 2009, **14**(6): p. 703–708.
12. von Lode, P., Point-of-care immunotesting: Approaching the analytical performance of central laboratory methods. *Clin Biochem*, 2005, **38**(7): p. 591–606.
13. Song, C. et al., Reversed-phase-reversed-phase liquid chromatography approach with high orthogonality for multidimensional separation of phosphopeptides. *Anal Chem*, **82**(1): p. 53–56.
14. Parsa, H. et al., Effect of volume- and time-based constraints on capture of analytes in microfluidic heterogeneous immunoassays. *Lab Chip*, 2008, **8**(12): p. 2062–2070.
15. Thorsen, T., Maerkl, S.J., Quake, S.R., Microfluidic large-scale integration. *Science*, 2002, **298**(5593): p. 580–584.
16. Herrmann, M. et al., Microfluidic ELISA on non-passivated PDMS chip using magnetic bead transfer inside dual networks of channels. *Lab Chip*, 2007, **7**(11): p. 1546–1552.
17. Addae-Mensah, K.A. et al., Actuation of elastomeric microvalves in point-of-care settings using handheld, battery-powered instrumentation. *Lab Chip*, **10**(12): p. 1618–1622.
18. Weibel, D.B. et al., Torque-actuated valves for microfluidics. *Anal Chem*, 2005, **77**(15): p. 4726–4733.
19. Ziegler, J. et al., High-performance immunoassays based on through-stencil patterned antibodies and capillary systems. *Anal Chem*, 2008, **80**(5): p. 1763–1769.
20. Gervais, L., Delamarche, E., Toward one-step point-of-care immunodiagnostics using capillary-driven microfluidics and PDMS substrates. *Lab Chip*, 2009, **9**(23): p. 3330–3337.
21. Moorthy, J. et al., Microfluidic tectonics platform: A colorimetric, disposable botulinum toxin enzyme-linked immunosorbent assay system. *Electrophoresis*, 2004, **25**(10–11): p. 1705–1713.
22. Linder, V., Sia, S.K., Whitesides, G.M., Reagent-loaded cartridges for valveless and automated fluid delivery in microfluidic devices. *Anal Chem*, 2005, **77**(1): p. 64–71.
23. Juncker, D. et al., Autonomous microfluidic capillary system. *Anal Chem*, 2002, **74**(24): p. 6139–44.
24. Weng, K.Y., Chou, N.J., Cheng, J.W., Triggering vacuum capillaries for pneumatic pumping and metering liquids in point-of-care immunoassays. *Lab Chip*, 2008, **8**(7): p. 1216–1219.
25. Cheng, C.M. et al., Paper-based ELISA. *Angew Chem Int Ed Engl*, **49**(28): p. 4771–4.
26. Meagher, R.J. et al., An integrated microfluidic platform for sensitive and rapid detection of biological toxins. *Lab Chip*, 2008, **8**(12): p. 2046–2053.
27. Herr, A.E. et al., Microfluidic immunoassays as rapid saliva-based clinical diagnostics. *Proc Natl Acad Sci U S A*, 2007, **104**(13): p. 5268–5273.
28. Sia, S.K. et al., An integrated approach to a portable and low-cost immunoassay for resource-poor settings. *Angew Chem Int Ed Engl*, 2004, **43**(4): p. 498–502.
29. Lee, B.S. et al., Fully integrated lab-on-a-disc for simultaneous analysis of biochemistry and immunoassay from whole blood. *Lab Chip*, 2011, **11**(1): p. 70–8.
30. Curtis, D. et al., Microfluidics-based diagnostics of infectious diseases in the developing world. *Nat Med* 2011 (In Press).
31. Dineva, M.A., Mahilum-Tapay, L., Lee, H., Sample preparation: A challenge in the development of point-of-care nucleic acidbased assays for resource-limited settings. *Analyst*, 2007, **132**(12): p. 1193–1199.

32. Chen, L., Manz, A., Day, P.J.R., Total nucleic acid analysis integrated on microfluidic devices. *Lab Chip*, 2007, **7**(11): p. 1413–1423.
33. Lee, T.M.H., Hsing, I.M., DNA-based bioanalytical microsystems for handheld device applications. *Anal Chim Acta*, 2006, **556**(1): p. 26–37.
34. Lui, C., Cady, N.C., Batt, C.A., Nucleic acid-based detection of bacterial pathogens using integrated microfluidic platform systems. *Sensors*, 2009, **9**(5): p. 3713–3744.
35. Ausubel, E.A., *Short Protocols in Molecular Biology: A Compendium of Methods from Current Protocols in Molecular Biology*. John Wiley & Sons, New York, 1992.
36. Crevillen, A.G. et al., Real sample analysis on microfluidic devices. *Talanta*, 2007, **74**(3): p. 342–357.
37. Marcus, J.S., Anderson, W.F., Quake, S.R., Microfluidic single-cell mRNA isolation and analysis. *Anal Chem*, 2006, **78**(9): p. 3084–3089.
38. Christel, L.A. et al., Rapid, automated nucleic acid probe assays using silicon microstructures for nucleic acid concentration. *J Biomech Eng-Trans ASME*, 1999, **121**(1): p. 22–27.
39. Zhang, C.S. et al., PCR microfluidic devices for DNA amplification. *Biotechnol Adv*, 2006, **24**(3): p. 243–284.
40. Cady, N.C. et al., Real-time PCR detection of Listeria monocytogenes using an integrated microfluidics platform. *Sens Actuators B*, 2005, **107**(1): p. 332–341.
41. Wang, F., Burns, M.A., Performance of nanoliter-sized droplet-based microfluidic PCR. *Biomed Microdevices*, 2009, from http://www.ncbi.nlm.nih.gov/pmc/articles/PMC2955802.
42. Holland, C.A., Kiechle, F.L., Point-of-care molecular diagnostic systems — past, present and future. *Curr Opin Microbiol*, 2005, **8**(5): p. 504–509.
43. *GeneOhm Products*. Retrieved March 20, 2004, from http://www.bd.com/geneohm/english/products/.
44. Braun, D., PCR by thermal convection. *Mod Phys Lett B*, 2004, **18**(16): p. 775–784.
45. Liao, C.S. et al., Micromachined polymerase chain reaction system for multiple DNA amplification of upper respiratory tract infectious diseases. *Biosens Bioelectron*, 2005, **20**(7): p. 1341–1348.
46. Lee, D.S. et al., Bulk-micromachined submicroliter-volume PCR chip with very rapid thermal response and low power consumption. *Lab Chip*, 2004, **4**(4): p. 401–407.
47. Van Ness, J., Van Ness, L.K., Galas, D.J., Isothermal reactions for the amplification of oligonucleotides. *Proc Natl Acad Sci U S A*, 2003, **100**(8): p. 4504–4509.
48. Tan, E. et al., Isothermal DNA amplification coupled with DNA nanosphere-based colorimetric detection. *Anal Chem*, 2005, **77**(24): p. 7984–7992.
49. Li Jeon, N. et al., Neutrophil chemotaxis in linear and complex gradients of interleukin-8 formed in a microfabricated device. *Nat Biotechnol*, 2002, **20**(8): p. 826–830.
50. Boehme, C.C. et al., Rapid molecular detection of tuberculosis and rifampin resistance. *N Engl J Med*, 2010, **363**(11): p. 1005–1015.
51. WHO, *Roadmap for Rolling Out Xpert MTB/RIF for Rapid Diagnosis of TB and MDR-TB*. World Health Organization, Geneva, 2010.
52. Taton, T.A., Mirkin, C.A., Letsinger, R.L., Scanometric DNA array detection with nanoparticle probes. *Science*, 2000, **289**(5485): p. 1757–1760.
53. Tsutsui, H., Ho, C.M., Cell separation by non-inertial force fields in microfluidic systems. *Mech Res Commun*, 2009, **36**(1): p. 92–103.

54. Osei-Bimpong, A. et al., Point-of-care method for total white cell count: An evaluation of the HemoCue WBC device. *Int J Lab Hematol*, 2008.
55. Rodriguez, W.R. et al., A microchip CD4 counting method for HIV monitoring in resource-poor settings. *PLoS Med*, 2005, **2**(7): p. E182.
56. Cheng, X.H. et al., A microchip approach for practical label-free CD4+ T-cell counting of HIV-infected subjects in resource-poor settings. *JAIDS*, 2007, **45**(3): p. 257–261.
57. Radisic, M., Iyer, R.K., Murthy, S.K., Micro- and nanotechnology in cell separation. *Int J Nanomed*, 2006, **1**(1): p. 3–14.
58. Mohamed, H. et al., Development of a rare cell fractionation device: Application for cancer detection. *IEEE Trans Nanobiosci*, 2004, **3**(4): p. 251–256.
59. Cho, B.S. et al., Passively driven integrated microfluidic system for separation of motile sperm. *Anal Chem*, 2003, **75**(7): p. 1671–1675.
60. Chen, X. et al., Continuous flow microfluidic device for cell separation, cell lysis and DNA purification. *Anal Chim Acta*, 2007, **584**(2): p. 237–243.
61. VanDelinder, V., Groisman, A., Separation of plasma from whole human blood in a continuous cross-flow in a molded microfluidic device. *Anal Chem*, 2006, **78**(11): p. 3765–3771.
62. Ma, N.N., Koelling, K.W., Chalmers, J.J., Fabrication and use of a transient contractional flow device to quantify the sensitivity of mammalian and insect cells to hydrodynamic forces. *Biotechnol Bioeng*, 2002, **80**(4): p. 428–437.
63. Doh, I., Cho, Y.H., A continuous cell separation chip using hydrodynamic dielectrophoresis (DEP) process. *Sens Actuators A* 2005, **121**(1): p. 59–65.
64. Wang, M.M. et al., Microfluidic sorting of mammalian cells by optical force switching. *Nat Biotechnol*, 2005, **23**(1): p. 83–87.
65. Sai, Y. et al., Continuous separation of particles using a microfluidic device equipped with flow rate control valves. *J Chromatogr, A*, 2006, **1127**(1–2): p. 214–220.
66. Petersson, F. et al., Free flow acoustophoresis: Microfluidic-based mode of particle and cell separation. *Anal Chem*, 2007, **79**(14): p. 5117–5123.
67. Cheng, X.H. et al., A microfluidic device for practical label-free CD4+ T cell counting of HIV-infected subjects. *Lab Chip*, 2007, **7**(2): p. 170–178.
68. Rao, L.V. et al., Evaluation of a new point of care automated complete blood count (CBC) analyzer in various clinical settings. *Clin Chim Acta*, 2008, **389**(1–2): p. 120–125.
69. Cheng, X. et al., Cell detection and counting through cell lysate impedance spectroscopy in microfluidic devices. *Lab Chip*, 2007, **7**(6): p. 746–755.
70. Jokerst, J.V. et al., Integration of semiconductor quantum dots into nano-bio-chip systems for enumeration of CD4+ T cell counts at the point-of-need. *Lab Chip*, 2008, **8**(12): p. 2079–2090.
71. Schafer, D. et al., Microfluidic cell counter with embedded optical fibers fabricated by femtosecond laser ablation and anodic bonding. *Opt Express*, 2009, **17**(8): p. 6068–6073.
72. Picard, F.J., Bergeron, M.G., Rapid molecular theranostics in infectious diseases. *Drug Discov Today*, 2002, **7**(21): p. 1092–1101.
73. Rye, P.D., Rittenhouse, H., Stigbrand, T., Up close and personal: Molecular diagnostics in oncology. *Tumor Biol*, 2004, **25**(4): p. 217–220.
74. Bissonnette, L., Bergeron, M.G., Next revolution in the molecular theranostics of infectious diseases: Microfabricated systems for personalized medicine. *Expert Rev Mol Diagn*, 2006, **6**(3): p. 433–450.

75. Cooper, J.W., Wang, Y.J., Lee, C.S., Recent advances in capillary separations for proteomics. *Electrophoresis*, 2004, **25**(23–24): p. 3913–3926.
76. Templin, M.F. *et al.*, Protein microarray technology. *Trends Biotechnol*, 2002, **20**(4): p. 160–166.
77. Sauer, S. *et al.*, Miniaturization in functional genomics and proteomics. *Nat Rev Genet*, 2005, **6**(6): p. 465–476.
78. Cordwell, S.J., Nouwens, A.S., Walsh, B.J., Comparative proteomics of bacterial pathogens. *Proteomics*, 2001, **1**(4): p. 461–472.
79. Sintchenko, V., Iredell, J.R., Gilbert, G.L., Opinion — pathogen profiling for disease management and surveillance. *Nat Rev Microbiol*, 2007, **5**(6): p. 464–470.
80. Situma, C., Hashimoto, M., Soper, S.A., Merging microfluidics with microarray-based bioassays. *Biomol Eng*, 2006, **23**(5): p. 213–231.

Chapter 7

Integrated Microfluidic Sample Preparation for Chip-based Molecular Diagnostics

*Jane Y. Zhang, Qingqing Cao, Madhumita Mahalanabis, and Catherine Klapperich**

*Departments of Biomedical and Mechanical Engineering
Boston University, Boston, MA 02215, USA*

Introduction to PCR

Around the world, the burden of infectious disease is high due to delayed disease detection, especially in developing countries and rural areas. Healthcare providers recognize the power of point-of-care testing (POCT) to reduce the cost and enhance the efficiency of total health management. Rapid results from POCT can enable better patient management decisions, improved patient outcomes, and a reduction in the overall cost of care. Therefore, new or newly adapted diagnostic tools that are developed must be rapid, sensitive, cost-effective, easy to use, and can be operated in areas without well-equipped and well-staffed clinical laboratories. Performing molecular assays in a microscale biochip has the potential to meet these requirements. Miniaturized nucleic acid-based diagnostics have the potential to achieve accurate and highly sensitive diagnostic results with a fast turn-around time and smaller sample and reagent volumes.[1]

The polymerase chain reaction (PCR) is the most commonly used nucleic acid amplification technology today due to its high sensitivity and specificity. Kary Mullis pioneered the PCR in 1983 and was awarded the Nobel Prize for

---

*Corresponding author. Email: catherin@bu.edu

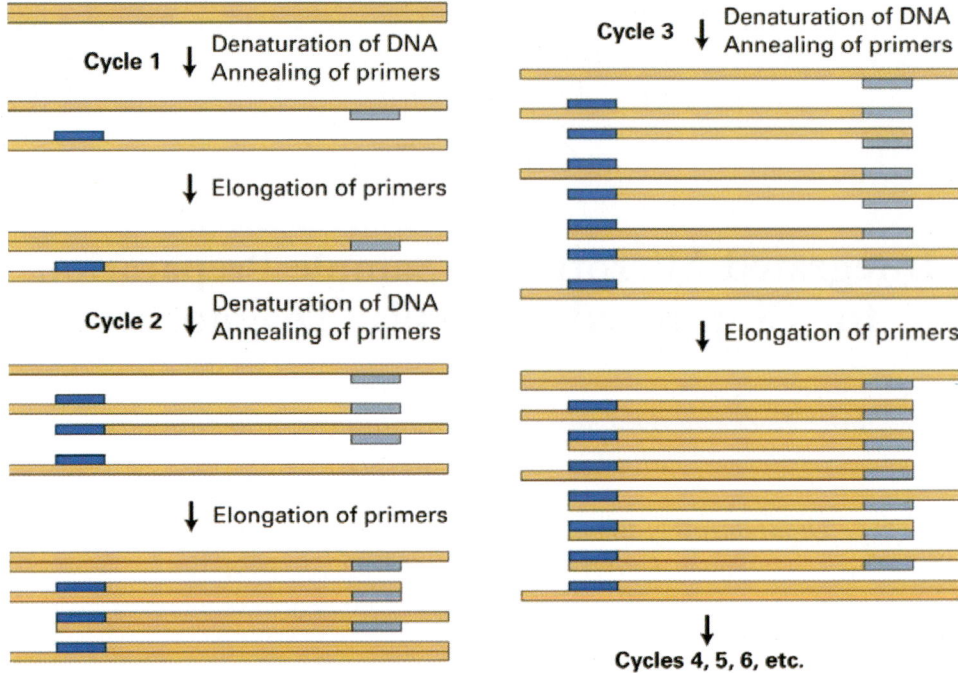

Fig. 1. Polymerase chain reaction (PCR).[8]

its discovery.[2–5] It is a cyclic process of the separation of double-stranded DNA by heat denaturation, specific hybridization, or annealing, of short oligonucleotide primers to single-stranded DNA, and elongation of the DNA strand.[4,6] Each cycle copies the region flanked by the two sets of primer sequences. By sequential iteration of the process, the PCR exponentially generates up to a billion copies of the target in less than 2 h (Fig. 1).[7,8]

History of PCR On-Chip

In 1993, Watson and colleagues first used the PCR in a microfabricated reaction chamber.[9] The temperature of the chamber was cycled between three different temperatures for the PCR amplification steps: denaturation, annealing, and extension. In 1998, Kopp *et al.* developed an innovative continuous-flow PCR (CF-PCR) chip that consists of a single fluidic channel where, by placing heating elements at regular intervals, fluid traveling downstream experiences the required thermal cycling to achieve amplification.[10] These two major types of microfluidic PCR devices (chamber PCR *vs.* continuous-flow PCR) have been extensively studied.

Contact and non-contact heat sources have been employed, including thin-film heaters,[11] metallic heating blocks,[12] thermo-electric heating blocks,[12] infrared (IR),[13] and lasers.[14] For global health applications, the thin-film heater is a good choice due to its small surface area, portability, relatively low cost, and ease of use. Different channel designs have been explored to achieve good performance: serpentine,[10] radial,[15] spiral,[16] and straight CF-PCR channels.[17,18] Various surface passivation methods have been applied to reduce nonspecific binding and inhibition.[19–21] Numerous materials have also been used as substrate materials for PCR chips, such as silicon,[22] glass,[23] cyclic polyolefins,[24] polydimethylsiloxane (PDMS), polymethyl methacrylate,[25] and polycarbonate.[26] The use of polymers has additional advantages for global health applications in terms of fabrication, as such materials can be fabricated using high-throughput manufacturing methods, including injection molding and heat sealing, which can greatly reduce unit cost in high-volume manufacturing.

Current Research in PCR On-Chip

Much recent work has been focused on simplifying the thermal and fluidic control of chip-based PCR to speed up the reaction rate,[10,27,28] increase sensitivity, and reduce sample and reagent consumption.[29–31] For example, Cao et al. worked toward simplified thermal control by using simulation to predict how fixed heater placement will affect temperature profiles in plastic CF-PCR chips.[24] Hashimoto et al. reported a fast amplification of a 500-bp λ-DNA fragment in 1.7 min on a 20-cycle CF-PCR chip, with an initial template concentration of 10 ng/µL.[28] Schaerli et al. presented a CF-PCR device to amplify single-copy DNA in water-in-oil droplets of nanoliter volumes, greatly reducing the amount of the sample and reagent.[15] Beer et al. built a PCR system to encapsulate a single copy of viral genomic DNA in a 10-pL droplet, and showed real-time PCR amplification curves with threshold cycles of 18, which is 20 cycles earlier than commercial instruments.[31] Microfluidic digital PCR represents the additional power of microfluidic PCR chips; PCR reagents were encapsulated in millions of picoliter droplets in a continuous-flow oil phase that flows through a PCR channel, and the amplification reactions were performed within individual droplets. The reactions consume a very small amount of sample and provide high-throughput analysis, since the PCR solutions can flow continuously through a reaction channel path in a droplet manner.[32–34] However, such systems could suffer from cross-contamination between samples and sample dispersion. Consequently, two-phase flow systems were often required, which introduced complicated fluidic control.[35]

Sample Preparation for Standard PCR

Although PCR is extremely efficient and can detect very low numbers of template molecules, the sensitivity of any given PCR depends on many factors. These include the purity and integrity of the input DNA/RNA template (e.g. minimal contaminants and potential PCR inhibitors) and assay design (primer optimization, target specificity, etc.) prior to amplification. For POC microfluidic diagnostics for infectious disease, assay developers will increasingly need to address the balance between reducing reaction volumes, while maintaining high sensitivity of the assay for clinical specimens that may often have low viral or bacterial loads, depending on the disease type and time course (acute *vs.* chronic types and phases of infection). For instance, a 25-µL PCR reaction volume typically only has 1–10 µL of template material. If the viral or bacterial load of the patient sample is very low, this will require volume reduction via enrichment or concentration of the sample before the sample is concentrated enough to provide results. In bacteremia for instance, a patient is already quite sick at bacterial loads of 1–10 colony-forming units/mL of blood; without sample preparation, it would be unlikely that one of those bugs would be present in any given 1–10 µL sample taken at random.[37] In addition, an optimized PCR requires not only optimized concentrations of specific primers, enzyme, divalent cations, and template molecules, but also well-controlled temperature, buffer pH, ionic strength, and other chemical agents. Any substances from the sample that inactivate the PCR reagent components or cause perturbation of the reaction conditions can potentially reduce PCR efficiency.[36] These PCR inhibitors can act by degrading the polymerase,[37] binding to the nucleic acid target which becomes unavailable to PCR,[38] or chelate or compete with the Mg^{2+} cofactor for *Taq* polymerases.[36,39,40] Therefore, a critical step to improve PCR performance is pre-PCR sample preparation. The goals of sample preparation are to reduce sample size to a workable volume while increasing pathogen concentration, to lyse pathogens efficiently to release the target DNA/RNA, and to remove PCR inhibitors. Extensive research and development has been carried out on each of the sample preparation steps for samples from different biological sources. Figure 2 outlines the sample preparation steps that may be necessary for PCR.

Tube methods for bacterial pathogens

Samples containing bacteria are usually concentrated by enrichment culture to increase the pathogen count and in some cases, to selectively remove the PCR inhibitor through controlled medium changes. Following enrichment culture, samples are further concentrated and purified by centrifugation, where cells

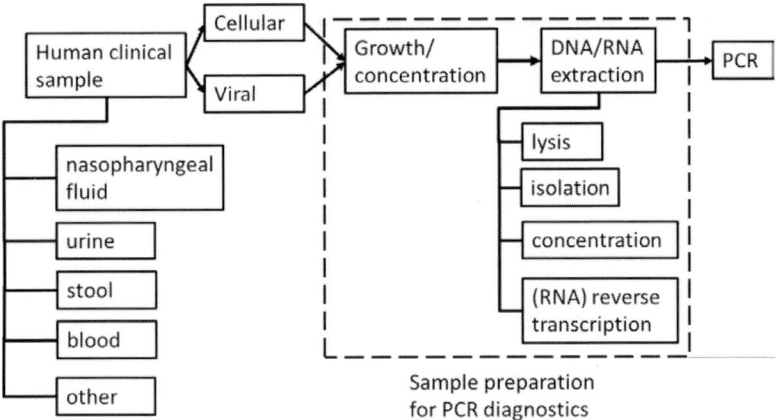

Fig. 2. Sample preparation of human samples for diagnostic PCR. Not all of the steps may be necessary for each sample type.

are pelleted in the tube, or by buoyant density centrifugation, where target cells and PCR inhibitors are separated in a density gradient medium, such as BactXtract, Ficoll, Percoll, sucrose, dextran, and cesium chloride via the difference in their buoyant densities.[41,42] Filtration-based methods can remove sample contaminants greater than the pore sizes of the filtration structure.[43,44] Immunocapture methods can also be used to purify and concentrate bacteria via antibody-conjugated magnetic beads that capture targeted cells expressing the corresponding antigen. Subsequent steps include lysis and washing to remove nonspecifically bound bacteria and then remove the magnetic beads.[45] This method can be easily automated but requires specific antibodies, which might be difficult to obtain or preserve in low-resource settings.

After cell-specific isolation, the next step in the nucleic acid recovery process is to perform cell lysis. The goal of cell lysis prior to PCR is to release the intracellular DNA/RNA by cell membrane disruption. For mammalian cells, lysis entails the destabilization and rupture of the lipid bilayer membrane. In the case of bacteria, both Gram-positive and Gram-negative cell types contain a covalent multilayered peptidoglycan structure of rigid glycan chains cross-linked by flexible peptide bridges,[46] with Gram-positive cells having a thicker and more extensively cross-linked cell wall than in Gram-negative cells.[47] Therefore, the lysis methods for Gram-positive bacteria must be more stringent to overcome a stronger peptidoglycan barrier. Cell lysis can be achieved at the bench with the following common techniques:

(1) mechanical, with glass-bead grinding or with a French pressure cell that shears or fractures the cell membrane.[47,48] These techniques require specialized laboratory equipment, and tend to be time-consuming.

(2) enzymatic, by muramidases such as lysozyme and mutanolysin to hydrolyze the glycoside linkages of the peptidoglycan backbone,[49] or proteinase K to denature proteins,[50]
(3) biochemical, with detergent and chaotropic salts. Detergents such as Triton X-100 and sodium dodecyl sulphate (SDS) destabilize the lipid bilayer and disrupt lipid–protein interactions. Chaotropic salts such as guanidinium thiocyanate (GuSCN), guanidinium hydrochloride (GuHCl), urea, and lithium perchloride are strong protein denaturants. They induce cell permeabilization by weakening the hydrophobic interactions of the membrane and by inhibiting the assembly of cross-linked peptidoglycans in the cell wall,[51] and
(4) physical, including thermal heating, sonication, osmotic shock with osmotic lytic agents such as deionized water, NaCl buffer, NH_4Cl–$KHCO_3$ buffer, electric field-induced electroporation, etc.

Depending on the sample type and the quantity of pathogen in the sample, one or more of the lysis methods may be used, and should be chosen carefully to minimize sample loss.

Nucleic acid isolation from cells can take place concurrently or subsequent to the lysis process. This step purifies the nucleic acid from the lysed organism and further removes PCR inhibitors from the sample. Two popular methods to achieve nucleic acid extraction are phenol–chloroform extraction and selective silica binding. Detailed by Sambrook *et al.* and Chomczynski and Sacchi, phenol–chloroform extraction takes advantage of the high solubility of DNA in the aqueous phase in a biphasic mixture of water-saturated phenol and chloroform.[52–54] DNA in the aqueous phase can be separated by centrifugation of the solution that separates the organic and aqueous phases. The protein remains dissolved in phenol and the lipids dissolved in chloroform, both of which remain in the organic phase. Although phenol–chloroform extraction has been optimized to be a highly efficient nucleic acid purification method, the drawbacks remain that both phenol and chloroform are hazardous to handle, and the process tends to be time-consuming. Premixed reagents are available commercially to reduce the duration and safety hazards involved in preparing the buffer.

Boom *et al.* derived the silica-binding method based on the observation that nucleic acid adsorbs to silica, glass beads, and unicellular algae, called "diatoms", in the presence of high concentrations of chaotropic salts.[55] Chaotropic salts destabilize hydrogen bonding, van der Waals forces, and hydrophobic interactions, inactivate proteins including nucleases, and dissociate nucleic acids from water. Lysis of the pathogen occurs in the presence of the

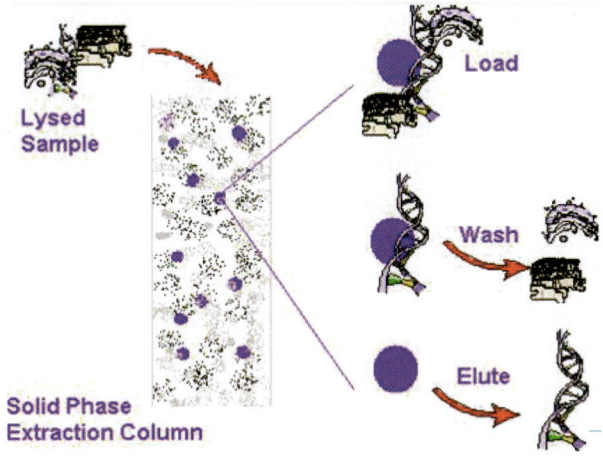

Fig. 3. DNA/RNA extraction process by solid phase extraction (SPE) column.[55]

chaotropic salt in combination with other chemical or mechanical methods. The lysate salt solution is mixed with silica to capture the DNA on the solid phase. The mechanism of binding is likely due to an energetically favorable interaction driven by combined intermolecular electrostatic forces, dehydration, and intermolecular hydrogen bonding.[56] The excess salts are then washed away with a polar protic solvent, such as 50–70% ethanol or isopropanol, or a polar aprotic solvent, such as 100% acetone or acetonitrile. Following washing, the nucleic acid is eluted into a low ionic strength buffer (Tris-EDTA or water) compatible with PCR. The silica binding method is later developed into solid phase extraction (SPE) of nucleic acids in a chromatographic setup, where the silica beads or another solid phase is packed into columns. The loading, washing, and elution steps through the columns are normally aided with centrifugation or vacuum. This process is illustrated in Fig. 3.

Tube methods for viral pathogens

Improved detection of viral pathogens traditionally requires viral culture in host cells,[57] chemical precipitation with ammonium sulfate or polyethylene glycol,[58–61] or filtration.[58,62–65] In the case that the purification of virus is required, depending on the nature of the different types of viruses and the ways they are released from the host cells, an ultracentrifuge or density gradient centrifuge is almost always required to pellet or band viral particles from the contaminating cell debris that contains lipids, carbohydrates, and proteins from the host cells.[66–70] Despite high efficiency, an ultracentrifuge is not portable and too costly and time-consuming to be used for rapid diagnostics. Cross-flow filtration methods are also used with the advantage of a simpler set up and less cost.[62–65] During cross-flow filtration, the viral sample flows in parallel to the filter membrane. Recirculating the fluid sample reduces viral particle build-up on the membrane. The retained fluid contains the concentrated and partially purified viral particles with the use of an appropriately sized filter membrane.

The procedures for extraction and purification of viral genomic DNA or RNA are similar to the procedures discussed earlier for bacteria. Since viruses can have genomes made up of either DNA or RNA, there are some variations depending on the type of virus being targeted. Compared to DNA, extraction and purification of RNA is more challenging due to ubiquitous RNases in the environment, which necessitates extra care during sample handling. Several standard benchtop procedures exist, and have been commercialized with various efficiencies, including phase separation, silica gel membrane columns, and magnetic bead isolation. In the phase separation reagent, in addition to the phenol–chloroform mixture, acidic guanidinium thiocyanate and β-mercaptoethanol are added to inactivate RNases and to precipitate genomic DNA in the organic phase along with unwanted proteins and lipids. RNA in the aqueous phase can then be recovered by centrifugation.[52] Similar to DNA extraction, commercial silica gel membrane columns are based on the binding of RNA to silica in a highly chaotropic environment. In this process, the RNA containing the pathogen is first lysed with a mixture of chaotropic salt, detergent and an enzyme such as proteinase K. In addition to achieving lysis, this buffer also denatures all nucleases, including RNases in the sample. The loading, washing, and elution steps are similar to DNA extraction.

Variations on tube methods for sample preparation from human body fluids

Clinical specimens are complex and difficult to process because of the proteins, fatty acids, salts, antibodies, collagen, enzymes, fiber, bacteria, and competitive nucleic acids present that will inhibit or reduce the sensitivity of the polymerase. Common human samples used in PCR diagnostics include nasopharyngeal fluid, urine, stool, and blood samples. Sample preparation methods differ for each kind of sample, depending on the content, composition, physical state, storage history, and other factors that can affect the sensitivity of PCR.

Many organisms that cause upper respiratory tract infections can be found in the nasopharynx. These include viruses such as influenza, parainfluenza, respiratory syncytial virus, adenovirus, pathogenic bacteria such as

Staphylococcus aureus, *Bordetella pertussis*, and *Neisseria meningitidis*, which can cause septicemia, whooping cough, and meningitis, respectively.[71] The nasopharyngeal fluid is collected from the nasopharynx by a swab, aspirate, or wash, and transferred to phosphate buffered saline or viral transport media immediately for storage. Conventional diagnosis is based on nasopharyngeal cultures to isolate these organisms. Selective bacterial culture and Gram staining are commonly used to enrich and identify bacteria. Viral cultures can take up to seven to 12 days, therefore, they are not commonly ordered.[72] Nasopharyngeal fluid samples are usually centrifuged to separate the cell components and debris from the viral-containing supernatant, before proceeding to conventional lysis and nucleic acid extraction.[73]

PCR on urinary samples reveals important information that current gold standard dipstick tests cannot provide, such as bacterial strain information, which can be critical prior to antibiotic treatment.[71] However, urine is a sample that places very high requirements on sample preparation prior to PCR. Urine contains potent PCR inhibitors such as urea. The chemical nature of the inhibitors in urine varies widely even from the same patient. Centrifugation and extraction with organic solvents have not been effective in removing the inhibitors extracted together with the nucleic acids. In addition, urinary bacteria can lyse during storage, causing nucleic acids to degrade. Bacteria can be separated from urine by repeated centrifugation or ion-exchange chromatography with Chelex® 100 resin (Sigma, St. Louis, MO, USA). The concentration of bacteria is achieved by filtration or centrifugation, followed by lysis with detergents.[73]

Stool is another particularly difficult diagnostic sample to prepare because it contains a large number of microorganisms that are part of the normal intestinal microflora. In addition, the physical states of stool samples vary widely from a semi-solid to a turbid liquid. Furthermore, stool is known to contain high concentrations of PCR inhibitors, such as bile salts, hemoglobin, degradation products, and complex polysaccharides.[74–76] Immunoprecipitation is one method to capture the target bacteria and to remove inhibitors within complex stool samples. Monteiro *et al.* demonstrated the use of a magnetic immunoassay to enable separation of *Helicobacter pylori* from potential PCR inhibitors in human stool samples.[77] Holland *et al.* evaluated three commercial kits for their efficiency in extracting *Escherichia coli* O157:H7 DNA directly from fecal samples for PCR.[78]

Blood contains rich genetic information for the indication of disease states. The ability to extract nucleic acids from whole blood during sample preparation is critical for molecular diagnostics based on PCR. The component of blood targeted by the PCR-based diagnostic determines the DNA or

RNA extraction procedure, beginning with the separation of red blood cells (RBCs), white blood cells (WBCs), and plasma or serum fractions. Human blood is composed of 45% highly specialized cells and 55% plasma. The RBCs comprise nearly 99% of the total cell population, while WBCs and platelets comprise less than 1%.[79] Conventional blood cell separation is achieved through gradient centrifugation, based on different rates of sedimentation. In addition, ammonium chloride and commercial cell lysis reagents including Erythrolyse buffer (ERYTHROLYSE, AbD Serotec, Oxford, UK), FACS lysing solution (FACS Lyse, Becton Dickinson, San Jose, CA), ZAP-OGLOBIN II lytic reagent, and Q-PREP/IMMUNOPREP lysing systems (Beckman Coulter, Brea, CA) are also used to lyse RBCs.[80] One can also use a filter to remove all blood cells to extract the plasma fraction if it is known to contain the pathogens of interest.

Sample Preparation Methods in Microfluidic Chips

Microfludics is a powerful solution for sample preparation because it can simplify and speed up the physical and chemical processes. The smaller volumes needed for on-chip analysis can save expensive reagents and precious samples, and can be made compatible with on-chip PCR.

For clinically relevant pathogen loads to be detectable in relatively small volumes of fluid manipulated in microfluidic chips, sample concentration or enrichment is often required, similar to culturing the pathogen from the clinical sample described earlier. Concentration is easily accomplished in a laboratory setting with a centrifuge. However, a benchtop centrifuge is not portable and is expensive. Overall, two general approaches are taken to concentrate samples in microfluidics: one is to capture or collect target pathogen from sample fluid flow, the second is to reduce fluid volume and conserve the number of pathogens. The current microfluidic sample concentration methods based on electrodynamics, filtration, and immunocapture follow the first methodology. On-chip dialysis and pervaporation-based methods follow the second.

Microfluidic electrodynamic methods include capillary electrophoresis (CE), dielectrophoresis (DEP), electroosmotic flow (EOF), and isoelectric focusing (IEF). All of the methods involve manipulating the sample flow using an electric field induced between two electrodes. Electrophoresis involves moving charged samples in an electric field.[80] CE implements electrophoresis in a capillary channel.[81] DEP moves charged samples in a varying electric field.[82] EOF utilizes fluid flow driven by a double layer of ions.[83] IEF is a kind of zoned EOF where sample movement is governed not only by its electric charge, but also by the pH of the surroundings.[84] The electric field-directed sample flow enables the effective concentration of the samples to a specified region. The negatively charged cell membrane or viral envelope can be conveniently manipulated to move toward the positive ionic region in electrophoresis. However, the electrodynamic methods often suffer from the drawbacks of complicated setup of external electrodes, power supply, and electric circuitry. The dependence on surface charge of the sample makes it difficult to adapt to different kinds of analytes. The addition of a conductive buffer to the sample to enhance EOF may also disturb the intrinsic surface properties required for subsequent detection.

Incorporating microfilters inside a microfluidic channel has been a popular method for trapping[85,86] or lysing[87,88] microbes. The filter can be made out of silicon or glass and has the shape of a weir or a collection of micropillars, allowing fluid to pass but trapping bacteria larger than the height of the weir or the gap between the micropillars.[85,86] The fabrication of such filters requires semi-conductor-processing technology, including deep reactive ion etching and silicon etching, which are batch processes. Each device has to be fabricated separately. One example of such a filter-based concentration device is shown in Fig. 4.

Immunocapture or surface-modified magnetic beads can be used to concentrate pathogens for subsequent detection by antibody–antigen binding.[45,89,90] Lien et al. developed a system to capture the dengue virus by an antibody conjugated on magnetic beads in a micromixer. The captured virus was then enriched by a magnetic field from a planar microcoil.[90]

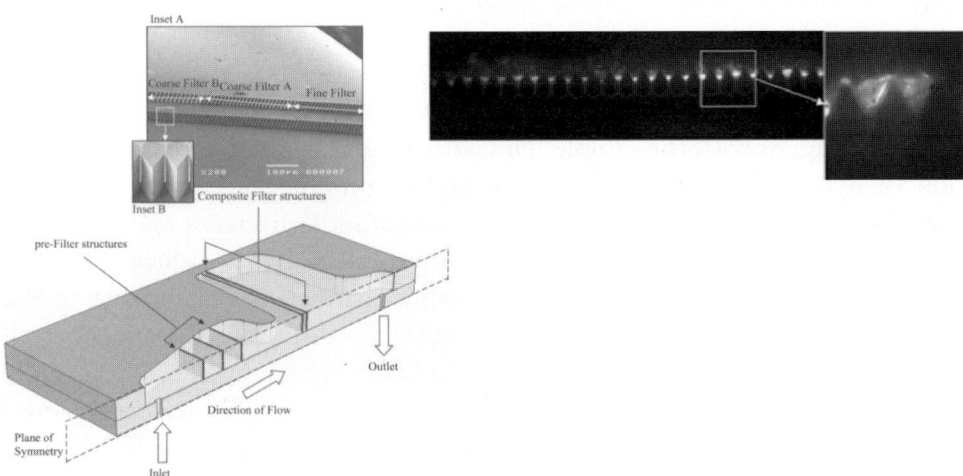

Fig. 4. Filtration-based microbial concentrator made of silicon.[85]

Dialysis or cross-flow filtration is the size-based separation of samples by selective diffusion through a semi-permeable membrane. On-chip dialysis separates the input sample constituents by size to concentrate pathogens or proteins larger than the molecular weight cut-off of the membrane. Song et al. used both a polymeric monolithic membrane and a laser-patterned nanoporous membrane, combined with electrophoresis to achieve on-chip cell lysis and dialysis to concentrate protein on-chip.[91,92]

Evaporation is used to concentrate biological samples in a simple and rapid fashion with minimal sample loss. Many microfluidic devices have utilized evaporation to concentrate microspheres, proteins, viruses, toxins, or electrolytes.[93–98] The device built by Walker et al.[97] is a simple channel with a "reservoir port" and a "collection port." The reservoir port and the channel are pre-filled with the liquid sample. As evaporation occurs at the collection port, water flows from the reservoir port to the collection port to compensate for liquid loss into the air. The analyte particles are dragged by the bulk fluid flow toward the collection port and is concentrated as liquid evaporates. This device is especially simple to fabricate and easy to use. The term "pervaporation" refers to the combined evaporation of the liquid and permeation of the vapor along a partial pressure gradient through a gas-permeable membrane. Sharma et al. developed an evaporation device composed of two layers, a fluid layer and a convective gas flow layer, with a hydrophobic porous membrane in between. They have built and tested a prototype for the rapid isothermal evaporation of water and bovine serum albumin.[95] Zhang et al. have developed pervaporation microfluidics with a surface tension-driven meniscus dragging effect to concentrate bacterial sample from 100 µL to sub-microliters within 15 min, with an efficiency greater than 90%.[99] The device design and configuration are shown in Fig. 5.

Following enrichment and concentration, sample lysis on chip can be achieved chemically or physically. Chemical lysis on chip requires the mixing of the lysis agent with the sample. Di Carlo et al. used on-chip electrogeneration of hydroxide ions to porate the cell membrane and lyse mammalian cells.[100] Stachowiak et al. demonstrated microfluidic bacterial spore lysis by solubilizing the coat protein with β-mercaptoethanol that reduces disulfide linkages.[101] The laminar flow in microfluidic channels is unfavorable to the mixing of individual cells and the lysis agent by diffusion. To enhance cell lysis, microfluidic mixing mechanisms are introduced. Sethu et al. successfully demonstrated a continuous-flow microfluidic system, where lysis is achieved by splitting the lysis buffer containing mainly ammonium chloride into two channels that meet to flank the cell suspension on both sides.[102] Bhattacharyya and Klapperich demonstrated continuous lysis by enhanced diffusive mixing

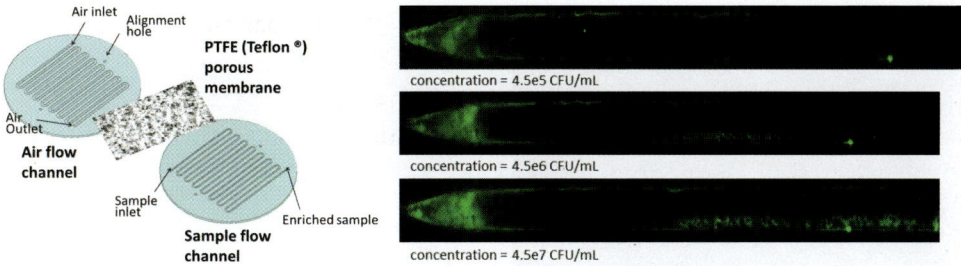

Fig. 5. Two-layered design of the evaporation microfluidics. Left: 3D view of each component, including the airflow channel layer, the PDMS sample flow channel layer, separated by the Teflon® (PTFE) porous membrane. Right: concentrated fluorescent *E. coli* from three different input concentrations, near the outlet terminal of the sample flow channel at the end of evaporation.[100]

of the chaotropic buffer (GuSCN) and cell suspension in a serpentine channel with sharp turns.[103]

Physical lysis on chip disrupts cell membranes mechanically, electrically, thermally, or by sonication. Microfluidic filters force cells through a barrier with openings too small for whole cells to pass. As a result, the shear force ruptures the cell membrane. Di Carlo *et al.* fabricated a microfluidic filter by the deep reactive ion etching process with nanostructured "nano-knives."[104] Kim *et al.* used a pneumatically controlled polydimethylsiloxane (PDMS) membrane to deform and crush cells to break their membranes.[105] Kulinski *et al.* and Mahalanabis *et al.* used combined chemical and mechanical methods in a polymeric porous monolith (PPM) inside a thermoplastic polymer chip to lyse Gram-negative *E. coli* in human urine and blood and Gram-positive *Bacillus subtilis* and *Enterococcus faecalis* in blood (Fig. 6). The fabrication procedure of a PPM involves delivering the mixture of a methacrylate-based monomer, a microporogen, a macroporogen, and an initiator to a sealed microchannel, followed by polymerization of the monomer by illuminating the microchannel with UV light.[49,87,88,106] The monolith pore size can be adjusted by the relative amount of the monomer and the porogens. The tortuous structure of the small open pore structure of the PPM shears the bacteria, which is enhanced in chaotropic or detergent lytic conditions.[49,106] In addition, cells can be lysed by the impact of collision caused by combined stirring and grinding. Kim *et al.* demonstrated the disruption of mammalian, bacterial, and yeast cells on a microfluidic compact disk platform using friction and collisions with granular beads in the lysis buffer.[107] Kido *et al.* developed a related CD-like device, where ferromagnetic disks and grinding matrices break down *E. coli* and *Saccharomyces cerevisiae* in

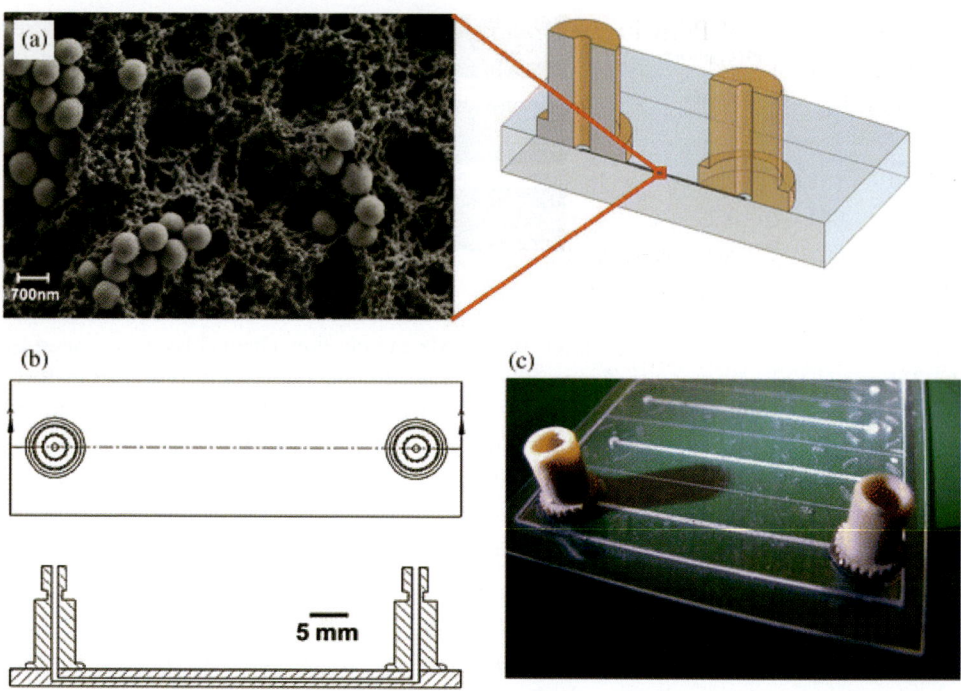

Fig. 6. (a) SEM image of a lysis/solid-phase extraction monolith taken at 10,000× magnification. The spheres in the image are silica particles, (b) CAD drawing of chip components, and (c) a photograph of an array of monolith filled channels.[106]

an oscillating magnetic field.[108] Furthermore, electroporation is adapted on chip for cell lysis. Jensen *et al.* demonstrated the electrolysis of human carcinoma cells in a micro-electroporation device.[109] Ramadan demonstrated lysis of WBCs and murine clonal cells by electroporation in a continuous-flow microfluidic device.[110] Ultrasonic agitation creates pressure waves to disrupt cells and is particular advantageous for lysing bacterial or fungal spores. Taylor *et al.* lysed bacterial spores trapped in a microfilter with ultrasonication. Although the ultrasonic horn was not incorporated in the microfluidic device, the noninvasive nature of the lysis process shows promise for integration with other on-chip modules.[111] Liu *et al.* demonstrated the thermal lysis of trapped *E. coli* cells using embedded heaters.[89] Lee *et al.* presented an integrated thermal cell lysis of *Streptococcus pneumonia* and PCR amplification in a microreactor.[112] Cheong *et al.* used an optothermal technique with on-chip gold nanoparticles, which generated enough thermal energy to lyse cells upon laser irradiation.[113] Many of the on-chip lysis methods introduced were integrated with downstream PCR in a single chip.[89,112–115]

The presence of PCR inhibitors in a complex biological sample will likely always require a nucleic acid extraction and purification step after lysis before PCR can be performed successfully on chip. Extraction and purification have been avoided in cases where a high concentration of the target molecule can be sufficiently diluted in the PCR reaction to also dilute and blunt the effects of any inhibitors, for example, using Phusion® Blood Direct PCR Kit (Finnzymes, Finland), KAPA Blood PCR Kits (KAPA Biosystems, Boston, MA, USA), and Extract-N-Amp® Blood PCR Kits (Sigma-Aldrich, St. Louis, MO, USA). The majority of the current microfluidic extraction methods are for DNA purification, although more attention is now being paid to RNA. RNA samples contain rich information about gene expression and genetic or pathogenic disease states because RNA is the intermediate material between DNA and the protein synthesis machinery of the cell. From a public health standpoint, the development of rapid, on-chip extraction of RNA could be beneficial to, for example, the response time, efficacy of exposure prevention, and treatment of outbreaks of highly infectious RNA viruses (such as influenza, Ebola, and HIV). Although many DNA extraction techniques can be directly applied, RNA extraction is a more complicated sample preparation process than DNA extraction. In addition, patient samples such as blood, stool, urine, and nasopharyngeal fluid add yet another layer of complexity to the extraction step.

The most widely studied on-chip DNA extraction is through micro-solid–phase extraction (microSPE), during which DNA binding to silica or glass fibers is facilitated in a microchannel. The large surface-to-volume ratio in a microfluidic channel greatly increases the exposed silica surface available to DNA binding, thus effectively enhancing the extraction efficiency. One of the earliest attempts to incorporate microSPE on chip was by Christel et al. using a microfabricated silicon chip with pillar-like structures using reactive ion etching to capture DNA.[116] This attempt inspired two different approaches for creating on chip microSPE — chemical formation and microfabrication. The chemical formation of microSPE requires silica to be packed or immobilized in a microfluidic channel or chamber. It is a simple and rapid process, but exhibits large variations in extraction efficiency between channels or runs due to the difference in the amount of silica immobilized. Therefore, each step has to be carried out in a carefully controlled fashion. Bhattacharyya and Klapperich used in situ UV polymerization to impregnate a PPM with silica particles on a disposable cyclic polyolefin chip.[117] The PPM is chemically grafted onto the plastic surface, providing a large surface area and good structural support and stability to the immobilized silica. Tian et al. retained silica beads in a polyethylene sleeve between two glass fiber frits on a glass chip.[118]

Later, improvement was demonstrated using a two-step formation of a hybrid microSPE of packed silica beads and silica sol-gel to replace packed silica alone. Together with lowered pH, this system reached well-controlled efficiency, stability, and repeatability. Tests on λDNA, *Salmonella typhimurium*, and *Bacillus anthracis* yielded PCR-amplifiable samples, with high efficiency and an overall process time of 15 min.[119,120] Furthermore, UV-induced photopolymerization of 3-(trimethoxysilyl)propyl methacrylate (TMSPM) monolith grafted with a tetramethyl orthosilicate (TMOS)-based sol-gel in a capillary was shown to significantly increase human genomic DNA extraction efficiency.[121] Another approach to achieve a large surface area for DNA binding is by microfabricating pillar-like structures. The disadvantage of microfabricating structures like these is that it is time-consuming and labor-intensive. However, they tend to have very consistent extraction efficiency. Cady *et al.* demonstrated the purification of bacteriophage λDNA and chromosomal DNA with a microfabricated device with silica-coated pillars.[122]

Besides microSPE based on affinity binding of DNA to silica, other DNA extraction methods utilize electrostatic interactions between DNA and chitosan- or carbonyl-/carboxyl-modified surfaces,[123–125] nanoporous aluminum oxide membranes (AOM) to trap and localize DNA,[126–128] and functionalized magnetic particles with silica, carboxyl groups, amine groups, or sequence-specific oligonucleotides on the surface.[129–133] A comparison between each of the extraction methods are reviewed and summarized by Kim *et al.* and Price *et al.*[134,135]

The extraction of RNA is difficult because of its susceptibility to degradation, contamination, and ubiquitous RNases. In addition, there are multiple types of RNA in cells that add to the complexity when purification of a particular type is required. RNA samples also have to undergo a reverse transcription (RT) step before PCR. The extraction step should have high efficiency to achieve high-sensitivity RT-PCR for RNA samples. One example of an on-chip RNA isolation device utilizes paramagnetic beads functionalized with oligo-dTs, to capture and manipulate eukaryotic mRNA or RNA from peripheral blood lysate that has a poly-A tail.[136,137] The challenge for magnetic bead-based systems to be easily implemented in a microfluidic setting for global health purposes is the extra steps and materials involved in the setup. RNA extraction is more readily achieved with silica-based methods. Bhattacharyya *et al.* used an SPE system that embedded silica particles in a PPM. RNA was separated through reversible binding of the nucleic acids to the silica particles in the monolith. This system allowed the isolation of intact influenza (H1N1) viral total RNA from a mixture of infected whole cell lysates and serum-supplemented cell culture supernatant.[138] Hagen *et al.* used chitosan-coated silica beads for capturing RNA without PCR inhibitory reagents such as guanidinium ions and isopropanol.[139] RNA extraction can also be achieved in a PPM without silica. Satterfield et al. modified the surface of PPM with oligo-dTs to immobilize and purify mRNA with up to 70% efficiency.[140] The limitation of such an approach, however, is that only eukaryotic mRNA with a poly-A tail can be extracted, but not prokaryotic RNA or tRNA. A recent method by Vulto et al. used neither PPM nor silica, but on-chip electrophoresis through a polyacrylamide gel to separate small, noncoding transfer-messenger RNA (tmRNA) from E. coli and Streptococcus thermophilus. The extraction limit was shown to be 1,000 times below that of commercial kits.[141]

Much recent development on miniaturized nucleic acid extraction on chip has been focused on processing real human patient samples, such as urine, stool, nasopharyngeal fluid, and blood, as this is the most relevant for POC diagnostics. On-chip nucleic acid extraction for real human samples must be portable, of low cost, rapid, and highly sensitive to overcome the equipment and personnel shortage at the POC. We have developed disposable polymeric nucleic acid extraction microfluidics for urine,[106] blood,[49] and stool.[142] The extraction was based on reversible silica binding in a PPM. The microfluidic chips were fabricated with low-cost thermoplastics undergoing thermal embossing and sealing processes. Thermoplastic sample preparation chips are less likely to fracture than glass or silicon and are more dimensionally stable than PDMS or PDMS hybrid chips, thus are more rugged for application in the developing world. The glass transition temperatures, optical properties, and other properties make this class of materials particularly attractive for mass-market applications.[142] It can be scaled up to large volume manufacturing to further reduce the cost.[143] Another direction of research is to further optimize assay performance and improve device sensitivity for human samples. Tian et al. evaluated different formats of on-chip silica immobilization to assess the efficiency of DNA extraction from whole blood.[118] They have then developed a two-stage technique on chip preparation, where proteins from blood are retained in a C18 phase, while DNA is extracted in the microSPE phase.[144] Recently, they have also used the aqueous phase created by chitosan-coasted silica to extract RNA from blood.[141]

Brief Review of Current Chips Integrating Sample Preparation and PCR

To expand the application of PCR for POC in a global health context, the integration of upstream sample preparation and downstream detection is necessary. Two decades ago, the micro total analysis system (μTAS) was proposed by Manz et al. to miniaturize total chemical analysis systems into a microfluidic platform, to reduce the reagent volume and process time during the sample preparation and analysis steps.[145] Subsequently, researchers started to miniaturize every single step, including sample preparation,[146,147] amplification,[10,148,149] and detection.[150] Further efforts have been put into partial or full integration of these steps.[21,89,109,151–157] Obeid et al. developed a system integrated continuous-flow DNA and RNA amplification with an in-house built laser-induced fluorescence (LIF) detection system that allows the analysis of amplification products within seconds. Lee et al. reported a microchip for DNA extraction and subsequent real-time detection by combining laser irradiation and magnetic beads.[154] Easley et al. built a glass microfluidic genetic analysis system capable of carrying out DNA purification, PCR-based amplification, separation, and detection to get sample-in–answer-out results in <30 min.[153] Zhang et al. developed a microfluidic device capable of fully integrated sample preparation and gene analysis for analyzing the *Rsf-1* gene from human whole blood.[157] We developed an integrated μTAS device capable of accepting the crude clinical sample, carrying out cell lysis, nucleic acid extraction, purification, reverse transcription, and amplification. This plastic microfluidic system is suitable for high-throughput fabrication by injection molding.[158] It uses the simple thermal and fluidic controls, and demonstrates the ability to detect clinical influenza patient samples from the 2008–2010 flu seasons.

Major Challenges Going Forward

While the development of μTAS for molecular diagnostics has evolved rapidly over the past 10 years, there are several challenges that remain for true application at the POC. First, many of the manufacturing processes needed to make the current devices are not appropriate for scale up to the numbers needed at very low unit production costs. A concerted effort is necessary on the part of the industry to address these scale-up issues. Injection molding of small plastic parts, heat sealing, and fluid input/output interfaces are but a few areas that need attention. There also remain significant challenges in working with real human sample matrices. Many demonstrations of integrated μTAS used either purified DNA or RNA[159,160] or cultured bacteria or viruses.[11,143,161] Fewer studies have addressed versatile samples for a broad spectrum of input species or clinical patient samples, due to their complex nature. Each different human sample type and each different disease target require specialized assay development, and assay development and chip development must be carried out simultaneously for the best chance of success. Finally, true POC devices require portability and true push button functionality. Many instruments and devices for protein diagnostics have surmounted these hurdles, and those of use working in molecular diagnostics for true POC applications can learn from those successes.

References

1. Sabella, S. *et al.*, Disposable plastic microreactors for genomic analyses. *Biomed Microdevices*, 2009, **11**(6): p. 1289–1295.
2. Saiki, R., Scharf, S., Faloona, F., Mullis, K., Horn, G., Erlich, H., Enzymatic amplification of beta-globin genomic sequences and restriction site analysis for diagnosis of sickle cell anemia. *Science*, 1985, **230**: p. 1350–1354.
3. Mullis, K., Faloona, F.A., Scharf, S., Saiki, R., Horn, G., Erlich, H., Specific enzymatic amplification of DNA *in vitro*: The polymerase chain reaction. *Cold Spring Harbor Symp Quant Biol*, 1986, **51**: p. 263–273.
4. Mullis, K.B., Faloona, F.A., Specific synthesis of DNA *in vitro* via a polymerase-catalyzed chain reaction. *Methods Enzymol*, 1987, **155**: p. 335–350.
5. Mullis, K., The unusual origin of the polymerase chain reaction. *Sci Am*, 1990.
6. Saiki RK, S.S., Faloona, F.A. *et al.*, Enzymatic amplification of beta-globin genomic sequences and restriction site analysis for diagnosis of sickle cell anaemia. *Science*, 1985, **230**: p. 1350–1354.
7. Metzker, A.L., Caskey, T.C. Polymerase chain reaction (PCR). In: *Encyclopedia of Life Sciences*, John Wiley & Sons, 2001.
8. Lodish, H.F., *Molecular Cell Biology*, 5th edn. W.H. Freeman and Company, 2003.
9. Northrup, M.A., Ching, M.T., White, R.M., Watson, R.T., DNA amplification in a microfabricated reaction chamber. *Transducer' 93: 7th Int Conf On Solid State Sens Actuators 1993*. Yokohama, Japan, 1993. p. 924–926.
10. Kopp, M.U. *et al.*, Chemical amplification: Continuous-flow PCR on a chip. *Science*, 1998, **280**(5366): p. 1046–1048.
11. Lagally, E.T. *et al.*, Integrated portable genetic analysis microsystem for pathogen/infectious disease detection. *Anal Chem*, 2004, **76**(11): p. 3162–3170.
12. Cady, N.C. *et al.*, Real-time PCR detection of *Listeria monocytogenes* using an integrated microfluidics platform. *Sens Actuators B*, 2005, **107**(1): p. 332–341.
13. Giordano, B.C. *et al.*, Polymerase chain reaction in polymeric microchips: DNA amplification in less than 240 seconds. *Anal Biochem*, 2001, **291**(1): p. 124–132.
14. Tanaka, Y. *et al.*, Non-contact photothermal control of enzyme reactions on a microchip by using a compact diode laser. *J Chromatogr A*, 2000, **894**(1–2): p. 45–51.
15. Schaerli, Y. *et al.*, Continuous-flow polymerase chain reaction of single-copy DNA in microfluidic microdroplets. *Anal Chem*, 2008, **81**(1): p. 302–306.
16. Mitchell, M.W. *et al.*, Modeling and validation of a molded polycarbonate continuous-flow polymerase chain reaction device, In: *Microfluidics, BioMEMS, and Medical Microsystems*, SPIE, San Jose, CA, USA, 2003.
17. Chiou, J. *et al.*, A closed-cycle capillary polymerase chain reaction machine. *Anal Chem*, 2001, **73**(9): p. 2018–2021.

18. Frey, O. et al., Autonomous microfluidic multi-channel chip for real-time PCR with integrated liquid handling. *Biomed Microdevices* 2007, **9**(5): p. 711–718.
19. Zhang, C. et al., PCR microfluidic devices for DNA amplification. *Biotechnol Adv,* 2006, **24**(3): p. 243–284.
20. Burns, M.A. et al., Microfabricated structures for integrated DNA analysis. *Proc Natl Acad Sci U S A,* 1996, **93**(11): p. 5556–5561.
21. Obeid, P.J., Christopoulos, T.K., Continuous-flow DNA and RNA amplification chip combined with laser-induced fluorescence detection. *Anal Chim Acta,* 2003, **494**(1–2): p. 1–9.
22. Chartier, I. et al. Fabrication of a hybrid plastic-silicon microfluidic device for high-throughput genotyping, In: *Microfluidics, BioMEMS, and Medical Microsystems,* SPIE, San Jose, CA, USA, 2003.
23. Shih, C-Y., Chen, Y., Tai, Y-C., Parylene-strengthened thermal isolation technology for microfluidic system-on-chip applications. *Sens Actuators A,* 2006, **126**(1): p. 270–276.
24. Cao, Q., Kim, M., Klapperich, C.M., Plastic microfluidic chip for continuous-flow polymerase chain reaction: Simulations and experiments. *Biotechnol J,* 2010.
25. Yao, L. et al., Micro flow-through PCR in a PMMA chip fabricated by KrF excimer laser. *Biomed Microdevices,* 2005, **7**(3): p. 253–257.
26. Wang, H. et al., Continuous flow thermal cycler microchip for DNA cycle sequencing. *Anal Chem,* 2006, **78**(17): p. 6223–6231.
27. Dettloff, R. et al., Nucleic acid amplification of individual molecules in a microfluidic device. *Anal Chem,* 2008, **80**(11): p. 4208–4213.
28. Hashimoto, M. et al., Ligase detection reaction/hybridization assays using three-dimensional microfluidic networks for the detection of low-abundant DNA point mutations. *Anal Chem,* 2005, **77**(10): p. 3243–3255.
29. Zhang, C., Xing, D., Single-molecule DNA amplification and analysis using microfluidics. *Chem Rev,* 2010, **110**(8): p. 4910–4947.
30. Wang, F., Burns, M., Performance of nanoliter-sized droplet-based microfluidic PCR. *Biomed Microdevices,* 2009, **11**(5): p. 1071–1080.
31. Beer, N.R. et al., On-chip, real-time, single-copy polymerase chain reaction in picoliter droplets. *Anal Chem,* 2007, **79**(22): p. 8471–8475.
32. Chabert, M. et al., Automated microdroplet platform for sample manipulation and polymerase chain reaction. *Anal Chem,* 2006, **78**(22): p. 7722–7728.
33. Chang, Y-H. et al., Integrated polymerase chain reaction chips utilizing digital microfluidics. *Biomed Microdevices,* 2006, **8**(3): p. 215–225.
34. Ottesen, E.A. et al., Microfluidic digital PCR enables multigene analysis of individual environmental bacteria. *Science,* 2006, **314**(5804): p. 1464–1467.
35. Zhang, C., Xing, D., Miniaturized PCR chips for nucleic acid amplification and analysis: Latest advances and future trends. *Nucleic Acids Res,* 2007, **35**(13): p. 4223–4237.
36. Wilson, I.G., Inhibition and facilitation of nucleic acid amplification. *Appl Environ Microbiol,* 1997, **63**(10): p. 3741–51.
37. Kornberg, A., *DNA Replication,* W.H. Freeman, San Francisco, 1980.
38. Al-Soud, W.A., Jonsson, L.J., Radstrom, P., Identification and characterization of immunoglobulin G in blood as a major inhibitor of diagnostic PCR. *J Clin Microbiol,* 2000, **38**(1): p. 345–350.
39. Kim, S., Labbe, R.G., Ryu, S., Inhibitory effects of collagen on the PCR for detection of *Clostridium perfringens. Appl Environ Microbiol,* 2000, **66**(3): p. 1213–1215.

40. Bickley, J. et al., Polymerase chain reaction (PCR) detection of *Listeria monocytogenes* in diluted milk and reversal of PCR inhibition caused by calcium ions. *Lett Appl Microbiol*, 1996, **22**(2): p. 153–158.
41. Jacobson, M. et al., Flotation-a new method to circumvent PCR inhibitors in the diagnosis of lawsonia intracellularis. *Int J Microbiol*, 2009, p. 410945.
42. Macey, M.G. et al., Effects of cell purification methods on CD11b and -selectin expression as well as the adherence and activation of leucocytes. *J Immunol Methods*, 1995, **181**(2): p. 211–219.
43. Chen, X. et al., Continuous flow microfluidic device for cell separation, cell lysis and DNA purification. *Anal Chim Acta*, 2007, **584**(2): p. 237–243.
44. Ji, H. et al., Silicon-based microfilters for whole blood cell separation. *Biomed Microdevices*, 2008, **10**(2): p. 251–257.
45. Furdui, V.I., Harrison, D.J., Immunomagnetic T cell capture from blood for PCR analysis using microfluidic systems. *Lab Chip*, 2004, **4**(6): p. 614–618.
46. Cabeen, M.T., Jacobs-Wagner, C., Bacterial cell shape. *Nat Rev Micro*, 2005, **3**(8): p. 601–610.
47. Vollmer, W., Holtje, J.-V., The architecture of the murein (peptidoglycan) in Gram-negative bacteria: Vertical scaffold or horizontal layer(s)? *J Bacteriol*, 2004, **186**(18): p. 5978–5987.
48. Middelberg, A.P., Process-scale disruption of microorganisms. *Biotechnol Adv*, 1995, **13**(3): p. 491–551.
49. Mahalanabis, M. et al., Cell lysis and DNA extraction of Gram-positive and Gram-negative bacteria from whole blood in a disposable microfluidic chip. *Lab Chip*, 2009, **9**(19): p. 2811–2817.
50. Rantakokko-Jalava, K., Jalava, J., Development of conventional and real-time PCR assays for detection of legionella DNA in respiratory specimens. *J Clin Microbiol*, 2001, **39**(8): p. 2904–2910.
51. Bhattacharyya, A., Klapperich, C.M., On-chip cell lysis. In: Li, D., ed. *Encyclopedia of Microfluidics and Nanofluidics*, Springer, 2008, p. 1513–1515.
52. Chomczynski, P., Sacchi, N., The single-step method of RNA isolation by acid guanidinium thiocyanate-phenol-chloroform extraction: Twenty-something years on. *Nat Protoc*, 2006, **1**(2): p. 581–585.
53. Chomczynski, P., Sacchi, N., Single-step method of RNA isolation by acid guanidinium thiocyanate-phenol-chloroform extraction. *Anal Biochem*, 1987, **162**(1): p. 156–159.
54. Sambrook, J., Russell, D.W., *Molecular Cloning: A Laboratory Manual*, 3rd edn. Cold Spring Harbor Laboratory Press, Cold Spring Harbor, New York, 2001.
55. Boom, R. et al., Rapid and simple method for purification of nucleic acids. *J Clin Microbiol*, 1990, **28**(3): p. 495–503.
56. Wen, J. et al., Purification of nucleic acids in microfluidic devices. *Anal Chem*, 2008, **80**(17): p. 6472–6479.
57. Imagawa, D.T. et al., Human immunodeficiency virus type 1 Infection in homosexual men who remain seronegative for prolonged periods. *N Engl J Med*, 1989, **320**(22): p. 1458–1462.
58. Trépanier, P., Payment P., Trudel, M., Concentration of human respiratory syncytial virus using ammonium sulfate, polyethylene glycol or hollow fiber ultrafiltration. *J Virol Methods*, 1981, **3**(4): p. 201–211.

59. Kohno, T. et al., A new improved method for the concentration of HIV-1 infective particles. *J Virol Methods*, 2002, **106**(2): p. 167–173.
60. Polson, A. et al., Polyethylene glycol purification of influenza virus with respect to aggregation and antigenicity. *J Hyg (Lond)*, 1972, **70**(2): p. 255–265.
61. Hamelin, C., Lussier, G., Concentration of human cytomegalovirus from large volumes of tissue culture fluids. *J Gen Virol*, 1979, **42**(1): p. 193–197.
62. Goyal, S.M., Hanssen, H., Gerba, C.P., Simple method for the concentration of influenza virus from allantoic fluid on microporous filters. *Appl Environ Microbiol*, 1980, **39**(3): p. 500–504.
63. Grzenia, D.L., Carlson, J.O., Wickramasinghe, S.R., Tangential flow filtration for virus purification. *J Membr Sci*, 2008, **321**(2): p. 373–380.
64. Yamagishi, T. et al., Simultaneous removal of phenol and ammonia by an activated sludge process with cross-flow filtration. *Water Res*, 2001, **35**: p. 3089–3096.
65. Pyke, A. et al., Sucrose density gradient centrifugation and cross-flow filtration methods for the production of arbovirus antigens inactivated by binary ethylenimine. *BMC Microbiol*, 2004, **4**(1): p. 3.
66. Arora, D.J.S. et al., Concentration and purification of influenza virus from allantoic fluid. *Anal Biochem*, 1985, **144**(1): p. 189–192.
67. Barry, R.D., Davies, P., The sedimentation of influenza virus and its RNA in sucrose density gradients. *J Gen Virol*, 1968, **2**(1): p. 59–69.
68. Brakke, M.K., Density-gradient centrifugation. *Methods Virol*, 1967, **2**: p. 93–118.
69. Neurath, A.R., Wiktor, T.J., Koprowski, H., Density gradient centrifugation studies on rabies virus. *J Bacteriol*, 1966, **92**(1): p. 102–106.
70. Reimer, C.B. et al., Purification of large quantities of influenza virus by density gradient centrifugation. *J Virol*, 1967, **1**(6): p. 1207–1216.
71. Byrne, J. et al., *Laboratory Tests: Implications for Nursing Care*, 2nd edn. Addison-Wesley Publication Co., Health Sciences Division, Menlo Park, CA, 1986.
72. Lee, H.H., Morse, S.A., Olsvik, O., *Nucleic Acid Amplification Technologies: Application to Disease Diagnosis*, Birkhauser Eaton Publications, Natick, MA, 1997.
73. Gupta, K., Hooton, T.M., Stamm, W.E., Increasing antimicrobial resistance and the management of uncomplicated community-acquired urinary tract infections. *Ann Intern Med*, 2001, **135**(1): p. 41–50.
74. Clarke, W., Dufour, D.R., eds. *Contemporary Practice in Clinical Chemistry*, AACC Press, Washington, DC, 2006.
75. Monteiro, L. et al., Complex polysaccharides as PCR inhibitors in feces: *Helicobacter pylori* model. *J Clin Microbiol*, 1997, **35**(4): p. 995–998.
76. Panaccio, M., Lew, A., PCR based diagnosis in the presence of 8% (v/v) blood. *Nucleic Acids Res*, 1991, **19**(5): p. 1151.
77. Monteiro, L., Gras, N., Megraud, F., Magnetic immuno-PCR assay with inhibitor removal for direct detection of *Helicobacter pylori* in human feces. *J Clin Microbiol*, 2001, **39**(10): p. 3778–3780.
78. Holland, J.L. et al., PCR detection of *Escherichia coli* O157:H7 directly from stools: Evaluation of commercial extraction methods for purifying fecal DNA. *J Clin Microbiol*, 2000, **38**(11): p. 4108–4113.
79. Vuorte, J., Jansson, S.E., Repo, H., Evaluation of red blood cell lysing solutions in the study of neutrophil oxidative burst by the DCFH assay. *Cytometry*, 2001, **43**(4): p. 290–296.

80. Castellanos, A. et al., Electrohydrodynamics and dielectrophoresis in microsystems: Scaling laws. *J Phys D* 2003, **36**: p. 2584–2597.
81. Boguslaw, B. et al., Separation of bacteria by capillary electrophoresis. *J Sep Sci*, 2003, **26**(11): p. 1045–1049.
82. Jie, W., Biased AC electro-osmosis for on-chip bioparticle processing. *IEEE Trans Nanotechnol*, 2006, **5**(2): p. 84–89.
83. Ocvirk, G., Electrokinetic control of fluid flow in native poly(dimethylsiloxane) capillary electrophoresis devices. *Electrophoresis*, 2000, **21**(1): p. 107–115.
84. Jacoby, G.A. and Han, P., Detection of extended-spectrum beta-lactamases in clinical isolates of *Klebsiella pneumoniae* and *Escherichia coli*. *J Clin Microbiol*, 1996, **34**(4): p. 908–911.
85. Lay, C. et al., Enhanced microfiltration devices configured with hydrodynamic trapping and a rain drop bypass filtering architecture for microbial cells detection. *Lab Chip*, 2008, **8**(5): p. 830–833.
86. Zhu, L. et al., Filter-based microfluidic device as a platform for immunofluorescent assay of microbial cells. *Lab Chip*, 2004, **4**(4): p. 337–341.
87. Kulinski, M. et al., Sample preparation module for bacterial lysis and isolation of DNA from human urine. *Biomed Microdevices*, 2009, **11**(3): p. 671–678.
88. Bhattacharyya, A., Klapperich, C.M., Thermoplastic microfluidic device for on-chip purification of nucleic acids for disposable diagnostics. *Anal Chem*, 2006, **78**(3): p. 788–792.
89. Liu, R.H. et al., Self-contained, fully integrated biochip for sample preparation, polymerase chain reaction amplification, and DNA microarray detection. *Anal Chem*, 2004, **76**(7): p. 1824–1831.
90. Lien, K.-Y. et al., Purification and enrichment of virus samples utilizing magnetic beads on a microfluidic system. *Lab Chip*, 2007, **7**(7): p. 868–875.
91. Song, S., Singh, A.K., Kirby, B.J., Electrophoretic concentration of proteins at laser-patterned nanoporous membranes in microchips. *Anal Chem*, 2004, **76**(15): p. 4589–4592.
92. Song, S. et al., Microchip dialysis of proteins using *in situ* photopatterned nanoporous polymer membranes. *Anal Chem*, 2004, **76**(8): p. 2367–2373.
93. Berthier, E. et al., Managing evaporation for more robust microscale assays Part 1. Volume loss in high throughput assays. *Lab Chip*, 2008, **8**(6): p. 852–859.
94. Berthier, E. et al., Managing evaporation for more robust microscale assays Part 2. Characterization of convection and diffusion for cell biology. *Lab Chip*, 2008, **8**(6): p. 860–864.
95. Sharma, N.R. et al., Development of an evaporation-based microfluidic sample concentrator, In: Woias, P. Papautsky, I., eds. *Microfluidics, BioMEMS, and Medical Microsystems VI*, SPIE, San Jose, CA, USA, 2008.
96. Timmer, B.H. et al., Micro-evaporation electrolyte concentrator. *Sens Actuators B*, 2003, **91**(1–3): p. 342–346.
97. Walker, G.M., Beebe, D.J., An evaporation-based microfluidic sample concentration method. *Lab Chip*, 2002, **2**(2): p. 57–61.
98. Zimmermann, M. et al., Continuous flow in open microfluidics using controlled evaporation. *Lab Chip*, 2005, **5**(12): p. 1355–1359.
99. Zhang, J.Y. et al., Rapid point-of-care concentration of bacteria in a disposable microfluidic device using meniscus dragging effect. *Lab Chip*, 2010, **10**(23): p. 3265–3270.

100. Di Carlo, D. et al., On-chip cell lysis by local hydroxide generation. *Lab Chip*, 2005, **5**(2): p. 171–178.
101. Stachowiak, J.C. et al., Autonomous microfluidic sample preparation system for protein profile-based detection of aerosolized bacterial cells and spores. *Anal Chem*, 2007, **79**(15): p. 5763–5770.
102. Sethu, P. et al., Continuous flow microfluidic device for rapid erythrocyte lysis. *Anal Chem*, 2004, **76**(21): p. 6247–6253.
103. Bhattacharyya, A., Klapperich, C.M., Polymeric microfluidic device for on-chip cell lysis and extraction of nucleic acids from biological samples. *9th Int Conf Miniaturized Syst Chem Life Sci*, Boston, 2005.
104. Carlo, D.D., Jeong, K-H., Lee, L.P., Reagentless mechanical cell lysis by nanoscale barbs in microchannels for sample preparation. *Lab Chip*, 2003, **3**(4): p. 287–291.
105. Kim, Y.C. et al., Microfluidic biomechanical device for compressive cell stimulation and lysis. *Sens Actuators, B*, 2007, **128**(1): p. 108–116.
106. Kulinski, M. et al., Sample preparation module for bacterial lysis and isolation of DNA from human urine. *Biomed Microdevices*, 2009, **11**(3): p. 671–678.
107. Kim, J. et al., Cell lysis on a microfluidic CD (compact disc). *Lab Chip*, 2004, **4**(5): p. 516–522.
108. Kido, H. et al., A novel, compact disk-like centrifugal microfluidics system for cell lysis and sample homogenization. *Colloids Surf, B*, 2007, **58**(1): p. 44–51.
109. Lu, H., Schmidt, M.A., Jensen, K.F., A microfluidic electroporation device for cell lysis. *Lab Chip*, 2005, **5**(1): p. 23–29.
110. Ramadan, Q. et al., Simultaneous cell lysis and bead trapping in a continuous flow microfluidic device. *Sens Actuators B*, 2006, **113**(2): p. 944–955.
111. Taylor, M.T. et al., Lysing bacterial spores by sonication through a flexible interface in a microfluidic system. *Anal Chem*, 2000, **73**(3): p. 492–496.
112. Lee, C-Y. et al., Integrated microfluidic systems for cell lysis, mixing/pumping and DNA amplification. *J Micromech Microeng*, 2005, **15**(6): p. 1215.
113. Cheong, K.H. et al., Gold nanoparticles for one step DNA extraction and real-time PCR of pathogens in a single chamber. *Lab Chip*, 2008, **8**(5): p. 810–813.
114. Ke, C. et al., Single step cell lysis/PCR detection of *Escherichia coli* in an independently controllable silicon microreactor. *Sens Actuators B*, 2007, **120**(2): p. 538–544.
115. Yeung, S-W. et al., A DNA biochip for on-the-spot multiplexed pathogen identification. *Nucleic Acids Res*, **34**(18): p. E118.
116. Christel, L.A. et al., Rapid, automated nucleic acid probe assays using silicon microstructures for nucleic acid concentration. *J Biomech Eng*, 1999, **121**(1): p. 22–27.
117. Bhattacharyya, A., Klapperich, C.M., Thermoplastic microfluidic device for on-chip purification of nucleic acids for disposable diagnostics. *Anal Chem*, 2005, **78**(3): p. 788–792.
118. Tian, H., Hühmer, A.F.R., Landers, J.P., Evaluation of silica resins for direct and efficient extraction of DNA from complex biological matrices in a miniaturized format. *Anal Biochem*, 2000, **283**(2): p. 175–191.
119. Breadmore, M.C. et al., Microchip-based purification of DNA from biological samples. *Anal Chem*, 2003, **75**(8): p. 1880–1886.
120. Wolfe, K.A. et al., Toward a microchip-based solid-phase extraction method for isolation of nucleic acids. *Electrophoresis*, 2002, **23**(5): p. 727–733.

121. Wen, J. et al., DNA extraction using a tetramethyl orthosilicate-grafted photopolymerized monolithic solid phase. *Anal Chem*, 2006, **78**(5): p. 1673–1681.
122. Cady, N.C., Stelick, S., Batt, C.A., Nucleic acid purification using microfabricated silicon structures. *Biosens Bioelectron*, 2003, **19**(1): p. 59–66.
123. Cao, W. et al., Chitosan as a polymer for pH-induced DNA capture in a totally aqueous system. *Anal Chem*, 2006, **78**(20): p. 7222–7228.
124. Witek, M.A. et al., 96-well polycarbonate-based microfluidic titer plate for high-throughput purification of DNA and RNA. *Anal Chem*, 2008, **80**(9): p. 3483–3491.
125. Witek, M.A. et al., Purification and preconcentration of genomic DNA from whole cell lysates using photoactivated polycarbonate (PPC) microfluidic chips. *Nucleic Acids Res*, **34**(10): p. E74.
126. Elgort, M.G. et al., Extraction and amplification of genomic DNA from human blood on nanoporous aluminum oxide membranes. *Clin Chem*, 2004, **50**(10): p. 1817–1819.
127. Kim, J., Gale, B.K., Quantitative and qualitative analysis of a microfluidic DNA extraction system using a nanoporous AlO membrane. *Lab Chip*, 2008, **8**(9): p. 1516–1523.
128. Kim, J. et al., A PCR reactor with an integrated alumina membrane for nucleic acid isolation. *Analyst*, **135**(9): p. 2408–2414.
129. Lee, J-G. et al., Microchip-based one step DNA extraction and real-time PCR in one chamber for rapid pathogen identification. *Lab Chip*, 2006, **6**(7): p. 886–895.
130. Hawkins, T.L. et al., DNA purification and isolation using a solid-phase. *Nucleic Acids Res*, 1994, **22**(21): p. 4543–4544.
131. Ramadan, Q. et al., Magnetic-based microfluidic platform for biomolecular separation. *Biomed Microdevices*, 2006, **8**(2): p. 151–158.
132. Yeung, S.W., Hsing, I.M., Manipulation and extraction of genomic DNA from cell lysate by functionalized magnetic particles for lab on a chip applications. *Biosens Bioelectron*, 2006, **21**(7): p. 989–997.
133. Duarte, G.R.M. et al., Characterization of dynamic solid phase DNA extraction from blood with magnetically controlled silica beads. *Analyst*, **135**(3): p. 531–537.
134. Kim, J. et al., Microfluidic sample preparation: Cell lysis and nucleic acid purification. *Integr Biol*, 2009, **1**(10): p. 574–586.
135. Price, C.W., Leslie, D.C., Landers, J.P., Nucleic acid extraction techniques and application to the microchip. *Lab Chip*, 2009, **9**(17): p. 2484–2494.
136. Jiang, G., Harrison, D.J., mRNA isolation in a microfluidic device for eventual integration of cDNA library construction. *Analyst*, 2000, **125**(12): p. 2176–2179.
137. Lee, H. et al., High-speed RNA microextraction technology using magnetic oligo-dT beads and lateral magnetophoresis. *Lab Chip*, **10**(20): p. 2764–2770.
138. Bhattacharyya, A., Klapperich, C.M., Microfluidics-based extraction of viral RNA from infected mammalian cells for disposable molecular diagnostics. *Sens Actuators B*, 2008, **129**(2): p. 693–698.
139. Hagan, K.A. et al., An integrated, valveless system for microfluidic purification and reverse transcription-PCR amplification of RNA for detection of infectious agents. *Lab Chip*, 2011, **11**(5): p. 957–961.
140. Satterfield, B.C. et al., Microfluidic purification and preconcentration of mRNA by flow-through polymeric monolith. *Anal Chem*, 2007, **79**(16): p. 6230–6235.
141. Vulto, P. et al., A microfluidic approach for high efficiency extraction of low molecular weight RNA. *Lab Chip*, 2010, **10**(5): p. 610–616.

142. Gillers, S. et al., Microscale sample preparation for PCR of C. difficile infected stool. *J Microbiol Methods*, 2009, 78(2): p. 203–207.
143. Sauer-Budge, A.F. et al., Low cost and manufacturable complete microTAS for detecting bacteria. *Lab Chip*, 2009, 9(19): p. 2803–2810.
144. Wen, J. et al., Microfluidic-based DNA purification in a two-stage, dual-phase microchip containing a reversed-phase and a photopolymerized monolith. *Anal Chem*, 2007, 79(16): p. 6135–6142.
145. Manz, A., Graber, N., Widmer, H.M., Miniaturized total chemical analysis systems: A novel concept for chemical sensing. *Sens Actuators B*, 1990, 1(1–6): p. 244–248.
146. Tan, A., Benetton, S., Henion, J.D., Chip-based solid-phase extraction pretreatment for direct electrospray mass spectrometry analysis using an array of monolithic columns in a polymeric substrate. *Anal Chem*, 2003, 75: p. 5504.
147. Bhattacharyya, A., Klapperich, C.M., Mechanical and chemical analysis of plasma and ultraviolet-ozone surface treatments for thermal bonding of polymeric microfluidic devices. *Lab Chip*, 2007, 7(7): p. 876–882.
148. Schneegass, I., Brautigam, R., Kohler, J.M., Miniaturized flow-through PCR with different template types in a silicon chip thermocycler. *Lab Chip*, 2001, 1(1): p. 42–49.
149. Hashimoto, M. et al., Rapid PCR in a continuous flow device. *Lab Chip*, 2004, 4(6): p. 638–645.
150. Auroux, P.A. et al., Miniaturised nucleic acid analysis. *Lab Chip*, 2004, 4(6): p. 534–546.
151. Pipper, J. et al., Clockwork PCR including sample preparation. *Angew Chem Int Edn*, 2008, 47(21): p. 3900–3904.
152. Easley, C.J., Karlinsey, J.M., Landers, J.P., On-chip pressure injection for integration of infrared-mediated DNA amplification with electrophoretic separation. *Lab Chip*, 2006, 6(5): p. 601–610.
153. Easley, C.J. et al., A fully integrated microfluidic genetic analysis system with sample-in-answer-out capability. *Proc Natl Acad Sci*, 2006, 103(51): p. 19272–19277.
154. Lee, J-G. et al., Microchip-based one step DNA extraction and real-time PCR in one chamber for rapid pathogen identification. *Lab Chip*, 2006, 6(7): p. 886–895.
155. Legendre, L.A. et al., A simple, valveless microfluidic sample preparation device for extraction and amplification of DNA from nanoliter-volume samples. *Anal Chem*, 2006, 78(5): p. 1444–1451.
156. Pipper, J. et al., Catching bird flu in a droplet. *Nat Med*, 2007, 13(10): p. 1259–1263.
157. Zhang, Y. et al., An all-in-one microfluidic device for parallel DNA extraction and gene analysis. *Biomed Microdevices*, 2010, 12(6): p. 1043–1049.
158. Cao, Q., Mahalanabis, M., Klapperich, C.M., An integrated plastic chip for PCR detection of influenza A. In: *42nd Annual AACC Oakridge Conference*, San Jose, CA, 2005.
159. Koh, C.G. et al., Integrating polymerase chain reaction, valving, and electrophoresis in a plastic device for bacterial detection. *Anal Chem*, 2003, 75(17): p. 4591–4598.
160. Hashimoto, M., Barany, F., Soper, S.A., Polymerase chain reaction/ligase detection reaction/hybridization assays using flow-through microfluidic devices for the detection of low-abundant DNA point mutations. *Biosen Bioelectron*, 2006, 21(10): p. 1915–1923.
161. Liu, R.H. et al., Self-contained, fully integrated biochip for sample preparation, polymerase chain reaction amplification, and DNA microarray detection. *Anal Chem*, 2004, 76(7): p. 1824–1831.

Chapter 8

Microfluidic Devices for Cellular Proteomic Studies

Yihong Zhan* and Chang Lu[†,‡]

*Birck Nanotechnology Center, Purdue University
West Lafayette, IN 47907, USA
†Department of Chemical Engineering, Virginia Tech
Blacksburg, VA 24061, USA

Introduction

The 23,000 human protein-coding genes introduce a large number of functional proteins due to alternative splicing and post-translational modifications. With the large progress made in the sequencing of the human genome, lots of efforts have been made to understand the complex biological processes, which generally involve the expression and interaction of multiple genes and proteins. Among this, the identification and quantification of various proteins constitute a critical procedure. In the following step, monitoring protein existence, quantifying protein expression levels, and detecting the dynamics in protein microheterogeneity comprise the major part of the detection and diagnosis of a wide range of diseases.[1,2]

The application of microfluidics technology in the field of molecular biology began with genomic studies. Two of the most widely used microfluidic tools include the DNA microarray[3–9] and capillary electrophoresis (CE) microchips.[10–12] The key success factors of applying microfluidics in these two areas are summarized to be the parallelization of hundreds of reactions on the same chip due to the advantages of the micrometer scale and precise control

---

[‡]Corresponding author. Email: changlu@vt.edu

of fluid and molecule movement in the microchannels. Compared to genomics, however, proteomics introduces certain challenges. First, sensitivity is very important for protein detection. Amplification is unsuitable for proteins as the polymerase chain reaction (PCR) is used for DNA.[13] The large variations in protein sizes and the complexity of protein mixtures are far beyond the current analytical capabilities. Some key characteristics of microfluidic systems make it promising for proteomics studies, including a short analysis time and a high surface-to-volume ratio, which is favorable for surface modifications.

In this chapter, we will limit our discussion to microfluidic devices with integrated cell pretreatment (starting from intact cells). There has been a tremendous amount of work done using microfluidic devices to conduct analysis based on cell lysates extracted off chip,[14–18] most notably utilizing mass spectrometry[19–25] and CE.[11,26–28]

This chapter describes the use of microfluidic systems in the proteomics field. We will focus on the area of cellular proteomics, since protein activity and characteristics in the cellular environment will more accurately reflect its role in realizing cellular functions. Several areas about the process will be discussed in the later parts, including trapped cells for time-lapse studies, cell culture and co-culture, flow cytometry, analysis after lysis, and some special analysis techniques.

Trapped Cells for Time-Lapse Studies

On-chip real-time detection of cellular signaling events is important for collecting information for studying signal transduction pathways. James et al.[29] developed a microfluidic device for real-time imaging of host–pathogen interactions. GFP–RelA translocates from the cytoplasm to the nucleus upon the activation of Toll-like receptor 4 (TLR4). The work imaged the host cell infection on the microfluidic device and quantitatively analyzed the initial 30-min of immune signaling events. The former methods for studying cell infection only provide an averaged values on the population level, which could not accurately reflect the real response since the cells in a population experience different initial physiological states and have different infection degrees.[30,31] Host cells are trapped in gaps formed by the channel's closely spaced triangular structures (Fig. 1), which reduces trap-to-trap fluid flow, minimizing chemical communication between cells.

Most conventional biochemical assays are performed based on a large number of cells, which only provide information in the analyzed population and often ignore the important information regarding their fluctuations among individual cells.[32,33] The ability to distinguish among individual cells is critical for studying several biological events such as cell differentiation. Thus, it is necessary to

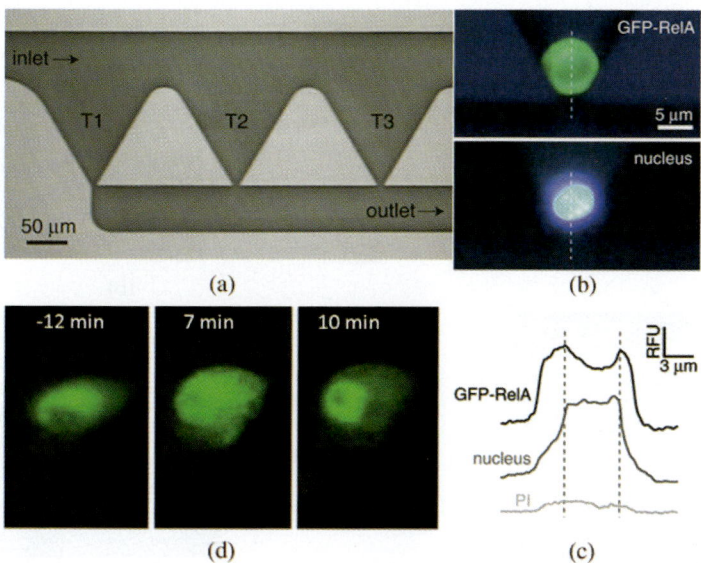

Fig. 1. (a) Bright-field image of the imaging chamber and three cell traps for trapping single cell separately. (b) Single cell captured showing the GFP–RelA and nucleus signals in the cytoplasm and nucleus, respectively. (c) Intensity line-plots at the position shown in (b) for GFP–RelA, the nucleus, and the PI signal. The location of the nucleus is noted using dashed lines. (d) GFP–RelA translocation progress with time. (Reproduced from Ref. 29 with permission from Springer Science + Business Media, LLC.)

compare the biochemical parameters of multiple single cells in parallel to obtain knowledge about genetic heterogeneity and heterotypic biological activities.[34] Sasuga et al.[35] described a simple microwell-based single-cell lysis method for the detection of proteins by antibody-conjugated microbeads. This goal was realized by three steps. First, cells were trapped in an array of picoliter-scale microwells; second, cells experienced chemical lysis in the closed microwell on a single-cell basis; finally, immunosignals were detected by a fluorescent microscope.

Figure 2 shows the overall procedures of the microwell-based, single-cell lysis method. Single-cell analysis was realized by confining each cell in the microwell and pressing the polydimethylsiloxane (PDMS) sheet for the closing/opening rounds. During the lysis step, the concentration of the lysis reagent was optimized to induce almost complete lysis while keeping the proper cell holding rate. Figures 2(b) and 2(c) demonstrate the application for the detection of a specific protein in a single cell by integrating the single-well lysis method with an antibody–microbead array reported previously.[36] Results in the paper also concluded that the protein production profiles vary widely from cell to cell.

Comparing with single-cell analysis, the high-throughput imaging system is also important. Many biomolecular networks exist in cells that process and

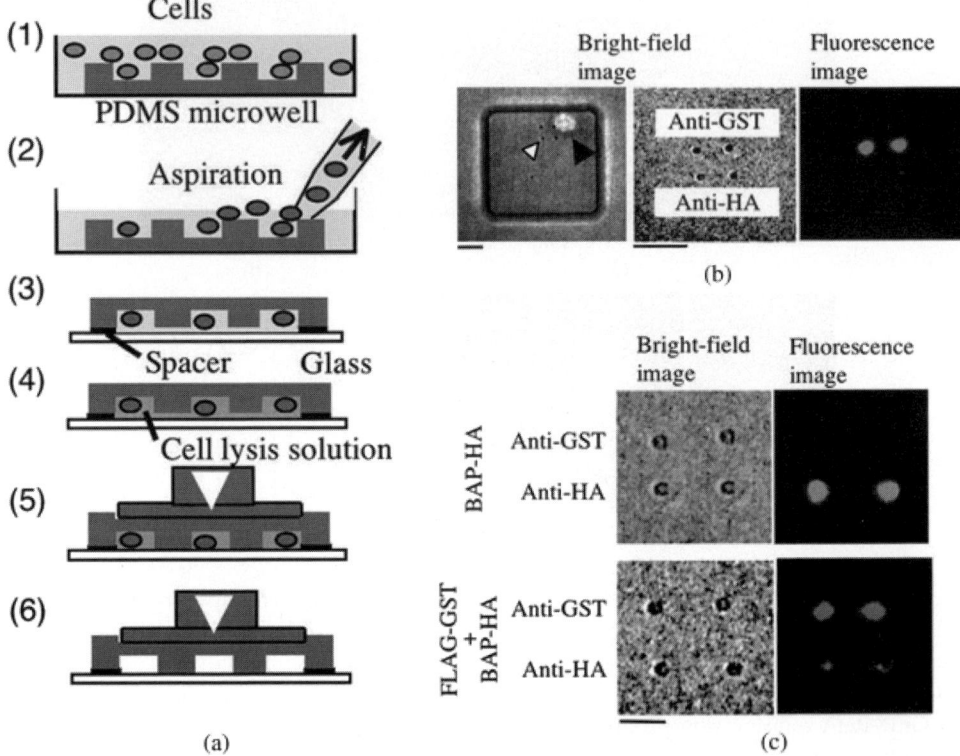

Fig. 2. (a) Schematic drawing of single-cell lysis using microwell array procedures: (1) Cells deposited onto the microwell array; (2) excess cells removed by aspiration; (3) the PDMS sheet inverted to form a flow cell with a bottom cover slip; (4) cell lysis reagent applied to the flow cell; (5) the PDMS sheet pressed against the bottom cover slip to close the open side of the microwell; (6) cells gradually lysed in the closed microwells. (b) Detection of FLAG–GST produced by a single PC12 cell. Left and middle panels show the bright-field image, while the right panel is a corresponding Cy3-fluorescence image. Cy3-fluorescence corresponded to the specific detection signal of FLAG–GST. Black and white arrowheads point to the trapped single cell and immobilized microbeads, respectively. (c) Analyses of the production of BAP–HA and coproduction of FLAG–GST and BAP–HA: left panel, bright-field image; right panel, Cy3-fluorescence image. (Reproduced with permission from Ref. 35. Copyright 2008, American Chemical Society.)

respond to dynamic chemical environments. It is important to understand how protein interactions affect biomolecular network properties. This requires the time-lapse analysis of cellular responses to many different genetic perturbations and chemical environments.[37] Taylor et al.[38] developed a high-throughput imaging system to trap single cells for studying network responses.

They used this system to investigate cellular memory by running >3,000 experiments to study the effect of gene deletions and changing stimulant conditions on mating response. As shown in Fig. 3, eight yeast strains were

Microfluidic Devices for Cellular Proteomic Studies 165

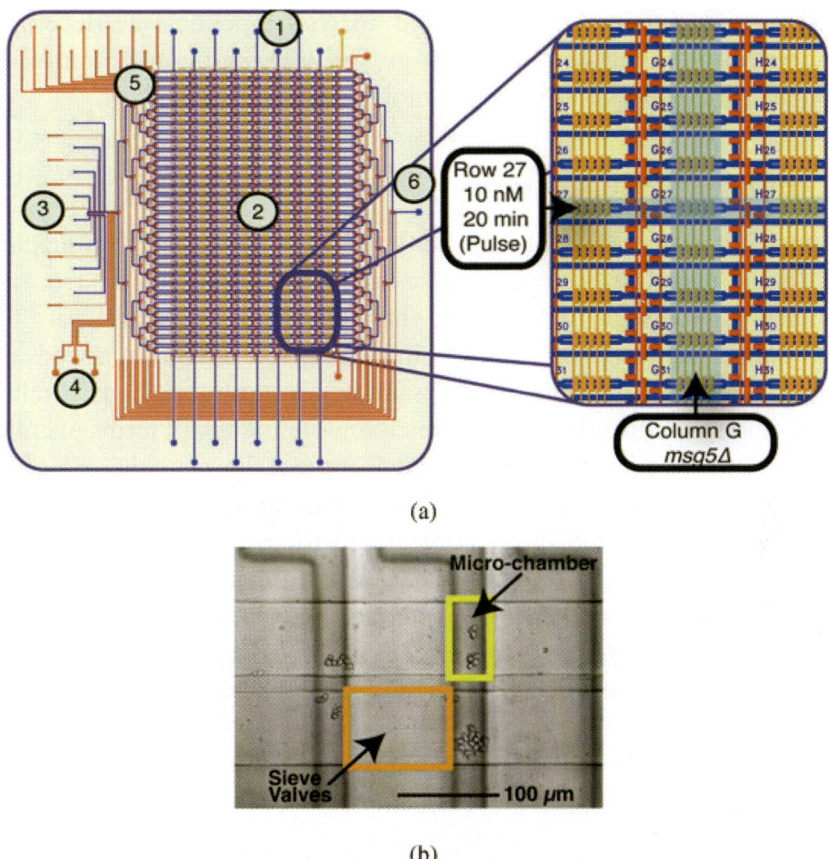

Fig. 3. Schematic of the microfluidic device for high-throughput dynamic analysis of MAPK signaling. (a) Layout of live-cell imaging matrix. Device features two layers of channels including a flow structure (blue) for introducing cells and reagents, and a control structure (red) for pneumatic valves. Regions are indicated by (1) cell-loading ports, (2) experiment matrix, (3) chemical inputs and control, (4) peristaltic pump, (5) fluidic multiplexer, and (6) waste outlet. Each column corresponded to a single yeast genotype while each row corresponded to a single experimental condition. (b) Differential interference contrast image of yeast cells in perfusion chambers. (Reproduced from Ref. 38. Copyright 2009, National Academy of Sciences, USA.)

tested against 32 different stimulant concentrations yielding a total of 256 simultaneous experiments. Genetic variations were created along the matrix columns, while the chemical variations were created along the rows. Valves were utilized to confine each yeast strain in a specific column, eliminating cross-contamination. Besides that, each cell trap contained a partially closed valve, allowing media exchange while retaining the cells. Throughout the whole experiment (12.5 h), high-resolution bright-field and fluorescence images of all 256 chambers were taken with a 15-min time resolution, generating >50,000 images. An image analysis pipeline was developed to record the cell number, cell morphology, cell size, and concentration of the fluorescent gene expression reporter molecule.

Cell Culture and Co-Culture

In the past few years, numerous studies have been conducted to utilize microfluidic devices to culture cells in diverse microenvironments.[39–41] The successful culturing of cells on microfluidic chips constitutes a crucial part for the screening of cellular responses in proteomics studies, arising from the advantages of microfluidic chips in terms of low-sample consumption, high parallelizing capability, and well-controlled culture microenvironments. In terms of culturing, cells respond to variations in the microenvironment. Particularly, cell–cell interactions can regulate the individual cell's fate and function. As cell–cell interactions occur mainly through direct contact and the exchange of soluble factors, by accurately positioning different cells, cell–cell interactions can be controlled. Hui et al.[42] developed a dynamic microfluidic platform to study cell–cell interactions between hepatocytes and stromal cells in the co-culture state. As is known, in vitro culturing, hepatocyte viability, and liver-specific functions need to be maintained by co-culturing with many different mesenchymal cell types, such as endothelial cells and fibroblasts.[43] The amount of liver-specific functions cells retained in vitro is decided by the interactions between the two cell types.[44]

As shown in Fig. 4, micromechanical reconfigurable cultures can be used to dynamically regulate cell–cell interactions through directly manipulating cell position. Generally, contact between different cell populations is controlled by positioning silicon parts together or apart. By imposing a small micrometer-scale distance between parts, there is no cell–cell contact while soluble signaling is maintained. If larger separation distances are applied, soluble signaling can also be eliminated. Also, different populations of cells can be exchanged by replacing plates.

Biochemical and biomechanical factors have been demonstrated to control endothelial cell migration and tube formation.[45–47] A three-dimensional (3D) environment with precise gradient and flow control for cell migration and capillary morphogenesis has been studied by many labs.[41,48,49] Chung et al.[50] recently demonstrated a new 3D microfluidic platform for controlling the biochemical and biomechanical forces within a 3D scaffold device. They used this system to monitor the endothelial cell response and quantify migration over time when the cells are co-cultured with physiologically relevant cell types, such as cancer cells and smooth muscle cells. This system was also used to study the cellular response to a growth factor gradient.

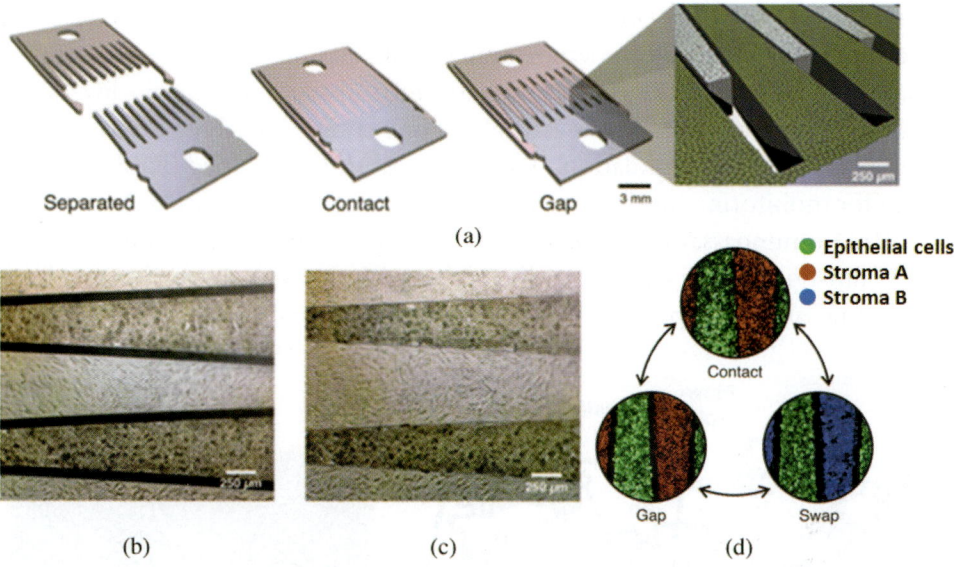

Fig. 4. Microfluidic device for micrometer resolution cell positioning using micromechanical substrates. (a) Different operation mode for positioning cells: fully separated (left), locked together with comb fingers in contact (center), or slightly separated (right). Cells are cultured on the top surfaces; manual scraping can be used to restrict cells to the comb fingers only (inset). The slope of the tapered comb fingers results in a 20:1 mechanical transmission ratio; (b) and (c) bright-field images of hepatocytes (darker cells) and 3T3 fibroblasts. (d) Reconfigurable cell culture. Cultures can be reversibly switched to initiate or to eliminate contact between different cell populations; individual populations can also be removed and replaced. Each cell type was prelabeled with an individual dye color. (Reproduced from Ref. 42. Copyright 2007, National Academy of Sciences, USA.)

In the system, cells were seeded in the central channel, contacting the collagen scaffolds on both sides. Cells would migrate outward in response to the biochemical and mechanical factors from the scaffold. The featured three-channel design facilitated the control and condition experiments to be simultaneously performed in the same device. In terms of co-culturing with different cell lines, with the MTLn3 cancer cell line, human dermal microvascular endothelial cell (HMVEC) migration is faster on the condition side than on the control side; with the U87MG cancer cell line, there is similar migration on both the condition and control sides. While for 10T 1/2 smooth muscle cells, HMVEC migration only occurs on the control side. This microfluidic system was shown to be an effective tool, providing a well-controlled culture microenvironment in mimicking *in vivo* physiological conditions.

Cell culturing on chip is convenient for mimicking cellular response *in vivo* for better understanding and evaluation of cell functions. The secretion of

168 *Microfluidic Technologies for Human Health*

insulin from pancreatic inlets employs complex kinetics.[51,52] The different secretion patterns have been demonstrated to be an important indicator of islet function and therefore, it is crucial to monitor insulin secretion in live islets. Single-islet pattern detection is crucial for preventing interference from other islets in a large cell population. Dishinger *et al.*[53] developed a microfluidic device for monitoring insulin secretion from individual islets with a throughput of 5,400 immunoassays per hour.

As illustrated in Fig. 5, the chip contains 15 channel networks capable of processing samples in parallel. In each unit, an inlet is placed in a chamber and

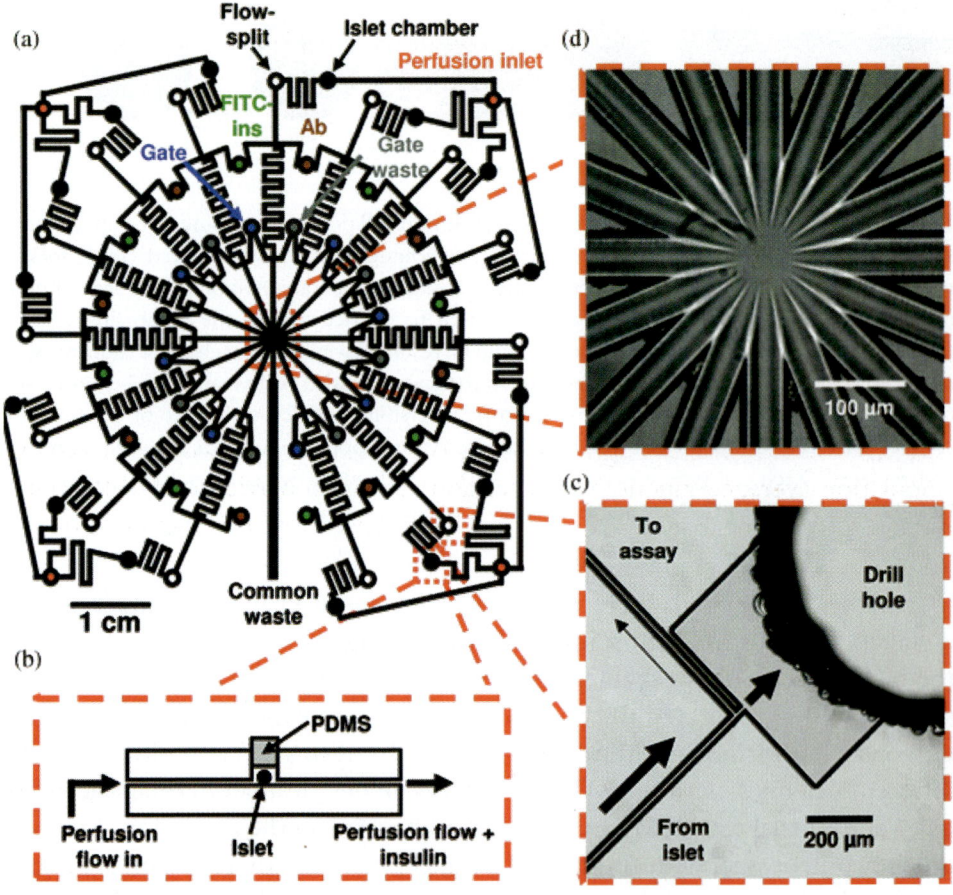

Fig. 5. Microfluidic chip for monitoring insulin secretion from islets. (a) Device's channel network. Channels are indicated by solid black lines. (b) Side view of an islet perfusion chamber. (c) CCD image of an on-chip flow-split with arrows indicating direction and estimated magnitude of flow. (d) Bright-field image of the detection area with a radial arrangement of channels for high-throughput assay. (Reproduced with permission from Ref. 53. Copyright 2009, American Chemical Society.)

superfused with media. The insulin secreted from the inlet is continuously sampled by electroosmotic flow. A flow-split is applied to control only a small part of the islet superfusate going to the assay region, where the secreted insulin is mixed with the immunoassay reagents (FITC-insulin and Ab) for a competitive binding reaction. The system is capable of obtaining dynamic information of inlet secretion on the single-inlet level with high-throughput, and is promising for application in evaluating islets for transplantation purposes.

Flow Cytometry

Electroporative flow cytometry (EFC) combines both advantages from electroporation and flow cytometry. Electroporation is applied in EFC as a tool to breach the cell membrane for probing intracellular molecules.[54–56] Flow cytometry provides single-cell resolution by sending cells one by one through a detection point in a carrier flow,[57,58] which facilitates high-throughput detection.

Nucleocytoplasmic transport is vital for many cellular functions. The kinetics study has been hindered by the problem of a lack of quantitative tools for measuring the nuclear and cytosolic fractions of an intracellular protein at the single-cell level for a large cell population. Wang et al.[59] developed a novel microfluidic electroporative flow cytometry method to study the kinetics of the nucleocytoplasmic transport of the transcription factor nuclear factor kappa B (NF-κB). Based on the single-cell data, they quantitatively characterized the population-averaged kinetic parameter and activation barrier for NF-κB transport. The channel design is shown in Fig. 6. Cells flow through the channel with a narrow middle section under hydrodynamic focusing. A DC power supply generates a constant high electric field exclusively in the narrow section for electroporating cells.[60] Electroporation triggers many nanoscale pores on the cell's plasma membrane, which allow intracellular molecules to be released into the surrounding solution.[61] The important feature of this design is that the released proteins (including GFP-labeled NF-κB) move into the upper reservoir driven by the electric field without interfering with the laser-induced fluorescence detection, while the remaining single cells flow to the downstream laser detection point and generate signals collected by a photomultiplier (PMT).

Due to the intensive electroporation used in the microfluidic EFC method, the cells lose their viability and biological function after the detection procedure. In order to develop a non-destructive method to detect protein translocation, Wang et al.[62] used a single-cell tool, microfluidic total internal reflection fluorescence flow cytometry (TIRF-FC) to detect both

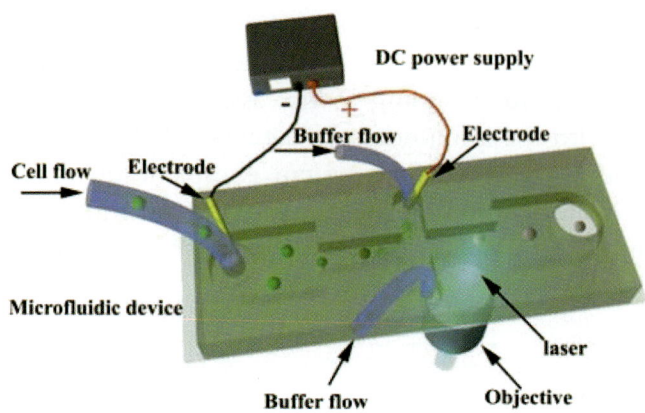

Fig. 6. Schematic of the microfluidic electroporative flow cytometry (EFC) device. The microfluidic device is mounted on a microscope's stage to allow incident laser from a 40× objective to focus in the channel to detect the fluorescent protein release. A DC power supply generates a constant high electric field exclusively in the narrow section for electroporating cells and release intracellular proteins. (Reproduced with permission from Ref. 59. Copyright 2010, Royal Society of Chemistry.)

cytosol-to-plasma membrane and cytosol-to-nucleus translocation using the tyrosine kinase Syk and NF-κB as models. The technique combined the advantages of high-throughput, single-cell analysis from microfluidics and sensitive, high signal-to-noise ratio from TIRF microscopy. This technique was able to quantitatively capture changes in the fluorescence density in the evanescent field associated with the translocation process for a large cell population with single-cell resolution.

As shown in the Figs. 7(a) and 7(b), the central part of the device employed a partially closed elastomeric two-layer valve[63] to force flowing cells to "squeeze through" a constriction with a depth smaller than the cell diameter. In this way, the part of the cell that is in the vicinity of the membrane is subjected to the evanescent field illumination and the fluorescence intensity from this region is recorded by a PMT. While passing through the constriction, clogging easily occurs due to cells being elastomeric and viscous. To solve this problem, a valve control circuit is used to respond promptly to clogging events by the rapid opening of the valve to remove the clogged cells. This device can reach a speed of ~200–300 cells/s for screening. As is well known, photobleaching is a critical limitation to quantitative time-lapse analysis of protein translocation by fluorescence imaging. TIRF-FC overcomes this problem by subjecting flowing cells only briefly to laser illumination. Additionally, the fluorescence of a much larger cell population is examined by TIRF-FC to reflect the characteristics of the population

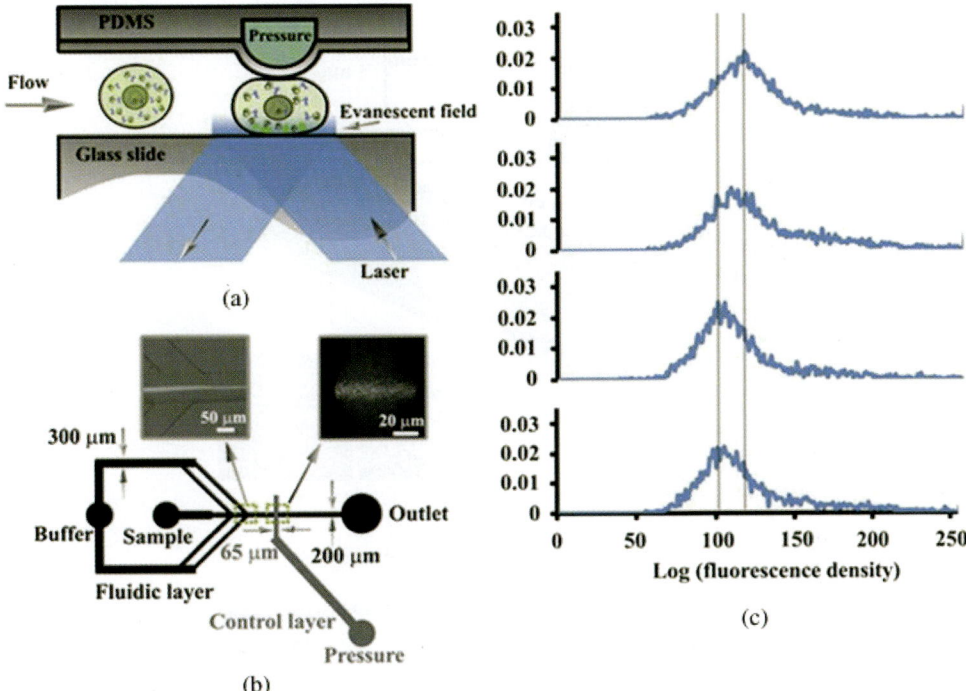

Fig. 7. (a) Schematic of microfluidic TIRF-FC device with an elastomeric valve, which forces flowing cells into the evanescent field for TIRF detection. (b) Microfluidic channel design. The buffer flow enters from one inlet and is then symmetrically divided into two pairs of streams for hydrodynamic focusing of the sample flow containing cells. Left inset: the hydrodynamic focusing. Right inset: TIRFM image of fluorescent trail detected by the evanescent wave illumination. (c) Nucleocytoplasmic translocation of NF-κB studied by microfluidic TIRF-FC. Histograms of fluorescence density collected from populations of ~5,000 cells at 0, 10, 60, and 120 min after stimulation by IL-1β at 22°C. (Reproduced with permission from Ref. 62. Copyright 2010, Royal Society of Chemistry.)

distribution. Figure 7(c) shows the histogram of the fluorescence density, which decreased substantially with the increased translocation of NF-κB into the nucleus as a function of time.

Most of the traditional biochemical analysis focus on the expression level of proteins from entire cells. However, the information from the protein's subcellular location has been gaining interest recently. Traditional methods for analyzing subcellular proteins include fractionation, which involves cell lysis and high-speed centrifugation. These steps are labor-intensive and not suitable for processing scarce cell samples. To overcome these problems, Zhan et al.[64] applied microfluidic flow-through electroporation to breach cell membranes and extract cytosolic proteins selectively in a single step. This approach

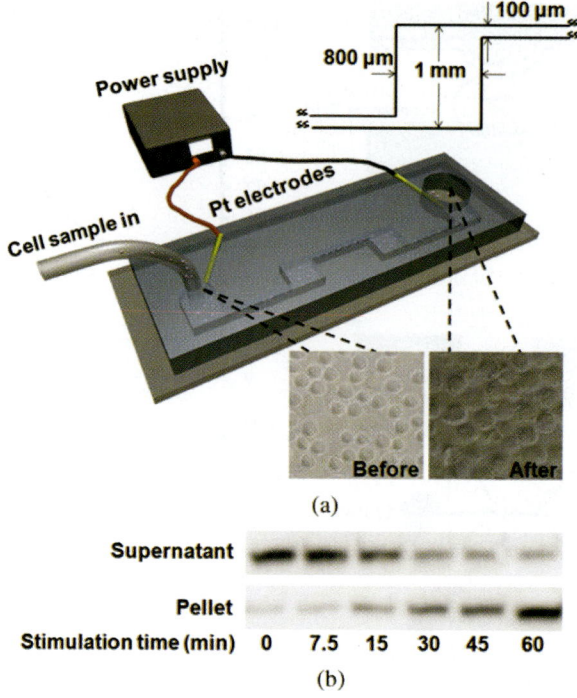

Fig. 8. (a) The layout of the electroporation device used for the selective release of intracellular proteins. The geometry of the wide sections is shown in the inset image. Each narrow section is 2.8-mm long and the channel has a depth of 60 μm. The inset images show that cells are mostly in one piece after electroporation. (b) Tracking of NF-κB translocation from the cytosol to the nucleus over time. The intracellular proteins were extracted by flow-through electroporation and the supernatant and pellet fractions were analyzed by Western blotting after stimulation of CHO/GFP–NFκBp65 cells by IL-1β for different periods of time. (Reproduced with permission from Ref. 64. Copyright 2010, Royal Society of Chemistry.)

has been successfully demonstrated to monitor the translocation of NF-κB from the cytosol to the nucleus without fractionation.

Figure 8 shows the flow-through electroporation device used in this study. The cell sample flowed through a microfluidic channel with alternating wide (~800-μm wide) and narrow (~100-μm wide) sections, while a constant DC voltage was established across the channel. Electroporation occurs exclusively in the narrow sections due to the significantly higher field intensity there. The supernatant and pellet fractions from the solution after electroporation were examined using SDS-PAGE and Western blotting analysis. The supernatant contained the intracellular proteins released into the solution by electroporation, while the pellet contained the cellular "remains." The experimental data showed that flow-through electroporation provides a significant differentiation in terms of its extraction for cytosolic and nuclear proteins, and could be a promising tool for bioanalytical analysis and spatial proteomics.

Enzymes, which control cytochemical balance and cell proliferation activity, are important biological components in cells. Electrochemical detection (ECD)[65,66] is an efficient tool for enzyme analysis. The strategy for determining the single cell's enzymatic activity is to detect electroactive products of the substrates of enzyme-catalyzed reactions. Gao et al.[67] developed a novel high-throughput method for analyzing enzymatic activity with single-cell resolution and without cytolysis, by combining chemical cell perforation and an intracellular enzyme-catalyzed reaction.

As shown in Fig. 9, neutrophils were first perforated with digitonin to form micropores on the cell membrane. After perforation, electrokinetic loading of a single cell was performed immediately to move the cell by electrokinetic flow. Cell movement was controlled by adjusting the liquid level in the buffer reservoir. Thereafter, the cell was incubated with the substrates for 5 min (c), forming a BQ zone around the cell (d) (PO converted H_2Q into BQ). An electric field of 100 V/cm was utilized to separate the cell and the BQ zone. The second BQ zone was formed by incubating substrates with released PO from rapidly lysed cells. Both BQ zones were electromigrated to the outlet for detection by applying another electric field of 100 V/cm. The concentration ratio of BQ with and without cytolysis could be determined from the electrophoretic peak areas corresponding to the two BQ zones respectively. This method can reach a detection rate of >1 cell/min. The advantage of this system is that the biological molecules remain inside the cells instead of adsorbing on the capillary wall or on the working electrode, and hence do not interfere with the following detection. Through this way, high-throughput analysis can be realized by eliminating the step of microchannel treatment between runs.

Analysis after Lysis

Most of the cell-based assays produce averaged data over large populations of cells, which will sometimes bury important information from individual cells. Individual cells may differ in many characteristics even though they seem identical in appearance.[32,68,69] Determining the individual cell's chemical composition and monitoring how the composition changes will affect the cell's response to the external environment, which is crucial in understanding basic cellular functions. Over the past few years, chemical cytometry has been demonstrated to be an efficient tool for analyzing single cells.[70,71] The most common among these techniques is CE, which has been demonstrated to

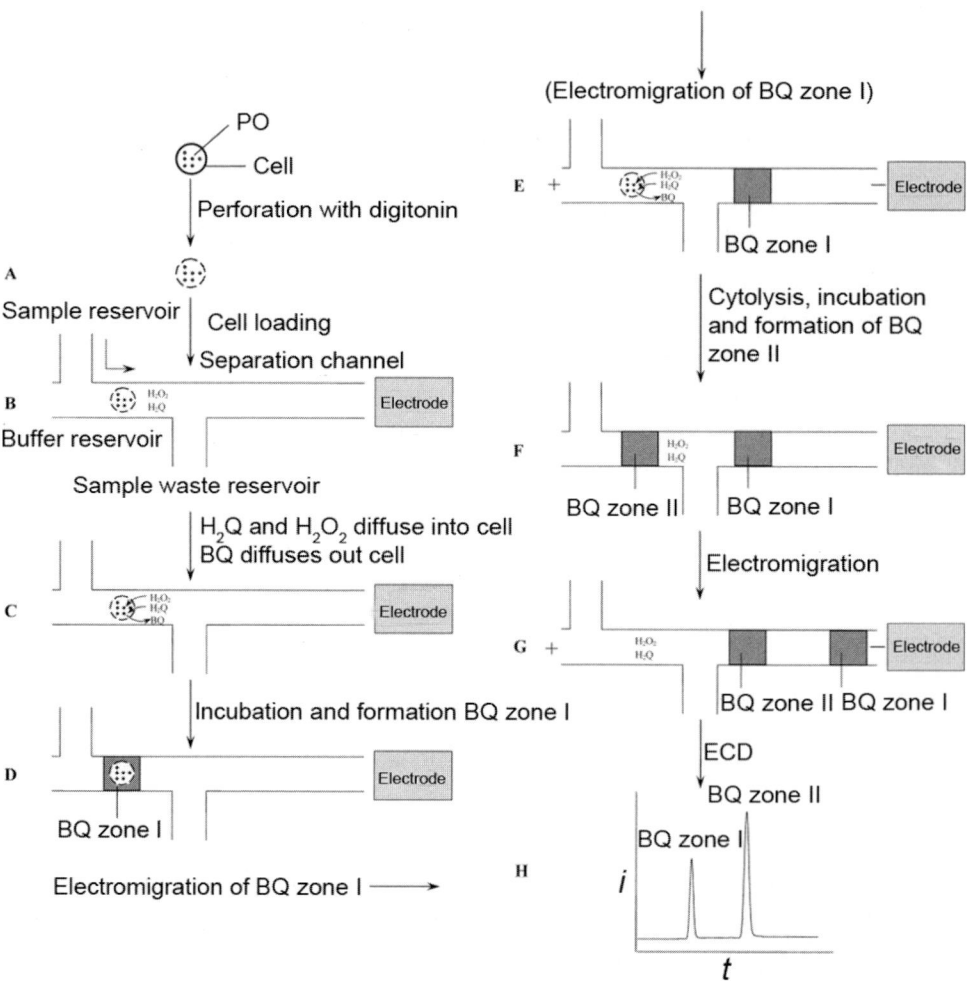

Fig. 9. Schematic diagram showing the process of microfluidic chip-based electrochemical detection of PO activity in single cells without and with cytolysis. (Reproduced with permission from Ref. 67. Copyright 2006, American Chemical Society.)

analyze various biomolecules.[72–75] The main drawback of CE is its relatively low throughput.[76,77] Integrated microfluidic platforms have been explored to improve CE performance, which uses sequential steps to acquire fast lysis and accurate on-chip analysis in a high-throughput manner. The key strategy for improving the throughput is by integrating and automating several cell handling and cell processing steps, followed by reducing the separation running time.

Among the multiple operations that need to be integrated, cell lysis is especially crucial. Enzymatic reactions usually occur in short time scales resulting in the concentrations of interested analytes changing by one order of magnitude in <1 s.[78] To accurately measure this concentration change, the cells must be rapidly lysed to stop the enzymatic reactions. Chemical lysis using surfactant and hypotonic media or mechanical lysis using sonic disruption is too slow for the influx of ions from outside the buffer to activate the cellular reactions and can thus change the concentrations of the analytes of interest. Electrical lysis is an efficient tool to rapidly destroy the cell membrane and in the meantime, dilute and denature the enzymes to stop enzymatic activity.

McClain et al.[55] demonstrated a microfluidic device for high-throughput chemical analysis of single cells by integrating cell handling, rapid cell lysis, efficient electrophoretic separation, and high sensitivity detection of fluorescent cytosolic dyes as shown in Fig. 10. This system has been used to rapidly lyse the cells pre-loaded with fluorogenic dyes with a high-voltage AC electric field. The dyes released from the cells were separated in a separation column by electrophoresis in less than 2.2 s. This high-throughput operation assures accurate measurements of the concentration change of the interested biomolecules, which can be used to detect the uncommon subpopulations of large cell populations.

Jiang et al.[79] developed a simple automated CE platform with buffer-switching and cell lysis features for rapidly analyzing adherent cell contents in a serial manner as shown in Fig. 11. The system employs a simple open, two-channel flow design, with one channel filled with a physiological buffer and the other channel filled with an electrophoretic buffer. The whole system is mounted on a microscope's motorized stage for fast and accurately control of the movement by the software. The flow rates of the two different buffers are carefully adjusted to prevent the electrophoretic buffer from contacting the cells in the microfabricated cell microwells. The individual cells are rapidly lysed by a laser[80] without affecting other nearby cells for realizing single-cell study. Then, the cell contents are loaded into the capillary, followed by moving to the channel containing the electrophoretic buffer for electrophoresis separation and detection. Following that, the chamber is repositioned for the next single-cell lysis and detection. As shown in Fig. 11, two dyes pre-loaded to the cells were clearly identified without any peak overlap and in the meantime, there were no overlaps between any two successive cells.

Other Analytical Techniques

Microfluidic techniques significantly reduce the costs due to reduced reagent and sample amounts, but are limited by the lack of suitable readouts.[81]

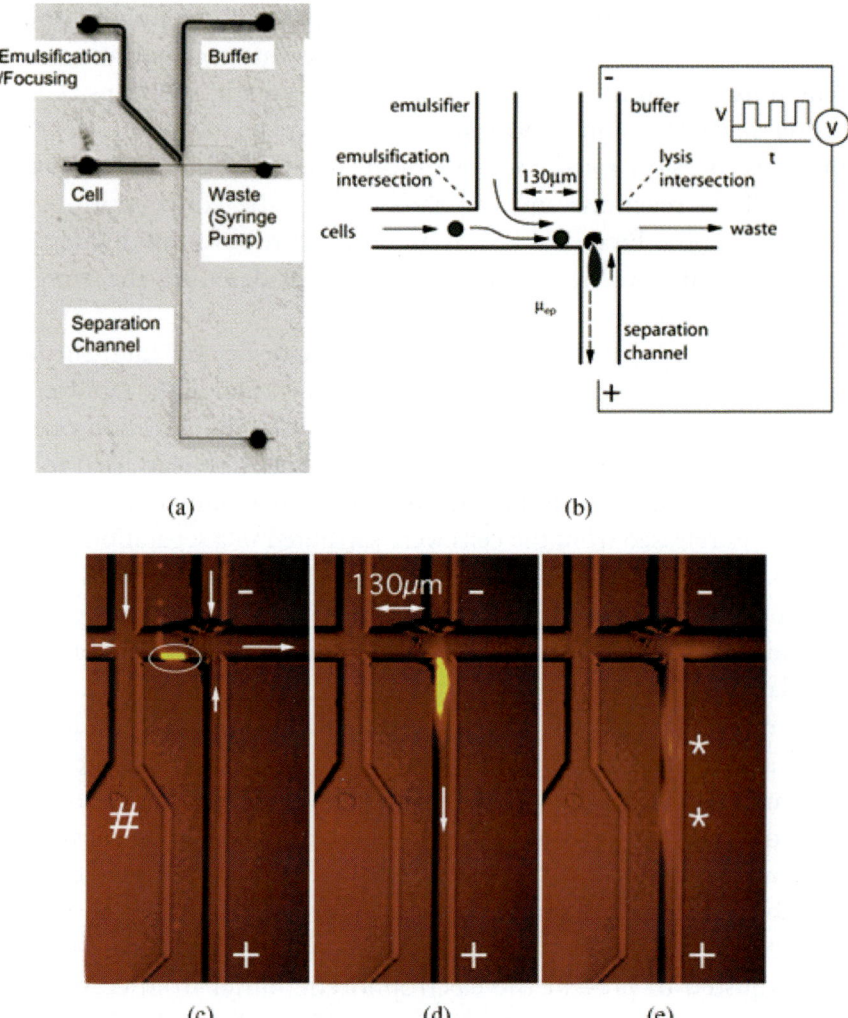

Fig. 10. (a) Design of the microchip used for the cell analysis experiments. (b) Schematic of intersections for the microchip design shown in (a). The solid arrows show the direction of bulk fluid flow and the dashed arrow shows the electrophoretic migration direction of the cell lysate. (c)–(e) CCD images of cell lysis. (c) The Jurkat cell loaded with Calcein AM (within the white oval) is being transported to the lysis intersection by hydrodynamic focusing. The arrows indicate the direction of fluid flow in the channels. (d) The cell has been lysed by the electric field. The fluorescently labeled contents are injected into the separation channel and migrate toward the anode. The arrow depicts the direction of lysate migration. (e) The separation of two of the fluorescently labeled components (marked by asterisks) is seen. (Rerproduced with permission from Ref. 55. Copyright 2003, American Chemical Society.)

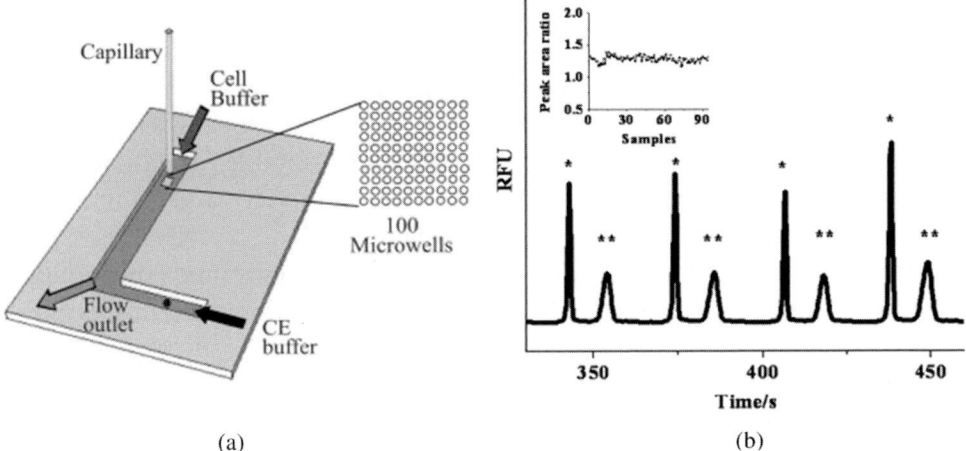

Fig. 11. Capillary-based platform containing two-channel chamber for fast serial analysis of single cells. (a) Channel with physiological and electrophoretic buffer flowing into the indicated channels. The circular region in the CE buffer channel marks the location of the capillary during electrophoresis. (b) Serial electrophoresis of fluorophore standards. Multi-peaks demonstrating the separation of fluorescein (*) and Oregon green (**). Inset: The ratio of the peak areas of fluorescein and Oregon green plotted against the sample number. (Reproduced with permission from Ref. 79. Copyright 2010, Wiley-VCH Verlag GmbH & Co.)

Staining proteins with dyes is common, but it requires manual analysis or image processing.[82–84] Manual analysis is subjected to experimental bias, while automatic microscopy image processing is usually affected by many environmental factors such as ambient light, excitation intensity, and difficulty in determining the cell edge. Immunostaining techniques such as Western blotting are the most common methods for protein qualification and quantification, but the amount of cell lysate needed is about 250-fold more than a microfluidic device can provide.[85] This situation gets even worse for specific cell types that produce fewer proteins per cell or sometimes, some specific proteins require a higher loading amount to be detected. To overcome these problems, the In-Cell Western™ (ICW) technique, integrating quantitative immunocytochemistry and laser scanner,[86,87] is used to perform *in situ* protein analysis in cells grown in microfluidic channels.[85]

The procedure for traditional ICW is that cells are grown in monolayer cultures before they are fixed and stained using a fluorescent secondary antibody. Then, the fluorescent signal for each well of a 96-well plate is quantified by commercially available laser scanners. The integration of ICW with a microfluidic platform generates a high-throughput screening method, which allows for a wide range of readouts to be studied simultaneously.

As mentioned earlier, single-cell–based assays are necessary for revealing the distinct responses by individual cells to the regulated activation of multiple proteins. One of the most critical requirements should be high throughput. Flow cytometry is one of most popular techniques for high-throughput cell screening.[88–90] However, flow cytometry is not suitable for adherent cells because cell detachment will destroy cell integrity and functions.[91] Immunocytochemistry can study the signaling of adherent cells on a single-cell level. Also, automatic imaging can be used to acquire high throughput. One main drawback for immunocytochemistry is the nonuniform treatment of each well.[92] The precise control on delivery can be easily realized on a microfluidic device. Cheong et al.[93] reported a device (in-chip immunostaining) which takes advantage of microfluidics to perform precise, high-throughput immunofluorescence analysis for signaling dynamics on a single-cell basis. The device featured in single-cell resolution, live-cell imaging, and was capable of visualizing dynamic signaling events without disturbing the cells. The device had a throughput comparable to flow cytometry; in the meantime, subcellular information and spatial distribution were acquired.

Figure 12 shows the schematic design of the device. In brief, the device is a compact monolithic two-layer chip, with the bottom layer containing an

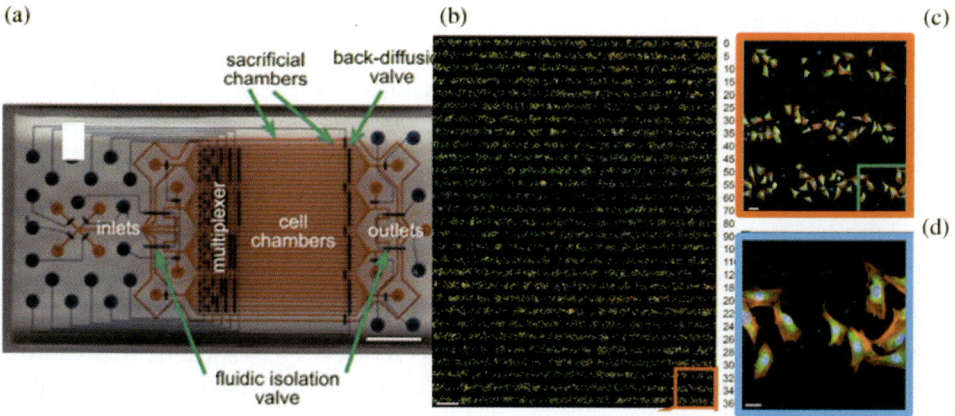

Fig. 12. (a) Layout of the device for high-throughput immunofluorescence analysis. The fluidic layer is filled with red food dye, while the valve layer is filled with blue food dye for demonstration. (b) Composite image of a single device used to measure the response of mouse NIH-3T3 fibroblasts exposed to TNF-α sampled at specific time points. Each row is an independent test chamber corresponding to a certain stimulation time point. Stains for p65 (antibody; green), actin (phalloidin; red), and DNA (Hoechst 33258; blue). (c) is a magnified picture of (b), and (d) is a magnified picture of (c). (Reproduced with permission from Ref. 93. Copyright 2009, American Society for Biochemistry and Molecular Biology.)

array of chambers for cell manipulation. The top layer is composed of valves for controlling the fluid flow in the bottom layer. The chip can be operated in various modes to facilitate sequential experiments. Proteins, DNA, and biomolecules of interest are stained for image analysis. All the images are quantified by the MATLAB program for fluorescence intensity and distribution changes.

Conclusions

Proteomics research is rapidly growing, generating new analytical challenges. The progress of microfluidics for studying cellular proteomics in laying proteins in their native conformation on chips and developing sensitive detection strategies facilitates the application of protein chips in many biomedical fields. The advantages of microfluidics in terms of low reagent consumption, high detection sensitivity, precise and highly automatic control, and rapid analysis make it a particularly promising tool in proteomics study. Also, the flexibility in the operation platform of microfluidic devices makes it convenient to couple it with many other advanced analysis techniques, such as microarray, flow cytometry, immunoassay, and mass spectroscopy. The integration strategy has drawn increasing interest in developing novel tools for biochemical and biological studies. Optimization in the integration interface can fully take advantage of different techniques, which will introduce revolutionary steps in cellular proteomic research.

References

1. Persidis, A., Proteomics — an ambitious drug development platform attempts to link gene sequence to expressed phenotype under various physiological states. *Nat Biotechnol*, 1998, **16**: p. 393–394.
2. Hood, L., A personal view of molecular technology and how it has changed biology. *J Proteome Res*, 2002, **1**: p. 399–409.
3. Cheek, B.J., Steel, A.B., Torres, M.P., Yu, Y.Y., Yang, H., Chemiluminescence detection for hybridization assays on the flow-thru chip, a three-dimensional microchannel biochip. *Anal Chem*, 2001, **73**: p. 5777–5783.
4. Adey, N.B., Lei, M., Howard, M.T., Jensen, J.D., Mayo, D.A., Butel, D.L., Coffin, S.C., Moyer, T.C., Slade, D.E., Spute, M.K., Hancock, A.M., Eisenhoffer, G.T., Dalley, B.K., McNeely, M.R., Gains in sensitivity with a device that mixes microarray hybridization solution in a 25-microm-thick chamber. *Anal Chem*, 2002, **74**: p. 6413–6417.
5. Wang, Y., Vaidya, B., Farquar, H.D., Stryjewski, W., Hammer, R.P., McCarley, R.L., Soper, S.A., Cheng, Y.W., Barany, F., Microarrays assembled in microfluidic chips fabricated from poly(methyl methacrylate) for the detection of low-abundant DNA mutations. *Anal Chem*, 2003, **75**: p. 1130–1140.

6. Zhou, X., Cai, S., Hong, A., You, Q., Yu, P., Sheng, N., Srivannavit, O., Muranjan, S., Rouillard, J.M., Xia, Y., Zhang, X., Xiang, Q., Ganesh, R., Zhu, Q., Matejko, A., Gulari, E., Gao, X., Microfluidic picoArray synthesis of oligodeoxynucleotides and simultaneous assembling of multiple DNA sequences. *Nucleic Acids Res*, 2004, **32**: p. 5409–5417.
7. Wei, C.W., Cheng, J.Y., Huang, C.T., Yen, M.H., Young, T.H., Using a microfluidic device for 1 μl DNA microarray hybridization in 500 s. *Nucleic Acids Res*, 2005, **33**.
8. Liu, R.H., Dill, K., Fuji, H.S., McShea, A., Integrated microfluidic biochips for DNA microarray analysis. *Expert Rev Mol Diagn*, 2006, **6**: p. 253–261.
9. Lee, H.H., Smoot, J., McMurray, Z., Stahl, D.A., Yager, P., Recirculating flow accelerates DNA microarray hybridization in a microfluidic device. *Lab Chip*, 2006, **6**: p. 1163–1170.
10. Paegel, B.M., Emrich, C.A., Weyemayer, G.J., Scherer, J.R., Mathies, R.A., High throughput DNA sequencing with a microfabricated 96-lane capillary array electrophoresis bioprocessor. *Proc Natl Acad Sci U S A*, 2002, **99**: p. 574–579.
11. Tran, N.T., Ayed, I., Pallandre, A., Taverna, M., Recent innovations in protein separation on microchips by electrophoretic methods: An update. *Electrophoresis*, 2010, **31**: p. 147–173.
12. Breadmore, M.C., Dawod, M., Quirino, J.P., Recent advances in enhancing the sensitivity of electrophoresis and electrochromatography in capillaries and microchips (2008–2010). *Electrophoresis*, 2011, **32**: p. 127–148.
13. Corthals, G.L., Wasinger, V.C., Hochstrasser, D.F., Sanchez, J.C., The dynamic range of protein expression: A challenge for proteomic research. *Electrophoresis*, 2000, **21**: p. 1104–1115.
14. Figeys, D., Adapting arrays and lab-on-a-chip technology for proteomics. *Proteomics*, 2002, **2**: p. 373–382.
15. Huber, D.L., Manginell, R.P., Samara, M.A., Kim, B.I., Bunker, B.C., Programmed adsorption and release of proteins in a microfluidic device. *Science*, 2003, **301**: p. 352–354.
16. Cooper, J.W., Wang, Y., Lee, C.S., Recent advances in capillary separations for proteomics. *Electrophoresis*, 2004, **25**: p. 3913–3926.
17. Schasfoort, R.B.M., Proteomics-on-a-chip: The challenge to couple lab-on-a-chip unit operations. *Expert Rev Proteomics*, 2004, **1**: p. 123–132.
18. Chen, H., Fan, Z.H., Two-dimensional protein separation in microfluidic devices. *Electrophoresis*, 2009, **30**: p. 758–765.
19. Zhang, S., Van Pelt, C.K., Chip-based nanoelectrospray mass spectrometry for protein characterization. *Expert Rev Proteomics*, 2004, **1**: p. 449–468.
20. Foret, F., Kusy, P., Microfluidics for multiplexed MS analysis. *Electrophoresis*, 2006, **27**: p. 4877–4887.
21. Devoe, D.L., Lee, C.S., Microfluidic technologies for MALDI-MS in proteomics *Electrophoresis*, 2006, **27**: p. 3559–3568.
22. Koster, S., Verpoorte, E., A decade of microfluidic analysis coupled with electrospray mass spectrometry: An overview. *Lab Chip*, 2007, **7**: p. 1394–1412.
23. Chen, G.D., Pramanik, B.N., Application of LC/MS to proteomics studies: Current status and future prospects. *Drug Discovery Today*, 2009, **14**: p. 465–471.
24. Lee, J., Soper, S.A., Murray, K.K., Microfluidics with MALDI analysis for proteomics-A review. *Anal Chim Acta*, 2009, **649**: p. 180–190.
25. Lee, J., Soper, S.A., Murray, K.K., Microfluidic chips for mass spectrometry-based proteomics. *J Mass Spectrom*, 2009, **44**: p. 579–593.

26. Dolnik, V., Capillary electrophoresis of proteins 2003–2005. *Electrophoresis*, 2006, **27**: p. 126–141.
27. Tia, S., Herr, A.E., On-chip technologies for multidimensional separations. *Lab Chip*, 2009, **9**: p. 2524–2536.
28. Fonslow, B.R., Yates, J.R., Capillary electrophoresis applied to proteomic analysis. *J Sep Sci*, 2009, **32**: p. 1175–1188.
29. James, C.D., Moorman, M.W., Carson, B.D., Branda, C.S., Lantz, J.W., Manginell, R.P., Martino, A., Singh, A.K., Nuclear translocation kinetics of NF-kappa B in macrophages challenged with pathogens in a microfluidic platform. *Biomed Microdevices*, 2009, **11**: p. 693–700.
30. Blake, W.J., Kaern, M., Cantor, C.R., Collins, J.J., Noise in eukaryotic gene expression. *Nature*, 2003, **422**: p. 633–637.
31. Swain, P.S., Elowitz, M.B., Siggia, E.D., Intrinsic and extrinsic contributions to stochasticity in gene expression. *Proc Natl Acad Sci U S A*, 2002, **99**: p. 12795–12800.
32. Ferrell, J.E., Jr., Machleder, E.M., The biochemical basis of an all-or-none cell fate switch in *Xenopus* oocytes. *Science*, 1998, **280**: p. 895–898.
33. Levsky, J.M., Singer, R.H., Gene expression and the myth of the average cell. *Trends Cell Biol*, 2003, **13**: p. 4–6.
34. Rubin, M.A., Tech. Sight. Understanding disease cell by cell. *Science*, 2002, **296**: p. 1329–1330.
35. Sasuga, Y., Iwasawa, T., Terada, K., Oe, Y., Sorimachi, H., Ohara, O., Harada, Y., Single-cell chemical lysis method for analyses of intracellular molecules using an array of picoliter-scale microwells. *Anal Chem*, 2008, **80**: p. 9141–9149.
36. Sasuga, Y., Tani, T., Hayashi, M., Yamakawa, H., Ohara, O., Harada, Y., Development of a microscopic platform for real-time monitoring of biomolecular interactions. *Genome Res*, 2006, **16**: p. 132–139.
37. Zamir, E., Bastiaens, P.I., Reverse engineering intracellular biochemical networks. *Nat Chem Biol*, 2008, **4**: p. 643–647.
38. Taylor, R.J., Falconnet, D., Niemisto, A., Ramsey, S.A., Prinz, S., Shmulevich, I., Galitski, T., Hansen, C.L., Dynamic analysis of MAPK signaling using a high-throughput microfluidic single-cell imaging platform. *Proc Natl Acad Sci U S A*, 2009, **106**: p. 3758–3763.
39. El-Ali, J., Sorger, P.K., Jensen, K.F., Cells on chips. *Nature*, 2006, **442**: p. 403–411.
40. Flaim, C.J., Teng, D., Chien, S., Bhatia, S.N., Combinatorial signaling microenvironments for studying stem cell fate. *Stem Cells Dev*, 2008, **17**: p. 29–39.
41. Vickerman, V., Blundo, J., Chung, S., Kamm, R., Design, fabrication and implementation of a novel multi-parameter control microfluidic platform for three-dimensional cell culture and real-time imaging. *Lab Chip*, 2008, **8**: p. 1468–1477.
42. Hui, E.E., Bhatia, S.N., Micromechanical control of cell-cell interactions. *Proc Natl Acad Sci U S A*, 2007, **104**: p. 5722–5726.
43. Bhatia, S.N., Balis, U.J., Yarmush, M.L., Toner, M., Effect of cell-cell interactions in preservation of cellular phenotype: Cocultivation of hepatocytes and nonparenchymal cells, *FASEB J*, 1999, **13**: p. 1883–1900.
44. Guguen-Guillouzo, C., Guillouzo, A., Modulation of functional activities in cultured rat hepatocytes. *Mol Cell Biochem*, 1983, **53–54**: p. 35–56.
45. Helm, C.L., Fleury, M.E., Zisch, A.H., Boschetti, F., Swartz, M.A., Synergy between interstitial flow and VEGF directs capillary morphogenesis *in vitro* through a gradient amplification mechanism. *Proc Natl Acad Sci U S A*, 2005, **102**: p. 15779–15784.

46. Zaman, M.H., Trapani, L.M., Sieminski, A.L., Mackellar, D., Gong, H., Kamm, R.D., Wells, A., Lauffenburger, D.A., Matsudaira, P., Migration of tumor cells in 3D matrices is governed by matrix stiffness along with cell-matrix adhesion and proteolysis. *Proc Natl Acad Sci U S A*, 2006, **103**: p. 10889–10894.
47. Yamamura, N., Sudo, R., Ikeda, M., Tanishita, K., Effects of the mechanical properties of collagen gel on the *in vitro* formation of microvessel networks by endothelial cells. *Tissue Eng*, 2007, **13**: p. 1443–1453.
48. Paguirigan, A., Beebe, D.J., Gelatin based microfluidic devices for cell culture. *Lab Chip*, 2006, **6**: p. 407–413.
49. Cheng, S.Y., Heilman, S., Wasserman, M., Archer, S., Shuler, M.L., Wu, M., A hydrogel-based microfluidic device for the studies of directed cell migration. *Lab Chip*, 2007, **7**: p. 763–769.
50. Chung, S., Sudo, R., Mack, P.J., Wan, C.R., Vickerman, V., Kamm, R.D., Cell migration into scaffolds under co-culture conditions in a microfluidic platform. *Lab Chip*, 2009, **9**: p. 269–275.
51. Nunemaker, C.S., Zhang, M., Wasserman, D.H., McGuinness, O.P., Powers, A.C., Bertram, R., Sherman, A., Satin, L.S., Individual mice can be distinguished by the period of their islet calcium oscillations: Is there an intrinsic islet period that is imprinted *in vivo*? *Diabetes*, 2005, **54**: p. 3517–3522.
52. Luciani, D.S., Misler, S., Polonsky, K.S., Ca^{2+} controls slow NAD(P)H oscillations in glucose-stimulated mouse pancreatic islets. *J Physiol*, 2006, **572**: p. 379–392.
53. Dishinger, J.F., Reid, K.R., Kennedy, R.T., Quantitative monitoring of insulin secretion from single islets of Langerhans in parallel on a microfluidic chip. *Anal Chem*, 2009, **81**: p. 3119–3127.
54. Han, F., Wang, Y., Sims, C.E., Bachman, M., Chang, R., Li, G.P., Allbritton, N.L., Fast electrical lysis of cells for capillary electrophoresis. *Anal Chem*, 2003, **75**: p. 3688–3696.
55. McClain, M.A., Culbertson, C.T., Jacobson, S.C., Allbritton, N.L., Sims, C.E., Ramsey, J.M., Microfluidic devices for the high-throughput chemical analysis of cells. *Anal Chem*, 2003, **75**: p. 5646–5655.
56. Wang, H.Y., Lu, C., Microfluidic chemical cytometry based on modulation of local field strength. *Chem Commun (Camb)*, 2006, p. 3528–3530.
57. Givan, A.L., *Flow Cytometry: First Principles*, 2nd edn. Wiley-Liss, New York, 2001.
58. Ormerod, M.G., *Flow Cytometry, A Practical Approach*, 3rd edn. Oxford University Press, Oxford, 2000.
59. Wang, J., Fei, B., Zhan, Y., Geahlen, R.L., Lu, C., Kinetics of NF-kappaB nucleocytoplasmic transport probed by single-cell screening without imaging. *Lab Chip*, 2010, **10**: p. 2911–2916.
60. Wang, H.Y., Lu, C., Electroporation of mammalian cells in a microfluidic channel with geometric variation. *Anal Chem*, 2006, **78**: p. 5158–5164.
61. Marc, P.J., Sims, C.E., Bachman, M., Li, G.P., Allbritton, N.L., Fast-lysis cell traps for chemical cytometry. *Lab Chip*, 2008, **8**: p. 710–716.
62. Wang, J., Fei, B., Geahlen, R.L., Lu, C. Quantitative analysis of protein translocations by microfluidic total internal reflection fluorescence flow cytometry. *Lab Chip*, 2010, **10**: p. 2673–2679.
63. Unger, M.A., Chou, H.P., Thorsen, T., Scherer, A., Quake, S.R., Monolithic microfabricated valves and pumps by multilayer soft lithography. *Science*, 2000, **288**: p. 113–116.

64. Zhan, Y., Martin, V.A., Geahlen, R.L., Lu, C., One-step extraction of subcellular proteins from eukaryotic cells. *Lab Chip*, 2010, **10**: p. 2046–2048.
65. Wightman, R.M., Jankowski, J.A., Kennedy, R.T., Kawagoe, K.T., Schroeder, T.J., Leszczyszyn, D.J., Near, J.A., Diliberto, E.J., Jr., Viveros, O.H., Temporally resolved catecholamine spikes correspond to single vesicle release from individual chromaffin cells. *Proc Natl Acad Sci U S A*, 1991, **88**: p. 10754–10758.
66. Bard, A.J., Li, X., Zhan, W., Chemically imaging living cells by scanning electrochemical microscopy. *Biosens Bioelectron*, 2006, **22**: p. 461–472.
67. Gao, N., Wang, W.L., Zhang, X.L., Jin, W.R., Yin, X.F., Fang, Z.L., High-throughput single-cell analysis for enzyme activity without cytolysis. *Anal Chem*, 2006, **78**: p. 3213–3220.
68. Teruel, M.N., Meyer, T., Parallel single-cell monitoring of receptor-triggered membrane translocation of a calcium-sensing protein module. *Science*, 2002, **295**: p. 1910–1912.
69. Marcus, J.S., Anderson, W.F., Quake, S.R., Parallel picoliter RT-PCR assays using microfluidics *Anal Chem*, 2006, **78**: p. 956–958.
70. Dovichi, N.J., Hu, S., Chemical cytometry, *Curr Opin Chem Biol*, 2003, 7: p. 603–608.
71. Lu, C., *Chemical Cytometry: Ultrasensitive Analysis of Single Cells*, Wiley-VCH, 2010.
72. Lee, C.L., Linton, J., Soughayer, J.S., Sims, C.E., Allbritton, N.L., Localized measurement of kinase activation in oocytes of *Xenopus laevis*. *Nat Biotechnol*, 1999, **17**: p. 759–762.
73. Luo, Y., Shoemaker, A.R., Liu, X., Woods, K.W., Thomas, S.A., de Jong, R., Han, E.K., Li, T., Stoll, V.S., Powlas, J.A., Oleksijew, A., Mitten, M.J., Shi, Y., Guan, R., McGonigal, T.P., Klinghofer, V., Johnson, E.F., Leverson, J.D., Bouska, J.J., Mamo, M., Smith, R.A., Gramling-Evans, E.E., Zinker, B.A., Mika, A.K., Nguyen, P.T., Oltersdorf, T., Rosenberg, S.H., Li, Q., Giranda, V.L., Potent and selective inhibitors of Akt kinases slow the progress of tumors *in vivo*. *Mol Cancer Ther*, 2005, **4**: p. 977–986.
74. Whitmore, C.D., Hindsgaul, O., Palcic, M.M., Schnaar, R.L., Dovichi, N.J., Metabolic cytometry. Glycosphingolipid metabolism in single cells. *Anal Chem*, 2007, **79**: p. 5139–5142.
75. Borland, L.M., Kottegoda, S., Phillips, K.S., Allbritton, N.L., Chemical analysis of single cells. *Annu Rev Anal Chem*, 2008, **1**: p. 191–227.
76. Meredith, G.D., Sims, C.E., Soughayer, J.S., Allbritton, N.L., Measurement of kinase activation in single mammalian cells. *Nat Biotechnol*, 2000, **18**: p. 309–312.
77. Hu, S., Le, Z., Krylov, S., Dovichi, N.J., Cell cycle-dependent protein fingerprint from a single cancer cell: Image cytometry coupled with single-cell capillary sieving electrophoresis. *Anal Chem*, 2003, **75**: p. 3495–3501.
78. Berridge, M.J., Inositol trisphosphate and calcium signalling. *Nature*, 1993, **361**: p. 315–325.
79. Jiang, D., Sims, C.E., Allbritton, N.L., Microelectrophoresis platform for fast serial analysis of single cells. *Electrophoresis*, 2010, **31**: p. 2558–2565.
80. Sims, C.E., Meredith, G.D., Krasieva, T.B., Berns, M.W., Tromberg, B.J., Allbritton, N.L., Laser-micropipet combination for single-cell analysis. *Anal Chem*, 1998, **70**: p. 4570–4577.
81. Paguirigan, A.L., Beebe, D.J., Microfluidics meet cell biology: Bridging the gap by validation and application of microscale techniques for cell biological assays. *BioEssays*, 2008, **30**: p. 811–821.
82. Toh, Y.C., Zhang, C., Zhang, J., Khong, Y.M., Chang, S., Samper, V.D., van Noort, D., Hutmacher, D.W., Yu, H., A novel 3D mammalian cell perfusion-culture system in microfluidic channels. *Lab Chip*, 2007, **7**: p. 302–309.

83. Jang, K., Sato, K., Igawa, K., Chung, U.I., Kitamori, T., Development of an osteoblast-based 3D continuous-perfusion microfluidic system for drug screening. *Anal Bioanal Chem*, 2008, **390**: p. 825–832.
84. Weigum, S.E., Floriano, P.N., Christodoulides, N., McDevitt, J.T., Cell-based sensor for analysis of EGFR biomarker expression in oral cancer. *Lab Chip*, 2007, **7**: p. 995–1003.
85. Paguirigan, A.L., Puccinelli, J.P., Su, X.J., Beebe, D.J., Expanding the available assays: Adapting and validating in-cell Westerns in microfluidic devices for cell-based assays. *Assay Drug Dev Technol*, 2010, **8**: p. 591–601.
86. Zhou, W.H., Du, M.R., Dong, L., Zhu, X.Y., Yang, J.Y., He, Y.Y., Li, D.J., Cyclosporin A increases expression of matrix metalloproteinase 9 and 2 and invasiveness *in vitro* of the first-trimester human trophoblast cells via the mitogen-activated protein kinase pathway. *Hum Reprod*, 2007, **22**: p. 2743–2750.
87. Du, Y.Q., Danjo, K., Robinson, P.A., Crabtree, J.E., In-cell Western analysis of *Helicobacter pylori*-induced phosphorylation of extracellular-signal related kinase via the transactivation of the epidermal growth factor receptor. *Microbes Infect*, 2007, **9**: p. 838–846.
88. Sachs, K., Perez, O., Pe'er, D., Lauffenburger, D.A., Nolan, G.P., Causal protein-signaling networks derived from multiparameter single-cell data. *Science*, 2005, **308**: p. 523–529.
89. Irish, J.M., Hovland, R., Krutzik, P.O., Perez, O.D., Bruserud, O., Gjertsen, B.T., Nolan, G.P., Single cell profiling of potentiated phospho-protein networks in cancer cells. *Cell*, 2004, **118**: p. 217–228.
90. Krutzik, P.O., Crane, J.M., Clutter, M.R., Nolan, G.P., High-content single-cell drug screening with phosphospecific flow cytometry. *Nat Chem Biol*, 2008, **4**: p. 132–142.
91. Krutzik, P.O., Irish, J.M., Nolan, G.P., Perez, O.D., Analysis of protein phosphorylation and cellular signaling events by flow cytometry: Techniques and clinical applications. *Clin Immunol*, 2004, **110**: p. 206–221.
92. Dunn, D.A., Feygin, I., Challenges and solutions to ultra-high-throughput screening assay miniaturization: Submicroliter fluid handling. *Drug Discovery Today*, 2000, **5**: p. 84–91.
93. Cheong, R., Wang, C.J., Levchenko, A., High content cell screening in a microfluidic device. *Mol Cell Proteomics*, 2009, **8**: p. 433–442.

Chapter 9

Microfluidics for Neuroscience: Novel Tools and Future Implications

Vivian M. Hernandez* and P. Hande Özdinler[*,†,‡,§]*

*Les Turner ALS Laboratory II, Department of Neurology
†Robert H. Lurie Comprehensive Cancer Center
‡Cognitive Neurology and Alzheimer's Disease Center
Feinberg School of Medicine, Northwestern University
Chicago, IL 60611, USA

Introduction

Challenges in the field of neuroscience

The brain is a complex system

The intricate neuronal networks in our cerebral cortex allow us to perform complex tasks such as thinking, reasoning, and voluntary movement. These networks determine how we perceive, analyze, and behave, and ultimately who we are. Therefore, the brain cannot be considered as just another organ; it is a combination of complex neural networks and it is a dynamic system that defines us as human beings.

In our brain, there are around 10 billion neurons in addition to an even greater number of non-neuronal cells, including astrocytes and microglia.[1] Thus, it is not surprising that there are thousands of different subtypes of neurons in the brain. This cellular heterogeneity is further complicated by the wide variation of neuronal spatial distribution, especially within the cerebral cortex.[2,3] Within a defined area in a single cortical layer, where the thickness

---

[§]Corresponding author. Email: ozdinler@northwestern.edu

is on the order of millimeters, one can find cell bodies of hundreds of different types of neurons. Yet despite the close proximity of neuronal cell bodies, the projections and circuitries they are involved in can vary widely. Together, these heterogeneous neurons and non-neuronal cells form an immeasurable number of connections, comprising an immense neuronal network, which control and initiate many of our behaviors and movements.

Understanding the brain's complexity requires an appreciation for the unique properties and complexities of the neuron as an entity. This chapter reviews the basic characteristics that set neurons apart from other cells, and how understanding these differences can help us reveal the neuron type-specific aspects of neuronal differentiation as well as selective neuronal vulnerability in diseases.

Neuronal polarity and heterogeneity

Neurons are different from other cells in our body because of their unique cellular morphology and their ability to communicate electrochemically with other cells. Neurons communicate by rapid changes in membrane potential, known as action potentials, that lead to the release of chemical and protein transmitters to other cells. These chemical and protein transmitters then signal to neighboring neurons through membrane specializations called synapses. This form of communication allows neurons to connect with other neurons to form functional circuitries. Not all circuitries are formed among neurons; in the motor neuron circuitry, for example, motor neurons in the spinal cord connect directly with muscle fibers via the neuromuscular junction and to allow muscle contraction and movement.

A neuron's ability to communicate over long distances is made possible by its unique cellular morphology (Fig. 1(a)). Similar to other cells in our body, all neurons have a cell body (soma) and a nucleus. However, different from other cells, neurons have axons, which are very long extensions that define their unique cellular polarity and allow long-distance connections (Fig. 1(a)). In addition, neurons have apical and basal dendrites, which are relatively short extensions that are specialized in receiving input from other neurons and cells in the environment. Dendrites play an important role in defining neuronal output, bursting potential, generation of action potentials, and neuronal activity.[4] Abnormalities in dendritic morphology are related to reduced neuronal activity and neurodegenerative diseases.[5,6] Once the action potential is initiated, it is carried through the axon in an all-or-none manner. In other words, the magnitude of an action potential is independent of the size of the stimulus, such that if the strength of stimulus is above the threshold, the response will be complete, and if it is below the threshold, there will be no response at all.

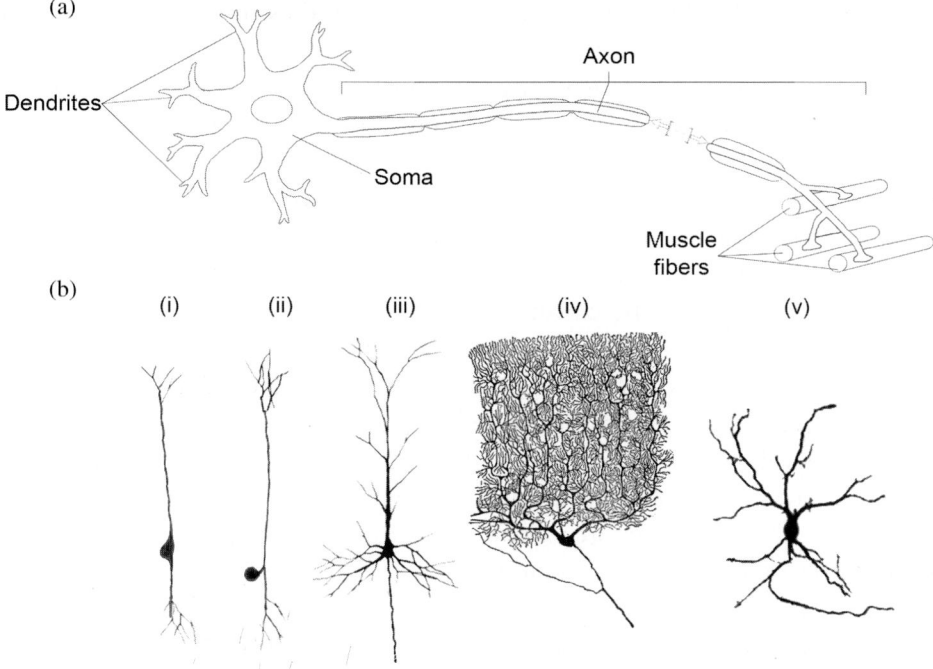

Fig. 1. Neurons have unique cellular characteristics and show a great level of cellular heterogeneity. (a) Schematic drawing of a motor neuron, indicating cell body (soma), dendrites, and a long axon that terminates at a muscle fiber. (b) Schematic drawing of different neuron types (adapted from drawings by neuroscientist Ramon y Cajal). (i) Bipolar neurons, (ii) pseudounipolar neurons, (iii) pyramidal projection neurons, (iv) Purkinje neurons of the cerebellum, and (v) interneurons (adapted from drawings by neuroscientist Ramón y Cajal).

Neurons are the most polarized cells in our body and can extend over long distances. For example, corticospinal motor neurons (CSMN) reside in layer V of the motor cortex but project their axons to spinal targets. Thus, the length of a CSMN axon, especially the one that innervates the sacral regions of the spinal cord, can be more than 1.5 m. In animals such as the giraffe and whale, this distance can reach 10–20 m. The ratio of a 20-µm cell body to a 1-m–long axon is 2×10^{-5}. This ratio is approximately equal to that of a person extending a 100-km–long arm to recognize and grab a target, which could be as small as a basketball. The remarkable extent of neuronal polarity gives a neuron its unique properties and is essential for its function. Yet, this high level of polarity limits our ability to study neurons in environments that mimic *in vivo* conditions.

Another important limitation to understanding the biology of neurons is their wide cellular heterogeneity (Fig. 1(b)). There are thousands of different neuron types in our brain. Neurons are classified based on their morphology, projection field, the type of neurotransmitter they express and secrete, the type of connections they make, and the type of activity they exert. For example, cochlear spiral ganglion neurons have bipolar morphology (Fig. 1(bi)), whereas dorsal root ganglion neurons have pseudounipolar morphology (Fig. 1(bii)) with two long axons and very limited arborization. The neurons of the central nervous system (CNS) show a high degree of heterogeneity and neuronal complexity (Fig. 1(biii–v)).

The variation even among neurons that are considered a "class" is immense. For example, the classification "projection neuron," used for neurons that send long axons to their targets, can refer to more than hundreds of different types of projection neurons[7–10] (Fig. 1(biii)). Corticospinal neurons are projection neurons that project distal axons from the cortex to the spinal cord. Likewise, corticocortical neurons project their axons from one hemisphere of the cortex to the other. As their names imply, the locations of their cell bodies and their projection fields differ greatly. Another classification is "interneuron" and is used for neurons that form local connections and induce, strengthen, monitor, and control local circuitries. Interneurons, like projection neurons, show a great level of heterogeneity both in their function and morphology (Fig. 1(bv)).[11–17] Even amacrine cells, a subset of interneurons in the retina that were once thought to constitute a single neuron type, are now classified into subtypes based on variations in their morphologies and the neurotransmitters they release.[18]

Interestingly, many projection neurons may look similar under the microscope, since they are located in very close proximity to each other, stain equally well with neuronal markers (i.e., NeuN, Map2); yet they are very different in nature. What makes a neuron different from other neurons is determined by a combination of extrinsic and intrinsic factors. Intrinsic factors such as gene expression, mRNA editing, and post-translational modifications of proteins in a cell type-specific manner are beginning to emerge as important determinants of neuronal identity.[20] Even though each neuron can code for the same genes, the level of expression and time of expression of distinct sets of genes may determine the identity of a neuron population and give them their unique characteristics. For example, the gene encoding the transcription factor, *Fezf2*, which has been reported to be necessary for the specification of several types of subcerebral cortical projection neurons, is sufficient to instruct them to become corticofugal neurons.[19–21] Upon *Fezf2* expression, neuronal progenitors acquire a glutamatergic neurotransmitter identity and express genes specific to corticofugal neurons in the cortex, including *Sox5*, which is expressed by all corticofugal neurons; *TBR1*, which is expressed by neurons in layer VI and a subset of layer V of the cortex; *BHLHB5*, which is expressed by subcerebral projection neurons during development; and *ZFPM2*, which is expressed by neurons in layer VI of the cortex. In addition, a significant fraction of the neuronal progenitors acquire an apical dendrite, a morphology characteristic of cortical projection neurons that project to corticofugal targets. Therefore, it is possible to instruct neuronal identity within the neural progenitors by directed gene expression. The level of neuronal heterogeneity is initiated in the cerebral cortex by the early stages of symmetric versus asymmetric division and by diversity among neural progenitor pools that give rise to new neurons.[22–24] At times, a neuron depends on cues from its environment to trigger these intrinsic changes. It is the interplay between these intrinsic and extrinsic factors that defines the full identity of a neuron.

Extrinsic factors, such as growth factors, and other inhibitory and excitatory molecules that are present in the environment can be very influential on neuronal differentiation, maturation, and response to injury.[3,25] Transplantation studies have the potential to reveal the importance of intrinsic properties of neurons and the role of the extracellular environment by introducing neurons to a different environment and thus test the potential role of the new environment on directed neuronal differentiation and function.[26–28] For example, the transplantation of inhibitory neurons enhances ocular dominance plasticity, a rapid reorganization of neuronal responses due to the occlusion of one eye, and suggests a role for inhibitory neurons in neuronal plasticity and repair.[29] Transplantation of early- and late-born interneurons revealed that fate switching is dependent on progenitor receptivity to environmental cues.[30] Another study by Gaillard *et al.*[29] found that the embryonic neuronal precursors isolated from the developing mouse cortex, when transplanted into injured adult cortex in the mouse, complete their neuronal differentiation and maturation in this new environment, and are able to respond to the cues and signals present to form connections even at distal targets. Investigators have used similar approaches to replace dying motor neurons in the spinal cord with neural precursor cells, in an attempt to devise new therapies for spinal cord injuries and diseases such as amyotrophic lateral sclerosis (ALS).[31,32] Such studies had limited success, likely because of the impurity of transplanted cells and the inability to control the environment they encounter. However, they revealed an important biology: the cellular dynamics between neurons and non-neuronal cells, their environment, and the intrinsic properties of the neurons are all important for proper neuronal function. Therefore, we need better tools to investigate both the intrinsic and extrinsic controls over neuronal differentiation, maturation, and function. Such studies will require novel platforms, where an individual neuron is investigated as an entity, and the results are observed in real time and with high precision.

Neuron type-specific vulnerability in neurodegenerative diseases

In neurodegenerative diseases, only a distinct neuron population shows initial vulnerability and progressive degeneration. For example, in Huntington's disease (HD), the extensive loss of medium spiny projection neurons of the caudate and putamen affects downstream target nuclei, the globus pallidus, and substantia nigra, leading to the dysregulation of motor control.[33,34] In Parkinson's disease (PD), mainly the dopaminergic neurons of the substantia nigra pars compacta degenerate with disease progression.[35,36] In spinal muscular atrophy, neuronal vulnerability and degeneration are mainly restricted to spinal motor neurons in the spinal cord.[37] We still do not know why only a distinct neuron population shows vulnerability in neurodegenerative diseases. Why are other neuron populations that are developmentally closely related, share common biology, and are located within the same microenvironment not affected? For example, in ALS, the corticospinal projection neurons located in the motor cortex are vulnerable and degenerate. Interestingly, corticocortical projection neurons, which are born together with corticospinal projection neurons from the same progenitor cells, migrate together during development, and settle together with corticospinal projection neurons in layer V of the motor cortex, are not primarily invloved in ALS. They do not show signs of vulnerability. This neuron type-specific vulnerability is one of the major mysteries in the field of neuroscience. It hampers the development of effective treatment strategies to neurodegeneration and further emphasizes the importance of understanding the biology of neuron populations as an entity.

Neurodegenerative diseases can either occur due to familial (passed by a genetic trait) or sporadic (occur either due to unknown, uncharacterized factors, or by genetic traits that are not evident) causes. About 90% of neurodegenerative diseases are due to uncharacterized mechanisms and effectors.[38] Furthermore, sometimes, genetics alone cannot explain neuronal vulnerability. For example, in ALS, mutations in the superoxide dismutase 1 (*SOD1*) gene are associated with familial forms of the disease.[39–41] Strikingly, the *SOD1* gene is expressed broadly in many different cells, and its expression is not restricted to one neuron type. However, mutations in this gene cause degeneration primarily in motor neurons, leaving other cell types largely unaffected. It is not easy to explain neuron type-specific vulnerability in neurodegenerative diseases with mutations that are present in the whole genome and in genes that are expressed in most cells. If the mutations are present in all cells, why is only a select neuron population affected? It is even harder to explain the sporadic occurrence of neurodegeneration with no known genetic or environmental predisposition.

Neuron type-specific vulnerability is one of the major challenges in the field of neuroscience. We are now beginning to realize the importance of studying neurons that are primarily vulnerable in diseases as a "pure" neuron population to reveal their biology. Previous investigations mostly ignored the importance of neuronal heterogeneity, mainly due to limitations in approaches and techniques. Novel applications that allow cell type-specific labeling, isolation, and purification of distinct neuron populations from the complex structure of the brain are beginning to emerge. Such studies have the potential to reveal the biology of distinct neuron populations. In one approach known as retrograde labeling, projection neurons are labeled by introducing fluorescent microspheres into their well-defined projection field. Microspheres are taken up by the axon and are transported by the cell's own transport machinery in the retrograde direction to their cell bodies. In this approach, the cell body gains a fluorescent character, and among many different neuron cell bodies in a defined area, only the neurons that project to the target area where microspheres were introduced become labeled. Retrograde labeling approaches have been used to label corticocortical projection neurons and corticospinal projection neurons.[42,43] Upon cellular labeling with a fluorescent marker, these cells can be purified via fluorescence-activated cell sorting (FACS). Another approach uses immunopanning to purify projection neuron populations. Rather than giving neurons a fluorescent character, cholera toxin beta (CTB) retrograde labeling induces neurons to express the cell surface epitope CTB, which is then used to purify neurons via immobilization with an anti-CTB antibody. This approach has been used to label and purify retinal ganglion cells[44] and corticospinal projection neurons.[45] Both retrograde labeling and immunopanning are very powerful for labeling and isolating neurons of interest from the cortex, but are mainly applicable to projection neurons and can yield only a limited number of neurons.

In addition to neuronal labeling techniques based on neuroanatomical properties and projection profiles, genetic approaches are gaining popularity as a means of visualizing and isolating neurons. Projects such as GENSAT (http://www.gensat.org/index.html) are using genetic labeling approaches to introduce a fluorescent character to a cell/neuron population based on their gene expression profile. Although this is a very powerful tool to label cells, it is still limited in its ability to verify specificity among cells/neurons that are labeled. Since the expression of a single gene may not always be strictly restricted to a neuron population, genetically labeled cells represent a heterogeneous population of cells, which happen to express the gene used to introduce a fluorescent character. As our knowledge and understanding of the molecular and genetic determinants of neuronal identity develop, these genetic labeling approaches will be very useful in labeling neuron populations based on their subtype-specific gene expression profile, and sources like GENSAT will be invaluable to studying neurons of interest.

As we develop better and more powerful approaches to label and isolate neurons that show vulnerability in diseases, we also need to develop improved platforms to study the biology of a single neuron population with high precision and accuracy. With the current tools and applications, it is not easy to reveal neuronal responses to extracellular factors and cell-intrinsic mechanisms that trigger neuronal vulnerability.

Microfluidics as Novel Tools and Solutions to Limitations

The spatial and temporal controls afforded by microfluidic applications allow the development of precisely controlled microenvironments that mimic many aspects of *in vivo* conditions, and permit the setting and control of parameters that are not easy to establish in *in vivo* experiments.[46–49] In addition, microfluidic devices are compatible with high-resolution imaging and microscopy, such that direct cellular responses can be monitored in real-time. They allow flexibility in multiple different applications by having the potential to generate different combinations of channels, grooves, and chambers. Microfluidic platforms can be prepared by using different polymers, including biodegradable ones that can be implanted into the human body. In addition, due to reduced volume requirements in experiments, they allow a significant reduction in cost. The many important characteristics of microfluidics and their utilization via microfluidic devices have improved experimental approaches in many different fields, including neuroscience.[50–54] We will focus our attention on five main characteristics of microfluidic devices: (1) their small size reduces the burden of culturing large number of neurons *in vitro* and allows the detailed investigation of a single neuron and its responses to a given stimuli with the highest level of precision; (2) the presence of chambers and channels allows spatial precision in plating, enabling the culture of very polarized neurons, and thus offers a distinct advantage over current culture approaches, in which the polarity of the neuron is neglected; (3) their capacity for spatial and temporal control of factors of interest enables the creation of realistic microenvironments that mimic the microenvironment of a neuron; (4) their compatibility with high-resolution live-cell imaging and microscopy allows the observation of direct cellular responses; and (5) in combination with biodegradable materials, they offer potential cellular therapeutic applications in the future for directed cellular/neuronal transplantation efforts.

Applications of microfluidic devices in neuroscience

One of the major advantages of using microfludic devices in neuroscience is their ability to bring about and reveal the biology of the neuron of interest by generating a controlled environment that is easy to manipulate[46–48] and monitor by real-time live imaging.[55] The oxygen levels,[56] the pH levels,[57] and the concentration gradients of extrinsic factors introduced[58–60] are strictly controlled and monitored. This is a great advantage over other *in vitro* applications, in which factors of interest are added homogeneously to both the cell body and the axon, neglecting the importance of neuronal polarity. In addition, having a strict control over extrinsic factors, which is not possible in *in vivo* applications, augments the importance of microfluidic devices and their proper use in the field of neuroscience.

Microfluidic devices have been used to culture a wide variety of cells, such as hepatocytes,[61] glial cells,[62] embryonic stem (ES) cells,[63] and myoblasts.[57] Their application in neuroscience is becoming increasingly evident.[48,64–66] Placing neurons specifically in distinct chambers, and positioning them with a distinct pattern on surfaces, enables a spatial resolution needed to study the neuron type-specific responses to given stimuli,[67] their interaction with other cells/neurons that are located nearby, as well as their directed differentiation under the influence of given factors.[68]

Neuron type-specific analysis

Since neurons show a high level of heterogeneity, and not all neurons can be accepted as one cell type, it seems obvious that taking a piece of brain for cellular and neuronal analysis of the cortex should not yield any reliable information. However, due to the lack of appreciation for neuronal heterogeneity and the absence of molecular markers and proper technologies to identify and purify neurons of interest from the complex structure of the brain, neuronal heterogeneity was mostly ignored in the past. There is a growing need to improve neuron type-specific analysis to enhance our understanding of differences among distinct neuron populations.

Attention to cell type-specific analysis has led to the development of diagnostic tools that take advantage of the surface protein profile of a cell. Microfluidic chambers use markers that are present in the cell membrane as a bait to identify and purify cells that are extremely rare in the blood.[69] This approach has revolutionized global health as it allows easy and cost-effective detection of diseases in large populations. Using a microfluidic chip, in which the channels are coated with antibodies that detect a certain epitope that is present only on the surface of a rare cell, we can now identify the presence of cancerous cells and cells transduced with viruses (e.g., HIV), by using only a small drop of blood.[70] It is possible to use similar approaches in the field of neuroscience as our knowledge of receptor expression profiles of distinct sets of neurons develops (Fig. 2).

As we uncover more on the proteins that become increasingly present either in the blood or in cerebrospinal fluid with respect to disease initiation and progression, we can begin to use approaches to detect these proteins even at very low concentrations.[71] This approach is used to detect prostate-specific antigen (PSA) in patients as an early detection marker for prostate cancer,[72] and amyloid-beta–derived diffusible ligands (ADDLs) as an early detection marker for Alzheimer's disease.[73] In the near future, as we identify proteins that become present in neurons that show cell type-specific vulnerability to diseases, such approaches will be valuable for generating tools to detect neurodegeneration at early stages of disease progression in patients.

Axon biology

Due to its polarity, a neuron is dependent on the health and stability of the axon for its connectivity and function. Various neurodegenerative diseases stem from defects in both retrograde (from distal end of the axonal toward cell body), and anterograde (from cell body to the distal part of the axon) axonal transport. These diseases are sometimes referred to as axonopathies, and include Alzheimer's disease,[74] hereditary spastic paraplegia (HSP),[75,76] spinal muscular atrophy (SMA),[77] and Rett syndrome.[78] Understanding the cellular aspects of axonal transport and maintenance of axonal health will have a huge impact in understanding the disease pathologies of different neurodegenerative diseases.

Injury can also affect axonal biology in an acute manner distinct from the more gradual deterioration in neurodegenerative diseases. For example, in spinal cord injury, multiple axon paths and neurocircuitries are damaged immediately upon impact, and the connections between the cell body and the axon terminals are lost. This eventually leads to neuron cell body degeneration and results in permanent paralysis in patients. It is therefore critically important to investigate axonal biology in detail. We need to develop novel platforms that allow the study of detailed cellular mechanisms related to axonal health and stability.

Since the diameter of the axon is very small, it is not easy to study the biology of axonal transport defects in neurons that are vulnerable in diseases. Investigators have taken advantage of the giant axon of the squid and developed *in vitro* axonoplasma cultures in which the cytoplasm of the axon is isolated

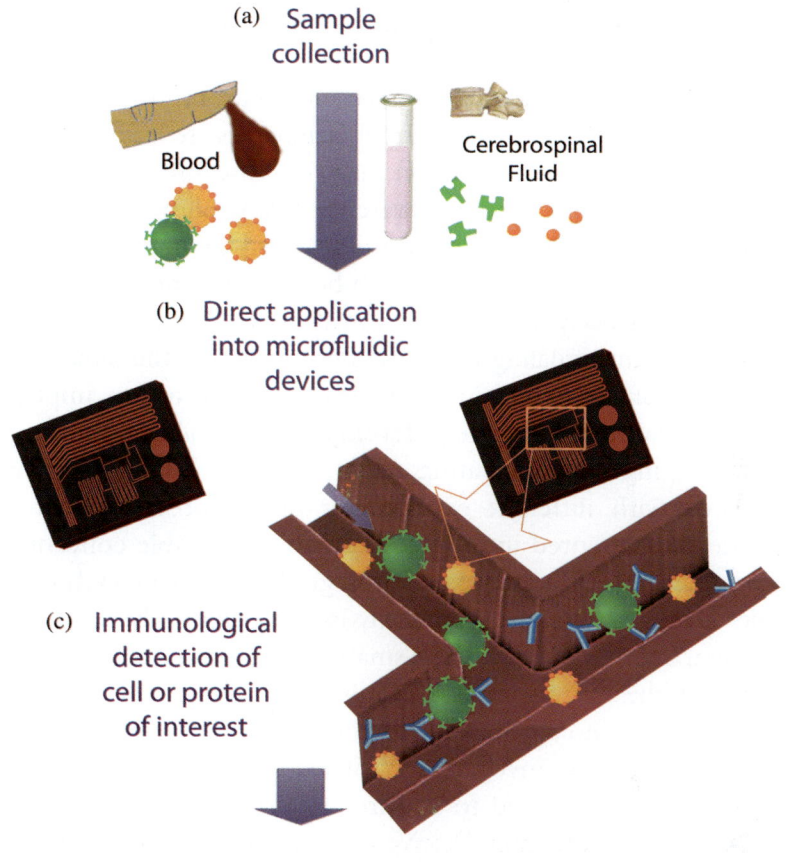

Fig. 2. Microfluidic devices can be used to detect very rare cells and proteins present in the blood and cerebrospinal fluid. (a) Disease-causing cells can be present at very low levels in the blood. Likewise, proteins that are related to disease pathogenesis can be present at low concentrations in the cerebrospinal fluid of a patient. (b) Both blood and cerebrospinal fluid can be directly applied to microfluidic devices that are coated with antibodies, which detect cells/proteins of interest. (c) Immunological detection and immobilization of cells/proteins of interest would help determine their presence even at very low concentrations and using very small volumes of starting biological material.

intact, and various parameters that would affect both anterograde and retrograde axonal transport are directly investigated on the axon.[79] These investigations have shed light on many biological phenomena such as fast axonal transport and slow axonal transport, and how these two different mechanisms are precisely manifested and controlled in the axon.[80] Even though the size of the squid giant axon offers advantages to the study of axonal biology, it comes with limitations. The squid is not an optimal model organism for neurodegenerative diseases and it is better to study axons of neurons that show primary vulnerability to disease.

In experimental setups in which dissociated neurons are cultured in large non-compartmentalized chambers, extracellular factors are homogeneously presented to the cell body and the axon, and information about axon-specific responses is lost. Campenot chambers were the first to allow a spatial separation between the neuron cell body and axon, and revealed the importance of location in neuronal responses to various stimuli.[81] It became evident that the application of factors to the cell body and to the axon do not exert the same effect, but rather initiate different signaling cascades of events within the same neuron.[82]

Compartmentalized microfluidic platforms offer major improvements over campenot chambers; they help develop a more elaborate microenvironment for developing and elongating axons, such as choice points, channels that are coated with different extracellular matrix proteins, and different factors of interest presented in different controlled variable concentrations.[83] One platform developed by the Jeon group guides axonal growth via multiple parallel microgrooves from one fluidically isolated cell body compartment to another compartment.[84] Due to their small dimensions (3-μm height), these grooves prevent the entrance of cell bodies, and thus generate a complete separation between the cell body and axon. Such platforms therefore allow the detailed analysis and investigation of axonal responses to stimuli and injury,[84,85] and have been used to culture and study peripheral neurons and their responses to axonopathy.[86,87]

Detailed biological investigations of movement within CNS axons were also made possible with the application of compartmentalized microfluidic devices. Recent studies applied single-molecule imaging techniques to visualize the movement of nerve growth factor (NGF)–quantum dot-tagged endosomes across microtubules during axonal transport process.[88] Using similar microfluidic platforms, axonal transport of distally applied alpha-herpes virus[89] and the retrograde transport of endosomes using either brain derived neutrophic factor (BDNF)–green fluorescent protein (GFP)[90] or NGF-conjugated quantum dots[91] have been studied. The role of axon–oligodendrocyte interactions in oligodendrocyte precursor maturation and the initiation of axon myelination have also been studied using microfluidic chambers that allow the co-culture of axons with oligodendrocyte precursors.[92]

Compartmentalized microfluidic platforms also enable axons to be harvested for further biochemical and genetic analysis. One important application is for the characterization of the mRNA pool in cortical axons after axotomy. Traditional non-compartmentalized chambers cannot isolate the distinct set of mRNA present in the axon from that in the soma.[93] Further development of such applications using pure populations of subtype-specific neuron populations would reveal potential differences between neurons to injury. Such platforms may prove to be very useful in the search of molecules and factors that enhance axon regeneration. By applying multiple factors to damaged axons in a controlled manner, their regrowth potential can be visually inspected and studied.

The role of the extracellular matrix on axonal outgrowth and choice points is an active area of research. We would like to know why distinct neuron populations use a particular path to reach their targets and whether we can manipulate other neuron populations to change their course during development or in relation to injury and neurodegeneration. This information would help us understand how connections are formed and maintained, and would set the stage for future attempts to rewire cortical or corticospinal connections that have been lost in injury and disease. Engineered micropatterned surfaces are beginning to reveal the interaction between the environment and the cell on determining cell shape, direction of axon outgrowth, and choice point decision.[68,94–96]

Directed neuronal differentiation

A new and exciting area in the field of neuroscience is the potential for generating different neuron types from neural precursors, stems cells, and dedifferentiated non-pluripotent cells. The generation of pluripotent stem cells artificially from terminally differentiated non-pluripotent cells by induced expression of a combination of specific genes, such as *Oct3/4, Sox2, cMyc,* and *Klf4*, created excitement in the field and opened up a new area of research.[97,98] Induced pluripotent stem (iPS) cells resemble pluripotent stem cells, such as ES cells in many ways. For example, they express certain stem cell markers similar to ES cells, and their chromatin methylation patterns and embryoid body formation are very similar to ES cells; however, the extent to which they emulate ES cells has not been fully characterized.

Even though the exact cellular and molecular mechanisms of directed differentiation of neurons toward more specific sub-types are not fully understood, the ability of a few groups to generate neurons from different sources such as skin biopsies of ALS patients[99] and fibroblasts isolated from PD patients[100] generated excitement for future cell therapy approaches in neurodegenerative diseases. The ability to generate neurons from stem cells,[101] and more broadly, the ability to have full control over the directed differentiation of stem cells into the cell lineage of interest, would give us huge power over the development of cell replacement therapies in neurodegenerative diseases and injury-induced neuronal loss. However, we still do not understand the intrinsic and extrinsic controls that shape and instruct differentiation in great detail. Such studies require the presence of the correct combination of intrinsic and extrinsic factors on precursor cells. Given the number of potential candidates that could direct neuronal differentiation, it is not easy to test their direct role in any given setting. However, microfluidic applications could prove to be useful for such studies in the near future. The ability to plate a single cell/neuron within a given spatial resolution, and the ability to apply a distinct set of factors with given concentrations, coupled with the ability to monitor cellular responses, gives researchers a unique opportunity to investigate the role of multiple extrinsic factors and their combinatorial effects on directed differentiation of neural precursors, and on cells that have differentiation potential.

Introduction of neuron type-specific analysis to drug discovery

To date, despite the amount of investment and emphasis given to drug discovery approaches for neurodegenerative diseases, not a single drug offers a sustainable cure to any neurodegenerative disease. This may be in part due to the fact that the majority of drug discovery approaches use cells that have very limited relevance to neurodegenerative diseases. Despite their limited relevance, cell lines and non-neuronal cells are used because they are easy to obtain and culture. However, although the drug discovery fields seem to have realized the importance of using neurons that matter, the technologies to incorporate cells/neurons of interest in cell-based assays are not yet fully defined.

This decade has seen rapid progress in the movement of drug discovery into the domain of neuron type-specific research. An early step of this direction was the introduction of mutant proteins, which induced neuronal vulnerability and degeneration, into cell lines that are used for drug discovery. One example is the introduction and expression of *SOD1* mutations in PC12 cells to screen for compounds that lower *SOD1* expression.[102] Another important step toward neuron type-specific analysis came with the generation of spinal motor neurons from ES cells and iPS cells via directed differentiation.[99,103,104] Even though these iPS cell-derived motor neuron cultures do not truly mimic all the aspects of spinal motor neurons and are not present as a pure neuron population within the culture, their use represents a positive development in the field of drug discovery. A third step in improvement is the generation of primary cell lines from dorsal root ganglion (DRG) neurons. Co-transfection of *SV40LargeT* and *hTERT* genes keeps neurons in the proliferative state, and neurons are taken out from this proliferative state by the addition of mitogens to their culture. These neurons are used for drug discovery efforts related to peripheral neuropathies.[105,106]

Since one of the major limitations in drug discovery is the requirement for a high cell number and easy culture conditions, microfluidic devices may offer a solution for bringing cell type specificity and direct readouts of neuronal survival in the presence of small molecules/compounds. Microfluidic chambers will allow culturing only a limited number of neurons and will permit the directed application of small molecules at given concentrations. Most importantly, these cultures can be performed on such a small scale that the whole culture can be maintained in a heated and CO_2-supplemented chamber located on top of an inverted microscope for live imaging and monitoring of neuronal health and degeneration. Since novel methodologies are being developed to plate cells of interest with high spatial resolution,[107,108] neurons that are related to a distinct neurodegenerative disease can be isolated and plated on microfluidic devices. Neurons of interest can be isolated either by retrograde labeling approaches that are coupled with FACS-mediated purification, or by using transgenic mice in which a distinct neuron population is genetically labeled using its neuron type-specific gene expression profile (Fig. 3). Another approach involves the immortalization of primary neuron cultures isolated from disease models. In the near future, similar approaches can be used to generate cell lines from neurons of interest, and direct cellular responses to drugs can be investigated by live/dead assay and real-time microscopy (Fig. 3). Such investigations would improve the analysis of potential drug candidates on neuron survival, and bring cellular relevance to disease.

The lack of a cure for any of the neurodegenerative diseases underlines the urgency with which drug discovery efforts need to transit into neuron type-specific analyses, despite greater difficulty in their implementation. Microfluidic devices can circumvent these limitations by minimizing the necessity for a large setup and high cell numbers required by traditional approaches. In addition, they can reduce cost since much less media and compounds/drugs are used for analysis. In the near future, the field of drug discovery will feel the need to adapt itself to accommodate analysis using neurons and cells that are directly relevant to disease.

Investigation of cellular interactions

Cell–cell interactions are very important for various biological responses and signaling cascades of events such as maturation, differentiation, and the initiation of novel cellular responses. Neurons also require proper extracellular stimuli to differentiate, mature, and act as functional neurons, and it is possible

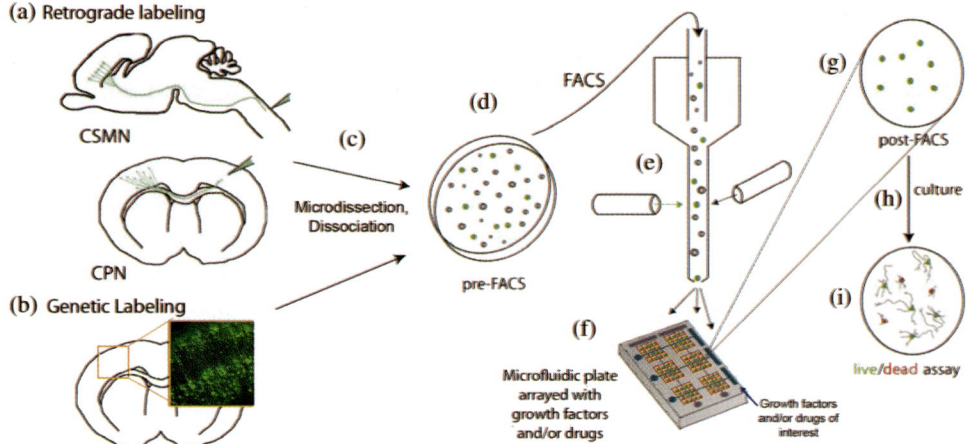

Fig. 3. Drug discovery approaches can be improved by using microfluidic devices and subtype-specific neuron populations that are vulnerable to disease. Neurons that show primary vulnerability can be either retrograde (a), or genetically (b) labeled with a fluorescent character, and dissected out (c)–(d). Pure populations of neurons can be obtained by FACS (fluorescence activated cell sorting)-mediated purification (e). These neurons can be cultured on microfluidic devices (f), where drugs of interest are directly applied. Using pure populations of neurons in culture (g), the efficacy of drugs on neuron survival can be directly investigated in culture (h), via live/dead assay (i).

that many aspects of neuronal diversity are in part due to different extracellular stimuli neurons receive from other cells in their microenvironment. What are these effectors and how do they exert their effect on distinct types of neurons? Which neurons or other cell types are involved in their neuronal development and maturation, and how is that interaction mediated? How do these extracellular effectors and cell–cell interactions affect neuronal vulnerability and degeneration in disease? These are some of the questions that await detailed answers in the field of neuroscience.

Microfluidic devices carry huge potential for revealing distinct cell type-specific responses to extracellular environments as well as cell–cell interactions. With their ability to generate controlled environments, whereby different cell types can be plated and cultured within defined chambers, in which the flow of medium and concentration gradient of extracellular factors are strictly controlled, microfluidic devices have the potential to revolutionize our understanding of neuron–neuron and neuron–non-neuronal cell interactions. Findings from these studies will have direct relevance to neuronal vulnerability in neurodegenerative diseases.

To study cellular interactions within a microenvironment, microfluidic channels can either be coated with different extracellular matrix proteins or can be enzymatically cross-linked, such as with cross-linked gelatin, to generate spatially distinct microenvironments. The cells' preferences over distinct extracellular matrix proteins to attach, to grow, and to differentiate are investigated and important information on cell type-specific responses are revealed.[109,110]

Co-culture applications using microfluidic devices are applied to a wide variety of studies to investigate the details of cellular interaction of numerous cell types, such as the host–pathogen interactions of bacteria and epithelial cells of the gastrointestinal tract.[111] Cell–cell interactions can be studied either by using microfluidic platforms, which have defined channels and chambers, or by generating 3D layered scaffolds, where cells of interest are trapped in engineered extracellular matrix environments.[96] These 3D layered structures can strongly mimic *in vivo* conditions, while providing a level of control impossible in *in vivo* conditions. In addition, patterned stencils can be used to generate different co-culture platforms, where cell–cell interactions can be studied at different spatial resolutions.[112,113] An example of high-resolution study uses a microfabricated cell-pair array in two different types of cells that are trapped in a small environment, and their ability to instruct each other to differentiate in a given and controlled environment is studied.[114]

The close interaction between oligodendrocyte precursors and axons can be investigated by using a specially designed microfluidic chamber in which the neuronal cell body and axon are spatially separated and oligodendrocyte precursors are plated only together with the axon (Fig. 4). Such studies would mimic *in vivo* conditions, since oligodendrocytes myelinate axons and are usually not in close contact with the neuron cell body, and they would reveal important biological aspects of myelination, which has direct relevance to diseases such as multiple sclerosis in which myelination is affected.[92]

The importance of the microenvironment and cellular interactions with non-neuronal cells are becoming more evident for neuron health. The activation of astrocytes (astrogliosis), is observed in numerous neurodegenerative diseases, especially toward the end-stage of disease progression. Astrogliosis has been suggested both to be detrimental and supportive of neuron health, and this may be due to cellular heterogeneity even among activated astrocytes.[115] The importance of the neuron–astrocyte interaction has been strongly implicated both in disease initiation and progression in ALS. It is now believed that via unknown mechanisms, the interaction between motor neurons and astrocytes may influence neuron health and vulnerability. Studies using chimeric transgenic mouse models[116,117] and *in vitro* applications[101] indicate the importance and relevance of astrocytes and other non-neuronal cells for spinal motor neuron health. However, to determine the molecules

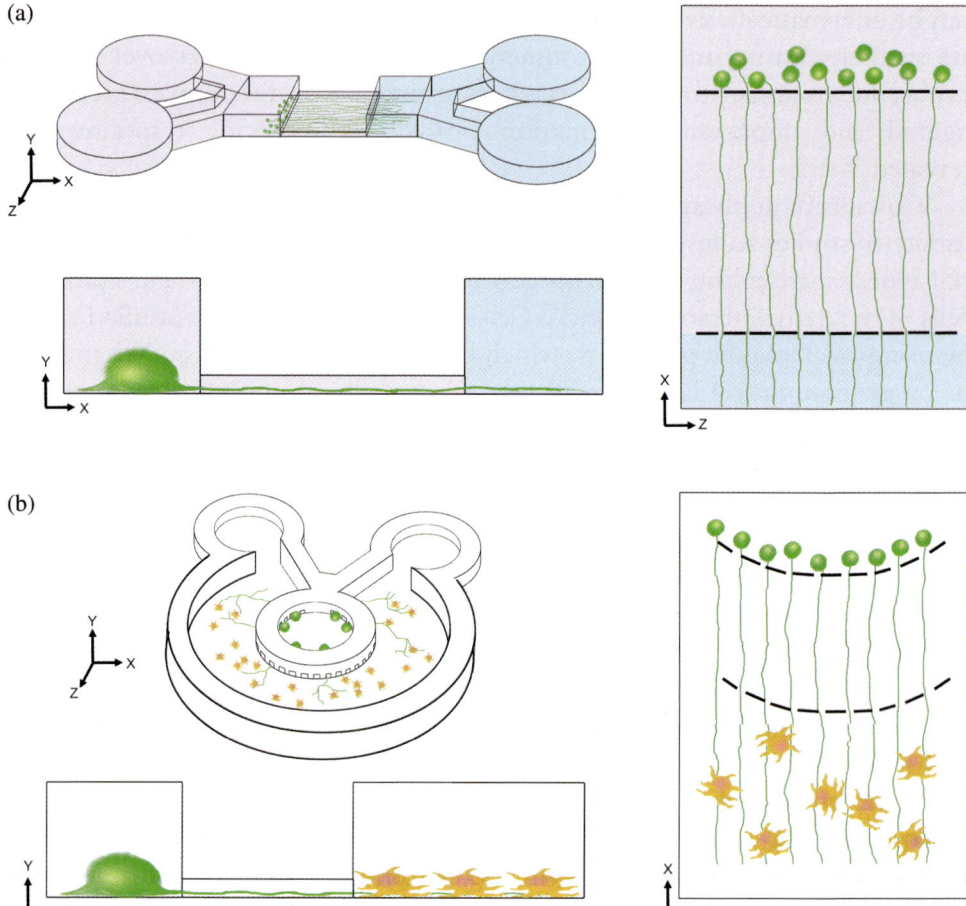

Fig. 4. New microfluidic culture platforms are developed to study axonal biology in detail. (a) Schematic drawing of a new microfluidic culture platform developed by the Jeon group to give a distinct spatial resolution to soma and axon of neurons. (b) Schematic drawing of a novel microfluidic culture platform developed by the Han group to investigate cell–axon interactions.

that mediate such responses and the conditions by which they exert their cellular effect, we need to build and use well-defined co-culture platforms (Fig. 5).

To further investigate the neuron and astrocyte interaction in relation to disease, it will be important to isolate and study pure populations of neurons and astrocytes. Identification of genes that are primarily expressed in developing spinal motor neurons, and generation of transgenic mouse models that express GFP under the promoter of neuron-type specific genes allow genetic labeling, visualization, and isolation of neurons of interest. One good example is the Hb9–GFP spinal motor neurons. The *Hb9* gene is expressed in spinal

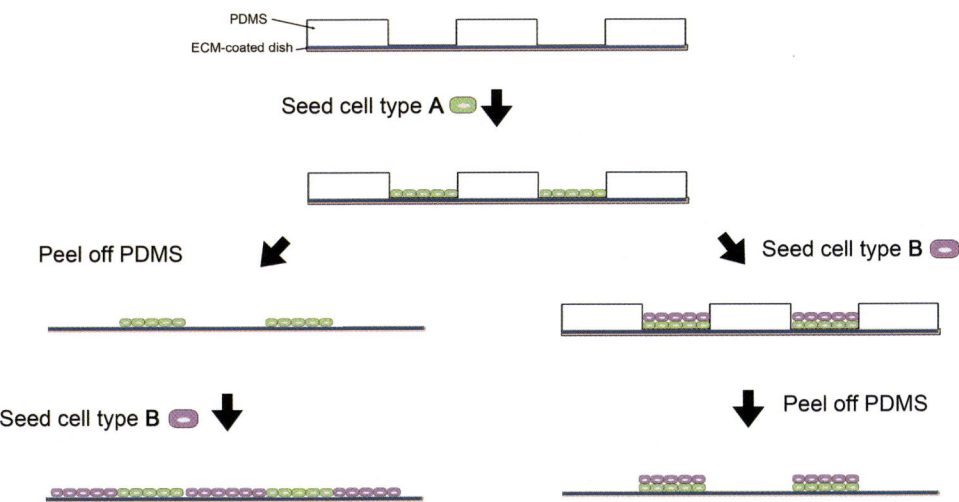

Fig. 5. Direct cell–cell interactions can be studied using different seeding approaches on microfluidic chambers. Cellular interaction of two different cell types can be investigated by plating cells sequentially in different spatial configurations. (Adapted from Ref. 111.)

motor neurons[118] and its expression in the Hb9–GFP transgenic mouse enables identification, purification, and culturing of GFP-labeled spinal motor neurons as a "pure" spinal motor neuron population. Non-neuronal cells, which adhere to noncharged substrates more readily than neurons, can be isolated as pure cell populations *in vitro*. Therefore, it would be possible to establish co-culture approaches using pure populations of astrocytes and spinal motor neurons, which are vulnerable in motor neuron diseases such as ALS and spinal muscular atrophy. Using microfluidic platforms for such cellular investigations would reveal cellular dynamics in a well-defined and controlled environment.

Summary

Many limitations to progress in the field of neuroscience can be overcome as we develop approaches allowing the investigation of neurons of interest in great detail. When the intrinsic and extrinsic factors that are important for neuronal differentiation, maturation, and sustainable health are revealed, we will begin to develop effective therapies for neurodegenerative diseases, and will have a better understanding of how the nervous system develops and functions.

Microfluidics offer unprecedented advantages over current *in vitro* applications, and many aspects of *in vivo* experimentation. Being able to manipulate, control, and engineer an environment to study the biology of a neuron, and to monitor cell type-specific responses in real time is revolutionary.

Acknowledgments

The authors declare no financial interest. This work is supported by Les Turner ALS and Wenske Foundation grants (to P.H.O), and V.M.H. is partially supported by NIH-T32 training grant. We would like to thank Dr. J. Jara, Dr. N. Khan, Dr. B. Genc, M. Tu, W. Weber, and other members of the Özdinler lab for critical reading of the manuscript.

References

1. Pakkenberg, B., Pelvig, D., Marner, L., Bundgaard, M.J., Gundersen, H.J., Nyengaard, J.R., Regeur, L., Aging and the human neocortex. *Exp Gerontol*, 2003, **38**: p. 95–99.
2. Anderson, S.A., Kaznowski, C.E., Horn, C., Rubenstein, J.L., McConnell, S.K., Distinct origins of neocortical projection neurons and interneurons *in vivo*. *Cereb Cortex*, 2002, **12**: p. 702–709.
3. Guillemot, F., Molnar, Z., Tarabykin, V., Stoykova, A., Molecular mechanisms of cortical differentiation. *Eur J Neurosci*, 2006, **23**: p. 857–868.
4. Branco, T., Hausser, M., The single dendritic branch as a fundamental functional unit in the nervous system. *Curr Opin Neurobiol*, 2010, **20**: p. 494–502.
5. Knobloch, M., Mansuy, I.M., Dendritic spine loss and synaptic alterations in Alzheimer's disease. *Mol Neurobiol*, 2008, **37**: p. 73–82.
6. Luebke, J.I., Weaver, C.M., Rocher, A.B., Rodriguez, A., Crimins, J.L., Dickstein, D.L., Wearne, S.L., Hof, P.R., Dendritic vulnerability in neurodegenerative disease: Insights from analyses of cortical pyramidal neurons in transgenic mouse models. *Brain Struct Funct*, 2010, **214**: p. 181–199.
7. Molyneaux, B.J., Arlotta, P., Macklis, J.D., Molecular development of corticospinal motor neuron circuitry. *Novartis Found Symp*, 2007, **288**: p. 3–15; discussion 15–20, 96–18.
8. Shoemaker, L.D., Arlotta, P., Untangling the cortex: Advances in understanding specification and differentiation of corticospinal motor neurons. *Bioessays*, 2010, **32**: p. 197–206.
9. Tomassy, G.S., Lodato, S., Trayes-Gibson, Z., Arlotta, P., Development and regeneration of projection neuron subtypes of the cerebral cortex. *Sci Prog*, 2010, **93**: p. 151–169.
10. Fame, R.M., MacDonald, J.L., Macklis, J.D., Development, specification, and diversity of callosal projection neurons. *Trends Neurosci*, 2011, **34**: p. 41–50.
11. Parra, P., Gulyas, A.I., Miles, R., How many subtypes of inhibitory cells in the hippocampus? *Neuron*, 1998, **20**: p. 983–993.
12. Buzsaki, G., Geisler, C., Henze, D.A., Wang, X.J., Interneuron diversity series: Circuit complexity and axon wiring economy of cortical interneurons. *Trends Neurosci*, 2004, **27**: p. 186–193.
13. Xu, Q., Cobos, I., De La Cruz, E., Rubenstein, J.L., Anderson, S.A., Origins of cortical interneuron subtypes. *J Neurosci*, 2004, **24**: p. 2612–2622.
14. Di Cristo, G., Development of cortical GABAergic circuits and its implications for neurodevelopmental disorders. *Clin Genet*, 2007, **72**: p. 1–8.
15. Batista-Brito, R., Fishell, G., The developmental integration of cortical interneurons into a functional network. *Curr Top Dev Biol*, 2009, **87**: p. 81–118.

16. Birke, G., Draguhn, A., No simple brake — the complex functions of inhibitory synapses. *Pharmacopsychiatry*, 2010, **43**(Suppl 1): p. S21–S31.
17. Tepper, J.M., Tecuapetla, F., Koos, T., Ibanez-Sandoval, O., Heterogeneity and diversity of striatal GABAergic interneurons. *Front Neuroanat*, 2010, **4**: p. 150.
18. MacNeil, M.A., Masland, R.H., Extreme diversity among amacrine cells: Implications for function. *Neuron*, 1998, **20**: p. 971–982.
19. Molyneaux, B.J., Arlotta, P., Fame, R.M., MacDonald, J.L., MacQuarrie, K.L., Macklis, J.D., Novel subtype-specific genes identify distinct subpopulations of callosal projection neurons. *J Neurosci*, 2009, **29**: p. 12343–12354.
20. Molyneaux, B.J., Arlotta, P., Hirata, T., Hibi, M., Macklis, J.D., Fezl is required for the birth and specification of corticospinal motor neurons. *Neuron*, 2005, **47**: p. 817–831.
21. Rouaux, C., Arlotta, P., Fezf2 directs the differentiation of corticofugal neurons from striatal progenitors in vivo. *Nat Neurosci*, 2010, **13**: p. 1345–1347.
22. Kriegstein, A., Noctor, S., Martinez-Cerdeno,. V., Patterns of neural stem and progenitor cell division may underlie evolutionary cortical expansion. *Nat Rev Neurosci*, 2006, **7**: p. 883–890.
23. Dehay, C., Kennedy, H., Cell-cycle control and cortical development. *Nat Rev Neurosci*, 2007, **8**: p. 438–450.
24. Noctor, S.C., Martinez-Cerdeno, V., Kriegstein, A.R., Neural stem and progenitor cells in cortical development. *Novartis Found Symp*, 2007, **288**: p. 59–73; discussion 73–58, 96–58.
25. Neary, J.T., Zimmermann, H., Trophic functions of nucleotides in the central nervous system. *Trends Neurosci*, 2009, **32**: p. 189–198.
26. MacLaren, R.E., Pearson, R.A., MacNeil, A., Douglas, R.H., Salt, T.E., Akimoto, M., Swaroop, A., Sowden, J.C., Ali, R.R., Retinal repair by transplantation of photoreceptor precursors. *Nature*, 2006, **444**: p. 203–207.
27. Sohur, U.S., Emsley, J.G., Mitchell, B.D., Macklis, J.D., Adult neurogenesis and cellular brain repair with neural progenitors, precursors and stem cells. *Philos Trans R Soc Lond B Biol Sci*, 2006, **361**: p. 1477–1497.
28. Gaillard, A., Prestoz, L., Dumartin, B., Cantereau, A., Morel, F., Roger, M., Jaber, M., Reestablishment of damaged adult motor pathways by grafted embryonic cortical neurons. *Nat Neurosci*, 2007, **10**: p. 1294–1299.
29. Southwell, D.G., Froemke, R.C., Alvarez-Buylla, A., Stryker, M.P., Gandhi, S.P., Cortical plasticity induced by inhibitory neuron transplantation. *Science*, 2010, **327**: p. 1145–1148.
30. Valcanis, H., Tan, S.S., Layer specification of transplanted interneurons in developing mouse neocortex. *J Neurosci*, 2003, **23**: p. 5113–5122.
31. Xu, L., Yan, J., Chen, D., Welsh, A.M., Hazel, T., Johe, K., Hatfield, G., Koliatsos, V.E., Human neural stem cell grafts ameliorate motor neuron disease in SOD-1 transgenic rats. *Transplant*, 2006, **82**: p. 865–875.
32. Hedlund, E., Hefferan, M.P., Marsala, M., Isacson, O., Cell therapy and stem cells in animal models of motor neuron disorders. *Eur J Neurosci*, 2007, **26**: p. 1721–1737.
33. Kremer, B., Weber, B., Hayden, M.R., New insights into the clinical features, pathogenesis and molecular genetics of Huntington disease. *Brain Pathol*, 1992, **2**: p. 321–335.
34. Bates, G.P., History of genetic disease: The molecular genetics of Huntington disease — a history. *Nat Rev Genet*, 2005, **6**: p. 766–773.

35. Del Tredici, K., Rub, U., De Vos, R.A., Bohl, J.R., Braak, H., Where does Parkinson disease pathology begin in the brain? *J Neuropathol Exp Neurol*, 2002, **61**: p. 413–426.
36. Braak, H., Del Tredici, K., Invited article: Nervous system pathology in sporadic Parkinson disease. *Neurology*, 2008, **70**: p. 1916–1925..
37. Murray, L.M., Comley, L.H., Thomson, D., Parkinson, N., Talbot, K., Gillingwater, T.H., Selective vulnerability of motor neurons and dissociation of pre- and post-synaptic pathology at the neuromuscular junction in mouse models of spinal muscular atrophy. *Hum Mol Genet*, 2008, **17**: p. 949–962.
38. Pasinelli, P., Brown, R.H., Molecular biology of amyotrophic lateral sclerosis: Insights from genetics. *Nat Rev Neurosci*, 2006, **7**: p. 710–723.
39. Jones, C.T., Swingler, R.J., Brock, D.J., Identification of a novel SOD1 mutation in an apparently sporadic amyotrophic lateral sclerosis patient and the detection of Ile113Thr in three others. *Hum Mol Genet*, 1994, **3**: p. 649–650.
40. Penco, S., Schenone, A., Bordo, D., Bolognesi, M., Abbruzzese, M., Bugiani, O., Ajmar, F., Garre, C., A SOD1 gene mutation in a patient with slowly progressing familial ALS. *Neurol*, 1999, **53**: p. 404–406.
41. Ricci, C.M., Benigni, M., Battistini, S., Greco, G., Torzini, A., Giannini, F. A novel exon 1 mutation (G10R) in the SOD1 gene in a patient with familial ALS. *Amyotroph Lateral Scler*, 2010, **11**: p. 481–485.
42. Catapano, L.A., Arnold, M.W., Perez, F.A., Macklis, J.D., Specific neurotrophic factors support the survival of cortical projection neurons at distinct stages of development. *J Neurosci*, 2001, **21**: p. 8863–8872.
43. Ozdinler, P.H., Macklis, J.D., IGF-I specifically enhances axon outgrowth of corticospinal motor neurons. *Nat Neurosci*, 2006, **9**: p. 1371–1381.
44. Barres, B.A., Silverstein, B.E., Corey, D.P., Chun, L.L., Immunological, morphological, and electrophysiological variation among retinal ganglion cells purified by panning. *Neuron*, 1988, **1**: p. 791–803.
45. Dugas, J.C., Mandemakers, W., Rogers, M., Ibrahim, A., Daneman, R., Barres, B.A., A novel purification method for CNS projection neurons leads to the identification of brain vascular cells as a source of trophic support for corticospinal motor neurons. *J Neurosci*, 2008, **28**: p. 8294–8305.
46. Atencia, J., Beebe, D.J., Controlled microfluidic interfaces. *Nature*, 2005, **437**: p. 648–655.
47. Kim, S., Kim, H.J., Jeon, N.L., Biological applications of microfluidic gradient devices. *Integr Biol (Camb)*, 2010, **2**: p. 584–603.
48. Taylor, A.M., Jeon, N.L., Micro-scale and microfluidic devices for neurobiology. *Curr Opin Neurobiol*, 2010, **20**: p. 640–647.
49. Velve-Casquillas, G., Le Berre, M., Piel, M., Tran, P.T., Microfluidic tools for cell biological research. *Nano Today*, 2010, **5**: p. 28–47.
50. Weibel, D.B., Garstecki, P., Whitesides, G.M., Combining microscience and neurobiology. *Curr Opin Neurobiol*, 2005, **15**: p. 560–567.
51. Whitesides, G.M., The origins and the future of microfluidics. *Nature*, 2006, **442**: p. 368–373.

52. Gross, P.G., Kartalov, E.P., Scherer, A., Weiner, L.P., Applications of microfluidics for neuronal studies. *J Neurol Sci*, 2007, **252**: p. 135–143.
53. Weibel, D.B., Diluzio, W.R., Whitesides, G.M., Microfabrication meets microbiology. *Nat Rev Microbiol*, 2007, **5**: p. 209–218.
54. Wang, J., Ren, L., Li, L., Liu, W., Zhou, J., Yu, W., Tong, D., Chen, S., Microfluidics: A new cosset for neurobiology. *Lab Chip*, 2009, **9**: p. 644–652.
55. Hsieh, C.C., Huang, S.B., Wu, P.C., Shieh, D.B., Lee, G.B., A microfluidic cell culture platform for real-time cellular imaging. *Biomed Microdevices*, 2009, **11**: p. 903–913.
56. Mehta, G., Mehta, K., Sud, D., Song, J.W., Bersano-Begey, T., Futai, N., Heo, Y.S., Mycek, M.A., Linderman, J.J., Takayama, S., Quantitative measurement and control of oxygen levels in microfluidic poly(dimethylsiloxane) bioreactors during cell culture. *Biomed Microdevices*, 2007, **9**: p. 123–134.
57. Ges, I.A., Ivanov, B.L., Werdich, A.A., Baudenbacher, F.J., Differential pH measurements of metabolic cellular activity in nl culture volumes using microfabricated iridium oxide electrodes. *Biosens Bioelectron*, 2007, **22**: p. 1303–1310.
58. Lin, F., Saadi, W., Rhee, S.W., Wang, S.J., Mittal, S., Jeon, N.L., Generation of dynamic temporal and spatial concentration gradients using microfluidic devices. *Lab Chip*, 2004, **4**: p. 164–167.
59. Saadi, W., Rhee, S.W., Lin, F., Vahidi, B., Chung, B.G., Jeon, N.L., Generation of stable concentration gradients in 2D and 3D environments using a microfluidic ladder chamber. *Biomed Microdevices*, 2007, **9**: p. 627–635.
60. Amadi, O.C., Steinhauser, M.L., Nishi, Y., Chung, S., Kamm, R.D., McMahon, A.P., Lee, R.T., A low resistance microfluidic system for the creation of stable concentration gradients in a defined 3D microenvironment. *Biomed Microdevices*, 2010, **12**: p. 1027–1041.
61. Leclerc, E., Sakai, Y., Fujii, T., Microfluidic PDMS (polydimethylsiloxane) bioreactor for large-scale culture of hepatocytes. *Biotechnol Prog*, 2004, **20**: p. 750–755.
62. Peterson, S.L., McDonald, A., Gourley, P.L., Sasaki, D.Y., Poly(dimethylsiloxane) thin films as biocompatible coatings for microfluidic devices: Cell culture and flow studies with glial cells. *J Biomed Mater Res A*, 2005, **72**: p. 10–18.
63. Kim, L., Vahey, M.D., Lee, H.Y., Voldman, J., Microfluidic arrays for logarithmically perfused embryonic stem cell culture. *Lab Chip*, 2006, **6**: p. 394–406.
64. Li, N., Tourovskaia, A., Folch, A., Biology on a chip: Microfabrication for studying the behavior of cultured cells. *Crit Rev Biomed Eng*, 2003, **31**: p. 423–488.
65. Morin, F., Nishimura, N., Griscom, L., Lepioufle, B., Fujita, H., Takamura, Y., Tamiya, E., Constraining the connectivity of neuronal networks cultured on microelectrode arrays with microfluidic techniques: A step towards neuron-based functional chips. *Biosens Bioelectron*, 2006, **21**: p. 1093–1100.
66. Bhattacharjee, N., Li, N., Keenan, T.M., Folch, A., A neuron-benign microfluidic gradient generator for studying the response of mammalian neurons towards axon guidance factors. *Integr Biol (Camb)*, 2010, **2**: p. 669–679.
67. Rhee, S.W., Taylor, A.M., Cribbs, D.H., Cotman, C.W., Jeon, N.L., External force-assisted cell positioning inside microfluidic devices. *Biomed Microdevices*, 2007, **9**: p. 15–23.

68. von Philipsborn, A.C., Lang, S., Loeschinger, J., Bernard, A., David, C., Lehnert, D., Bonhoeffer, F., Bastmeyer, M., Growth cone navigation in substrate-bound ephrin gradients. *Development*, 2006, **133**: p. 2487–2495.
69. Toner, M., Irimia, D., Blood-on-a-chip. *Annu Rev Biomed Eng*, 2005, **7**: p. 77–103.
70. Kim, Y.G., Moon, S., Kuritzkes, D.R., Demirci, U., Quantum dot-based HIV capture and imaging in a microfluidic channel. *Biosens Bioelectron*, 2009, **25**: p. 253–258.
71. Nam, J.M., Thaxton, C.S., Mirkin, C.A., Nanoparticle-based bio-bar codes for the ultrasensitive detection of proteins. *Science*, 2003, **301**: p. 1884–1886.
72. Thaxton, C.S., Elghanian, R., Thomas, A.D., Stoeva, S.I., Lee, J.S., Smith, N.D., Schaeffer, A.J., Klocker, H., Horninger, W., Bartsch, G., Mirkin, C.A., Nanoparticle-based bio-barcode assay redefines "undetectable" PSA and biochemical recurrence after radical prostatectomy. *Proc Natl Acad Sci U S A*, 2009, **106**: p. 18437–18442.
73. Georganopoulou, D.G., Chang, L., Nam, J.M., Thaxton, C.S., Mufson, E.J., Klein, W.L., Mirkin, C.A., Nanoparticle-based detection in cerebral spinal fluid of a soluble pathogenic biomarker for Alzheimer's disease. *Proc Natl Acad Sci U S A*, 2005, **102**: p. 2273–2276.
74. Stokin, G.B., Lillo, C., Falzone, T.L., Brusch, R.G., Rockenstein, E., Mount, S.L., Raman, R., Davies, P., Masliah, E., Williams, D.S., Goldstein, L.S., Axonopathy and transport deficits early in the pathogenesis of Alzheimer's disease. *Science*, 2005, **307**: p. 1282–1288.
75. Miura, H., Oda, K., Endo, C., Yamazaki, K., Shibasaki, H., Kikuchi, T., Progressive degeneration of motor nerve terminals in GAD mutant mouse with hereditary sensory axonopathy. *Neuropathol Appl Neurobiol*, 1993, **19**: p. 41–51.
76. Pirozzi, M., Quattrini, A., Andolfi, G., Dina, G., Malaguti, M.C., Auricchio, A., Rugarli, E.I., Intramuscular viral delivery of paraplegin rescues peripheral axonopathy in a model of hereditary spastic paraplegia. *J Clin Invest*, 2006, **116**: p. 202–208.
77. Frequin, S.T., Gabreels, F.J., Gabreels-Festen, A.A., Joosten, E.M., Sensory axonopathy in hereditary distal spinal muscular atrophy. *Clin Neurol Neurosurg*, 1991, **93**: p. 323–326.
78. Wakai, S., Kameda, K., Ishikawa, Y., Miyamoto, S., Nagaoka, M., Okabe, M., Minami, R., Tachi, N., Rett syndrome: Findings suggesting axonopathy and mitochondrial abnormalities. *Pediatr Neurol*, 1990, **6**: p. 339–343.
79. Brady, S.T., Richards, B.W., Leopold, P.L., Assay of vesicle motility in squid axoplasm. *Methods Cell Biol*, 1993, **39**: p. 191–202.
80. Schroer, T.A., Brady, S.T., Kelly, R.B., Fast axonal transport of foreign synaptic vesicles in squid axoplasm. *J Cell Biol*, 1985, **101**: p. 568–572.
81. Campenot, R.B., Independent control of the local environment of somas and neurites. *Methods Enzymol*, 1979, **58**: p. 302–307.
82. Heerssen, H.M., Segal, R.A., Location, location, location: A spatial view of neurotrophin signal transduction. *Trends Neurosci*, 2002, **25**: p. 160–165.
83. Park, J.W., Kim, H.J., Byun, J.H., Ryu, H.R., Jeon, N.L., Novel microfluidic platform for culturing neurons: Culturing and biochemical analysis of neuronal components. *Biotechnol J*, 2009, **4**: p. 1573–1577.
84. Taylor, A.M., Blurton-Jones, M., Rhee, S.W., Cribbs, D.H., Cotman, C.W., Jeon, N.L., A microfluidic culture platform for CNS axonal injury, regeneration and transport. *Nat Methods*, 2005, **2**: p. 599–605.

85. Vahidi, B., Park, J.W., Kim, H.J., Jeon, N.L., Microfluidic-based strip assay for testing the effects of various surface-bound inhibitors in spinal cord injury. *J Neurosci Methods*, 2008, **170**: p. 188–196.
86. Hengst, U., Deglincerti, A., Kim, H.J., Jeon, N.L., Jaffrey, S.R., Axonal elongation triggered by stimulus-induced local translation of a polarity complex protein. *Nat Cell Biol*, 2009, **11**: p. 1024–1030.
87. Yang, I.H., Siddique, R., Hosmane, S., Thakor, N., Hoke, A., Compartmentalized microfluidic culture platform to study mechanism of paclitaxel-induced axonal degeneration. *Exp Neurol*, 2009, **218**: p. 124–128.
88. Mudrakola, H.V., Zhang, K., Cui, B., Optically resolving individual microtubules in live axons. *Struct*, 2009, **17**: p. 1433–1441.
89. Liu, W.W., Goodhouse, J., Jeon, N.L., Enquist, L.W., A microfluidic chamber for analysis of neuron-to-cell spread and axonal transport of an alpha-herpesvirus. *PLoS One*, 2008, **3**: p. E2382.
90. Poon, W.W., Blurton-Jones, M., Tu, C.H., Feinberg, L.M., Chabrier, M.A., Harris, J.W., Jeon, N.L., Cotman, C.W., Beta-amyloid impairs axonal BDNF retrograde trafficking. *Neurobiol Aging*, 2009.
91. Cui, B., Wu, C., Chen, L., Ramirez, A., Bearer, E.L., Li, W.P., Mobley, W.C., Chu, S., One at a time, live tracking of NGF axonal transport using quantum dots. *Proc Natl Acad Sci U S A*, 2007, **104**: p. 13666–13671.
92. Park, J., Koito, H., Li, J., Han, A., Microfluidic compartmentalized co-culture platform for CNS axon myelination research. *Biomed Microdevices*, 2009, **11**: p. 1145–1153.
93. Taylor, A.M., Berchtold, N.C., Perreau, V.M., Tu, C.H., Li Jeon, N., Cotman, C.W., Axonal mRNA in uninjured and regenerating cortical mammalian axons. *J Neurosci*, 2009, **29**: p. 4697–4707.
94. Chen, C.S., Mrksich, M., Huang, S., Whitesides, G.M., Ingber, D.E., Micropatterned surfaces for control of cell shape, position, and function. *Biotechnol Prog*, 1998, **14**: p. 356–363.
95. Walker, G.M., Zeringue, H.C., Beebe, D.J., Microenvironment design considerations for cellular scale studies. *Lab Chip*, 2004, **4**: p. 91–97.
96. Li, N., Folch, A., Integration of topographical and biochemical cues by axons during growth on microfabricated 3-D substrates. *Exp Cell Res*, 2005, **311**: p. 307–316.
97. Aoi, T., Yae, K., Nakagawa, M., Ichisaka, T., Okita, K., Takahashi, K., Chiba, T., Yamanaka, S., Generation of pluripotent stem cells from adult mouse liver and stomach cells. *Science*, 2008, **321**: p. 699–702.
98. Okita, K., Nakagawa, M., Hyenjong, H., Ichisaka, T., Yamanaka, S., Generation of mouse induced pluripotent stem cells without viral vectors. *Science*, 2008, **322**: p. 949–953.
99. Dimos, J.T., Rodolfa, K.T., Niakan, K.K., Weisenthal, L.M., Mitsumoto, H., Chung, W., Croft, G.F., Saphier, G., Leibel, R., Goland, R., Wichterle, H., Henderson, C.E., Eggan, K., Induced pluripotent stem cells generated from patients with ALS can be differentiated into motor neurons. *Science*, 2008, **321**: p. 1218–1221.
100. Soldner, F., Hockemeyer, D., Beard, C., Gao, Q., Bell, G.W., Cook, E.G., Hargus, G., Blak, A., Cooper, O., Mitalipova, M., Isacson, O., Jaenisch, R., Parkinson's disease patient-derived induced pluripotent stem cells free of viral reprogramming factors. *Cell*, 2009, **136**: p. 964–977.

101. Di Giorgio, F.P., Boulting, G.L., Bobrowicz, S., Eggan, K.C., Human embryonic stem cell-derived motor neurons are sensitive to the toxic effect of glial cells carrying an ALS-causing mutation. *Cell Stem Cell*, 2008, **3**: p. 637–648.
102. Broom, W.J., Auwarter, K.E., Ni, J., Russel, D.E., Yeh, L.A., Maxwell, M.M., Glicksman, M., Kazantsev, M.A.G., Brown, R.H., Jr., Two approaches to drug discovery in SOD1-mediated ALS. *J Biomol Screen*, 2006, **11**: p. 729–735.
103. Hu, B.Y., Zhang, S.C., Directed differentiation of neural-stem cells and subtype-specific neurons from hESCs. *Methods Mol Biol*, 2010, **636**: p. 123–137.
104. Nizzardo, M., Simone, C., Falcone, M., Locatelli, F., Riboldi, G., Comi, G.P., Corti, S., Human motor neuron generation from embryonic stem cells and induced pluripotent stem cells. *Cell Mol Life Sci*, 2010, **67**: p. 3837–3847.
105. Chen, W., Mi, R., Haughey, N., Oz, M., Hoke, A., Immortalization and characterization of a nociceptive dorsal root ganglion sensory neuronal line. *J Peripher Nerv Syst*, 2007, **12**: p. 121–130.
106. Melli, G., Hoke, A., Dorsal root ganglia sensory neuronal cultures: A tool for drug discovery for peripheral neuropathies. *Expert Opin Drug Discovery*, 2009, **4**: p. 1035–1045.
107. Song, Y.S., Adler, D., Xu, F., Kayaalp, E., Nureddin, A., Anchan, R.M., Maas, R.L., Demirci, U., Vitrification and levitation of a liquid droplet on liquid nitrogen. *Proc Natl Acad Sci U S A*, 2010, **107**: p. 4596–4600.
108. Tasoglu, S., Kaynak, G., Szeri, A.J., Demirci, U., Muradoglu, M., Impact of a compound droplet on a flat surface: A model for single cell epitaxy. *Phys Fluids*, 1994, **2010**: p. 22.
109. Turner, A.M., Dowell, N., Turner, S.W., Kam, L., Isaacson, M., Turner, J.N., Craighead, H.G., Shain, W., Attachment of astroglial cells to microfabricated pillar arrays of different geometries. *J Biomed Mater Res*, 2000, **51**: p. 430–441.
110. Kam, L., Shain, W., Turner, J.N., Bizios, R., Selective adhesion of astrocytes to surfaces modified with immobilized peptides. *Biomaterials*, 2002, **23**: p. 511–515.
111. Kim, J., Hegde, M., Jayaraman, A., Co-culture of epithelial cells and bacteria for investigating host-pathogen interactions. *Lab Chip*, 2010, **10**: p. 43–50.
112. Cho, C.H., Park, J., Tilles, A.W., Berthiaume, F., Toner, M., Yarmush, M.L., Layered patterning of hepatocytes in co-culture systems using microfabricated stencils. *Bio Techniques*, 2010, **48**: p. 47–52.
113. Okuyama, T., Yamazoe, H., Mochizuki, N., Khademhosseini, A., Suzuki, H., Fukuda, J., Preparation of arrays of cell spheroids and spheroid-monolayer cocultures within a microfluidic device. *J Biosci Bioeng*, 2010, **110**: p. 572–576.
114. Parekkadan, B., Berdichevsky, Y., Irimia, D., Leeder, A., Yarmush, G., Toner, M., Levine, J.B., Yarmush, M.L., Cell-cell interaction modulates neuroectodermal specification of embryonic stem cells. *Neurosci Lett*, 2008, **438**: p. 190–195.
115. Hoke, A., Silver, J., Heterogeneity among astrocytes in reactive gliosis. *Perspect Dev Neurobiol*, 1994, **2**: p. 269–274.
116. Clement, A.M., Nguyen, M.D., Roberts, E.A., Garcia, M.L., Boillee, S., Rule, M., McMahon, A.P., Doucette, W., Siwek, D., Ferrante, R.J., Brown, R.H., Jr., Julien, J.P., Goldstein, L.S., Cleveland, D.W., Wild-type nonneuronal cells extend survival of SOD1 mutant motor neurons in ALS mice. *Science*, 2003, **302**: p. 113–117.

117. Boillee, S., Yamanaka, K., Lobsiger, C.S., Copeland, N.G., Jenkins, N.A., Kassiotis, G., Kollias, G., Cleveland, D.W., Onset and progression in inherited ALS determined by motor neurons and microglia. *Science,* 2006, **312**: p. 1389–1392.
118. Arber, S., Han, B., Mendelsohn, M., Smith, M., Jessell, T.M., Sockanathan, S., Requirement for the homeobox gene *Hb9* in the consolidation of motor neuron identity. *Neuron,* 1999, **23**: p. 659–674.

Chapter 10

Microfluidics: On-Chip Platforms as *In Vitro* Disease Models

Shan Gao*, Erkin Şeker†, and Martin L. Yarmush*,‡,**

*Center for Engineering in Medicine, Massachusetts General Hospital Harvard Medical School and Shriners Hospitals for Children 51 Blossom Street, Boston, MA 02114, USA
†Department of Electrical and Computer Engineering University of California, Davis Davis, CA 95616, USA
‡Department of Biomedical Engineering, Rutgers University Piscataway, NJ 08854, USA

Introduction

Significance of in vitro platforms to study diseases

Tissue engineering, since unsuccessfully proposed by Y. C. Fung to the National Science Foundation in 1985,[1] has offered new possibilities for the study of complex physiological and pathological processes *in vitro*. Later defined by Langer and Vacanti[2] in 1993, tissue engineering is "an interdisciplinary field that applies the principles of engineering and life sciences toward the development of biological substitutes that restore, maintain, or improve tissue function." While the ultimate goal of tissue engineering is to replace non-functional organs and tissues with artificial replicas, the goal of "off-the-shelf replacement parts for the human body" may not become a reality as soon as what was predicted a decade ago.[3-5] However, there is a great need for creating more realistic tissue models *in vitro* for drug development and the

---

**Corresponding author. Email: ireis@sbi.org

study of human tissue physiology and pathology, instead of aiming for direct implantation in patients. These attempts can also assist scientists in identifying new therapeutic targets and offering novel strategies for tissue regeneration so that it actually decreases the need for organ transplantations.[6]

The development of more physiologically relevant *in vitro* tissue models is also driven by the significant achievements in molecular biology and tissue engineering. First of all, while animal models are highly useful for preclinical tests of novel therapeutics, they do not fully predict therapeutic and toxic behaviors of new drugs in humans, because they fail to capture some of the essential human physiological mechanisms. For instance, only 1% of the medicinal products that progress to clinical testing ever reach the registration stage.[7] Approximately half of all drug failures are attributed to problems with efficacy and toxicity that preclinical studies failed to reveal.[7] Considering the ~1 billion-dollar per compound development cost of new pharmaceuticals,[8,9] human cell-based *in vitro* models are likely to play an increasingly important role in evaluating the efficacy and toxicity of new drugs before clinical trials. In addition, some pathogens are species-specific (e.g., hepatitis C), for which animal models are not applicable for preclinical studies. Second, *in vitro* studies have shown that the interactions between cells (different types or the same type), as well as the extracellular matrix (ECM) and cells, are crucial to induce specific cellular responses existing *in vivo*.[10–12] Traditional monocell-type culture system that lacks ECM–cell and/or cell–cell interactions may lack desired cellular functions present *in vivo*. For example, isolated primary hepatocytes rapidly lose their liver-specific functions in culture, unless they are co-cultured with endothelial cells or fibroblasts and/or in a collagen-gel sandwich[12–14] compared to monoculture. Finally, as much as biochemical stimuli, mechanical forces play an essential role in mediating cellular functions.[15–19] As a result, it is necessary to precisely engineer the mechanical microenvironment of cells and tissues *in vitro*. For instance, the exertion of shear forces on endothelial cells is crucial to the formation of proper cytoskeletal alignment.[20] Therefore, the combination of optimizing the ECM and scaffold as well as controlling biochemical and mechanical cues on cell cultures is expected to be integrated in a culture system at microscales relevant to physiological performance. There is an ever-increasing demand for more realistic *in vitro* tissue models that better capture mechanisms underlying tissue physiology and pathology than conventional animal models and cell culture models, in order to accelerate the development of therapeutics, including pharmaceuticals.

The challenge in creating high-fidelity tissue models *in vitro* is to recapitulate the *in vivo* cellular microenvironment as closely as possible, with a platform that is user-friendly to enable experimental replicates in a reasonable amount of time. A technology that addresses these challenges equally well will be of great utility to the industry and academia alike.

Microfluidics for an in vitro disease model: Significance and advantages

Unlike artificial organs developed for transplantation, *in vitro* tissue models are not required to be at a similar scale of a whole organ to recapitulate essential tissue functions. The microenvironment in which cells reside, including soluble factors, cell–matrix interactions, cell–cell contacts, and specific physicochemical properties, dictates the specific structure and function of specific tissue (Fig. 1). This microenvironment can, in part, be mimicked by microfluidic structures. The prerequisite of creating realistic disease models *in vitro* is to manipulate and monitor cell responses by spatially and temporally controlling the concentrations of soluble factors, as well as controlling mechanical environment of cell growth *in vitro* similar to that *in vivo*. Microfluidic technologies can precisely control fluid flow with a volume of nano- and

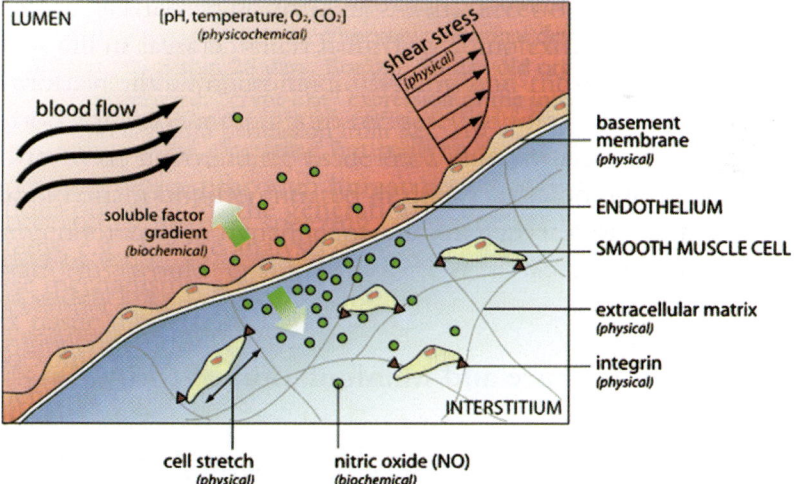

Fig. 1. The cell microenvironment consists of physical, biochemical, and physicochemical factors. For example, the endothelium that lines blood vessels is exposed to hemodynamic shear stress (external physical force) that stimulates a biochemical response, releasing nitric oxide (NO). NO diffuses to neighboring smooth muscle cells (SMCs), where it regulates cell contraction and relaxation. The gradient of diffused NO affects nearby SMCs more than distant SMCs. Endothelial cells are anchored to the basement membrane, while SMCs are anchored to the extracellular matrix (ECM) of the interstitium, both via integrins that act as sensors and transducers of physical force. Local physicochemical properties ensure proper regulation of both physical and biochemical mechanisms.[32]

picoliters in channels with dimensions of tens to hundreds of micrometers, thereby establishing a tightly regulated microenvironment.

Adopted directly from microelectronics, the first generation of microfluidic devices fabricated in silicon and glass using photolithography and etching techniques was expensive and rigid. In the 1980s and 1990s,[21] the adoption of certain synthetic polymers (e.g., polydimethylsiloxane (PDMS)) exhibited better biocompatibility and versatility in micropatterning. These materials processed with a specialized technique of molding, known as soft lithography, greatly reduced the time and expense to fabricate microfluidic devices.[22–25] Meanwhile, demands to control fluid flow in microfluidic channels led to the development of novel fluid flow techniques,[26] such as special valves,[27,28] mixers,[28–30] and pumps.[31] Integrated microfluidic systems have found use in various applications, especially in biological and medical fields. The advantages of microfluidics, including miniaturization, high throughput, and automation,[32] allow dynamic control of the cell microenvironment, including medium compositions, oxygen or CO_2 level, ECM, chemical and physical stimuli, or insults delivered in various spatial and temporal fashions, at physiologically relevant spatial and temporal scales. Although microfluidic chips have yet to become "off-the-shelf" supplies, various designs and prototypes have been created in research groups aimed at serving a myriad of research areas in life sciences.

Here, we will review the recent advances in microfluidic platforms, with an emphasis on the integrated platforms of *in vitro* models for normal and pathological physiological states. We will focus on analyzing proof-of-concept examples of *in vitro* models of the liver, neuron networks, the blood–brain barrier (BBB), and vascular systems and their potential impacts on drug discovery and physiology research.

Microfluidic Cell Culture and Manipulation of the Microenvironment

Controlling the cellular microenvironment with microfluidics

Cell culture has been a fundamental tool for constructing disease models. For the last 50 years, significant effort has been devoted to optimize the soluble biochemical environment of *in vitro* two-dimensional (2D) cell culture, including medium composition, temperature, CO_2 level, and surface hydrophilicity, in order to maintain *in vivo*-like cell function. However, it was not until the major development of three-dimensional (3D) cell cultures on various scaffolds and the ECM in the last 25 years that a more realistic model evolved, resembling the physical microenvironment existing *in vivo* (Fig. 1), which has led to a number of novel discoveries and applications, such as the discovery of breast cancer development and the mammary gland physical microenvironment in 3D breast epithelial cell culture.[10,33] Three-dimensional culture has combined physical and biochemical cues applied from the ECM and soluble factors in the medium to the cells in order to induce proper cell responses for the formation of desired tissue structures and functions *in vivo*, which is difficult to achieve in traditional cell culture systems. However, the scale and architecture of the culture as well as transport properties in 3D culture cannot be effectively controlled at physiologically relevant microenvironments. Microfabricated culture systems, on the other hand, can offer better controlled culture architectures in both 2D and 3D cultures by (1) micropatterning a single type or multiple types of cells for static culture conditions and (2) micropatterning flow channels and cell chambers for continuous perfusion conditions. The static 2D cell culture systems created by the first method have been mainly used for studying cell–matrix[34] or cell–cell interactions.[35] In the perfusion cell culture systems created by the second method, great efforts have been taken to develop continuous perfusion microfluidic culture systems[36–39] that can be used for high-throughput cell array culture in 2D or 3D with each population of cells cultured under dependent or independent conditions, including contents or concentrations of soluble factors in the medium and various ECM. Although there are challenges involved in surface treatment, cell/scaffold seeding, and the user-friendliness of these systems,[40] the current technologies constitute an effective starting point for developing new ideas and techniques in order to establish robust microfluidic perfusion culture systems for various applications. For instance, complex systems with tube-based inlets (Fig. 2) may require a specially designed loading interface or program that limits their accessibility to users such as biologists that are accustomed to conventional biological techniques. An alternative tubeless microfluidic culture system was developed by adopting a loading strategy of using automated liquid handling systems,[41] which may promote the acceptance of microfluidics for cell culture in the life sciences community.

One of the challenges of microfluidics-based cell culture is selecting the material and the corresponding fabrication method for the devices. It has been well recognized that the introduction of the elastomer PDMS, rather than traditional silicon or glass as the key material for preparing microfluidic systems in late 1980s[42] and 1990s,[43] has led to the extensive applications of microfluidics in biological and medical studies.[21] With its excellent optical transparency, low toxicity, high gas permeability, and simple fabrication process using soft lithography, PDMS has become the most widely used material in microfluidic systems. Although little attention has been paid to potential

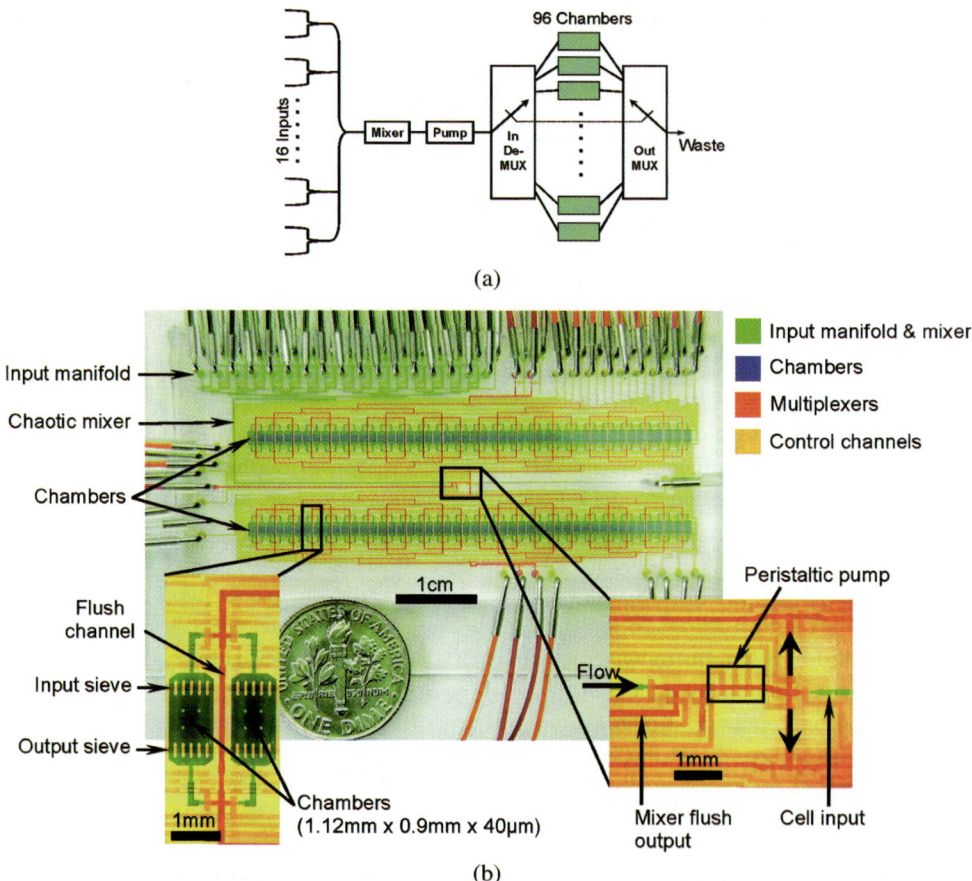

Fig. 2. Design of fully automated microfluidic cell culture chip. (a) Simplified schematic diagram of the fluidic path in the chip (MUX, multiplexer). (b) Annotated photograph of a chip with the channels filled with a color dye to indicate different parts of the device. The left inset gives a closer view of two culture chambers, with the multiplexer flush channel in between them. The right inset shows the root of the input multiplexer, with the peristaltic pump, a waste output for flushing the mixer, and the cell input line.[37]

experimental artifacts introduced by the chemical properties of PDMS on long-term cell cultures for which the aforementioned continuous perfusion systems have been developed, it is worth noting the effects of PDMS on cell properties and functions. For instance, the leaching of uncured oligomers from the polymer into the media,[44] the evaporation of water through the PDMS,[45] and the absorption of small, hydrophobic molecules (culture supplements or key chemicals used in bio-assays, e.g., estrogen) from the media into the polymer bulk[44,46] may change the biochemical environment of cells. Various surface treatment or ECM coating[47] methods as well as alternative materials (e.g. polystyrene,[40] poly(methylmethacrylate) (PMMA),[40] polycarbonate,[48] and polyolefin[48]) have been developed to overcome these challenges related to PDMS. However, a perfect material of microfluidics for biological application has yet to be identified.[48]

Primary cell culture in microfluidics: Toward in vitro tissue models

Compared to immortalized cell lines, primary cells tend to retain more desirable cellular function, which is especially desirable in creating *in vitro* disease models. Therefore, incorporating primary cells into microfluidics while preserving their functionality is of great importance as well as a significant challenge. Primary hepatocytes are one of the primary cell types that have gained a great deal of attention due to the major metabolic and detoxification function of liver, which has significant application on drug screening and understanding human liver diseases, e.g., tumors and hepatitis A, B, and C infections. These infectious diseases are becoming increasingly serious in public health, while related studies are hampered by the difficulties in creating disease models in animals or even a cell culture system (e.g., HCV). Recently, the collaboration between Dr. Sangeeta N. Bhatia's microfluidic research group and Dr. Charles M. Rice's virology group has led to a significant achievement in developing persistent hepatitis C infection in micropatterned primary human hepatocyte cultures,[49] through which the infection of human hepatocytes, viral propagation, and host responses[50] can be studied in a more physiologically relevant system instead of a cell line-based system.[51,52] Although the infection efficiency is expected to be improved, this work has laid the foundation for the preclinical assessment of antiviral drugs in a more realistic system, thereby demonstrating the value of applying microfabricated systems in life sciences.

One major challenge in primary cultures *in vitro* is to maintain the viability and specific phenotype of the primary cells for extended time periods. Primary cultures of stem cells and hepatocytes are two of the primary cultures most studied due to their significant application in regenerative medicine and drug toxicity. Readers interested in the application of microfluidics- and microfabrication-based devices in culturing stem cells *in vitro* are referred to a tutorial review published recently.[53] The application of primary cultures of hepatocytes in liver toxicity studies *in vitro* is hampered due to the low throughput of traditional culture methods and uncontrollable co-culture format.[54] Microfluidics, on the other hand, can provide greater control of the co-culture condition and the parameter variation in a high-throughput and scalable manner. For example, a 64 (8×8) element array of microfluidic wells was developed to support micropatterned primary rat hepatocytes in co-culture with 3T3-J2 fibroblasts continuously perfused with medium and oxygen.[54] More importantly, this culture system exhibits stable liver-specific function in the form of relatively constant albumin and urea production rates for up to 32 days. Primary human hepatocytes, instead of their rodent counterparts, present a promising *in vitro* hepatic culture system for physiologically relevant toxicology studies. In a commercialized microfluidic device, the HµREL® device, cryopreserved human hepatocytes from a single donor were cultured up to 24 h to predict the *in vivo* hepatic clearance of six marketed model compounds.[55] Although the reference compounds exhibiting different degrees of *in vivo* clearance were metabolized at relatively similar clearance rates within the HµREL® device, the absolute values of these estimated clearance rates did not seem to reflect the values from *in vivo* experiments better than those estimated using the cell culture in traditional Biocoat™ plates. To improve the *in vitro* and *in vivo* clearance correlation, co-culture of the primary hepatocytes and non-parenchymal cells was established in the same device to study clearance and metabolite generation of a battery of molecular entities. The flow-based co-culture system yielded improved metabolite generation and prediction of drug clearance *in vivo*.[56]

Complex culture systems

Cell culture analog systems, better known as "animal-on-a-chip"[57] and "human-on-a-chip"[58] (which are thoroughly reviewed elsewhere[59]), attempt to compartmentalize the functional components of key organs involved in drug metabolism and transport in an on-chip setup by using cells derived from those organs.[60] The compartments, corresponding to various organs including the lung, liver, and adipose tissue, are integrated in a microfabricated culture system and interconnected based on their sequence *in vivo*, which allows the study of drug action and metabolism in different organs in a systematic way *in vitro*. However, the attempt to realizing this concept is yet to replace animal models. The prototype "animal-on-a-chip" was created by integrating two or three types of mammalian cells on multi-chamber microfluidic devices to represent key functions of liver, lung, and fat compartments.[61,62] The systems were evaluated by studying biodistribution and toxicity of select compounds[61] and the toxicity in the lung compartment caused by the metabolized products of compounds in the liver.[62] Recently, a proof-of-principle study toward human-on-a-chip has demonstrated the impact of a cytokine on four compartments of human-derived cells cultured in a multi-channel 3D microfluidic cell culture system.[63] However, the function of the multicell-type culture system is dictated by the specific cells seeded in the system. It is important to distinguish the physiologically relevant cellular responses from *in vitro* artifacts of cultured cells. For example, the C3A cells, representing liver-on-a-chip in the above-described cell culture platform, may not be an ideal physiological relevant cell type, because it has been shown that the urea cycle present in normal liver is not functional in C3A cells in that the urea production in C3A does not reflect ammonia detoxification.[64] In addition, the types of cells studied in this culture system are still limited. For instance, a central neuron system has yet to be represented and may not be properly integrated into the culture system without a robust BBB culture *in vitro*. Finally, it could be also challenging to optimize a culture medium suitable for maintaining specific functions of various types of cells. Although still at its infancy, multi-tissue culture systems have the potential of constituting a more physiologically relevant *in vitro* model for the pre-clinical assessment of drug candidates, not only for direct liver metabolism, but also for secondary effects of the metabolites from the liver on other tissues.

Microfluidic Platforms for Bioanalytical Assays

Microfluidic control of the cell microenvironment is an effective way to "input" stimuli into the *in vitro* model (Fig. 3). In order to read the "output" of the *in vitro* model, it is necessary to monitor the molecular or cellular functions of the model (Fig. 3). Cell responses can be divided into three fundamental categories: (1) direct molecular responses, i.e., soluble biochemical factors consumed or produced by the cells; (2) indirect genetic responses, i.e., dynamics of transcription factors and gene expression profiles; (3) cell phenotypes, i.e., morphological and developmental properties. Coincidentally, as engineers and biologists developed microfluidics-based cell culture systems, chemists simultaneously scaled down conventional analytical assays for microfluidics.[65]

Molecular detection: Proteomic and metabolic analyses

For the direct biochemical detection of small molecules and proteins, most traditional analytical assays, including immunoassays and enzymatic assays, have been developed as microfluidic analytic modules that can be used independently or integrated with cell culture modules as a downstream detection unit. Significant reductions in cell culture scale and sample volume (from 10–100 µL used in microtittering plate to nanoliters in microfluidics) lead to the decreased consumption of detection proteins (antibodies or enzymes),

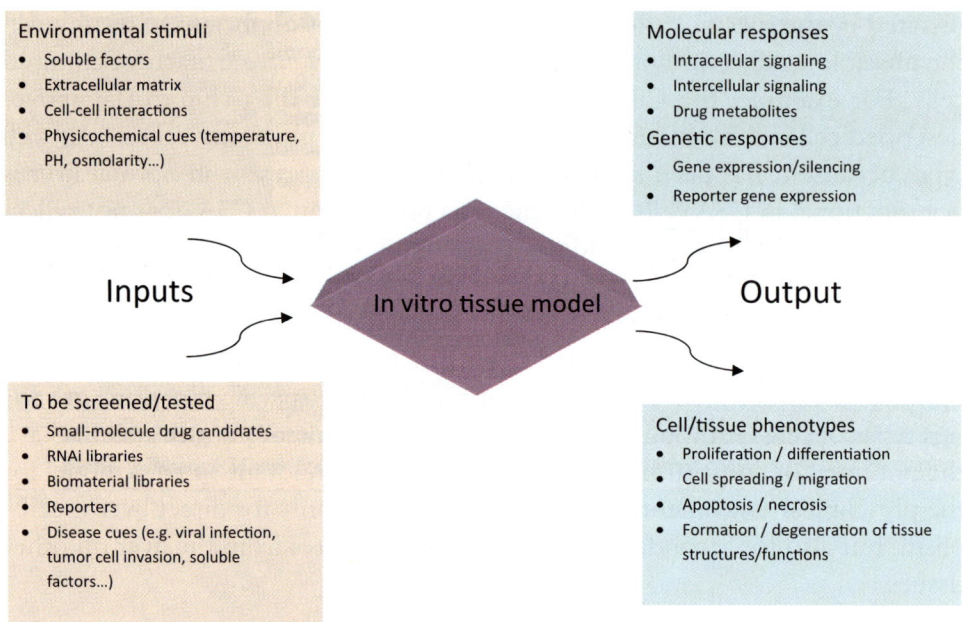

Fig. 3.   Attributes of an *in vitro* tissue model.

but require higher detection sensitivity to measure desired metabolites without a pre-concentrating step. While PDMS is a suitable material for cell culture, its surface properties (e.g., unstable hydrophilicity and non-reliable surface bioconjugate chemistry) have rendered it a difficult material for use in biochemical assays. Great efforts have been taken to explore alternative materials[66] or to reproducibly functionalize PDMS surfaces[67,68] and silicon surfaces[69] with detecting antibodies. However, the specifically functionalized and fabricated device cannot be reconfigured to measure other analytes of interests. Alternatively, protein- or desired functional molecule-conjugated microbeads have become a promising tool to detect biochemical molecules in microfluidics,[70,71] which offers an easy approach to multiplex detection and eliminates the difficulties of tailoring the PDMS surface for specific detection. For example, a multiplexed protein detection strategy was developed using a set of optically distinguishable microparticles on an immunoassay chip, each of which is functionalized with an antibody against a protein of interest.[71] It was demonstrated that four standard cytokines were simultaneously measured at picogram/milliliter concentrations in nanoliter volumes using this system, which may be potentially used for the detection of protein secretion of individual cells.

Genomic analyses

For the detection of genetic responses, endpoint analysis of direct genetic information requires cell lysis and DNA or RNA extraction for either sequencing or hybridization with target probes. Portable microfluidic systems have been developed to extract DNA and RNA from blood samples[72] and then to introduce them into DNA microarray or microfluidics-based sequencing modules,[73] which is not the focus of this chapter. To obtain real-time information on microfluidic cell assays at the gene expression level, non-destructive detection strategies are desired to monitor cellular response to dynamic stimuli. Engineered plasmids for controlled fluorescent protein expression by engineered promoters have been developed to measure transcriptional activity,[74] which enables data collection using high-throughput imaging methods. These methods have already been widely used in next-generation DNA sequencing, systems biology, and drug discovery studies with well-established image acquisition and data analysis programs.[75,76] However, in order to obtain quantitative kinetic data from the imaging results, additional image processing is expected to correct the time-lapse fluorescent baseline from the culture background (including medium and untreated cells), adjust dynamic fluorescent intensity with the half-life of the fluorescent protein, and identify cell movement as well as cell boundary.

In general, microfluidics-based molecular detection is one of the best developed microfluidic systems with successful applications and commercialization. As a self-functioned module, it can be easily coupled with cell culture modules to evaluate cell responses indicated by soluble factors. Future development will more likely focus on the handling of the fluid sample directly from an upstream cell culture and improvement of detection sensitivity. For measuring genetic responses of live cells in real time, however, more challenges are expected to be solved in creating efficient reporters, introducing reporters into the cells, and analyzing the data from the high-throughput experiments.

Integrated Microfluidic Platforms to Study Diseases

Microfluidic cell arrays

The integration of microfabricated arrays, micropatterned ECM, and cell culture under the precise temporal and spatial control of chemical mediators has enabled high-throughput living cell arrays, which provides powerful tools for studying cellular responses to massive combinations of stimuli,[77,78] screening small-molecule libraries,[79] RNAi libraries,[80,81] ECM libraries,[82,83] and studying intercellular communications.[84] For example, 2D microfluidic cell arrays have already been created to systematically vary one parameter across a row of elements containing small adherent cell populations (Figs. 4(a) and

4(b)).[85] In this device, various doses of a cytokine stimulus were generated through an upstream diffusive mixing microfluidic network and delivered to downstream pre-seeded NFκB–green fluorescent protein (GFP) transcriptional reporter cells. The resulting gene expression dynamics were measured

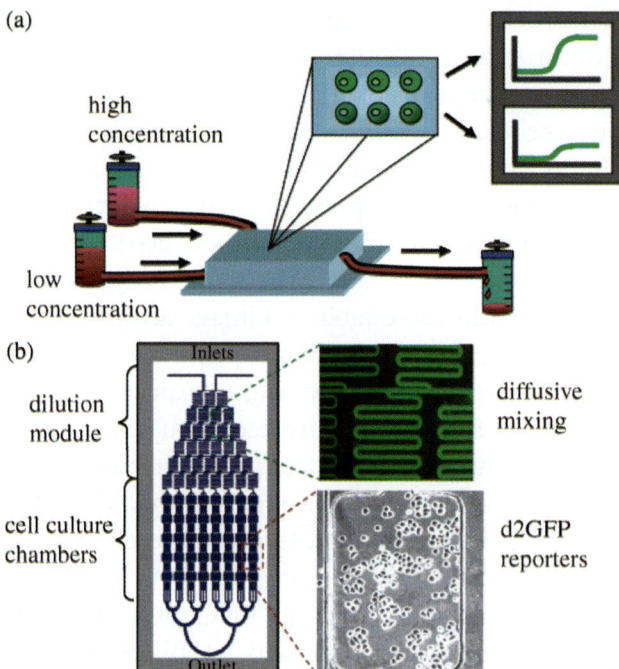

Fig. 4. Microfluidic cell arrays. (a) Schematic representation of the living cell array (LCA) device. Medium containing a soluble mediator enters the device, delivers nutrients, stimulates cells cultured in the cell chambers, and exits into a waste stream. Gene expression dynamics are obtained by time-lapse imaging and quantified using image analysis software.[85] (b) The microfluidic network design and micrographs of the dilution and cell cultivation modules are shown. The transparent PDMS device allows the visualization of the TNF-α gradient (doped with fluorescein dye) and EGFP reporter cells. The microfluidic channels in the upstream dilution module generate several concentrations of the stimulus by continuous-flow diffusive mixing of adjacent laminar flow streams. The various concentrations are delivered to the downstream array of culture chambers.[85] (c) A microfluidic device uses flow-encoded switching to control parallel delivery of different stimulus timing regimens to pre-seeded GFP reporter cells. The figure illustrates device design (top left), dye filled channels at different time points (top middle), channels seeded with fluorescent GFP reporter cells (top right), example excitation sequences (black) and the resulting dynamic exposures of four channels (green) varying pulse duration, width, and frequency.[86] (d) A microfluidic valve-controlled array enables dynamic gene expression profiling. The figure illustrates device design, valve design, reporter cell seeding, stimulation strategy (left), phase and fluorescence images of GFP reporter cells in an array element (top right), and a heat map quantifying dynamic GFP responses across the array (bottom right).[77]

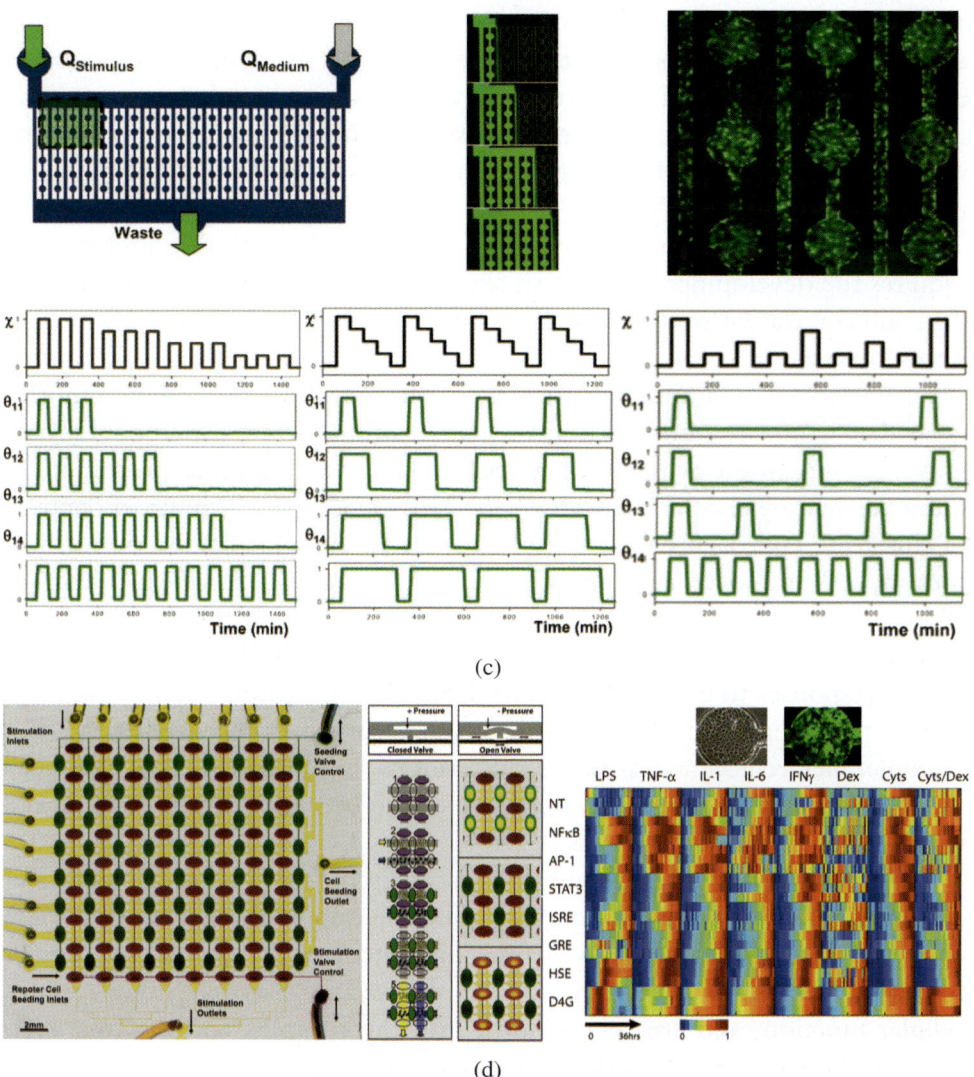

Fig. 4. (*Continued*)

using time-lapse fluorescence microscopy. Using a novel flow-encoded switching with a comparatively small number of inputs, the controlled delivery of many different dynamic stimulus patterns to downstream GFP reporter cells was achieved (Fig. 4(c)).[86] Stimulation patterns included variable pulse widths, pulse train lengths, and pulse frequencies. Besides varying the temporal aspects of the soluble environment, a facile spatial control of soluble stimuli can also be achieved by creating a valve-controlled 2D microfluidic array that allows independent isolation of rows and columns for cell seeding and stimulation. In a proof-of-principle experiment, the dynamic responses of eight stable monoclonal GFP reporter cell lines were seeded in rows, and eight stimuli, including inflammatory stimuli, inhibitors, and modulators, were introduced into columns to acquire ~5,000 single time-point measurements over 24 h (Fig. 4(d)).[77]

The abundance of the acquired data from these high-throughput systems requires the development of analysis algorithms to quantify dynamic imaging data and correlate it to desired cellular function in order to reveal the mechanisms underlying the cell behaviors. For example, in the living cell array (LCA) device[77] containing hepatocytes transfected with a fluorescent reporter gene that would respond to a given transcription factor, experimental data indicated significant cross-talk between inducers and specific promoter. In order to identify the mechanism of the cross-talk, a novel non-overlapping bi-clustering approach was developed to determine which transcription factors are activated over a variety of stimulating conditions,[87] which essentially allows the deciphering of the information obtained from the LCA to reveal transcription factor interaction networks that give rise to the observed experimental responses to injury, disease, or drug administration.[88] Other bioinformatics techniques are also valuable in analyzing live cell array data. For example, hierarchical and *k*-means clustering was used to visualize single-cell array data and provide a detailed picture of the heterogeneity present in cell responses to administered stress.[89]

The advances in LCA design, fluorescent reporter constructs, and LCA data analysis have offered unique and promising opportunities for using these emerging LCA platforms as a valuable component of *in vitro* disease models to rapidly identify genetic determinants of disease, discover modulators of cellular function, and create a window into the complex and dynamic relationships between cells and their local environments.

Liver bioreactors

Miniaturized devices using microfabrication and microfluidic techniques have also offered opportunities to develop bioreactors with potentially feasible clinical applications. For example, a radial flow bioreactor was developed through optimizing the flow environment for the cultured hepatocytes in the bioreactor with stacked substrates that served to protect the seeded hepatocytes from high shear stresses (Fig. 5(a)).[90] The seeded hepatocytes exhibited stable liver-specific function over five days of perfusion with better oxygen

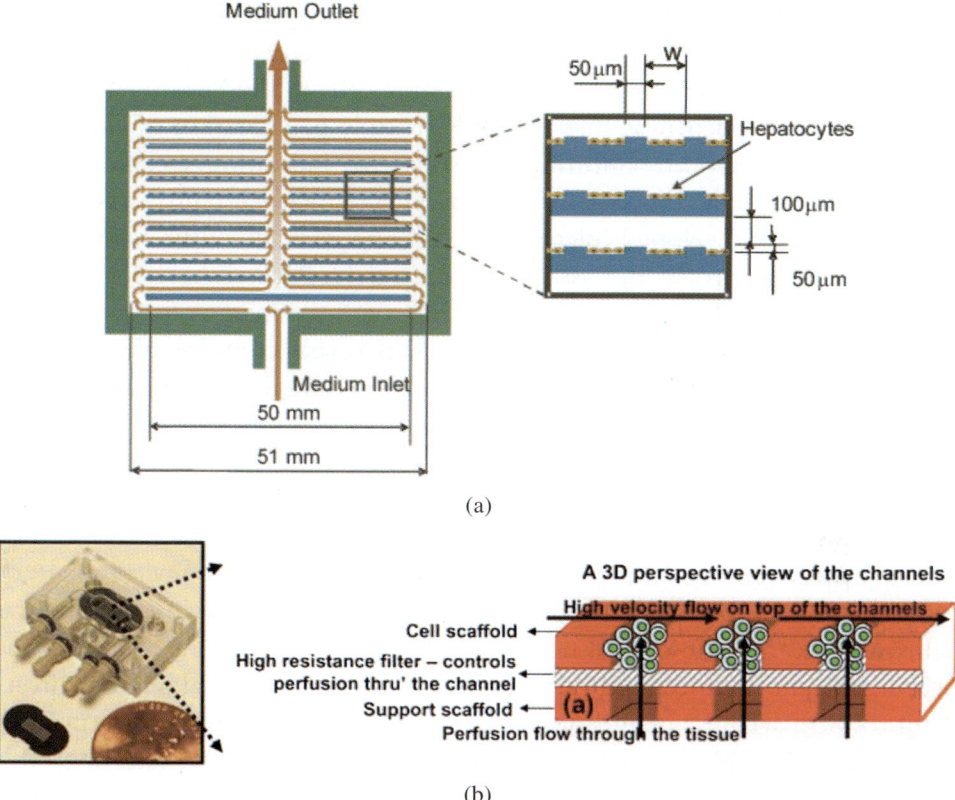

Fig. 5. Microfludics-based liver bioreactors. (a) Radial flow bioreactor with stacked microgrooved substrates. The flow network in the bioreactor shows the microgrooved substrates. The channel height between the substrates was 100 mm, and the groove width, w, was based on the shear stress level at a particular location on the substrate.[90] (b) 3D-perfused microtissue units seeded with rat hepatocytes. The assembled macroreactor that has 1,000 channels comprising one million cells. The schematic illustrates a side-view cross-section of a single channel in the silicon chip, indicating the locations of the scaffold, filter, and direction of fluid flow.[94]

delivery to the cells than a flat-plate bioreactor.[91,92] The calculation based on albumin secretion suggests this design is scalable to a clinically feasible bioartificial liver (BAL) devices in the treatment of liver failure. In addition, an array-based microfluidic culture system was also developed for creating an *in vitro* liver tissue model by co-culturing hepatocytes and liver sinusoidal endothelial cells (LSECs) on each scaffold in the bioreactor (Fig. 5(b)).[93] The liver-specific functions of a similar system, i.e., the expression levels of cytochrome P450 and hepatic transcription factors up to 20 days, were verified and compared to that in a 2D collagen sandwich culture as well as that in tissues *in vivo*.[94] By co-culturing carcinoma cells in the hepatocyte-seeded bioreactor, this system was further used to study proliferation of carcinoma cells in a 3D hepatocyte-culturing environment, indicating specific interactions between carcinoma cells and hepatocytes in metastatic progression.[95]

These newly developed microfluidic 2D and 3D cell culture systems are transforming traditional bio-assays carried out in Petri dishes. The precise control of each patterned cell population with defined combinations of stimuli in a versatile temporal fashion would reveal complex physiological pathways and mechanisms of diseases that are discouragingly difficult to study either in conventional *in vitro* bio-assays or *in vivo* animal models.

Microfluidics in neurosciences

The ability to control the physical and biochemical microenvironment of *in vitro* cell cultures using microfluidics is especially valuable in experimental neuroscience.[96] In the central nervous system (CNS), neurons extend axons over significant distances and varying physical and biochemical microenvironment to form synaptic connections. Axonal damage or activation is crucial to the etiology of CNS injuries, development of brain circuitry, and neurodegenerative diseases. However, it is challenging to pattern neuritis and axons, to construct desired neuronal networks, and to create distinct somal and axonal microenvironments using traditional *in vitro* culture techniques in order to gain insights in axonal biology and synaptogenesis.

Thirty years ago, Campenot *et al.* first described a multicompartment culturing method for studying axonal biology in the peripheral nervous system (PNS).[97,98] Recently, a microfluidic device was developed for the long-term culture and compartmentalization of primary CNS neurons that are traditionally more difficult to culture and exhibit different axonal biology from PNS neurons.[99,100] This microfluidics-based culture platform directs the axonal growth of CNS neurons and creates the isolated chemical microenvironment between the somal and axonal sides generated by the hydrostatic pressure due to a medium volume difference between the two sides. As a proof-of-concept, this culture system was successfully used as an *in vitro* model for axonal injury and regeneration within the CNS, which demonstrated (1) the changes of somal gene expression upon selective axotomy and vacuum aspiration in the axonal compartment and (2) neuronal regeneration of axonally localized neurotrophin treatment of lesioned axons.[100] This system was further developed into a microfluidic local perfusion chamber for culturing two populations of neurons to access and manipulate synaptic regions of presynaptic and postsynaptic compartments *in vitro*, allowing spatial and temporal control over the duration and frequency of stimulation.[101] In addition to a cell culture-based system, organotypic culture-based platforms were developed to capture neural electrophysiological activity and axonal pathways through functional connections between fluidically isolated co-cultures of tissue slices from different regions of brain, e.g., cortex and hippocampus.[102] These *in vitro* culture platforms have provided powerful tools in experimental neurosciences to easily build and manipulate various specific circuits that could be involved in neurodegenerative diseases, psychiatric disorders, epilepsy, and the development of brain circuitry in order to reveal the underlying physiological and pathological pathways and mechanisms that may not be identified using conventional *in vitro* and *in vivo* experimental techniques.

One of the open fields in building *in vitro* models in neuroscience is the microfluidics-based *in vitro* BBB, a unique physiological structure involved in CNS that is formed by specialized BBB capillary endothelia and the tight junctions between them. This physical and enzymatic barrier segregates the brain from the systemic circulation with the selective discrimination of substance transcytosis based on lipid solubility, molecular size, and charge. In addition, specific transporters present at the BBB regulate the transport of nutrients to the brain and toxicant/waste products from the brain. Therefore, it is especially difficult to study the pathological pathways of the BBB and quantify BBB permeability changes which can be caused by local/direct (e.g., mechanical brain injury) and remote/indirect insults *in vivo*. Furthermore, the presence of the BBB has been the biggest challenge of predicting pharmacological properties for a given compound in CNS therapeutic development. Many transwell-based static *in vitro* BBB models and hollow fiber-based dynamic models have been developed to evaluate pharmacological properties of neuroactive compounds. Although the studies have generally focused on characterizing the endothelial monolayer (co-cultured with astrocytes in some cases), especially in transwell-based models, the permeability obtained from the *in vitro* models tends to be higher than that obtained *in vivo*. The use of these models for trans-BBB drug screening would potentially overestimate the drug permeability across the BBB. In addition, the presence of the supporting membrane and non-transparent hollow fiber does not permit the microscopic evaluation of the tight junction formation. The low throughput also hinders its application as a feasible drug screening assay in the pharmaceutical industry. In general, no model has been developed that fully recapitulates the properties of the BBB as found *in vivo*. With physiologically relevant scale and dynamic fluidic components, microfluidics has the potential to be a platform with advanced cell culture methods for building more realistic *in vitro* BBB models. The use of fluidics not only provides continuous medium replenishment, but also applies shear forces to cultured endothelial cells that are subjected to the continuous pulsatile flow of blood along their luminal surface. Flow along the luminal surface has been shown to induce angiogenesis and the polarity of transporters.

Surprisingly, few studies have focused on developing microfluidic systems for highly specialized vascular cultures as a BBB model even though a number of general vascular systems, which will be discussed in the next section, have been developed for potentially various applications, including the BBB. A simple microflow channel was created within cross-linked collagen by applying a focused laser beam.[103] Isolated rat brain capillary endothelial cells were successfully cultured on the inner tunnel wall with the formation of two typical tight junction markers. However, the complete characterization of BBB function and refining design, such as the co-culture of endothelial cells and astrocytes and quantification of permeability, are expected to evaluate the feasibility and effectiveness of this model in BBB pathology and compound pharmacology study. For instance, a 3D astrocyte–endothelial tissue construct, consisting of astrocytes cultured within the hydrogel and endothelial cells cultured on the top of the hydrogel,[60] may be integrated in a microfabricated system for creating a more structural and physiological relevant BBB model. For more information on microfluidics for neuroscience, see Chapter 9.

Microfluidics in Vascular Research

Vascular cell biology is an active research field in tissue engineering with significant biomedical relevance and strong translational focus. Vascular dysfunction appears in major diseases such as atherosclerosis,[104] cancer,[105] and diabetes.[106] *In vitro* models of blood vessels are essential tools in revealing the molecular and cellular mechanisms of the vascular system in order to gain insight into vascular physiology and treatment of disease. Similar to the BBB model, microfluidic technology allows the creation of a vascular model by integrating vascular cell culture, under shear stress from the fluidics, in a physiologically relevant dimension, e.g., the scale of capillaries with diameters at microsizes. Vascular endothelial cells are highly sensitive to the changes of shear stress caused by the flow of the fluid. The biological response, known as the endothelial mechanoresponse, has been shown as a key process in preventing vascular disease.[107] The mechanoresponse is usually studied in conventional parallel plate flow chambers, in which the flow rates are in the order of hundreds of milliliters per hour to resemble the level of shear force that endothelial cells would be subjected to *in vivo*. The consumption of large amounts of culture media and other reagents, including proteins and drug candidates to be tested, has been a major limitation of this testing device. In microfluidics, however, it has been shown that a flow rate of less than 200 μL/h is already enough to make the sheared endothelial cells elongate and orient in the direction of the flow, which is a prominent feature of the endothelial mechanoresponse that is also found *in vivo*.[108] The significantly reduced volumes of reagents provide an ideal and feasible *in vitro* platform for studying dynamic pathological effects of changes in flow patterns on endothelial cell gene expression, cytoskeletal arrangement, wound repair, leukocyte adhesion,[109–112] tumor cell adhesion[113,114] as well as the vasoreactive, oxidative,[115–117] and inflammatory states of the vascular wall. The insights gained from the *in vitro* models would suggest potential therapeutic methods or compounds for treating related vascular diseases through biochemical or physical intervention. The 3D culture environment is also an important factor in vascular physiology. Hydrogel is the most common ECM used in microfluidic channels to support cell culture. For example, the endothelial cells grown in or into hydrogel developed complex interconnected multicellular capillary-like structures in the gel with the presence of pro-angiogenic factors or interstitial flow, whilst cells grown on a 2D gel surface tend to form circular and flattened lumen-like structures.[118] Although microinjection was used to load the gel into the chamber in this studies, the gel can also be hydrodynamically loaded into channels with specially designed inlets on the chip.[119] Barkefors *et al.*[120] have taken a step further to culture *ex vivo* kidney tissue and induced directional angiogenesis within the tissue by growth factor gradients that are created by the microfluidic device. These proof-of-concept studies have offered not only valuable and feasible tools to facilitate vascular research, but also new research areas using more physiologically relevant *in vitro* vascular models or even tissue with vessel models.

Conclusion and Outlook

In the late 1800s, the scale up of benchtop chemical reactions to massive processing operations in plants has revolutionized the way we harvest and utilize natural recourses. A century later, scale down has become the new fashion led by the microelectronics industry, which has revolutionized the way we communicate with one another. Microfluidics is expected to become the next revolutionary technology in changing the way we understand and utilize molecular/cellular/tissular biology and chemistry in human health care. However, its potential values are yet to be fully recognized by biologists and clinicians, and the microfluidic devices have yet to reach mass production of off-the-shelf supplies. As every new technology evolves, it is a process of building sufficient trust between the developers and users. On one hand, microfluidics research is mainly performed by academic experts in mechanics, biomedical engineering, material science, or analytical chemistry, who have established powerful devices for potential applications in biology, life sciences, or even in clinical settings. On the other hand, biologists and clinicians may not be ready to actively seek microfluidic devices to solve their problems until they experience the advantages of the new devices over traditional methods or instruments. More importantly, this technology may not flourish until the users, e.g., biologists and clinicians, participate in the development by requesting desired functions of devices that cannot be achieved by conventional devices.

As it has been debated whether a single "killer application"[48] or a wide range of life science applications[121] would be more important for microfluidics to thrive as new technology, creating microfluidics-based tissue or disease models *in vitro* may offer opportunities in both situations. The basic functions of *in vitro* models require culturing cells in a controlled microenvironment with a combination of physical, biochemical, and physicochemical cues, and a set of detection modules to measure cell responses upon exposure to specific stimuli, which involve the major applications of microfluidic devices under development. In addition, a successful *in vitro* model will be desired by biologists for basic cellular and molecular physiology studies and the discovery of new therapeutic targets; clinicians for pathology studies and engineering therapies; and the pharmaceutical industry for new drug development. The key issues are (1) the standardization of the functional modules and their interfaces; (2) the validation of the reliability, reproducibility, and versatility of the device models; (3) the mass production of the devices; and (4) the acceptance of the devices by users. The collaboration between research groups in the academic community certainly leads to the fundamental development of new devices. For example, the collaboration between Quake's group and Chen's group on a project of microfluidics-based screening biomaterials for mesenchymal stem cell differentiation[122] has resulted in the development of the aforementioned automated microfluidic cell culture system with tube-based inlets[37] that may completely change the biomaterial screening process if it can be widely used in other biomaterial or biological laboratories. However, the collaborations in academia have their limitations, such as the limited production of the devices, custom designing for specific projects, and the requirement of specially trained personnel to use the devices. In this case, the complexity of the sophisticated system may lack user-friendliness to biologists for general cell assay purposes. Therefore, widely used tissue-in-/on-a-chip may not likely become reality until it can be successfully commercialized. It has become a trend of collaboration between the academia and industry in developing microfluidics-based systems. For instance, the Micro/Nano Fluidics Fundamentals Focus (MF3) Center based at the University of California, Irvine, is partnering academic research groups and commercial developers in an attempt to speed up the transition from the module-based design to large-scale manufacturing and commercialization. Another good example of bridging the gap between fundamental research and clinical applications is the BioMEMS Resource Center, formed through partnerships between interdisciplinary research groups and institutes in engineering, science, and medicine at the Massachusetts General Hospital. This research center has not only initiated new biomedical applications of bio-microelectromechanical systems, but also promoted the clinician's awareness of using microfluidics to solve problems. The achievement obtained from both fundamental development and translational technology in microfluidics-based cell culture and biomolecular detection will contribute to the development and recognition of microfluidics-based *in vitro* tissue or disease models in the academic, industrial, and health care communities.

Acknowledgments

The authors which to acknowledge support from NIH grants R01 AI063795 and P41 EB002503 and support from the Shriners Hospitals for Children.

References

1. Viola, J., Lal, B., Grad, O., The emergence of tissue engineering as a research field. The National Science Foundation, from http://www.nof.gov/pubs/2004/nsf0450/start.htm, 2004.
2. Langer, R., Vacanti, J.P., Tissue engineering. *Science*, 1993, **260**(5110): p. 920–96.
3. Griffith, L.G., Naughton, G., Tissue engineering — current challenges and expanding opportunities. *Science*, 2002, **295**(5557): 1009–1014.
4. Lysaght, M.J., Hazlehurst, A.L., Tissue engineering: The end of the beginning. *Tissue Eng*, 2004, **10**(1–2): p. 309–320.
5. Griffith, L.G., Swartz, M.A., Capturing complex 3D tissue physiology *in vitro*. *Nat Rev Mol Cell Biol*, 2006, **7**(3): p. 211–224.
6. Suuronen, E.J., Sheardown, H. *et al.* Building *in vitro* models of organs. *Int Rev Cytol*, 2005, **244**: p. 137–173.
7. Bhogal, N., Balls, M., Translation of new technologies: From basic research to drug discovery and development. *Curr Drug Discovery Technol*, 2008, **5**(3): p. 250–262.
8. DiMasi, J.A., Hansen, R.W. *et al.*, The price of innovation: New estimates of drug development costs. *J Health Econ*, 2003, **22**(2): p. 151–185.

9. Adams, C.P., Brantner, V.V., Estimating the cost of new drug development: Is it really 802 million dollars? *Health Aff (Millwood)*, 2006, **25**(2): p. 420–428.
10. Debnath, J., Brugge, J.S., Modelling glandular epithelial cancers in three-dimensional cultures. *Nat Rev Cancer*, 2005, **5**(9): p. 675–688.
11. Janmey, P.A., McCulloc, C.A., Cell mechanics: Integrating cell responses to mechanical stimuli. *Annu Rev Biomed Eng*, 2007, **9**: p. 1–34.
12. Khetani, S.R., Bhatia, S.N., Microscale culture of human liver cells for drug development. *Nat Biotechnol*, 2008, **26**(1): 120–126.
13. Dunn, J.C., Tompkins, R.G. et al., Long-term *in vitro* function of adult hepatocytes in a collagen sandwich configuration. *Biotechnol Prog*, 1991, 7(3): 237–245.
14. Zinchenko, Y.S., Culberson, C.R. et al., Contribution of non-parenchymal cells to the performance of micropatterned hepatocytes. *Tissue Eng*, 2006, **12**(8): 2241–1251.
15. Ingber, D.E., Folkman, J., Mechanochemical switching between growth and differentiation during fibroblast growth factor-stimulated angiogenesis *in vitro*: Role of extracellular matrix. *J Cell Biol*, 1989, **109**(1): 317–330.
16. Ingber, D.E., Folkman, J., Tension and compression as basic determinants of cell form and function: utilization of a cellular tensegrity mechanism. In: Stein, B.F., Orland, W., eds. *Cell Shape: Determinants, Regulation and Regulatory Role*, Academic Press, 1989.
17. Roskelley, C.D., Desprez, P.Y. et al., Extracellular matrix-dependent tissue-specific gene expression in mammary epithelial cells requires both physical and biochemical signal transduction. *Proc Natl Acad Sci U S A*, 1994, **91**(26): p. 12378–12382.
18. Ingber, D.E., Extracellular matrix: A solid-state regulator of cell form, function, and tissue development. In: Jamieson, J.D.H., ed., *Handbook of Cell Physiology*, Oxford University Press, New York, New York, 1997.
19. Fu, J., Wang, Y.K. et al., Mechanical regulation of cell function with geometrically modulated elastomeric substrates. *Nat Methods*, 2010, 7(9): p. 733–736.
20. van der Meer, A.D., Poot, A.A. et al., Microfluidic technology in vascular research. *J Biomed Biotechnol*, 2009, **2009**: p. 823148.
21. Mukhopadhyay, R., When PDMS isn't the best. *Anal Chem*, 2007, **79**(9): p. 3248–3253.
22. Xia, Y., Whiteside, G., Soft lithography. *Annu Rev Mater Sci*, 1998, **28**: p. 153–184.
23. Jo B, Motsegood, L.L., K., Beebe, D., Three-dimensional micro-channel fabrication in polydimethylsiloxane (PDMS) elastomer, *J Microelectromech Syst*, 2000, **9**: P. 76–81.
24. Quake, S. R., Scherer, A., From micro- to nanofabrication with soft materials. *Science*, 2000, **290**(5496): 1536–1540.
25. Whitesides, G.M., Ostuni, E. et al., Soft lithography in biology and biochemistry. *Annu Rev Biomed Eng*, 2001, **3**: p. 335–373.
26. Beebe, D.J., Mensing, G.A. et al., Physics and applications of microfluidics in biology. *Annu Rev Biomed Eng*, 2002, **4**: 261–286.
27. Hong, J.W., Quake, S.R., Integrated nanoliter systems. *Nat Biotechnol*, 2003, **21**(10): p. 1179–1183.
28. Weibel, D.B., Kruithof, M. et al., Torque-actuated valves for microfluidics. *Anal Chem*, 2005, 77(15): p. 4726–4233.
29. Gunther, A., Jhunjhunwala, M. et al., Micromixing of miscible liquids in segmented gas-liquid flow. *Langmuir*, 2005, **21**(4): p. 1547–1555.

30. Garstecki, P., J.F.M. et al., Mixing with bubbles: A practical technology for use with portable microfluidic devices. *Lab Chip*, 2006, **6**(2): p. 207–212.
31. Laser, D., Santiago, J.G., A review of micropumps. *J Micromech Microeng*, 2004, **14**: p. R35–R64.
32. Young, E.W., Beebe, D.J., Fundamentals of microfluidic cell culture in controlled microenvironments. *Chem Soc Rev*, 2010, **39**(3): p. 1036–1048.
33. Bissell, M.J., Radisky, D.C. et al., The organizing principle: Microenvironmental influences in the normal and malignant breast. *Differentiation*, 2002, **70**(9–10): p. 537–546.
34. Tweedie, C., Anderson, D. et al., Cover picture: Combinatorial material mechanics: High-throughput polymer synthesis and nanomechanical screening. *Adv Mater*, 2005, **17**(21).
35. Jinno, S., Moeller, H.C. et al., Microfabricated multilayer parylene-C stencils for the generation of patterned dynamic co-cultures. *J Biomed Mater Res A*, 2008, **86A**(1): 278–288.
36. Hung, P.J., Lee, P.J. et al., Continuous perfusion microfluidic cell culture array for high-throughput cell-based assays. *Biotechnol Bioeng*, 2005, **89**(1): p. 1–8.
37. Gomez-Sjoberg, R., Leyrat, A.A. et al., Versatile, fully automated, microfluidic cell culture system. *Anal Chem*, 2007, **79**(22): p. 8557–8563.
38. Yu, Z.T., Kamei, K. et al., Integrated microfluidic devices for combinatorial cell-based assays. *Biomed Microdevices*, 2009, **11**(3): p. 547–555.
39. Park, E.S., Brown, A.C. et al., Continuously perfused, non-cross-contaminating microfluidic chamber array for studying cellular responses to orthogonal combinations of matrix and soluble signals. *Lab Chip*, 2010, **10**(5): p. 571–580.
40. Kim, L., Toh, Y.C. et al., A practical guide to microfluidic perfusion culture of adherent mammalian cells. *Lab Chip*, 2007, **7**(6): p. 681–694.
41. Meyvantsson, I., Warrick, J.W. et al., Automated cell culture in high density tubeless microfluidic device arrays. *Lab Chip*, 2008, **8**(5): p. 717–724.
42. Masuda, S.W., Nanba, M., T. Novel method of cell fusion in field constriction area in fluid integration circuit. *IEEE Trans Ind Appl*, 1989, **25**: p. 732–737.
43. Duffy, D.C., McDonald, J.C. et al., Rapid prototyping of microfluidic systems in poly(dimethylsiloxane). *Anal Chem*, 1998, **70**(23): p. 4974–4984.
44. Regehr, K. J., Domenech, M. et al., Biological implications of polydimethylsiloxane-based microfluidic cell culture. *Lab Chip* 2009, **9**(15): p. 2132–2139.
45. Heo, Y.S., Cabrera, L.M. et al., Characterization and resolution of evaporation-mediated osmolality shifts that constrain microfluidic cell culture in poly(dimethylsiloxane) devices. *Anal Chem*, 2007, **79**(3): p. 1126–1134.
46. Toepke, M.W., Beebe, D.J., PDMS absorption of small molecules and consequences in microfluidic applications. *Lab Chip*, 2006, **6**(12): p. 1484–1486.
47. Puleo, C.M., McIntosh Ambrose, W. et al., Integration and application of vitrified collagen in multilayered microfluidic devices for corneal microtissue culture. *Lab Chip*, 2009, **9**(22): p. 3221–3227.
48. Whitesides, G.M, The origins and the future of microfluidics. *Nature*, 2006, **442**(7101): p. 368–373.
49. Ploss, A., Khetani, S.R. et al., Persistent hepatitis C virus infection in microscale primary human hepatocyte cultures. *Proc Natl Acad Sci U S A*, 2010, **107**(7): p. 3141–3145.

50. Jones, C.T., Catanese, M.T. et al., Real-time imaging of hepatitis C virus infection using a fluorescent cell-based reporter system. *Nat Biotechnol*, 2010, **28**(2): p. 167–171.
51. Nagao, K., Ohyashiki, J.H. et al., Expression of hTERT mRNA in a mortal liver cell line during S phase without detectable telomerase activity. *Int J Mol Med*, 2005, **15**(4): p. 683–688.
52. Durantel, D., Zoulim, F., Going towards more relevant cell culture models to study the *in vitro* replication of serum-derived hepatitis C virus and virus/host cell interactions? *J Hepatol*, 2007, **46**(1): p. 1–5.
53. Gupta, K., Kim, D.H. et al., Lab-on-a-chip devices as an emerging platform for stem cell biology. *Lab Chip*, 2010, **10**(16): p. 2019–2031.
54. Kane, B.J., Zinner, M.J. et al., Liver-specific functional studies in a microfluidic array of primary mammalian hepatocytes. *Anal Chem*, 2006, **78**(13): p. 4291–4298.
55. Chao, P., Maguire, T. et al., Evaluation of a microfluidic based cell culture platform with primary human hepatocytes for the prediction of hepatic clearance in human. *Biochem Pharmacol*, 2009, **78**(6): p. 625–632.
56. Novik, E., Maguire, T.J. et al., A microfluidic hepatic coculture platform for cell-based drug metabolism studies. *Biochem Pharmacol*, 2010, **79**(7): p. 1036–1044.
57. Sin, A., Baxter, G.T., Shuler, M.L., Animal on a chip: A microscale cell culture analog device for evaluating toxicological and pharmacological profiles. *Proc SPIE, Int Soc Opt Eng*, Bellingham, WA, 2010.
58. Khamsi, R., Labs on a chip: Meet the stripped down rat. *Nature* 2005, **435**(7038): p. 12–13.
59. Shuler, M.L., Cropek, D.M. et al., Animal/human on a chip. *Annu Rev Biomed Eng*, 2011, **13**(1).
60. Frampton, J.P., Shuler, M.L. et al., Biomedical technologies for *in vitro* screening and controlled delivery of neuroactive compounds. *Cent Nerv Syst Agents Med Chem*, 2008, **8**(3): p. 203–219.
61. Viravaidya, K., Shuler, M.L., Incorporation of 3T3-L1 cells to mimic bioaccumulation in a microscale cell culture analog device for toxicity studies. *Biotechnol Prog*, 2004, **20**(2): p. 590–597.
62. Viravaidya, K., Sin, A. et al., Development of a microscale cell culture analog to probe naphthalene toxicity. *Biotechnol Prog*, 2004, **20**(1): p. 316–323.
63. Zhang, C., Zhao, Z. et al., Towards a human-on-chip: Culturing multiple cell types on a chip with compartmentalized microenvironments. *Lab Chip*, 2009, **9**(22): p. 3185–3192.
64. Mavri-Damelin, D., Damelin, L.H. et al., Cells for bioartificial liver devices: The human hepatoma-derived cell line C3A produces urea but does not detoxify ammonia. *Biotechnol Bioeng*, 2008, **99**(3): p. 644–651.
65. Lion, N., Rohner, T.C. et al., Microfluidic systems in proteomics. *Electrophoresis*, 2003, **24**(21): p. 3533–3662.
66. Vreeland, W.N., Barron, A.E., Functional materials for microscale genomic and proteomic analyses. *Curr Opin Biotechnol*, 2002, **13**(2): p. 87–94.
67. Dodge, A., Fluri, K. et al., Electrokinetically driven microfluidic chips with surface-modified chambers for heterogeneous immunoassays. *Anal Chem*, 2001, **73**(14): p. 3400–3409.
68. Bange, A., Halsall, H.B. et al., Microfluidic immunosensor systems. *Biosens Bioelectron*, 2005, **20**(12): p. 2488–2503.

69. Yakovleva, J., Davidsson, R. et al., Microfluidic enzyme immunoassay using silicon microchip with immobilized antibodies and chemiluminescence detection. *Anal Chem*, 2002, **74**(13): p. 2994–3004.
70. Derveaux, S., Stubbe, B.G. et al., Synergism between particle-based multiplexing and microfluidics technologies may bring diagnostics closer to the patient. *Anal Bioanal Chem*, 2008, **391**(7): p. 2453–2467.
71. Diercks, A.H., Ozinsky, A. et al., A microfluidic device for multiplexed protein detection in nano-liter volumes. *Anal Biochem*, 2009, **386**(1): p. 30–35.
72. Hui, W.C., Yobas, L. et al., Microfluidic systems for extracting nucleic acids for DNA and RNA analysis. *Sens Actuators*, 2007, **133**(2): p. 335–339.
73. Liu, P., Mathies, R.A., Integrated microfluidic systems for high-performance genetic analysis. *Trends Biotechnol*, 2009, **27**(10): p. 572–581.
74. Wieder, K.J., King, K.R. et al., Optimization of reporter cells for expression profiling in a microfluidic device. *Biomed Microdevices*, 2005, **7**(3): p. 213–222.
75. Metzker, M.L., Sequencing technologies — the next generation. *Nat Rev Genet* **11**(1): p. 31–46.
76. Rettig, J.R., Folch, A., Large-scale single-cell trapping and imaging using microwell arrays. *Anal Chem*, 2005, **77**(17): p. 5628–5634.
77. King, K.R., Wang, S. et al., A high-throughput microfluidic real-time gene expression living cell array. *Lab Chip*, 2007, **7**(1): p. 77–85.
78. Yarmush, M.L., King, K.R., Living-cell microarrays. *Annu Rev Biomed Eng*, 2009, **11**: p. 235–257.
79. Einav, S., Gerber, D. et al., Discovery of a hepatitis C target and its pharmacological inhibitors by microfluidic affinity analysis. *Nat Biotechnol*, 2008, **26**(9): p. 1019–1027.
80. Erfle, H., Neumann, B. et al., Reverse transfection on cell arrays for high content screening microscopy. *Nat Protoc*, 2007, **2**(2): p. 392–399.
81. Sharma, S., Rao, A., RNAi screening: Tips and techniques. *Nat Immunol*, 2009, **10**(8): p. 799–804.
82. Anderson, D. G., Levenberg, S. et al., Nanoliter-scale synthesis of arrayed biomaterials and application to human embryonic stem cells. *Nat Biotechnol*, 2004, **22**(7): 863–866.
83. Flaim, C.J., Chien, S. et al., An extracellular matrix microarray for probing cellular differentiation. *Nat Methods*, 2005, **2**(2): p. 119–125.
84. Chen, Z., Li, Y. et al., Patterning mammalian cells for modeling three types of naturally occurring cell-cell interactions. *Angew Chem Int Ed Engl*, 2009, **48**(44): p. 8303–8305.
85. Thompson, D.M., King, K.R. et al., Dynamic gene expression profiling using a microfabricated living cell array. *Anal Chem*, 2004, **76**(14): p. 4098–4103.
86. King, K.R., Wang, S. et al., Microfluidic flow-encoded switching for parallel control of dynamic cellular microenvironments. *Lab Chip*, 2008, **8**(1): p. 107–116.
87. Yang, E., Foteinou, P.T. et al., A novel non-overlapping bi-clustering algorithm for network generation using living cell array data. *Bioinformatics*, 2007, **23**(17): p. 2306–2313.
88. Yang, E., Yarmush, M.L. et al., Transcription factor network reconstruction using the living cell array. *J Theor Biol*, 2009, **256**(3): p. 393–407.
89. Roach, K.L., King, K.R. et al., High throughput single cell bioinformatics. *Biotechnol Prog*, 2009, **25**(6): p. 1772–1779.

90. Park, J., Li, Y. et al., Radial flow hepatocyte bioreactor using stacked microfabricated grooved substrates. *Biotechnol Bioeng*, 2008, **99**(2): p. 455–467.
91. Tilles, A.W., Baskaran, H. et al., Effects of oxygenation and flow on the viability and function of rat hepatocytes cocultured in a microchannel flat-plate bioreactor. *Biotechnol Bioeng*, 2001, **73**(5): p. 379–389.
92. Park, J., Berthiaume, F. et al., Microfabricated grooved substrates as platforms for bio-artificial liver reactors. *Biotechnol Bioeng*, 2005, **90**(5): p. 632–644.
93. Domansky, K., Inman, W. et al., Perfused multiwell plate for 3D liver tissue engineering. *Lab Chip*, 2010, **10**(1): p. 51–58.
94. Sivaraman, A., Leach, J.K. et al., A microscale *in vitro* physiological model of the liver: Predictive screens for drug metabolism and enzyme induction. *Curr Drug Metab*, 2005, **6**(6): p. 569–591.
95. Yates, C., Shepard, C.R. et al., Novel three-dimensional organotypic liver bioreactor to directly visualize early events in metastatic progression. *Adv Cancer Res*, 2007, **97**: p. 225–246.
96. Wang, J., Ren, L. et al., Microfluidics: A new cosset for neurobiology. *Lab Chip*, 2009, **9**(5): p. 644–652.
97. Campenot, R.B., Local control of neurite development by nerve growth factor. *Proc Natl Acad Sci U S A*, 1977, **74**(10): p. 4516–4519.
98. Campenot, R.B., Development of sympathetic neurons in compartmentalized cultures. II Local control of neurite growth by nerve growth factor. *Dev Biol*, 1982, **93**(1): p. 1–12.
99. Taylor, A.M., Rhee, S.W. et al., Microfluidic multicompartment device for neuroscience research. *Langmuir*, 2003, **19**(5): p. 1551–1556.
100. Taylor, A.M., Blurton-Jones, M. et al., A microfluidic culture platform for CNS axonal injury, regeneration and transport. *Nat Methods*, 2005, **2**(8): p. 599–605.
101. Taylor, A.M., Dieterich, D.C. et al., Microfluidic local perfusion chambers for the visualization and manipulation of synapses. *Neuron*, 2010, **66**(1): p. 57–68.
102. Berdichevsky, Y., Staley, K.J. et al., Building and manipulating neural pathways with microfluidics. *Lab Chip*, 2010, **10**(8): 999–1004.
103. Shibata, K., Terazono, H., Hattori, A., Yasuda, K., Collagen micro-flow channels as an for *in vitro* blood-brain barrier model. *Jpn J Appl Phys*, 2008, **47**(8): p. 5208–5211.
104. Davignon, J., Ganz, P., Role of endothelial dysfunction in atherosclerosis. *Circulation*, 2004, **109**(23 Suppl 1): p. III27–III32.
105. Hicklin, D.J., Ellis, L.M., Role of the vascular endothelial growth factor pathway in tumor growth and angiogenesis. *J Clin Oncol*, 2005, **23**(5): p. 1011–1027.
106. Cosentino, F., Luscher, T.F., Endothelial dysfunction in diabetes mellitus. *J Cardiovasc Pharmacol*, 1998, **32**(Suppl 3): p. S54–S61.
107. Cunningham, K.S., Gotlieb, A.I., The role of shear stress in the pathogenesis of atherosclerosis. *Lab Invest*, 2005, **85**(1): p. 9–23.
108. Song, J.W., Gu, W. et al., Computer-controlled microcirculatory support system for endothelial cell culture and shearing. *Anal Chem*, 2005, **77**(13): p. 3993–3999.
109. Schaff, U.Y., Xing, M.M. et al., Vascular mimetics based on microfluidics for imaging the leukocyte--endothelial inflammatory response. *Lab Chip*, 2007, **7**(4): 448–456.
110. Gutierrez, E., Petrich, B.G. et al., Microfluidic devices for studies of shear-dependent platelet adhesion. *Lab Chip*, 2008, **8**(9): p. 1486–1495.

111. Ku, C.J., D'Amico Oblak, T. et al., Interactions between multiple cell types in parallel microfluidic channels: Monitoring platelet adhesion to an endothelium in the presence of an anti-adhesion drug. *Anal Chem*, 2008, **80**(19): p. 7543–7548.
112. Neeves, K.B., Diamond, S.L., A membrane-based microfluidic device for controlling the flux of platelet agonists into flowing blood. *Lab Chip*, 2008, **8**(5): p. 701–709.
113. Chaw, K.C., Manimaran, M. et al., Multi-step microfluidic device for studying cancer metastasis. *Lab Chip*, 2007, **7**(8): p. 1041–1047.
114. Song, J.W., Cavnar, S.P. et al., Microfluidic endothelium for studying the intravascular adhesion of metastatic breast cancer cells. *PLoS One*, 2009, **4**(6): p. e5756.
115. Spence, D.M., Torrence, N.J. et al., Amperometric determination of nitric oxide derived from pulmonary artery endothelial cells immobilized in a microchip channel. *Analyst*, 2004, **129**(11): p. 995–1000.
116. D'Amico Oblak, T., Root, P. et al., Fluorescence monitoring of ATP-stimulated, endothelium-derived nitric oxide production in channels of a poly(dimethylsiloxane)-based microfluidic device. *Anal Chem*, 2006, **78**(9): p. 3193–3197.
117. Genes, L.I., V Tolan, N., Hulvey, M.K., Marth, R.S., Spence, D.M., Addressing a vascular endothelium array with blood components using underlying microfluidic channels. *Lab Chip*, 2007, **7**(10): p. 1256–1259.
118. Vickerman, V., Blundo, J. et al., Design, fabrication and implementation of a novel multi-parameter control microfluidic platform for three-dimensional cell culture and real-time imaging. *Lab Chip*, 2008, **8**(9): p. 1468–1477.
119. Kim, M.S., Yeon, J.H. et al., A microfluidic platform for 3-dimensional cell culture and cell-based assays. *Biomed Microdevices*, 2007, **9**(1): p. 25–34.
120. Barkefors, I., Thorslund, S. et al., A fluidic device to study directional angiogenesis in complex tissue and organ culture models. *Lab Chip*, 2009, **9**(4): p. 529–535.
121. Blow, N., Microfluidics: The great divide. *Nat Methods*, 2009, **6**(9): p. 683–686.
122. Griffiths, J., Research profile: A completely automated, microfluidic cell-culture chip. *Anal Chem*, 2007, **79**(23): p. 8829–8829.

Chapter 11

Application of Microfluidics in Stem Cell and Tissue Engineering

Sasha H. Bakhru*, Christopher Highley†, and Stefan Zappe†,‡

**Department of Molecular Pharmacology, Physiology, and Biotechnology
Brown University, Providence, RI 02912, USA*
*†Department of Biomedical Engineering, Carnegie Mellon University
Pittsburgh, PA 15213, USA*

Introduction

Stem cells have immense potential for use in therapeutic strategies.[1,2] Ever since the first indications of the existence of hematopoietic stem cells (HSCs) in the 1960s,[3] and more intensely after the isolation and culture of human embryonic stem and germ cells in 1998,[4,5] researchers have strived to identify new stem cell types to understand their biology, as well as to manipulate and use them in order to achieve specific therapeutic aims. In 2007, the successful creation of induced pluripotent stem (iPS) cells through the reprogramming of human somatic cells generated hope for easy access to patient-specific stem cells that bypass ethical concerns related to human embryonic stem cells (hESCs) and can be used for autologous therapies.[6,7] Stem cells find applications, for example, in cell-based gene therapies, therapies for the replacement of specific cell types, or in more complex tissue engineering strategies.[2] Similar to stem cell research, early tissue engineering experiments were already carried out in the 1960s, for example, with artificial skin for burn victims. However, it was not until the end of the 1980s that tissue engineering was recognized and discussed by the broader research community as a

---

[‡]Corresponding author. Email: zappe@cmu.edu

new scientific discipline,[8] based on seminal papers that described, for example, the successful isolation of parenchymal cells, their seeding onto biodegradable scaffolds *in vitro*, and their functional engraftment *in vivo* after surgical transplantation of cell-seeded scaffolds.[9] In 1993, tissue engineering was most prominently defined as an "interdisciplinary field that applies the principles of engineering and life sciences toward the development of biological substitutes that restore, maintain, or improve tissue function or a whole organ."[10] Today, many FDA-approved tissue engineering products are commercially available,[11] including acellular products such as Pro Osteon® for bone tissue regeneration as well as cellular products such as Carticel® for the repair of articular cartilage in the knee or the skin substitute Apligraf®. Additionally, while hematopoietic stem cells have been used, for example, in bone marrow transplant therapies for decades and many other adult human stem cell types in clinical trials for years, the first clinical trial with oligodendrocyte progenitor cells derived from hESCs for spinal cord repair is currently underway.[12]

Stem cell and tissue engineering are closely related fields. Stem cells can be used in tissue engineering approaches to generate cells of therapeutic relevance. Due to their immature nature and capacity for proliferation and differentiation on cue, stem cells may be better suited than more specialized cells to let a tissue engineering construct mature into functional tissue. In the ideal tissue engineering approach, some tissue-specific adult stem cells would actually remain undifferentiated in their respective niches and would help to maintain generated tissue through continued renewal of cells. The tissue engineering community has generated processes and materials designed to elicit specific responses in cells in support of the formation of new tissue. Such cell-instructive materials can be applied, for example, to the expansion of stem cells to clinically relevant numbers or to their efficient, lineage-specific differentiation into therapeutic cell types. Further development of tissue engineering and stem cell-based regenerative concepts, however, appears to be limited by our ability to analyze and understand complex biological systems as well as to replicate such complexity in engineered systems. While progress will depend on highly interdisciplinary solutions, microfluidic technologies, as discussed in this chapter, are uniquely suited to address complex challenges in a systematic manner and to contribute to the further advancement of stem cell and tissue engineering-based approaches in regenerative medicine.

Established in the 1990s, microfluidic technologies have enabled experimentation at the molecular and cellular levels in unprecedented ways.[13] Microfluidic devices can generally be mass-fabricated at low cost per device, can reliably and efficiently handle the smallest sample volumes, and are extremely versatile with respect to the integration of sensors, actuators, and even electronic circuits for control and signal processing. The term *microTAS* was coined in 1990, referring to the generation and application of miniaturized total chemical analysis systems.[14] Soon after, the concept of *lab-on-a-chip* was developed, aimed at processing not only chemical, but also biological samples in complex microfluidic systems. Initially, microfluidic devices were fabricated through the application of silicon technologies to the patterning of silicon and glass wafers. For economic reasons, experimental and especially commercial devices were soon made of rigid polymers and more recently, using elastic polymers such as polydimethylsiloxane (PDMS), e.g., in a process termed *soft lithography*.[15] In 2000, an enhanced, two-layer soft lithography process was presented that enabled the integration of pumps and valves at very high density,[16] paving the road for *microfluidic large-scale integration*,[17,18] analogous to the large-scale integration of electronic devices. More recently, researchers have also emphasized on the opportunity to combine microfluidic and micro-optic devices to integrated *optofluidic* systems.[19,20] During the past 15 years, complex microfluidic systems have been developed for automated handling, manipulation, and analysis of mammalian cells[21–25] and in support of systems biology research.[26] However, it was only in very recent years that the engineering community started to apply microfluidic technologies to research in stem cell biology[27–29] and tissue engineering.[30–32] While this chapter focuses on microfluidic applications, other microtechnologies will also be briefly discussed based on their importance or their potential for use in combination with microfluidic systems.

Microfluidic technologies can contribute to advancements in stem cell and tissue engineering at several levels. Microfluidic systems can help to isolate relatively rare adult stem cells from tissue biopsies, to screen for microenvironmental factors that govern the fate of cultured stem cells, to automatically expand small numbers of stem cells to clinically relevant quantities, to differentiate stem cells into therapeutic cell types, as well as to support high-throughput and single-cell analysis strategies, for example, for the identification of novel stem cell markers or measurement of gene expression profiles, in an attempt to understand stem cell fate decisions in more detail. As applied to tissue engineering, the use of microfluidic devices is particularly compelling either for on-chip complexes, and three-dimensional cell cultures under the precise control of culture conditions in space and time, or for the generation of cell-containing building blocks that can be assembled off chip to complex, cell-instructive environments that mature into functional tissue. Microfluidics can either support the generation of actual tissue engineering constructs for use *in vivo* or the generation of tissue models for *in vitro* assays. Such assays could attempt to mimic specific *in vivo* conditions, e.g., for study of disease mechanisms, evaluation of therapeutic strategies, prediction of the behavior of transplanted stem cells *in vivo*, or for testing of drugs.

Properties of Stem Cells

The following section briefly reviews basic properties of stem cells and characteristics of the various tissue-specific niches in which adult stem cells reside. It further discusses microenvironmental cues that stem cells respond to *in vivo* as well as some microtechnologies, applied to demonstrate the effect of single, cell-instructive stimuli *in vitro*.

Origin, properties, and isolation of stem cells

Stem cell types can be distinguished from one another based on their potency, or in other words, the types of mature cells they can yield. The zygote, or fertilized egg, is classified as *totipotent* since it will form not only the entire organism, with all its somatic cell types, but also placental tissue. As a zygote matures over several days, a blastocyst is formed, from which cells of the inner cell mass, as shown in Fig. 1 (left), are harvested to yield ESCs. In contrast to totipotent cells, ESCs are termed *pluripotent* as they are capable of generating all cells of the developing organism, but not those found within supporting placental structures.

iPS cells are created through nuclear reprogramming of more specialized, mature cells, for example, by the permanent or transient expression of transgenes that encode key transcriptional factors active in ESCs. iPS

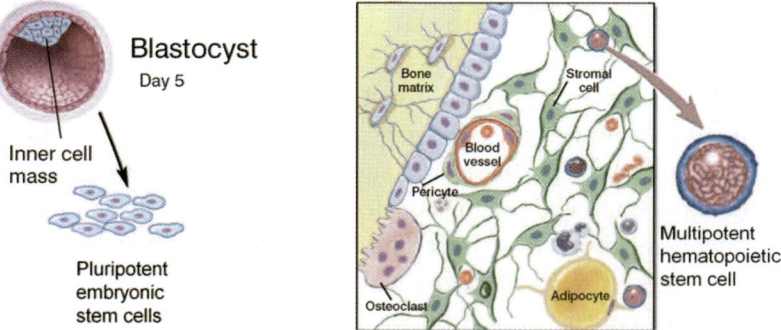

Fig. 1. Origin of embryonic and adult stem cells. Left: Embryonic stem cells are harvested from a blastocyst stage embryo. They form the inner cell mass of the blastocyst and give rise to an entire organism. Right: Adult stem cells (here: hematopoietic stem cells as an example) reside in specific locations in virtually any tissue of the developed (adult) organism and help to maintain and to some extent, repair tissue. (Reprinted with permission from Ref. 1.)

cells are thought to have properties nearly identical to ESCs. Adult stem cells, on the other hand, are found in virtually all tissues of the developed organism and help to maintain and to some extent, repair the adult tissue. The example in Fig. 1 (right) shows HSCs, the first adult stem cell type described, residing in the bone marrow. Adult stem cells are classified as *multipotent* as they give rise typically only to lineages found in their native tissue. However, some research results suggested that at least *in vitro*, adult stem cells are able to cross their lineage boundaries, or transdifferentiate.[33]

Common to all stem cell types is the defining property of indefinite self-renewal. Significant differences exist, however, between embryonic and adult stem cells regarding their tumorigenicity. Upon transplantation into a host, ESCs typically give rise to teratomas, tumors which display cell types that are found in tissues derived from all three embryonic germ layers. Therapeutic use of ESCs therefore requires their reliable pre-differentiation into therapeutic cell types.[34] In contrast, adult stem cells are typically non-tumorigenic and may be transplanted either directly or upon pre-differentiation, especially if the specific targeted clinical condition may not be expected to elicit differentiation along desired lineages.

While ESCs are harvested with relative ease, the isolation of relatively rare adult stem cells, e.g., from tissue biopsies of an adult person, typically requires enzymatic tissue digestion and separation of stem cells from more mature cell types. In some cases, this can be achieved through relatively simple *in vitro* culture procedures. For example, mesenchymal stem cells (MSCs) readily adhere to tissue culture plastic within 24–48 h and can be easily separated from other bone marrow cells that are nonadherent.[35] Other cell types require more sophisticated cell cytometry and sorting strategies such as fluorescence-activated cell sorting (FACS). Microfluidics has been applied to produce miniaturized cytometry, sorting, and FACS systems for cost-effective analysis and enrichment of cells,[36,37] but only a few systems have been specifically designed for the isolation or enrichment of stem cells. Mouse neural stem/precursor cells, for example, were successfully isolated in microfluidic systems through dielectrophoresis (DEP).[38] Similarly, a microfluidic system for DEP field-flow fractionation was developed for enrichment of up to 14-fold of putative stem cells from adipose tissue.[39] Other microfluidic cell sorting systems with potential for applications to stem cell isolation include sorting based on size exclusion,[40] as well as immobilization on microchannel surfaces through surface-bound antibodies,[41] or peptide sequences that bind stem cell-specific surface receptors.[42]

In vivo adult stem cell niches

Controlling stem cell fate, e.g., for *in vitro* expansion of stem cells to clinically relevant numbers or for lineage-specific differentiation into therapeutically relevant cell types is of key interest to the research community. In general, stem cell quiescence, proliferation, differentiation, apoptosis, and migration are tightly governed by a variety of environmental cues such as soluble factors (e.g., growth factors and cytokines secreted by neighboring cells), physical cues (e.g., extracellular matrix (ECM) interactions, material stiffness, nanoscale topography, temperature), and both hetero- and homotypic cell–cell interactions (Fig. 2).[43] Understanding the specific microenvironment in which stem cells reside *in vivo* can yield important insights into factors that regulate stem cell fate decisions and enable the generation of cell-instructive materials, designed to elicit specific cellular responses or to ensure that stem cell-seeded scaffolds mature into functional tissue.[44,45]

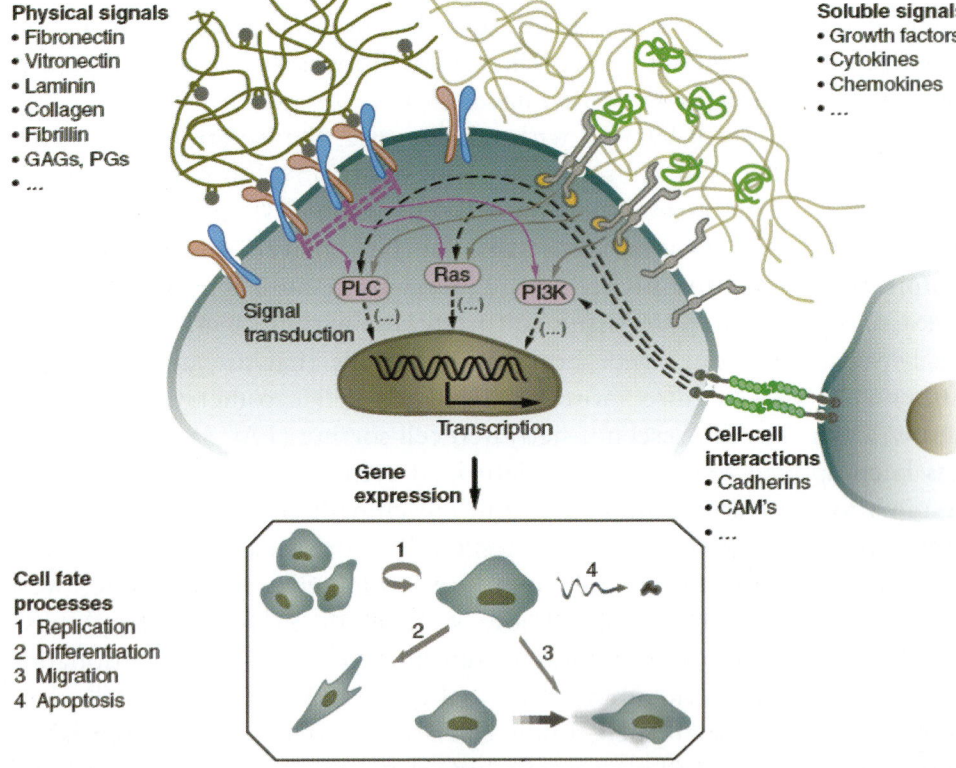

Fig. 2. The cellular microenvironment. Cells experience a combination of physical, soluble, and juxtracrine (cell–cell) cues *in vivo*, which in combination, influence cellular fate. (Reprinted with permission from Ref. 44.)

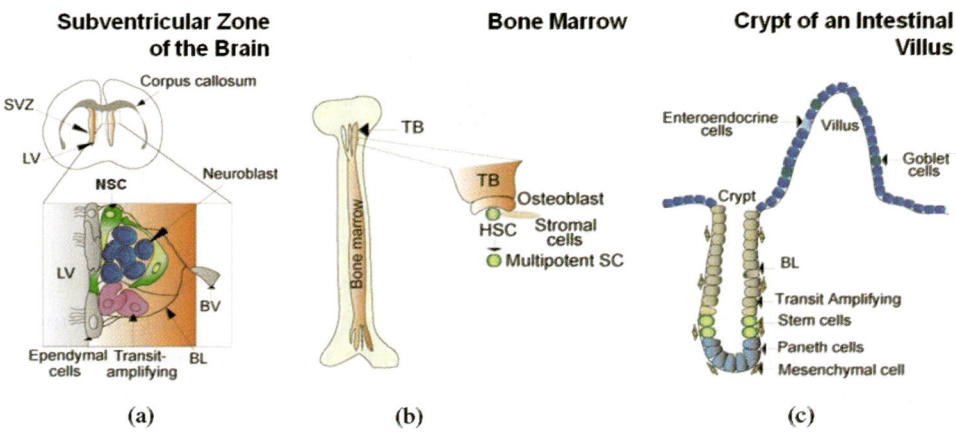

Fig. 3. Examples of *in vivo* somatic stem cell niches. (a) Neural stem cell (NSC) niche in the subventricular zone. (b) Hematopoietic stem cell (HSC) niche in the bone marrow. (c) Intestinal stem cell niche in the crypt of an intestinal villus. (Reprinted with permission from Ref. 56.)

In vivo, the microenvironment immediately surrounding stem cells is termed the stem cell *niche* and is a complex, well-defined venue for signals that regulate stem cell fate.[46–51] The concept of a stem cell niche dates back to early experiments with *in vitro* hematopoiesis,[52,53] in which cell fate decisions were observed to depend on combinations of soluble factors presented, presumed *in vivo* to be provided by neighboring cells. Since that time, a host of *in vivo* stem cell niches in a diverse array of tissues have been described (Fig. 3).

In the case of soluble cues, not only the concentration, but also the mode of presentation, influenced by the cellular microenvironment, are important. A prime example of this phenomenon is captured in fibroblast growth factor (FGF) signaling, common to almost all stem cell types. Heparan sulfate, a member of the glycosaminoglycan family of carbohydrates, is a cofactor for FGF signaling. Heparan sulfate binds, clusters, and stabilizes members of the FGF family,[54] increasing their receptor-binding affinities by more than an order of magnitude. Modeling studies have demonstrated a wide range of flexibility in stoichiometry, but most typically describe a dimer of FGF ligands bound to a dimer of FGF receptors stabilized by a cell surface-anchored heparan sulfate proteoglycan through its heparan sulfate chain.[55] Important to virtually all stem cell types, heparan sulfate proteoglycan interacts with a diverse array of growth factors and cytokines, influencing a myriad of biological processes. For these reasons, delivering FGF and other heparan sulfate-binding soluble factors in solution *in vitro* is inefficient and the original presentation mode is poorly recapitulated in standard cell culture methodologies. While soluble factors have proven to be potent regulators of cell fate, other classes of environmental cues play equally important roles.

Physical signaling, for example, involves interactions between the cell and ECM or matrix constituents. The nature of the extracellular milieu varies dramatically among stem cell types; however, in some cases, various aspects or themes are conserved. For example, a number of adult stem cell types occupy a *perivascular* niche. That is, these cells reside adjacent to the tissue microvasculature and interact directly with constituents of the basement membrane overlying the vessel surface, including matrix glycoproteins, proteoglycans, and collagen. Adhesion glycoproteins and collagens engage dimeric, transmembrane receptors known as integrins on the cell surface. Neural stem cells (NSCs)[57] and HSCs[58] are prime examples of cells occupying a perivascular niche, receiving physical signals from the vessel surface, paracrine signals from the cells composing the vessel, and endocrine (hormonal) signals from the blood carried within the vessel lumen (Fig. 3(a) and 3(b)). A second example of a highly ordered *in vivo* stem cell niche architecture is that of the intestinal crypt. In this case, the structure contains a populations of roughly 250 cells (murine), all derived from a multipotent cell population at its base, serving to populate each individual crypt[59] — i.e., each crypt is a clonal population of cells, derived from stem cells differentiating in an orderly fashion along the crypt–villus axis (Fig. 3(c)).[60] The niche is thought to comprise the proliferating and differentiating epithelial cells, surrounding MSCs, subepithelial fibroblasts and myoblasts, and the interpenetrating matrix in which they reside.[59,61] This matrix again contains integrin-binding glycoproteins directing cellular organization within this highly ordered cellular structure.

The concept of the ECM influencing stem and progenitor cell fate is well established. Two prime examples are found for β1 integrin engagement influencing stem cell localization to the niche in two very different tissues. The first is in the skin, where β1 integrins are differentially expressed on stem and progenitor cells and participate in constrained localization of stem cells via interaction with ECM glycoprotein ligands. In this case, β1 integrin serves two functions — if expression is downregulated via a dominant-negative mutation, stem cells behave like transit-amplifying cells, differentiating within a few proliferative cycles.[62,63] Thus, β1 integrins presumably serve to maintain stem cell phenotype. Second, high β1 integrin expression maintains the patterned distribution of stem cells within this ependymal niche, rendering them less motile than transit-amplifying cells and leaving them mostly clustered within the epidermal basal layer.[49,62] The second example, involving one of the best studied integrin-mediated homing mechanisms for a stem cell, is HSC migration from the circulating blood to the perivascular bone marrow niche. In this case, HSCs are recruited via systemic blood circulation, locate (e.g., via ECM-bound stromal cell-derived factor-1 (SDF-1)), and attach to their perivascular niche in a process known as stem cell *homing*. This process of attachment involves integrins expressed on the circulating HSCs, upregulated by SDF-1.[64] According to the literature, migrating HSCs use β1 integrin-mediated adhesion to bind to vascular cell adhesion molecule 1 (VCAM-1) on the surface of endothelial cells comprising the bone marrow vasculature.[65,66] This receptor (integrin) engagement initiates the migration of the HSCs through the vasculature into the bone marrow, where they may repopulate their niche.

While this brief discussion of selected factors that govern stem cell fate *in vivo* is by no means exhaustive, it may indicate both the challenges of understanding the complexity of stem cell niches as well as the opportunities for the implementation of research findings in the design of processes or materials toward directing stem cell fate.

Responses of stem cells to single environmental cues

Microtechnologies, including microfluidic technologies discussed later in this chapter, are well suited for investigating individual cues that stem cells respond to, and for verifying hypotheses derived from the analysis of *in vivo* stem cell niches. For example, toward fabricating a culture substrate affording control over cell–cell interaction, a microfabricated polymeric chip containing several thousands of microwells was developed, each sized to trap and seed a single stem cell (Fig. 4).[67]

This platform enabled efficient generation of a square array of murine ESCs onto a variety of substrates, ranging from a gelatin-patterned culture dish, to a three-dimensional substrate, to an underlying layer of supporting cells. It was similarly possible to pattern small groups of cells both with and without cell contact, allowing "incremental control" over contact-mediated signaling. This platform enabled studies demonstrating a role for cell–cell contact in depressing ESC colony formation, and implicated E-cadherin in this negative regulatory pathway.

Interaction with for example, surface-bound ECM constituents enables a cell to indirectly gather information, e.g., on surface topography and substrate stiffness, and to respond to such cues. Based on microtechnologies for the nanopatterning of cell culture substrates, studies toward elucidating the influence of topographical cues on stem cell fate have recently gained

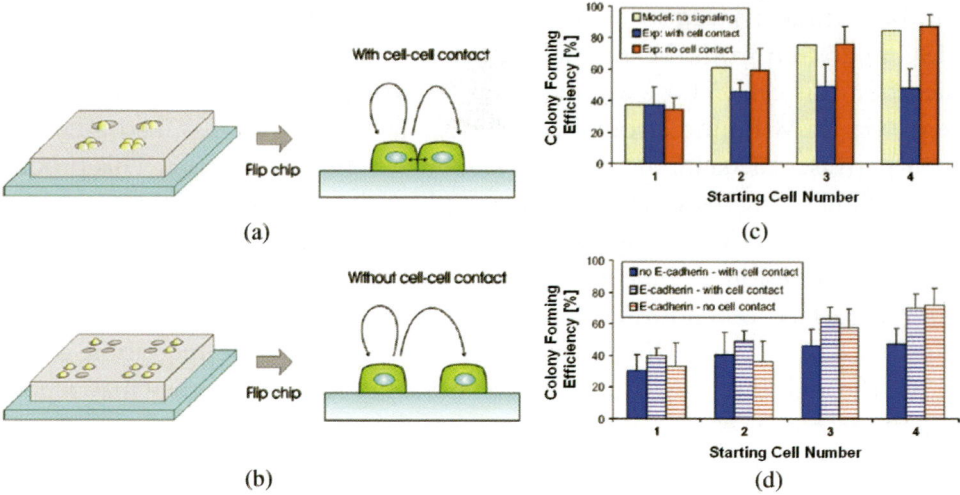

Fig. 4. A microfabricated polymeric chip affording control over cell–cell contact. Embryonic stem cells (ESCs) can be patterned in (a) groups with cell–cell contact, enabling controlled juxtacrine interaction, or (b) without cell–cell contact, limiting interaction to paracrine signaling between neighboring cells. (c) and (d) Colony-forming efficiency of patterned cells was quantified as degree of cell contact and the presence of the tight-junction constituent E-cadherin were varied, demonstrating the importance of cell–cell contact to the colony-forming potential of cultured ESCs. (Reprinted with permission from Ref. 67.)

attention. Among these are seminal reports providing early evidence for substrate-mediated control of stem cell fate, without the need for the use of biochemical signals. In 2005, empirical evidence that nanotopography halts cellular proliferation was reported, based on a bromodeoxyuridine (BrdU) incorporation assay with human smooth muscle cells grown on nano-imprinted substrates.[68] This finding prompted further investigation into the influence of nanotopographical cues on stem cell fate, including studies probing the influence of similar two-dimensional nanopatterns on MSC differentiation.[69] The directed growth along the long axis of the nanopattern elicited preferential neuronal differentiation of cultured MSCs, as shown in Fig. 5.

These studies provided early evidence that surface geometry or texture alone is sufficient to influence stem cell proliferation and differentiation in culture. More recently, the same group showed that human MSCs (hMSCs) cultured on 350-nm gratings of tissue culture polystyrene (TCPS) and PDMS yielded decreased expression of integrin subunits $\alpha 2$, $\alpha 6$, αV, $\beta 2$, $\beta 3$, and $\beta 4$ compared to planar controls. On gratings, hMSCs exhibited an aligned, elongated actin cytoskeleton, while on unpatterned controls, hMSCs exhibited a random but denser actin cytoskeleton network.[70] The expression of cytoskeleton and focal adhesion components was also altered by the nanotopography,

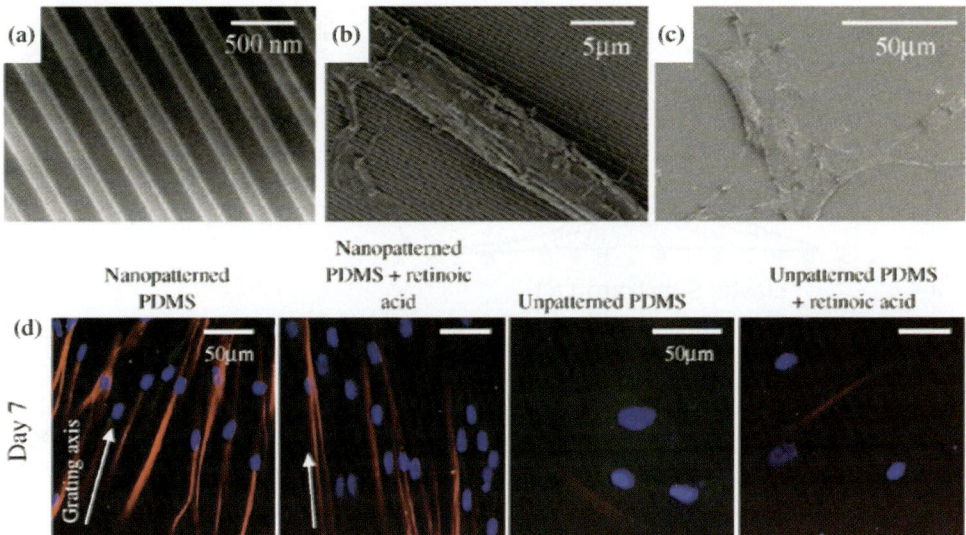

Fig. 5. Culture substrate nanotopography influences MSC differentiation. Shown are scanning electron microscope (SEM) images of (a) PDMS nanopatterned by replica molding; human MSCs (hMSCs) cultured on (b) nano-patterned PDMS and (c) unpatterned PDMS. (d) Cells are indirectly immunofluorescently stained for TuJ1, an early marker of neuronal lineage, expressed most prominently in MSCs cultured on aligned nanoscale gratings. (Reprinted with permission from Ref. 69.)

as reflected in mechanical properties measured by atomic force microscopy. For example, on rigid TCPS, hMSCs on gratings exhibited lower Young's moduli and apparent viscosity. On the softer, elastomeric PDMS, hMSCs exhibited further reduced cellular mechanical properties compared to those on more rigid substrates, regardless of topography. These data suggested that not only nanotopography, but also substrate stiffness could be important in determining mechanical cell properties, while nanotopography may be more dominant in determining intracellular cytoskeleton and focal adhesion organization.

In another landmark study, MSCs were again employed to demonstrate the direct influence of substrate stiffness on stem cell lineage specification (Fig. 6). MSCs were cultured in suspension and on hydrogel substrates of varied stiffness. It was observed that MSCs in suspension culture gave rise to blood cells; those on the softest hydrogel matrices gave rise to neural cells, those on moderately stiff hydrogels gave rise to muscle cells, and those on stiff substrates gave rise to collagenous bone, within the range of 0 to 100 kPa substrate elastic modulus.[71] Moreover, using blebistatin, an inhibitor of non-muscle myosin 2 function, non-muscle myosin 2 was identified as a key player in these cells' ability to sense their environment. In the presence of blebistatin, cultured MSCs

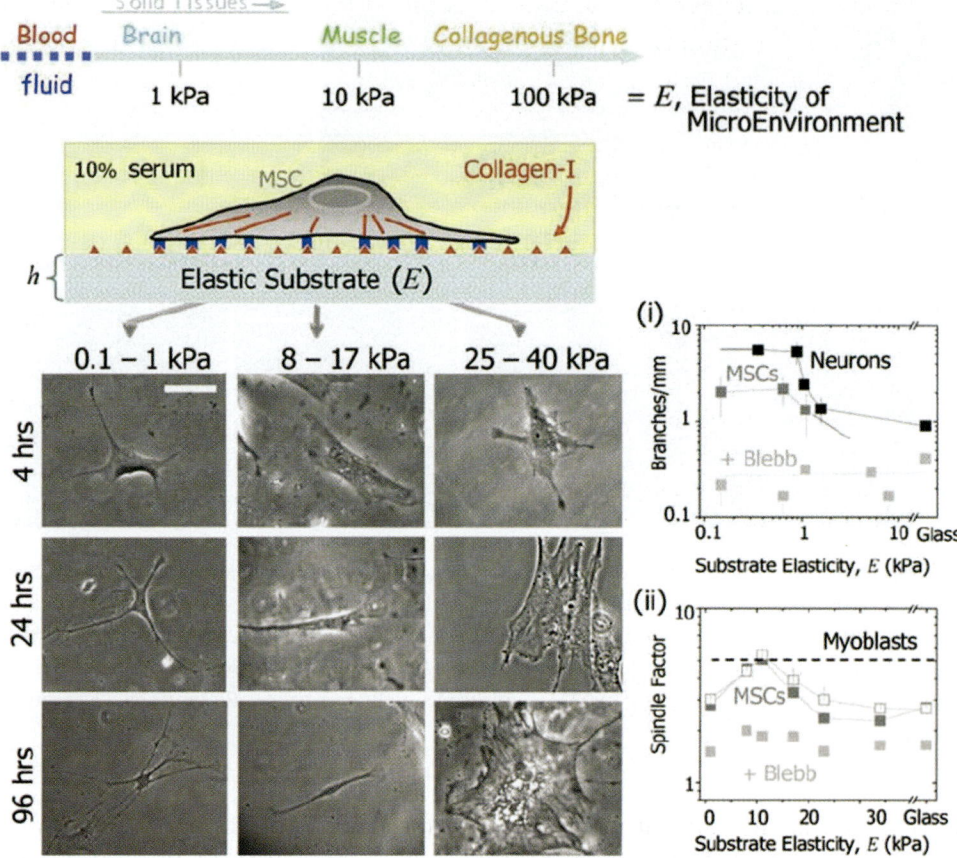

Fig. 6. Substrate stiffness influences stem cell lineage specification. MSCs in fluid differentiate into blood cells; those on softest substrates into brain (1 kPa), then muscle (10 kPa), then bone (100 kPa) as substrate stiffness increases over orders of magnitude. (Reprinted with permission from Ref. 71.)

demonstrated little response to substrate stiffness. Following this study, a host of others investigating the influence of substrate rigidity on stem cell fate have followed with embryonic and a variety of adult stem cell types.[72–82]

Microfluidics for Stem Cell Culture and Analysis

The previously discussed experiments based on microtechnologies serve to illustrate how single microenvironmental cues can influence stem cell fate. Microfluidic systems, in particular, are uniquely suited for manipulating and culturing stem cells not only in the presence of single microenvironmental cues, but also within more complex settings, enabling the combination of several classes of cues in a static or dynamic fashion. The following section reviews the application of microfluidic systems to stem cell culture for expansion and controlled differentiation *in vitro*, as well as for more detailed analysis of cellular processes.

Microfluidics for stem cell culture

Each stem cell type requires unique procedures and reagents for expansion and differentiation. ESCs are mainly cultured either in adherent monolayer culture on fibroblast feeder layers for expansion, or in suspension culture as cell aggregates called embryoid bodies (EBs), typically employed for research on the generation of lineage-specific derivatives of ESCs. Numerous devices have been developed in support of both culture modes. For example, design considerations for a glass slide-sized bioreactor array for adherent ESC cultures have been discussed and preliminary data have been presented regarding the influence of cell density and exerted shear stress generated by flow of culture media on cell maintenance and differentiation into cells positive for smooth muscle actin.[83] A soft lithography system for ESC expansion, differentiation, and analysis was developed.[84] hESC clusters that were introduced into the system were microscopically analyzed for shape and size. Selected clusters were sorted into one of six culture chambers for co-culture on mouse embryonic fibroblast feeder layers under varying conditions. The device enabled screening for optimized culture conditions. Proliferation and pluripotency of cultured ESCs was monitored through *Oct-4* promoter-driven expression of enhanced green fluorescent protein (EGFP) as well as on-chip immunocytochemistry.

Very recently, the same group presented a similar technological approach to analyze serum/feeder free conditions for culture of hESCs and iPS cells based on Matrigel and various chemically defined culture media (StemPro, mTeSR7, and N2/B27), supplemented with various concentrations of bFGF.[86] Analysis of cell fate at the single-cell level was performed through an image cytometry approach. In another example, a microfluidic system for the arrayed culture of adherent murine ESCs over a logarithmic range of flow rates was used to optimize conditions for cell survival and expansion, as shown in Fig. 7.[85] Cells grown under slowest flow rate condition did not proliferate, while ESC colonies grown in higher flow rates exhibited healthy, round morphology. The co-culture of hESCs with murine embryonic fibroblasts was demonstrated on microwell-patterned substrates for the expansion of ESCs and compared to traditional co-culture on flat surfaces.[87] Seeding conditions resulted in ESCs settling preferentially in microwells (Fig. 8). A

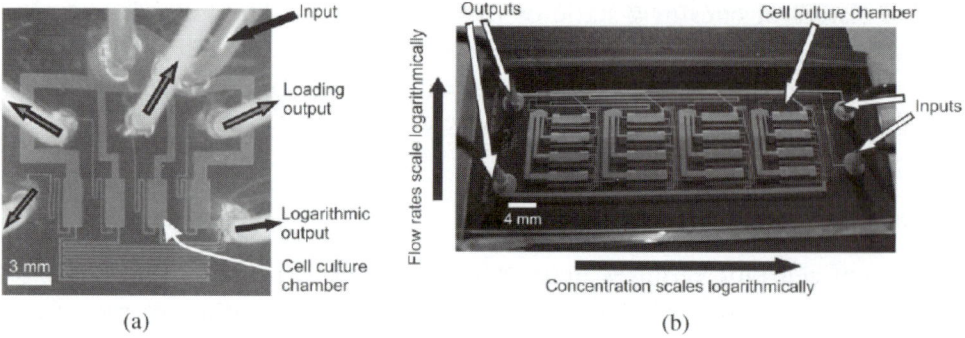

Fig. 7. Microfluidic systems with chamber arrays for the culture of murine ESCs over a logarithmic range of flow rates. (Reprinted with permission from Ref. 85.)

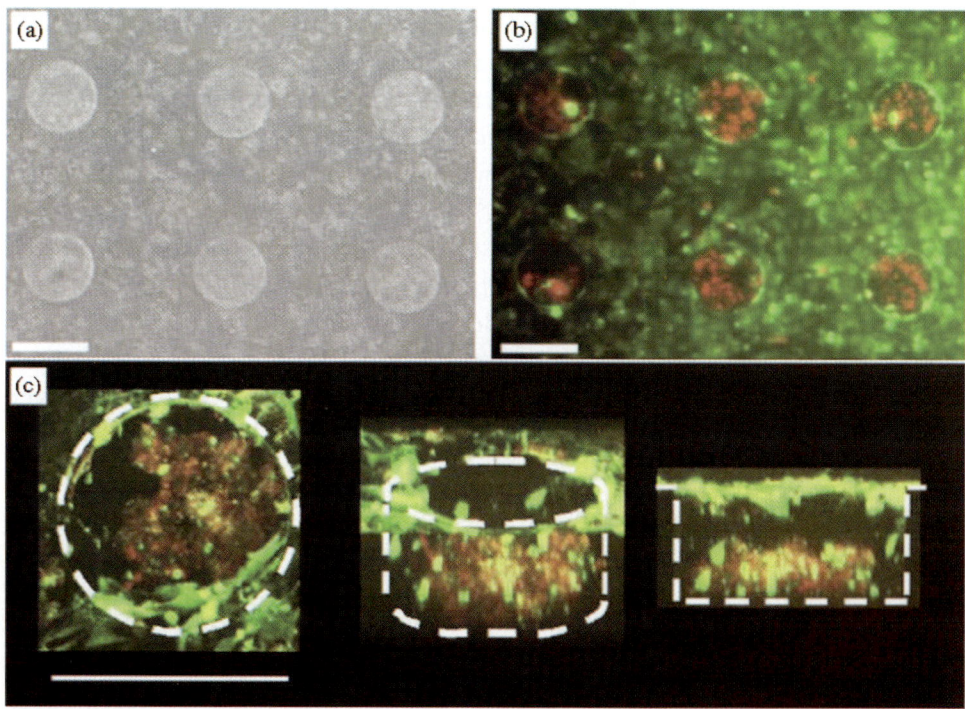

Fig. 8. Microwell platform for the co-culture of human ESCs (red) on a murine fibroblast feeder layer (green) for generation of uniformly sized EBs. (a) Light image. (b) Fluorescent image. (c) Confocal z-stack projections at various angles. (Reprinted with permission from Ref. 87.)

comparison to traditional co-culture on flat surfaces with regard to ESC aggregate size, viability, and expression of the ESC markers Oct-4 and alkaline phosphatase showed that viability and self-renewal profiles were similar, but that ESCs grown on microwell-patterned substrates showed a greater level of homogeneity in aggregate size. The application of the system to generate EBs with uniform size for further experimentation was suggested and later on demonstrated.[88]

Similarly, a microfluidic channel system with integrated microwell arrays was built to seed ESCs in wells to produce uniformly sized EBs and to generate neural progenitor cells in a subsequent differentiation step.[89] Another approach to the generation of uniformly sized EBs used a PDMS stencil with arrays of holes with varying diameters for the seeding of ESCs onto a collagen-coated substrate.[90] Initially growing as monolayer, ESCs soon proliferated and formed EB-like aggregates. Aggregates with 300-µm diameter showed similar differentiation profiles compared to EBs made using the hanging drop technique. Aggregates with a diameter of 100 µm exhibited increased expression of ectodermal markers, while 500-µm aggregates expressed increased levels of mesodermal and endodermal markers.

As illustrated in Fig. 9, a microfluidic system enabled the exposure of a differentiating EB to two laminar streams of different culture media at the same time,[91] resulting in the sub-differentiation of a single EB into more than one lineage. In another study, a microfluidic system was used to culture ESCs under the periodic removal of cell-conditioned media, leading to observations of differential cell survival that were in agreement with theoretical models.[92] In a related study, a model for oxygen and glucose consumption by ESCs cultured in a microfluidic system was developed and experimentally verified.[93]

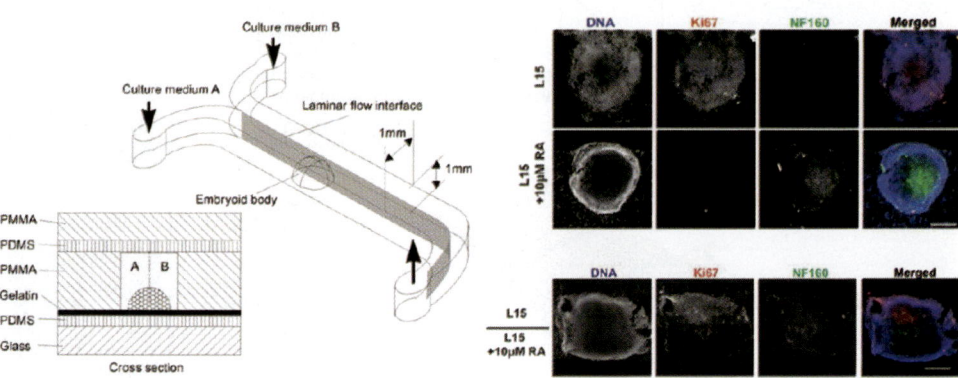

Fig. 9. Single EB culture in a microfluidic device. Left: Microfluidic system that enables exposure of an EB to two different culture media A and B. Top right: Culture of the entire EB in complete L15 media did not result in neuronal differentiation, while the addition of retinoic acid (RA) yielded neuronal differentiation throughout the EBs. Bottom right: Split media flow resulted in neuronal differentiation only in the half of the EBs that was exposed to media with RA. (Reprinted with permission from Ref. 91.)

In order to systematically screen for ECM components that may support the differentiation of ESCs along hepatogenic lineages, an ECM microarray was generated using a standard DNA microarray spotting system.[94] A total of 32 different combinations of five ECM components (collagen I, collagen III, collagen IV, laminin, and fibronectin) were deposited, and the effect on differentiation of mouse ESCs toward an early hepatic fate was assessed, using a β-galactosidase reporter fused to the fetal liver-specific gene *Ankrd17* (also known as *gtar*). An approximately 140-fold difference in β-galactosidase signal intensity was determined between the least efficient condition (laminin alone) and the most efficient condition (laminin + collagen I + fibronectin). In a follow-up experiment, a multiwell microarray platform for 1,200 simultaneous experiments using 240 unique signaling environments was applied to screen for the differentiation of ESCs along cardiac lineages based on a myosin heavy chain-a (MHCα)–GFP reporter, as illustrated in Fig. 10.[95] In addition to previous applied mixtures of ECM components, various soluble factors such as Wnt3a, activin A, bone morphogenetic protein 4 (BMP4), and FGF4 were used for the screen. A DNA microarray scanner was used for fast experimental readout.

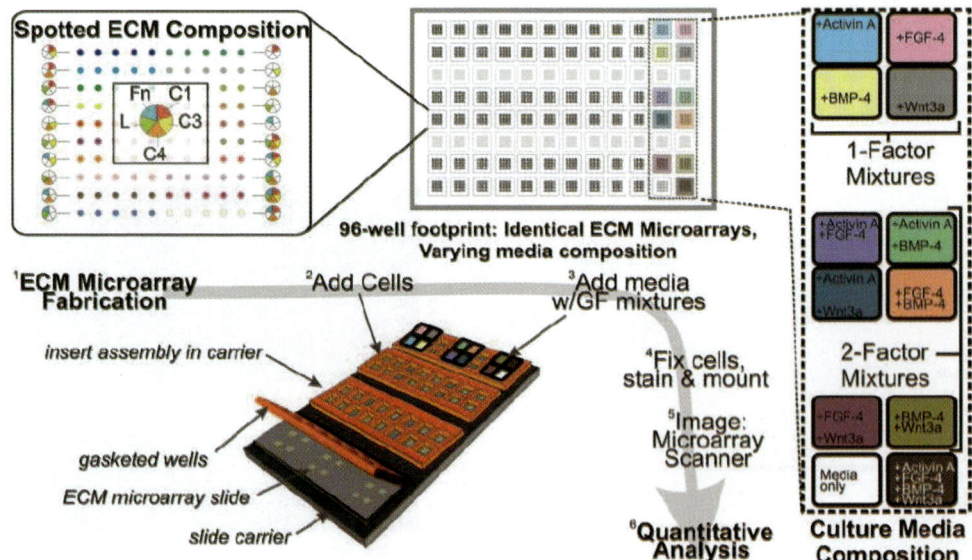

Fig. 10. Multiwell cell culture microarray with spotted ECM islands of various compositions (using collagen I, collagen III, collagen IV, laminin, and fibronectin) for the assessment of the differentiation of ESCs toward cardiac lineages. (Reprinted with permission from Ref. 95.)

Further array technologies for high-throughput analysis of signals regulating stem cell fate and function have been reviewed.[96] Described research includes the differentiation of neural precursor cells on printed microenvironment arrays composed of mixtures of ECM components, morphogens, and other signaling proteins.[97] It was found that Wnt and Notch co-stimulation could maintain the cells in an undifferentiated-like, proliferative state, whereas BMP4 induced an "indeterminate" differentiation phenotype characterized by the simultaneous expression of glial and neuronal markers. The review also emphasized the potential of RNAi-based screens, for example, applied to the identification of genes that are essential for mouse ESC self-renewal.[98]

The design of bioreactors for MSC culture has recently been reviewed, emphasizing the promise for the application of microfluidic and soft lithography-based devices.[99] A soft lithography screening system with an array of 96 individual culture chambers has been developed for the quantification of the effects of cell seeding density, composition of culture medium, and transient feeding schedules on the proliferation, osteogenic differentiation, and motility of human primary MSCs, as documented in Fig. 11.[100] The differentiation of hMSCs toward a hepatogenic lineage in a microfluidic culture chamber was compared to parallel differentiation in 96-well plates.[101] The expressions of hepatoblast markers α-feroprotein, tyrosine aminotransferase, and

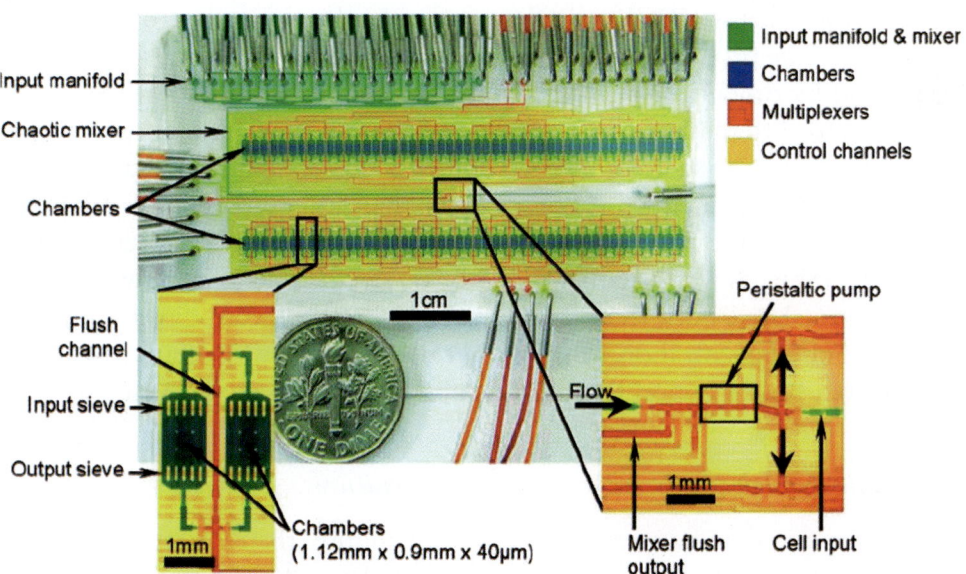

Fig. 11. Microfluidic, soft lithography chip with an array of 96 culture chambers for the screening of optimized culture conditions for proliferation and differentiation, respectively, of human primary MSCs. (Reprinted with permission from Ref. 100.)

albumin was detected in both culture systems through RT-PCR and immunocytochemistry after four weeks. However, MSCs that differentiated on chip showed superior functionality through the increased uptake of light-density lipoprotein (LDL), likely due to constant media exchange. In another microfluidic system, MSCs were seeded on predefined regions on a polystyrene substrate by either partially oxidizing polystyrene to enable MSC attachment or by partially masking the surface with the triblock copolymer Pluronic to prevent MSCs from attaching.[102] Oil Red-O staining of lipids accumulated in adipocytes and Alizarin Red S staining of calcium deposited by osteoblasts were used to verify adipogenic as well as osteogenic differentiation upon exposure to specific differentiation protocols. In a further study, an osmotic pump was applied to generate a gradient of fetal bovine serum concentration across a microfluidic culture chamber with seeded MSCs, enabling the optimization of culture conditions.[103]

In addition, adult hippocampal progenitor cells were cultured in arrays of 10,000 microwells on a glass cover slip, with all wells exposed to the same culture media.[104] The degree of cell proliferation heterogeneity was analyzed using automated microscopy. The approach enabled the initial analysis of survival, proliferation, or differentiation in dependence on number of cells per well and potential cell–cell interactions. NSCs were also cultured in a microfluidic system that generated a gradient of a mixture containing epidermal growth factor (EGF), FGF2, and platelet-derived growth factor (PDGF).[105] The proliferation and differentiation of NSCs into astrocytes were monitored by time-lapse microscopy and immunocytochemistry. NSCs proliferated and differentiated in a graded and proportional fashion that varied directly with growth factor concentrations. In another study, neural progenitor cells were generated from ESCs and further differentiated in a microfluidic system that created oppositely oriented gradients of the signals sonic hedgehog (Shh) as well as either FGF8 or BMP4.[106] The average numbers of both neuronal cell body clusters and neurite bundles were directly proportional to Shh concentration. A significant decrease in TuJ1-positive neuronal cells at high BMP4 concentrations was observed.

Microscale technologies for advanced stem cell culture

Apart from microfluidic systems for basic cell culture, the research community has demonstrated the generation of more complex cell arrangements through the application of microtechnologies for patterning substrates or modifying surfaces for spatial control over seeded cell types.[107] For example, systematic research of the effects of various parameters that influence cell docking and retention within grooved, micromachined substrates was performed.[108] In another study, microwells at the bottom of reversibly sealed channels were used to capture cells and demonstrate the generation of precisely defined cell arrays, consisting of ESCs, osteoblasts (Saos-2), hepatocytes (AML12), fibroblasts (NIH-3T3), and human prostate cells (PC3).[109] Potential applications include the analysis of cell–cell interactions and effects of paracrine signaling. In a study on the generation of micropatterned co-cultures, photolithography was applied to mask some areas of a substrate and to coat exposed areas with proteins that allowed cells to adhere.[110] After the removal of photoresist and exposure to one cell type in serum-free media, cells adhered to patterned regions. Subsequent seeding of a second cell type enabled the deposition of that cell type in the field, enabling defined co-culture for the investigation of hetero- and homotypic cell–cell interactions. Similarly, micromachined devices were made for separate seeding of cells and placement of cells at a defined, variable distance after microdevice assembly, enabling the separation of effects of cell–cell interactions from paracrine signaling.[111] Microwell surfaces were generated for the capture of single NSCs for the assessment of heterogeneity regarding survival, proliferation, and differentiation patterns.[104] Parylene-C stencils were used to generate patterned co-cultures of ESCs with other cell types.[112] Layer-by-layer deposition of hyaluronic acid (HA) and poly-L-lysine was applied for patterned co-culture of hepatocytes or ESCs with fibroblasts.[113] A soft embossing process was developed for the patterning of hydrogels and applied to a culture of HSCs in hydrogel wells.[114] Microcontact printing is another important technique for patterned cell growth.[115] For example, patterns of fibronectin on PDMS-coated cover slips were generated by microcontact printing, and non-printed areas were rendered nonadhesive after incubation with Pluronic F-127.[116] The influence of cell patterns and corresponding gradients in mechanical forces on the differentiation of hMSCs along adipogenic and osteogenic lineages, respectively, was assessed. Micromolding of cells in hydrogel microstructures and transfer to a new substrate were applied to a patterned co-culture of fibroblasts with human hepatoblastoma cells in chitosan hydrogels,[117] patterned three-dimensional culture of ESCs or fibroblasts in photocrosslinkable HA gels,[118] as well as the generation of multiphenotype cell microarrays.[119] A photosensitive pCLLA acrylate polymer was used to create biodegradable microstructured culture surfaces and bioreactors.[120] Another study demonstrated the generation of patterned HA substrates through a microfluidic patterning approach for the formation of contractile tissue-like structures from cardiomyocytes.[121] A combination of cell clustering by DEP in a microfluidic culture chamber, incorporation of patterned cell clusters in a hydrogel through photopolymerization, and stacking of cell cluster-containing gel slabs was demonstrated, as shown in Fig. 12.[122]

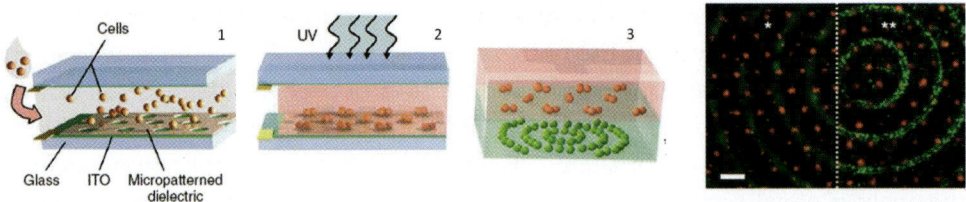

Fig. 12. Patterned cell co-culture in photopolymerizable hydrogels. Left: Microelectrodes inside microfluidic reservoirs are used for patterned cell seeding by dielectrophoresis (DEP) (1). Photopolymerization embeds positioned cell clusters in a hydrogel (2). Removal of the gel slab, repetition of the process with a second cell type, and layering of gel slabs enable a patterned, three-dimensional co-culture (3). Right: Confocal image of a hydrogel that contains red labeled cell clusters at the top (*) and concentric rings of green labeled cells at the bottom (**). (Reprinted with permission from Ref. 122.)

Furthermore, many culture systems based on electrospun micro- and nanofibers have been demonstrated, including the use of polyurethane fibers for the proliferation and neuronal differentiation of hESCs,[123] investigations of the effects of fiber diameter on the differentiation of NSCs,[124] the enhancement of adhesion and expansion of human umbilical cord blood HSCs on surface-aminated nanofibers,[125] as well as the expansion of MSCs on type I collagen nanofibers.[126] In combination with hole arrays in polymer substrates, electrospinning was applied to create microwell arrays with electrospun fiber networks at the bottom for the high-throughput assembly of three-dimensional cell clusters.[127] Microfluidic devices themselves have been used for the creation of fibers from poly(lactic-*co*-glycolic acid) and chitosan through a hydrodynamic focusing technique.[128,129] The fibers were evaluated for use outside of the microfluidic devices for the culture of neural and hepatic cell types.

Microfluidics for stem cell analysis

While microfluidic technologies are well suited for stem cell culture for efficient expansion or specific differentiation, they can also play an important role in genetically engineering cells as well as in analyzing stem cell biology in more detail, e.g., through gene expression profiling. Microfluidic devices have been developed in the past for cellomics,[22] systems biology,[26] high-throughput cell-based screens,[130,131] as well as for measuring gene network dynamics in single cells.[132] Similar to FACS, significant efforts have been spent on the development of systems that contain cells in single droplets for further analysis.[133,134] Capillary electrophoresis has been integrated on chip for single-cell analysis based on laser-induced fluorescence readout,[135] typically combined with integrated single-cell docking and lysis.[136] Capacitance cytometry is another single-cell analysis modality that has been implemented in microfluidic systems.[137] Regarding high-throughput analysis of gene expression in response to soluble stimuli, soft lithography has, for example, been applied to generate a scalable bioreactor array for the seeding of fluorescent reporter cell lines that are exposed to different stimuli.[138] Fluorescence time-lapse microscopy was applied to record the dynamics of gene expression in real time. These examples already indicate the significant potential of microfluidic systems for the analysis of cells in general. The stem cell community, however, has just started to implement such technologies for systematic research.

In support of the delivery of DNA, RNA, and other biological reagents, numerous electroporation strategies have been implemented in microfluidic devices.[140] For example, a device for the immobilization of single cells in microchannels by suction and electroporation with integrated electrodes was applied to deliver DNA encoding a green fluorescent–Erk1 fusion protein to single mouse myoblastic C2C12 cells and single hMSCs for cell signaling studies, as shown in Fig. 13.[139]

Microfluidic systems were also developed in support of the analysis of single stem cells, for example, in order to take into account phenotypic differences within a stem cell population. Soft lithography has enabled the development of integrated fluidic circuits for gene expression analysis by digital PCR,[141] a technique that is based on a limiting dilution approach and enables the precise quantification of mRNA copy numbers. As indicated in Fig. 14, using digital PCR for the analysis of hematopoietic progenitor cells, it was possible to define two distinct subpopulations among common myeloid

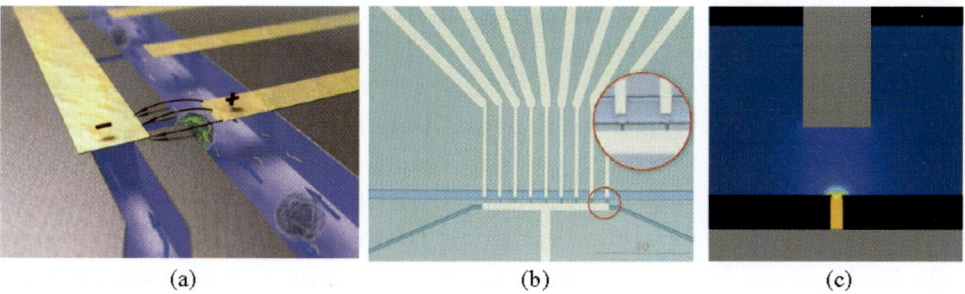

Fig. 13. Microfluidic system for single-cell gene delivery through electroporation. (a) Schematic of the microfluidic system with cell trapping site and integrated electrodes. (b) Photograph of the system. (c) Simulation of the electric field strength at the trapping site. The system was applied to deliver, e.g., DNA encoding a green fluorescent–Erk1 fusion protein to hMSCs for cell signaling studies. (Reprinted with permission from Ref. 139.)

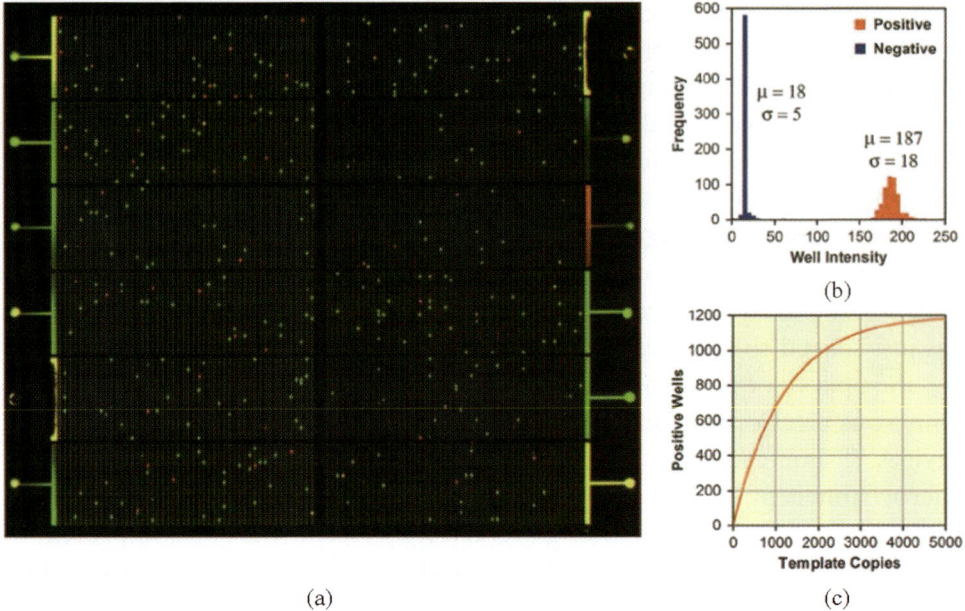

Fig. 14. Microfluidic chip with chamber arrays for on-chip digital PCR for the quantification of mRNA transcript copy numbers of single hematopoietic progenitor cells. (a) Fluorescence image of the array system. (b) Counts of positive and negative wells after 40 cycles of PCR. (c) Digital PCR response characteristic. (Reprinted with permission from Ref. 142.)

progenitors, a cell type that was previously thought to be homogeneous in phenotype.[142]

In another study, digital PCR was applied to comparatively analyze the expression levels of free radical scavenging systems in non-tumorigenic as well as epithelial cancer stem cells.[143] Results indicated that, similar to normal tissue stem cells, subsets of cancer stem cells in some tumors contain lower reactive oxygen species (ROS) levels and enhanced ROS defenses, compared to their non-tumorigenic progeny, which may contribute to tumor radioresistance and may be relevant for the design of cancer therapies.

In order to investigate the chemoresistance of individual CD34$^+$ HSCs and chronic myeloid leukemia (CML) cells, a microfluidic system was developed with 440 mechanical traps for single cells within a single culture chamber, as shown in Fig. 15.[144] Automated microscopy was applied to assess single-cell proliferation, migration, and apoptosis under normal culture conditions and upon the administration of the drug dasatinib. Both cell types proliferated as normal in the constant flow-through environment; CML cells showed an unexpected increased resistance to apoptosis in response to dasatinib, but were compromised in their migratory capacity.

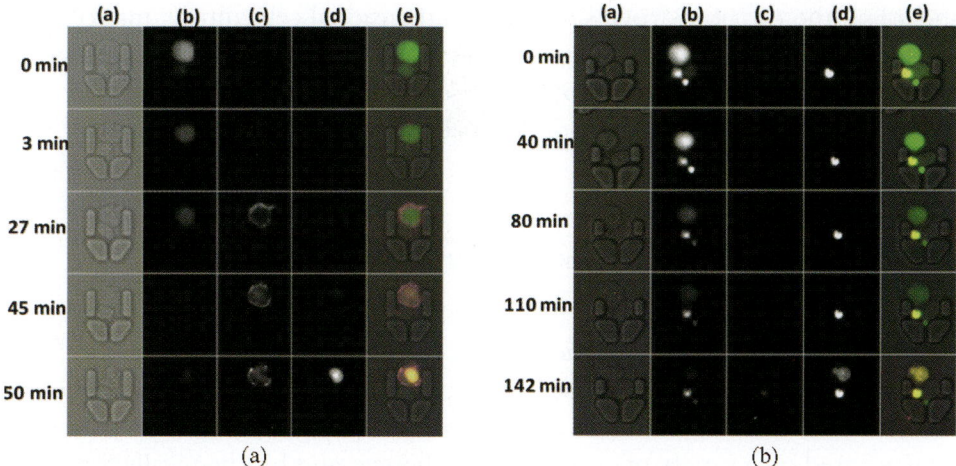

Fig. 15. Microfluidic chip with an array of 440 mechanical traps (only single traps are shown) for positioning of normal HSCs (a) and chronic myeloid leukemia (CML) stem cells (b), respectively. Administration of the drug dasatinib elicited different apoptosis responses over time. Lane (a): bright field image, (b): cell labeling with Fluo-3 calcium indicator, (c): annexin-V Alexa-fluor 647 conjugate for apoptosis detection, (d): propidium iodide for apoptosis detection, and (e): colored overlay of the bright field and fluorescence channels. (Reprinted with permission from Ref. 144.)

In another report, a soft lithography-based microfluidic processor was presented for single-cell on-chip mRNA extraction and reverse transcription for subsequent off-chip PCR.[145] Absolute numbers of mRNA molecules of three genes (*B2M, Nodal*, and *Fzd4*) in single hESCs were measured, indicating significant diversity in expression levels among individual hESCs.

Microfluidics for Tissue Engineering

The microfluidic systems discussed so far were mainly designed to yield cells as final products, e.g., after the expansion or differentiation of stem cells for their subsequent use in therapeutic approaches, or after the manipulation of stem cells for subsequent further analysis. Microfluidic systems in support of tissue engineering concepts require the realization of more complex approaches to cell culture that eventually yield cells that are embedded in cell-instructive biomaterials[146] and are arranged in three dimensions to complex tissue engineering constructs.[30–32] These constructs may subsequently be used either in therapeutic approaches or as experimental test beds, e.g., for the development of specific aspects of a therapeutic approach, analysis of specific physiological aspects, or for drug screening. The following section discusses opportunities and challenges related to the application of microfluidics to tissue engineering, including the use of hydrogels for three-dimensional cell culture, microfluidic strategies for ensuring mass transport or vascularization in three-dimensional constructs, the generation of complex tissue engineering constructs based on assembly of cell-containing building blocks, as well as the generation of tissue models.

Hydrogels in microfluidic systems for three-dimensional tissue engineering

Hydrogels find widespread applications in tissue engineering approaches[147] and are particularly well suited for use in microfluidic systems. Hydrogels can easily be introduced into microfluidic systems as liquid pre-gels, gelled and patterned *in situ* based on a variety of cross-linking schemes, and biochemically functionalized in support of specific therapeutic or biological aims.[148] Researchers have in the past emphasized opportunities for the design of cell-instructive hydrogels through the synthesis of biomaterials, for example, with incorporated cell adhesion ligands, protein binding sites, e.g., for the immobilization of growth factors, specific degradation sites as a target for cell-secreted proteases, and under the application of design strategies that either support self-assembly or enable cross-linking.[44] Such materials show tremendous promise for use in tissue engineering applications as well as for related control of stem cell fate.[45]

For example, a multifunctional protein-based, self-assembling triblock hydrogel for NSC culture was reported.[149] This construct presented an interesting advancement because it not only could serve as a soft culture substrate itself, but could also be used to make potentially any hydrophobic polymeric substrate cell-interactive. Furthermore, the active component of the construct is modularly interchangeable. The protein construct is composed of a pair of flanking alpha-helices, between which a random coil domain contains an ECM-inspired minimal binding peptide sequence (e.g., RGDS from fibronectin). Depending on the concentration of the protein in solution, either self-assembly into a soft hydrogel occurs spontaneously at room temperature, or the protein may coat the underlying culture substrate (e.g., polystyrene) by hydrophobic interactions between the polymer substrate and a leucine zipper motif on the flanking protein helices. The random coil region presents the binding sequence to cells in culture and has been shown to shift the growth mode of rat NSCs in culture from the neurosphere to adherent monolayer culture as the percentage of RGDS-incorporated protein increased from 0 to 50 mol%.

In another prominent example, self-assembling peptide amphiphile-based, high-epitope density nanofiber matrices have been described.[151] Applications include the selective neuronal differentiation of NSCs.[150] As shown in Fig. 16,

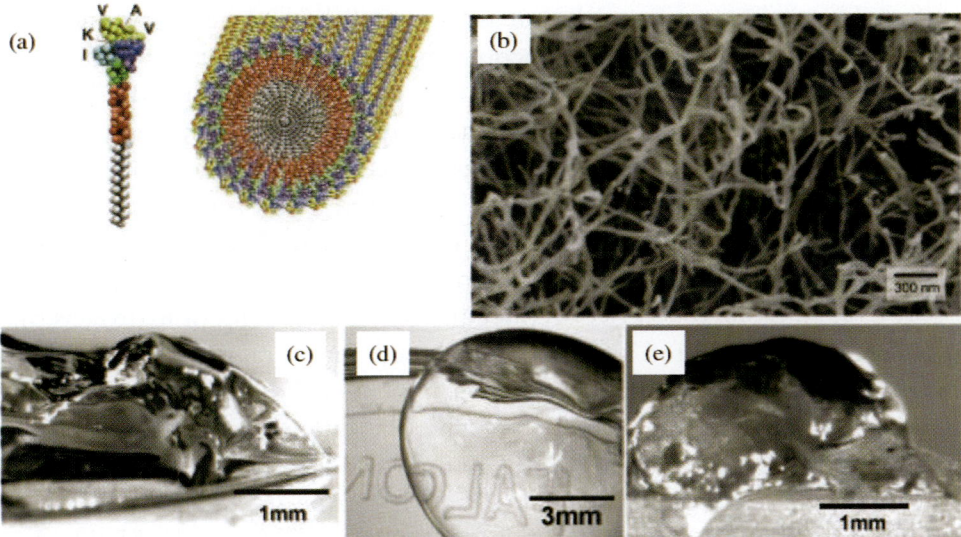

Fig. 16. High-epitope density peptide amphiphiles self-assemble to yield cell-instructive, encapsulating, and nanofiber-based hydrogels. (a) Schematic of peptide amphiphile and cross-section of the self-assembled nanofiber. (b) SEM of nanofiber mesh. (c)–(e) Photographs of hydrogels. The hydrogels supported specific neuronal differentiation of neural progenitor cells. (Reprinted with permission from Ref. 150.)

the constituent amphiphile in this case contained the pentapeptide epitope IKVAV, which is found in laminin and is known to promote neurite sprouting and to direct neurite growth. Adjacent to the binding epitope, the molecule contains a Glu residue, resulting in a net negative charge at physiological pH so that cations in the culture media can screen electrostatic repulsion among the amphiphilic molecules and drive self-assembly. The remaining sequence contains four Ala and three Gly residues (A/G), followed by a hydrophobic, 16-carbon alkyl tail. The A/G and alkyl segments result in increasing hydrophobicity along the molecule length, opposite the binding epitope. Once electrostatic repulsions are screened, the molecules are driven to assemble by hydrogen bond formation and hydrophobic interactions. The nanofibers that self-assemble in aqueous media present the binding epitopes on their surfaces at "van der Waals packing distances." These nanofibers aggregate to form three-dimensional, interpenetrating networks and produce a hydrated, gel-like solid. Neural progenitor cells cultured in these hydrogels showed a strong tendency to undergo neuronal differentiation. *In vivo*, similar amphiphile-based materials supported the repair of spinal cord injuries in mouse and rat models.[152] Of note is also the generation of an HA-based gel which was designed to be degradable by cell-secreted proteases.[153] In a secondary photo-initiated UV cross-linking process using a photomask, parts of the hydrogel were modified to inhibit cellular remodeling. Functionality was validated by controlling the outgrowth from chick aortic arches or the spreading of encapsulated MSCs, where only UV-untreated regions permitted arch outgrowth and MSC spreading. In another seminal work, photodegradable hydrogels for dynamic tuning of physical and chemical properties were developed.[154] Post-gelation control of the gel properties was demonstrated to introduce temporal changes, creation of arbitrarily shaped features, and on-demand pendant functionality release.

A basic microfluidic device architecture for the handling of hydrogels that was already demonstrated in the 1990s is based on several input channels that merge in a larger microfluidic chamber.[156,157] Due to laminar flow at low Reynolds numbers in microfluidic systems, the different input flows do not mix, enabling, e.g., the deposition of pre-gels at the chamber center, while, after the induction of gelation, side channels can be used to deliver culture media or to establish gradients of soluble factors across the embedded hydrogel. Integrated micropillars can be used to keep the hydrogel in place.[158] This basic architecture also enables the generation of gels with embedded gradients, e.g., of concentrations of embedded RGDS adhesive ligands.[159] A variant of such an architecture that also makes use of surface tension and hydrophobic interactions for definition of pre-gel locations is shown in Fig. 17.[155] An application of such an architecture to liver tissue engineering is illustrated in Fig. 18, demonstrating the formation of vascular sprouts by endothelial cells through a collagen hydrogel toward seeded hepatocytes.[160] The same paper also presented an assay for tumor cell invasion and endothelial cell layer sprouting in a system that again separates the two cell types with a hydrogel. Another related study discussed endothelial cell migration into hydrogel

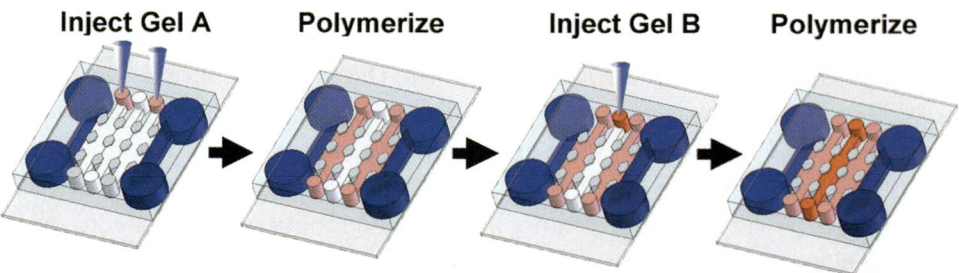

Fig. 17. Properties of fluids and structures at the microscale can be used to pattern three-dimensional, cell-containing hydrogels in multiple discrete regions (red) in a sequence of simple steps. Side channels (blue) can be used, for example, for supply with culture media or for the creation of concentration gradients of soluble factors across the embedded hydrogels. (Reprinted with permission from Ref. 155.)

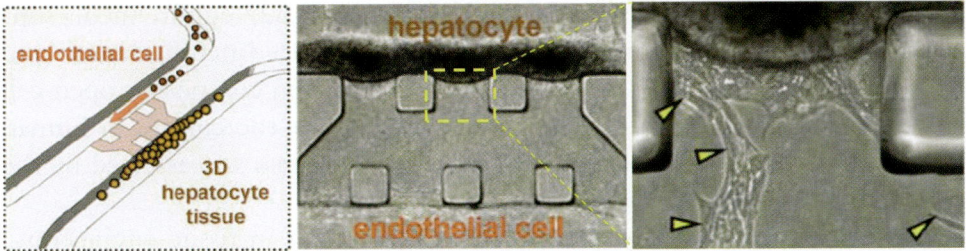

Fig. 18. Microfluidic system for liver tissue engineering research. Left: A collagen hydrogel (red) is created between two microfluidic channels and endothelial cells as well as hepatocytes are seeded on opposing sides. Center: Photo of seeded cells on day 0. Right: On day 5, some endothelial cells formed sprouts that extended toward the three-dimensional hepatocyte tissue. (Reprinted with permission from Ref. 160.)

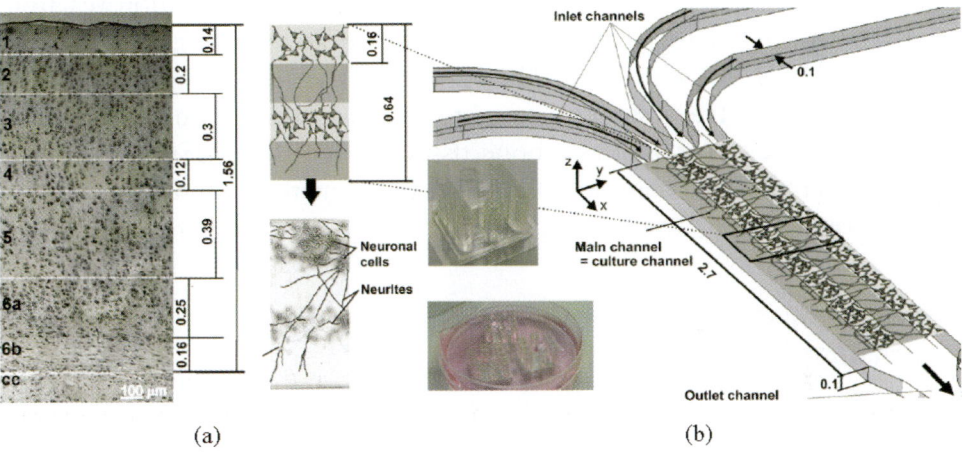

Fig. 19. Microfluidic system for recreation of cortical neural tissue architecture. (a) Image of a cortical brain slice with distinct cell layers and interconnections (small inserts). (b) Schematic of the microfluidic chip for layered culture of neuronal cells with photographs (small inserts) of the chip package and culture in a Petri dish. (Reprinted with permission from Ref. 162.)

scaffolds, e.g., in response to a concentration gradient of angiogenic vascular endothelial growth factor (VEGF).[161]

Using a similar microfluidic system, researchers attempted to mimic the architecture of cortical neural tissue as shown in Fig. 19.[162] An alginate-enriched agarose scaffold was used to create side by side four different hydrogel layers, containing primary cortical neurons in the first and third layers. Gelation in the microfluidic system was thermally induced. Initial chemical loading conditions of the different hydrogels were varied and the effect on neurite outgrowth was investigated. Gel loading with B27 culture media supplement in various configurations significantly enhanced neurite length and density. A similar system was used to compare collagen composite, open-cell polylactic acid (OPLA), and Puramatrix scaffolds for the culture of human hepatocellular carcinoma cells (HepG2).[163] Effectiveness was assessed by the measurement of albumin secretion.

While hydrogels serve as extremely versatile tools in tissue engineering, mass transport issues can arise when dense hydrogels with large dimensions are used that are hard to effectively perfuse. An interesting alternative approach is based on the three-dimensional seeding of cells in a microfluidic channel that is lined by pillars that physically constrain cells to their location.[165] The porosity of the cell aggregate enables the deposition of an ECM-like environment by the layer-by-layer deposition of oppositely charged polyelectrolytes, as illustrated in Fig. 20.[164] For this complex coacervation reaction, positively charged methylated collagen and the negatively charged terpolymer HEMA–MMA–MAA were used. The porosity of the cell aggregate facilitated perfusion culture. The system was used to establish a liver tissue model using primary hepatocytes as well as a bone tissue model using MSCs. Of note is also the attempt to recreate in a fibrin gel the hematon, a fundamental unit of hematopoiesis.[166]

Microfluidic strategies for efficient mass transport or vascularization

As already indicated, the creation of complex, voluminous constructs for tissue engineering brings the challenge of providing embedded cells with nutrients and soluble signals as well as removing metabolic waste products to ensure cell survival and specific function. Significant efforts were spent on creating either large constructs with embedded microchannels for perfusion *in vitro*, with microchannels for seeding of endothelial and supporting cells for the formation of vasculature, or with embedded angiogenic signals for the recruitment of endothelial cells and induction of vasculature formation *in vivo*.[167,168]

For example, poly(glycerol sebacate) (PGS) was engineered to be both biodegradable and flexible, and as a biocompatible elastomer, it is well suited for tissue engineering applications. In early experiments, the culture of endothelial cells was demonstrated in PGS microfluidic systems, created by molding using a silicon micromold and the attachment of a patterned to a flat PGS substrate.[169] Endothelial cells were successfully seeded in the channel system and grew under perfusion to confluency. In another experiment, PGS was used to form multilayered structures by stacking individual,

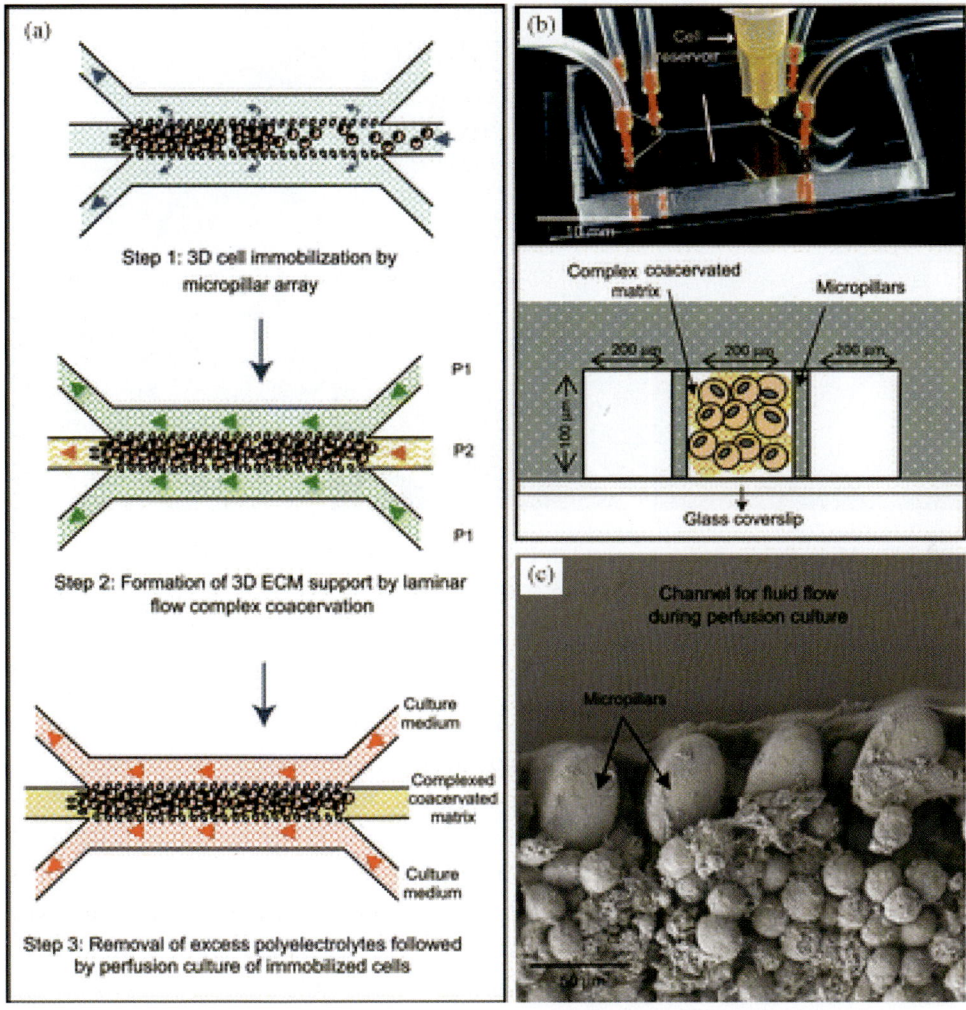

Fig. 20. Microfluidic system for three-dimensional culture of cells embedded in an ECM-like environment, applied to bone and liver tissue engineering. (a) Micropillars constrain cells to the center portion of a culture chamber. Alternating perfusion with solutions of oppositely charged polyelectrolytes with intermittent washing enables layer-by-layer deposition of cell-instructive materials. (b) Top: Photograph of the microfluidic device. Bottom: Schematic of the device cross-section. (C) SEM of seeded cells next to micropillars. (Reprinted with permission from Ref. 164.)

micropatterned PGS sheets as shown in Fig. 21.[170] Circular pillars acted as spacers between sheets and defined microfluidic networks that were seeded with hepatocytes and perfused for one week. Albumin production was measured to assess liver-specific cell function. The system enables the co-culture of different cell types and, due to its biodegradable nature, might support more complex strategies for cellular reorganization into functional tissue.

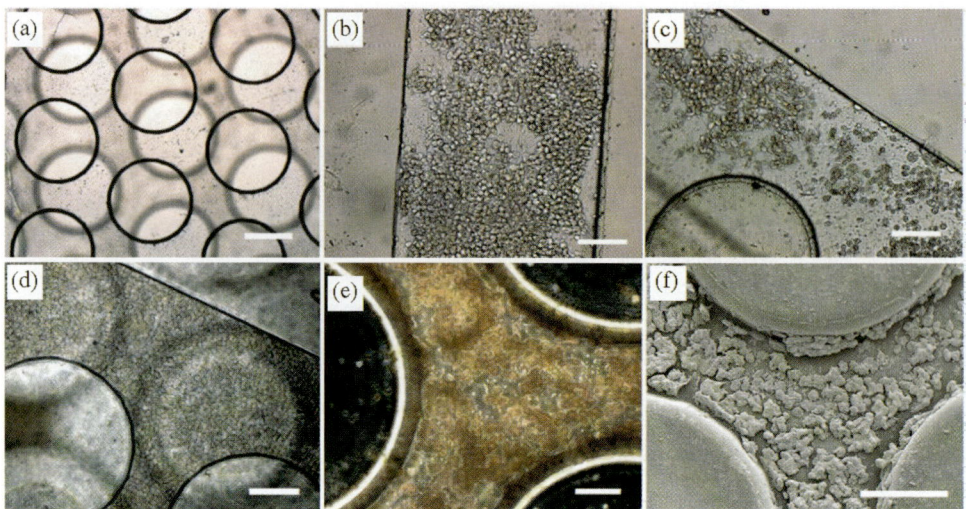

Fig. 21. Microfluidic devices created in biodegradable PGS for long-term perfusion culture of hepatocytes. (a) Layered PGS sheets with circular posts before seeding. (b)–(e) Device with seeded hepatocytes. (f) SEM of seeded hepatocytes. (Reprinted with permission from Ref. 170.)

Other promising approaches are based on the use of the cell-containing hydrogels for the formation of microfluidic channel systems for perfusion. In this approach, cells can not only be seeded onto the surfaces of the channels, but also within the bulk of the device itself. Hydrogels based on alginate, a material broadly used in tissue engineering, have been used for the demonstration of such microfluidic biomaterials.[171] A PDMS mold was used to cast a patterned, cell-containing alginate hydrogel. The transfer to another patterned gel created a device with microfluidic channels embedded at the interface.[172] Fluorescent dye solutions were used to characterize transport properties, and live/dead staining revealed excellent cell survival (Fig. 22).

In a similar approach, agarose, a polymer that naturally forms a hydrogel upon cooling from a liquid state, has been used to create cell-containing gels with embedded microchannels.[173] Extensive efforts were spent on analyzing cell survival in dependence on media perfusion and distance from embedded channels. In support of the better design of such microfluidic biomaterial geometries, mathematical models were developed for the analytical description of mass transport, with applications including not only the analysis and prediction of cell survival,[174] but also the potential analysis of vasculature formation and tumor angiogenesis,[175] as well as the analysis of a biomimetic, microfluidic model of liver vasculature.[176]

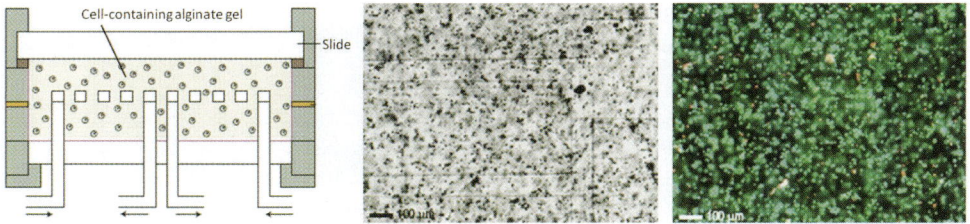

Fig. 22. Microfluidic biomaterial based on cell-containing alginate gel slabs. Left: Layering of two patterned hydrogels produces a microfluidic biomaterial with embedded flow channels. Center: Image of the gel, seeded with primary chondrocytes (embedded channels are faintly visible). Right: Live/dead (green/red) staining reveals excellent cell survival. (Reprinted with permission from Ref. 172.)

In order to increase the porosity of and mass transport through hydrogels, leachable materials such as sucrose crystals, can be incorporated within pre-gels, resulting in pores within the gel after gelation once the leachable material is removed — by dissolution in the case of sucrose.[177] Materials that are derived from mammalian sources have also been used to implement microfluidic networks within hydrogels. HA, a polysaccharide which is a native component of the ECM and is often used in tissue engineering scaffolds, can be modified with methacrylate groups to allow photocross-linking. This modified material has been cast over molds and used for the creation of a hydrogel-based system of microfluidic channels similar to the ones previously discussed.[178] In another study, gelatin, a derivative of the ECM constituent collagen, has been used to create a microfluidic device by casting it over a silicon mold and cross-linking the gelatin enzymatically.[179] Gelatin can also thermally cross-link to form a hydrogel upon cooling, and this property has been leveraged to create gelatin meshes in the shape of the vasculature using PDMS molds. These meshes, kept cool, can then be embedded in another material, which can be gelled around them. Upon heating to physiologic temperatures, the gelatin mesh will melt, leaving microfluidic channels throughout the bulk material, as documented in Fig. 23.[180]

Microfluidic networks through a gel can also serve as a template into which a cell-containing pre-gel is introduced and gelled, resulting in a system in which multiple materials and cells can be arranged relative to one another. One difficulty that may have to be addressed in such a construct is the potential for two neighboring gels to delaminate from each other, particularly if cells that remodel their environment are inducing mechanical stresses. One study that addressed this problem used collagen in both, the hydrogel comprising the device and the hydrogel filling the channels. During the gelation of the

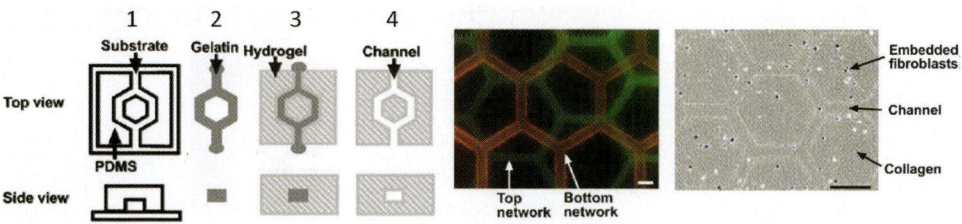

Fig. 23. Microfluidic networks created within hydrogels through the use of gelatin channel templates. Left: A PDMS mold (1) is used to create a gelatin template (2). A second, cell-containing gel is cast around the gelatin template (3). At elevated temperatures, the gelatin melts and can be flushed away (4). Center: Color-labeled stack of two channel networks in gel. Right: Photograph of channels in a fibroblast-containing collagen gel. (Reprinted with permission from Ref. 180.)

second gel, collagen fibers in the first gel started to extend into the second gel across the interface, mechanically linking and stabilizing the two gels.[181]

Microfluidic systems also play an important role in studying details of vasculogenesis and angiogenesis with the aim of developing corresponding tissue engineering strategies. Examples include a microfluidic system that was designed for eliciting the appearance and guidance of vascular sprouts in differentiating EBs,[185] as well as for endothelial cell sprouting assays.[186] Furthermore, microsystems for complex, perfusable cell culture have not only been used in studies *in vitro*, but also in animal experiments. Based on a design that was aided by simulation and verified *in vitro*,[182] a microfluidic system in support of liver function was built, consisting of two layers of channels, allowing primary hepatocytes to be cultured in one layer which was separated from a vessel-like layer by a porous membrane.[183] This microfluidic system was integrated via a shunt with systemic circulation in an animal model,[184] enabling interactions between hepatocytes and blood through the porous membrane (Fig. 24). This step toward a major tissue engineering goal of supporting, and eventually replacing, a dysfunctional liver was the first example of a microfluidic system integrated with systemic circulation.

Microfluidics for generation of modular tissue engineering constructs

While the approaches discussed so far of the generation of cell-containing hydrogels show great promise for tissue engineering research, they offer limited control over the composition of tissue engineering constructs with high spatial resolution. In order to create more complex constructs that may better replicate the anatomy and physiology of native tissue, researchers have strived

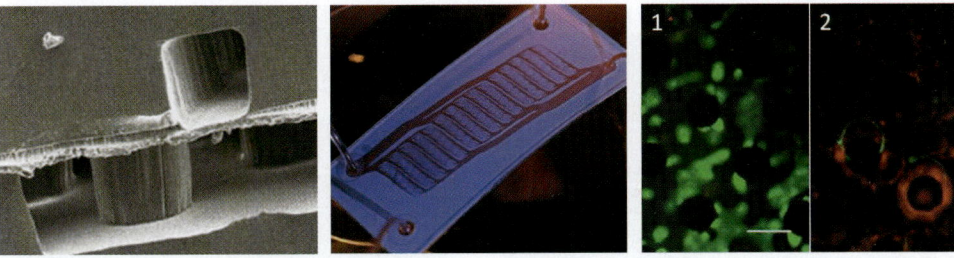

Fig. 24. Microfluidic liver assist device for use in animal models. Left: Bilayer PDMS channel system, separated by a nanoporous polyethersulfone membrane. Hepatocytes are cultured in the lower channel system and blood is perfused through the upper channel system. Center: Perfused device, showing upper blood channel system. Right: Live/dead stain of hepatocytes, 24 h after device implantation with (1) and without blood flow (2). (Reprinted with permission from Refs. 182–184.)

to use cell-containing building blocks for construct assembly with high spatial control over seeded cell types and their microenvironments. Microtechnologies and microfluidic devices have enabled the creation of individual building blocks as well as their assembly on or off chip,[187] potentially with a resolution down to the single-cell level.

For example, submillimeter-sized collagen gel rods seeded with endothelial cells were randomly assembled to three-dimensional clusters with high cell densities and perfusion culture was successfully demonstrated.[188] Results suggested that a functional non-thrombogenic layer of endothelial cells was generated on the gel rod surfaces and that it was sufficient to maintain continuous blood flow through the engineered modular tissue without significant platelet loss.

Similarly, rat aortic endothelial cells were seeded onto submillimeter-sized modules made of type I bovine collagen supplemented with Matrigel embedded with cardiomyocyte-enriched neonatal rat heart cells and assembled into a contractile, macroporous, sheet-like construct.[190] Muscle bundles co-expressing cardiac troponin I and connexin-43 were evident near the perimeter of modules and at intermodule junctions. In a different approach, a photocross-linkable poly(ethylene glycol) pre-gel was patterned in a photolithographic process into individual, cell-containing microgel units with various shapes and with dimensions in the range of several hundred micrometers.[189] Droplets of microgels in the pre-gel solution were suspended in mineral oil and mechanically agitated to support self-assembly. The fixation of assembled units through secondary cross-linking was demonstrated (Fig. 25). Microfluidic devices can be conveniently used to create such microgel building blocks with arbitrary shapes through continuous-flow lithography.[191] Microfluidic devices

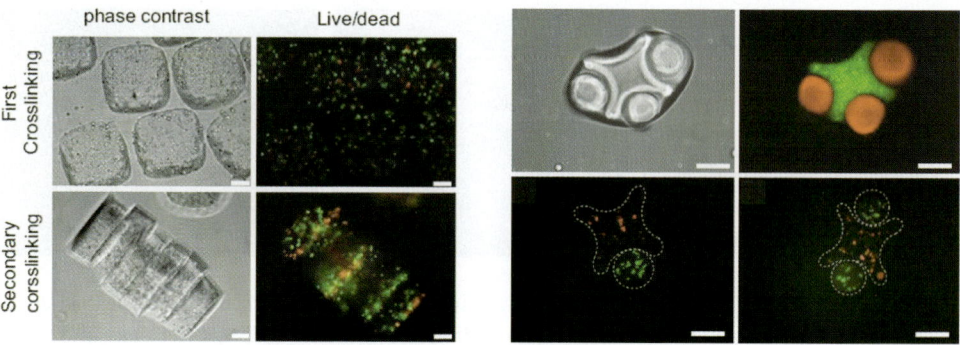

Fig. 25. Self-assembly of patterned, cell-containing microgels. Left: Phase contrast and live/dead fluorescence images of microgels after primary cross-linking as well as after self-assembly and secondary cross-linking. Right: Self-assembled star and rod microgels. (Reprinted with permission from Ref. 189.)

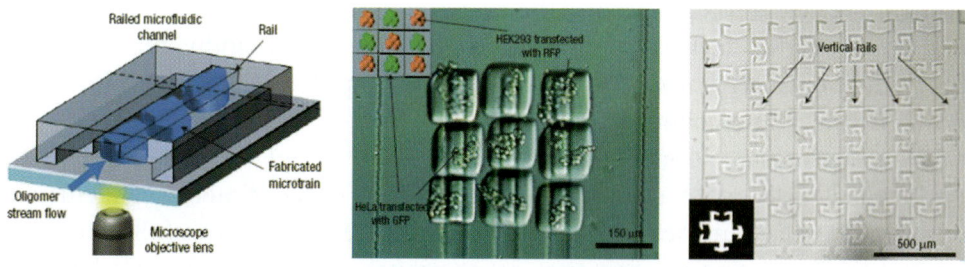

Fig. 26. Microfluidic system for photolithographic patterning of microgel units and their rail-guided assembly. Left: Using UV light and a microscope objective, a shape is projected into the pre-gel solution for microgel unit generation. Center: Using a rail guidance system, generated microgels can be assembled into larger constructs, containing, for example, blocks of alternating cell types. Right: Assembled, latched microgel units. (Reprinted with permission from Ref. 193.)

are also well suited to support the assembly of construct elements that were fabricated on or off chip. For example, microfluidic systems with culture chambers were used to assemble single layers of stacked particles.[192] Integrated micropillars kept elements in place while enabling perfusion culture with high cell viability. In order to enable precise unit assembly by design, microfluidic systems were used for on-chip photolithographic generation of microgels, which were then moved to specific locations in the microfluidic system using an embedded rail system, as illustrated in Fig. 26.[193]

Rather than generating movable microgel units on chip, the assembly of complex constructs can be based on units that stay in place after

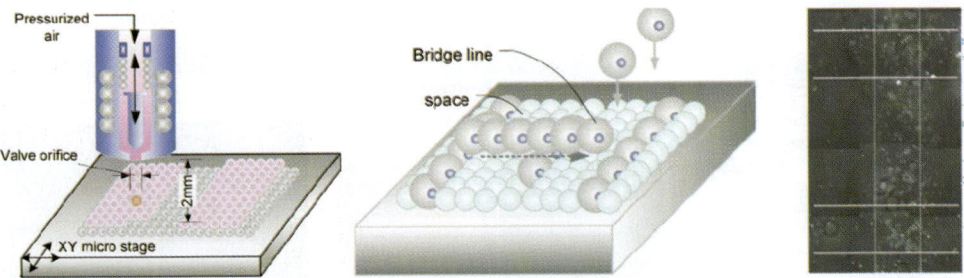

Fig. 27. Bioprinting of empty or cell-containing collagen pre-gel droplets for the generation of three-dimensional tissue engineering constructs. Left: Scheme of droplet ejector and substrate with printed cell layers. Center: Scheme of printed, cell-containing lines in two stacked layers. Right: Image of lines of seeded cells in different layers of the construct. (Reprinted with permission from Ref. 194.)

generation.[195] In this approach, a microfluidic chamber is filled with photocurable pre-polymer solution, and a scanning micromirror is used to expose the solution and to create a defined building block. The replacement of the pre-polymer solution and repetition of the exposure process in a different location enables serial generation of complex constructs. In a similar approach, heterogeneous side-by-side hydrogels were formed through photopolymerization in microfluidic channels.[196] Another related technology includes the layer-by-layer deposition of cell-containing gels in microchannels for the creation of heterogeneous constructs in three dimensions.[197] Of note are also microfluidic bioprinting technologies for the generation of complex structures, e.g., through the patterned deposition of growth factors on substrates,[198] deposition of cell-containing droplets,[199,200] or through the deposition of cell-containing tissue spheroids.[201,202] In the approach shown in Fig. 27, cooled, collagen-based pre-gel droplets, containing either smooth muscle cells or no cells, were printed onto a surface to form a three-dimensional pattern layer by layer.[194] Printed layers were gelled by incubation for 5 min at 37°C. An excellent long-term cell viability of >90% after 14 days of culture was reported.

Microfluidics for the generation of tissue models

Microfluidic technologies, such as those discussed here, not only advance stem cell research and the generation of cell/material systems toward regenerative approaches, but can also be central to the development of *in vitro* tissue models for drug development, study of disease mechanisms, and replication of aspects of embryonic development as well as adult physiology.[203,204] For example, a miniaturized, multiwell culture system for

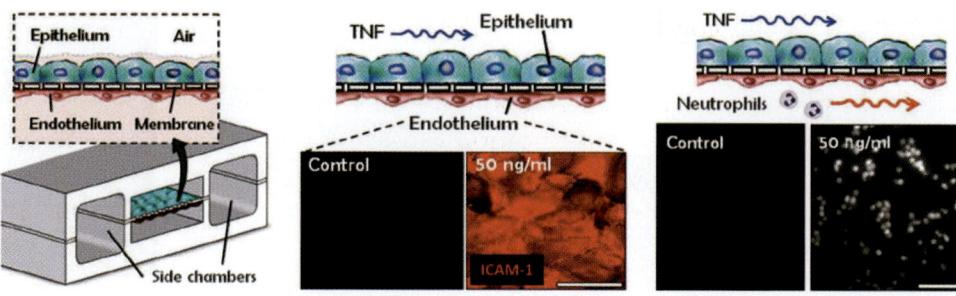

Fig. 28. Microfluidic system for replication of the alveolar–capillary interface of lung tissue. Left: Epithelial cells and endothelial cells are seeded on opposite sides of a porous polyester membrane that is embedded in a PDMS microfluidic system. Center: Exposure of the epithelial cell layer to the immunostimulant TNF causes the upregulation of the inter-cellular adhesion molecule ICAM-1 (red) in the endothelium. Right: Upon ICAM-1 upregulation, neutrophils (white dots) attach to the endothelial layer and transmigrate through endo- and epithelium. (Reprinted with permission from Ref. 206.)

human liver cells was developed with optimized microscale architecture that maintained phenotypic functions for several weeks.[205] The system enabled the assessment of gene expression profiles, phase I/II metabolism, canalicular transport, secretion of liver-specific products, and susceptibility to hepatotoxins. The need for such a liver model was underscored by citing the high rate of pre-launch and post-market attrition of pharmaceuticals due to liver toxicity.

In another recent example of the replication of tissue features, a microfluidic system was developed that recreates aspects of the alveolar/capillary interface in lung tissue, enabling the generation and investigation of injurious mechanical stresses in the lungs.[207] Endothelial and epithelial cells were cultured on opposite sides of a porous polyester membrane which separates channels in PDMS for air and liquid flows. This system was further developed to contain side chambers parallel to the main channels that enabled, upon the application of vacuum, the stretching of the membrane and mechanical loading of seeded cells.[206] Apart from mechanical investigations, the system also enabled studies of the toxicity of nanoparticles as well as the demonstration of the attachment of neutrophils to the endothelial layer and their transmigration through the endothelium and epithelium upon exposure of the epithelium to the immunostimulant tumor necrosis factor (TNF) (Fig. 28).

Conclusion and Outlook

This overview on the various applications of microfluidic systems in stem cell and tissue engineering not only indicates the tremendous potential that microfluidic technologies have for further advancements in these fields, but also reveals that these efforts are still at an early stage. Opportunities exist for the further exploitation of microfluidic devices for the analysis of stem cell biology, e.g., through the identification of novel stem cell markers through sorting and lineage analysis procedures. While such markers are well defined, for example, for HSCs,[208,209] other adult stem cell types lack such clear definition, hindering systematic, repeatable approaches to stem cell research. There is significant potential in applying microfluidics to develop a systematic understanding of gene expression profiles under well-defined experimental conditions, both in a high-throughput fashion on stem cell populations as well as at the single-cell level, for example, to assess heterogeneity among a population of cells. Microfluidics will also continue to play an important role in screens for environmental, fate-controlling factors that stem cells respond to and signaling pathways that are regulated in response to such stimuli. A better understanding of the exact signals that stem cells are exposed to within their niches *in vivo* may help to identify candidate factors for such screens. Pioneering work, e.g., regarding screens for ECM components and soluble factors that influence ESC fate, can be expanded in microfluidic systems to systematic combinatorial screens, e.g., with direct optical readout using suitable reporter transgenes. Increasing the complexity of engineered cellular microenvironments in microfluidic systems and their influence on stem cell fate may even include attempts to recreate the *in vivo* stem cell niche and its interactions with surrounding tissue on a chip. In the ideal case, adult stem cells would remain undifferentiated in their niche-like environment and generate progenitor cells in dynamic interaction with surrounding mature tissue. Such platforms would enable the systematic analysis of niche signaling dynamics and could help to develop advanced strategies for the design of tissue engineering constructs. Furthermore, stem cells in microfluidic systems have the potential for extensive use in drug screens. Apart from basic stem cell research, microfluidic systems are also well suited for the creation of bioreactors for the production of large numbers of stem cells or stem cell derivatives, directly for clinical use.

The recreation of biological complexity in the context of stem cell research and especially in the context of tissue engineering will largely depend on the development of advanced, synthetic biomaterials, with hydrogels being especially promising for use in microfluidic systems. A particularly compelling approach is to program advanced materials systems to induce a chain of events, leading for example to vascularization as suggested in Ref. 44, through embedded cues for the recruitment and differentiation of endothelial progenitors, formation of capillary structures, and stabilization through recruited supporting cells. Microfluidic systems can serve as versatile platforms for the development of assays for testing such advanced material concepts. For *in vitro* tissue engineering assays, microfluidic strategies have been effectively used to ensure mass transport and cell survival in complex, larger volume constructs to compensate for the lack of functional vascular structures. As briefly discussed, microfluidic systems can also play an important role in the development of strategies for the vascularization of tissue engineering constructs *in vivo*. Toward further increasing of the complexity of tissue engineering constructs, approaches based on cell-containing building blocks that can be assembled to precisely defined, three-dimensional structures are very promising. The use of building blocks offers precise control over the composition of cells as well as the microenvironments that the cells are exposed to, with spatial resolution potentially down to the single-cell level. Several microfluidic approaches for the creation and assembly of such cell-containing building blocks on chip have been discussed in the last part of this chapter. In addition, microfluidic bioprinting technologies become increasingly important for the generation of complex tissue engineering constructs, e.g., based on cells embedded in pre-gel droplets or tissue spheroids as building blocks. Such approaches will play important roles in the creation of constructs that better mimic tissue anatomy and physiology.

Another very exciting potential use of building block-based constructs is the recreation of specific conditions that are observed during embryonic development in order to instruct embedded ESCs to form specific tissues or organs. In analogy to creating environments that are inspired by the adult stem cell niche for advanced adult stem cell culture, human embryonic development can inspire tissue engineering approaches that are based on the recreation of developmental events, rather than on the synthesis of adult tissues.

An excellent example for such a strategy has been published recently. Researchers have managed to direct the differentiation of iPS cells into complex, three-dimensional intestinal tissue *in vitro*.[210] They applied a multistep differentiation protocol using factors that were identified through embryonic research, e.g., using gene expression analysis of embryonic endodermal tissue. The resulting intestinal "organoids" consisted of a polarized, columnar epithelium that was patterned into villus-like structures and crypt-like proliferative zones that expressed intestinal stem cell markers (Fig. 29). The epithelium contained functional enterocytes, as well as goblet, Paneth, and enteroendocrine cells. Recreating such complexity through traditional synthetic tissue engineering approaches is currently not possible. Microfluidic technologies could help to further develop the concept of development-inspired tissue and organ engineering. Additionally, there is

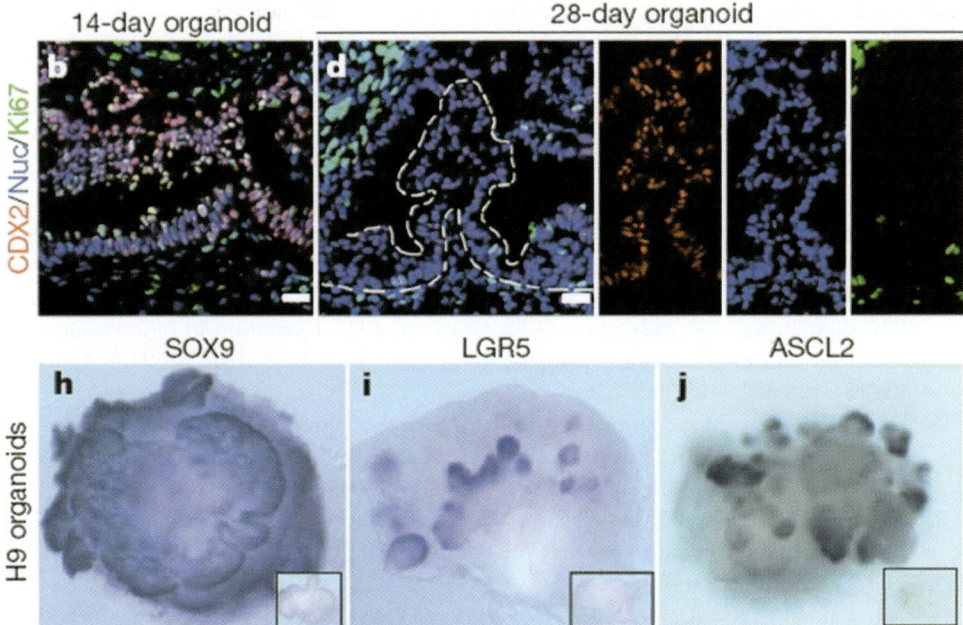

Fig. 29. Stimulation of iPS cells for the formation of intestinal organoids. Top: iPS cells that are stimulated with factors that govern embryonic intestinal development form structures that resemble intestinal villi within 28 days. Bottom: 56-day-old organoids show epithelial expression of SOX9 and restricted, "crypt-like" expression of the intestinal stem cell markers LGR5 and ASCL2. (Reprinted with permission from Ref. 210.)

tremendous potential for the generation of further tissue models in microfluidic systems, e.g., for the analysis of certain aspects of tissue physiology, study of disease mechanisms, screens for drugs, or analysis of stem cell behavior in complex tissue models, such as stem cell migration, homing, or differentiation. To some extent, work with microfluidic tissue models could replace experiments in animal models.

Overall, responding to challenges and opportunities in stem cell and tissue engineering will require highly interdisciplinary approaches. While the use of microfluidic systems in these fields is still in its infancy, exciting opportunities exist to massively apply microfluidic technologies ranging from basic stem cell research to supporting the creation of advanced functional tissues and possibly entire organs, toward improving human health and quality of life.

References

1. Department of Health and Human Services, Regenerative medicine, Retrieved August 22, 2012, from http://www.stemcells.nih.gov/info/scireport/2006report.htm, 2006.

2. Lanza, R., Gearhart, J., Hogan, B., Melton, D., Pederson, R., Thomas, E.D, Thomson, J., Wilmut, S.I., *Essentials of Stem Cell Biology*, 2nd edn. Academic Press, Burlington, MA, 2009.
3. Becker, A., McCulloch, E., Till, J., Cytological demonstration of the clonal nature of spleen colonies derived from transplanted mouse marrow cells. *Nature*, 1963, **197**: p. 452–454.
4. Thomson, J.A., Itskovitz-Eldor, J., Shapiro, S.S., Waknitz, M.A., Swiergiel, J.J., Marshall, V.S., Jones, J.M., Embryonic stem cell lines derived from human blastocysts. *Science*, 1998, **282**(5391): p. 1145–1147.
5. Shamblott, M.J., Axelman, J., Wang, S., Bugg, E.M., Littlefield, J.W., Donovan, P.J., Blumenthal, P.D., Huggins, G.R., Gearhart, J.D., Derivation of pluripotent stem cells from cultured human primordial germ cells. *Proc Natl Acad Sci U S A*, 1998, **95**(23): p. 13726–13731.
6. Yu, J., Vodyanik, M.A., Smuga-Otto, K., Antosiewicz-Bourget, J., Frane, J.L., Tian, S., Nie, J., Jonsdottir, G.A., Ruotti, V., Stewart, R. *et al.*, Induced pluripotent stem cell lines derived from human somatic cells. *Science*, 2007, **318**(5858): p. 1917–1920.
7. Takahashi, K., Tanabe, K., Ohnuki, M., Narita, M., Ichisaka, T., Tomoda, K., Yamanaka, S., Induction of pluripotent stem cells from adult human fibroblasts by defined factors. *Cell*, 2007, **131**(5): p. 861–872.
8. Lanza, R., Langer, R., Vacanti, J., *Principles of Tissue Engineering*, 3rd edn. Academic Press, Burlington, MA, 2007.
9. Vacanti, J.P., Morse, M.A., Saltzman, W.M., Domb, A.J., Perez-Atayde, A., Langer, R., Selective cell transplantation using bioabsorbable artificial polymers as matrices. *J Pediatric Surg*, 1988, **23**(1): p. 3–9.
10. Langer, R., Vacanti, J.P., Tissue engineering. *Science*, 1993, **260**(5110): p. 920–926.
11. Place, E.S., Evans, N.D., Stevens, M.M., Complexity in biomaterials for tissue engineering. *Nat Mater*, 2009, **8**(6): p. 457–470.
12. Alper, J., Geron gets green light for human trial of ES cell-derived product. *Nat Biotechnol*, 2009, **27**(3): p. 213–214.
13. Whitesides, G.M., The origins and the future of microfluidics. *Nature*, 2006, **442**(7101): p. 368–373.
14. Manz, A., Graber, N., Widmer, H.M., Miniaturized total chemical analysis systems: A novel concept for chemical sensing. *Sens Actuators B*, 1990, **1**(1–6): p. 244–248.
15. Xia, Y., Whitesides, G.M., Soft lithography. *Angewandte Chemie Int Edn*, 1998, **37**(5): p. 550–575.
16. Unger, M.A., Chou, H-P., Thorsen, T., Scherer, A., Quake, S.R., Monolithic microfabricated valves and pumps by multilayer soft lithography. *Science*, 2000, **288**(5463): p. 113–116.
17. Melin, J., Quake, S.R., Microfluidic large-scale integration: The evolution of design rules for biological automation. *Annu Rev Biophys Biomol Struct*, 2007, **36**(1): p. 213–231.
18. Thorsen, T., Maerkl, S.J., Quake, S.R., Microfluidic large-scale integration. *Science*, 2002, **298**(5593): p. 580–584.
19. Psaltis, D., Quake, S.R., Yang, C., Developing optofluidic technology through the fusion of microfluidics and optics. *Nature*, 2006, **442**(7101): p. 381–386.
20. Monat, C., Domachuk, P., Eggleton, B.J., Integrated optofluidics: A new river of light. *Nat Photon*, 2007, **1**(2): p. 106–114.

21. El-Ali, J., Sorger, P.K., Jensen, K.F., Cells on chips. *Nature*, 2006, **442**(7101): p. 403–411.
22. Andersson, H., van den Berg, A., Microfluidic devices for cellomics: A review. *Sens Actuator B: Chem*, 2003, **92**(3): p. 315–325.
23. Yi, C., Li, C-W., Ji, S., Yang, M., Microfluidics technology for manipulation and analysis of biological cells. *Anal Chim Acta*, 2006, **560**(1–2): p. 1–23.
24. Voldman, J., Electrical forces for microscale cell manipulations. *Annu Rev Biomed Eng*, 2006, **8**(1): p. 425–454.
25. Beebe, D.J., Mensing, G.A., Walker, G.M., Physics and applications of microfluidics in biology. *Annu Rev Biomed Eng*, 2002, **4**: p. 261–286.
26. Breslauer, D.N., Lee, P.J., Lee, L.P., Microfluidics-based systems biology. *Mol Biosyst*, 2006, **2**(2): p. 97–112.
27. Toh, Y-C., Blagovic, K., Voldman, J., Advancing stem cell research with microtechnologies: Opportunities and challenges. *Integr Biol*, 2010, **2**(7–8): p. 305–325.
28. van Noort, D., Ong, S.M., Zhang, C., Zhang, S., Arooz, T., Yu, H., Stem cells in microfluidics. *Biotechnol Progress*, 2009, **25**(1): p. 52–60.
29. Gupta, K., Kim, D., Ellison, C., Smith, A., Kundu, A., Tuan, J. Suh, K-Y., Levchenko, A., Lab-on-a-chip devices as an emerging platform for stem cell biology. *Lab Chip*, 2010, **10**(16): p. 2019–2031.
30. Khademhosseini, A., Langer, R., Borenstein, J., Vacanti, J.P., Microscale technologies for tissue engineering and biology. *Proc Natl Acad Sci U S A*, 2006, **103**(8): p. 2480–2487.
31. Liu Tsang, V., Bhatia, S.N., Three-dimensional tissue fabrication. *Adv Drug Delivery Rev*, 2004, **56**(11): p. 1635–47.
32. Park, H., Cannizzaro, C., Vunjak-Novakovic, G., Langer, R., Vacanti, C.A., Farokhzad, O.C., Nanofabrication and microfabrication of functional materials for tissue engineering. *Tissue Eng*, 2007, **13**(8): p. 1867–1877.
33. Anderson, D.J., Gage, F.H., Weissman, I.L., Can stem cells cross lineage boundaries? *Nat Med*, 2001, **7**: p. 393–395.
34. Hwang, N.S., Varghese, S., Elisseeff, J., Controlled differentiation of stem cells. *Adv Drug Delivery Rev*, 2008, **60**(2): p. 199–214.
35. Alhadlaq, A., Mao, J.J., Mesenchymal stem cells: Isolation and therapeutics. *Stem Cells Dev*, 2004, **13**(4): p. 436–448.
36. Dongeun, H., et al., Microfluidics for flow cytometric analysis of cells and particles. *Physiol Meas*, 2005, **26**(3): p. R73.
37. Fu, A.Y., Chou, H-P., Spence, C., Arnold, F.H., Quake, S.R., An integrated microfabricated cell sorter. *Anal Chem*, 2002, **74**(11): p. 2451–2457.
38. Flanagan, L.A., Lu, J., Wang, L., Marchenko, S.A., Jeon, N.L., Lee, A.P., Monuki, E.S., Unique dielectric properties distinguish stem cells and their differentiated progeny. *Stem Cells*, 2008, **26**(3): p. 656–665.
39. Vykoukal, J., Vykoukal, D.M., Freyberg, S., Alt, E.U., Gascoyne, P.R.C., Enrichment of putative stem cells from adipose tissue using dielectrophoretic field-flow fractionation. *Lab Chip*, 2008, **8**(8): p. 1386–1393.
40. Murthy, S., Sethu, P., Vunjak-Novakovic, G., Toner, M., Radisic, M., Size-based microfluidic enrichment of neonatal rat cardiac cell populations. *Biomed Microdevices*, 2006, **8**(3): p. 231–237.
41. Plouffe, B.D., Kniazeva, T., Mayer, J.E., Murthy, S.K., Sales, V.L., Development of microfluidics as endothelial progenitor cell capture technology for cardiovascular tissue engineering and diagnostic medicine. *FASEB J*, 2009, **23**(10): p. 3309–3314.

42. Plouffe, B.D., Njoka, D.N., Harris, J., Liao, J., Horick, N.K., Radisic, M., Murthy, S.K., Peptide-mediated selective adhesion of smooth muscle and endothelial cells in microfluidic shear flow. *Langmuir*, 2007, **23**(9): p. 5050–5055.
43. Discher, D.E., Mooney, D.J., Zandstra, P.W., Growth factors, matrices, and forces combine and control stem cells. *Science*, 2009, **324**(5935): p. 1673–1677.
44. Lutolf, M.P., Hubbell, J.A., Synthetic biomaterials as instructive extracellular microenvironments for morphogenesis in tissue engineering. *Nat Biotechnol*, 2005, **23**(1): p. 47–55.
45. Lutolf, M.P., Gilbert, P.M., Blau, H.M., Designing materials to direct stem-cell fate. *Nature*, 2009, **462**(7272): p. 433–441.
46. Moore, K.A., Lemischka, I.R., Stem cells and their niches. *Science*, 2006, **311**(5769): p. 1880–1885.
47. Scadden, D.T., The stem-cell niche as an entity of action. *Nature*, 2006, **441**(7097): p. 1075–1079.
48. Sujata, L., Chaudhuri, S., Stem cell niche, the microenvironment and immunological crosstalk. *Cell Mol Immunol*, 2008, **5**(2): p. 107–112.
49. Watt, F.M., Hogan, Brigid, L.M., Out of Eden: Stem cells and their niches. *Science*, 2000, **287**(5457): p. 1427–1430.
50. Raymond, K., Deugnier, M-A., Faraldo, M.M., Glukhova, M.A., Adhesion within the stem cell niches. *Curr Opin Cell Biol*, 2009, **21**(5): p. 623–629.
51. Blank, U., Karlsson, G., Karlsson, S., Signaling pathways governing stem-cell fate. *Blood*, 2008, **111**(2): p. 492–503.
52. Quesenberry, P.J., Becker, P.S., Stem cell homing: Rolling, crawling, and nesting. *Proc Natl Acad Sci U S A*, 1998, **95**(26): p. 15155–15157.
53. Morrison, S.J., Shah, N.M., Anderson, D.J., Regulatory mechanisms in stem cell biology. *Cell*, 1997, **88**(3): p. 287–298.
54. Ornitz, D.M., FGFs, heparan sulfate and FGFRs: Complex interactions essential for development. *Bioessays*, 2000, **22**(2): p. 108–112.
55. Levenstein, M.E., Berggren, W.T., Lee, J.E., Conard, K.R., Llanas, R.A., Wagner, R.J., Smith, L.M., Thomson, J.A., Secreted proteoglycans directly mediate human embryonic stem cell-basic fibroblast growth factor 2 interactions critical for proliferation. *Stem Cells*, 2008, **26**(12): p. 3099–3107.
56. Fuchs, E., Tumbar, T., Guasch, G., Socializing with the neighbors: Stem cells and their niche. *Cell*, 2004, **116**(6): p. 769–778.
57. Shen, Q., Wang, Y., Kokovay, E., Lin, G., Chuang, S.M., Goderie, S.K., Roysam, B., Temple, S., Adult SVZ stem cells lie in a vascular niche: A quantitative analysis of niche cell-cell interactions. *Cell Stem Cell*, 2008, **3**(3): p. 289–300.
58. Kopp, H.G., Avecilla, S.T., Hooper, A.T., Rafii, S., The bone marrow vascular niche: Home of HSC differentiation and mobilization. *Physiology (Bethesda)*, 2005, **20**(5): p. 349–356.
59. Mills, J.C., Gordon, J.I., The intestinal stem cell niche: There grows the neighborhood. *Proc Natl Acad Sci U S A*, 2001, **98**(22): p. 12334–12336.
60. Bjerknes, M., Cheng, H., Clonal analysis of mouse intestinal epithelial progenitors. *Gastroenterology*, 1999, **116**(1): p. 7–14.
61. Whitehead, R.H., Demmler, K., Rockman, S.P., Watson, N.K., Clonogenic growth of epithelial cells from normal colonic mucosa from both mice and humans. *Gastroenterology*, 1999, **117**(4): p. 858–865.

62. Watt, F.M., Role of integrins in regulating epidermal adhesion, growth and differentiation. *EMBO J*, 2002, **21**(15): p. 3919–3926.
63. Zhu, A.J., Haase, I., Watt, F.M., Signaling via beta1 integrins and mitogen-activated protein kinase determines human epidermal stem cell fate *in vitro*. *Proc Natl Acad Sci U S A*, 1999, **96**(12): p. 6728–6733.
64. Kopp, H.G., Avecilla, S.T., Hooper, A.T., Rafii, S., The bone marrow vascular niche: Home of HSC differentiation and mobilization. *Physiology (Bethesda)*, 2005, **20**: p. 349–56.
65. Levesque, J.P., Takamatsu, Y., Nilsson, S.K., Haylock, D.N., Simmons, P.J., Vascular cell adhesion molecule-1 (CD106) is cleaved by neutrophil proteases in the bone marrow following hematopoietic progenitor cell mobilization by granulocyte colony-stimulating factor. *Blood*, 2001, **98**(5): p. 1289–1297.
66. Papayannopoulou, T., Current mechanistic scenarios in hematopoietic stem/progenitor cell mobilization. *Blood*, 2004, **103**(5): p. 1580–1585.
67. Rosenthal, A., Macdonald, A., Voldman, J., Cell patterning chip for controlling the stem cell microenvironment. *Biomaterials*, 2007, **28**(21): p. 3208–3216.
68. Yim, E.K., Reano, R.M., Pang, S.W., Yee, A.F., Chen, C.S., Leong, K.W., Nanopattern-induced changes in morphology and motility of smooth muscle cells. *Biomaterials*, 2005, **26**(26): p. 5405–5413.
69. Yim, E.K.F., Pang, S.W., Leong, K.W., Synthetic nanostructures inducing differentiation of human mesenchymal stem cells into neuronal lineage. *Exp Cell Res*, 2007, **313**(9): p. 1820–1829.
70. Yim, E.K., Darling, E.M., Kulangara, K., Guilak, F., Leong, K.W., Nanotopography-induced changes in focal adhesions, cytoskeletal organization, and mechanical properties of human mesenchymal stem cells. *Biomaterials*, 2010, **31**(6): p. 1299–1306.
71. Engler, A.J., Sen, S., Sweeney, H.L., Discher, D.E., Matrix elasticity directs stem cell lineage specification. *Cell*, 2006, **126**(4): p. 677–689.
72. Buxboim, A., Rajagopal, K., Brown, A.E., Discher, D.E., How deeply cells feel: Methods for thin gels. *J Phys Condens Matter*, 2010, **22**(19): p. 194116.
73. Chowdhury, F., Li, Y., Poh, Y.C., Yokohama-Tamaki, T., Wang, N., Tanaka, T.S., Soft substrates promote homogeneous self-renewal of embryonic stem cells via downregulating cell-matrix tractions. *PLoS One*, 2010, **5**(12): p. e15655.
74. Evans, N.D., Minelli, C., Gentleman, E., LaPointe, V., Patankar, S.N., Kallivretaki, M., Chen, X., Roberts, C.J., Stevens, M.M., Substrate stiffness affects early differentiation events in embryonic stem cells. *Eur Cell Mater*, 2009, **18**: p. 1–13; discussion 13–14.
75. Kim, T.J., Seong, J., Ouyang, M., Sun, J., Lu, S., Hong, J.P., Wang, N., Wang, Y., Substrate rigidity regulates Ca^{2+} oscillation via RhoA pathway in stem cells. *J Cell Physiol*, 2009, **218**(2): p. 285–293.
76. Leong, W.S., Tay, C.Y., Yu, H., Li, A., Wu, S.C., Duc, D.H., Lim, C.T., Tan, L.P., Thickness sensing of hMSCs on collagen gel directs stem cell fate. *Biochem Biophys Res Commun*, 2010, **401**(2): p. 287–292.
77. Saha, K., Keung, A.J., Irwin, E.F., Li, Y., Little, L., Schaffer, D.V., Healy, K.E., Substrate modulus directs neural stem cell behavior. *Biophys J*, 2008, **95**(9): p. 4426–4438.
78. Poh, Y.C., Chowdhury, F., Tanaka, T.S., Wang, N., Embryonic stem cells do not stiffen on rigid substrates. *Biophys J*, 2010, **99**(2): p. L19–L21.

79. Hanjaya-Putra, D., Yee, J., Ceci, D., Truitt, R., Yee, D., Gerecht, S., Vascular endothelial growth factor and substrate mechanics regulate *in vitro* tubulogenesis of endothelial progenitor cells. *J Cell Mol Med*, 2010, **14**(10): p. 2436–2447.
80. Leipzig, N.D., Shoichet, M.S., The effect of substrate stiffness on adult neural stem cell behavior. *Biomaterials*, 2009, **30**(36): p. 6867–6878.
81. Teixeira, A.I., Ilkhanizadeh, S., Wigenius, J.A., Duckworth, J.K., Inganas, O., Hermanson, O., The promotion of neuronal maturation on soft substrates. *Biomaterials*, 2009, **30**(27): p. 4567–4572.
82. Seidlits, S.K., Khaing, Z.Z., Petersen, R.R., Nickels, J.D., Vanscoy, J.E., Shear, J.B., Schmidt, C.E., The effects of hyaluronic acid hydrogels with tunable mechanical properties on neural progenitor cell differentiation. *Biomaterials*. **31**(14): p. 3930–3940.
83. Cimetta, E., Figallo, E., Cannizzaro, C., Elvassore, N., Vunjak-Novakovic, G., Microbioreactor arrays for controlling cellular environments: Design principles for human embryonic stem cell applications. *Methods*, 2009, **47**(2): p. 81–89.
84. Kamei, K-I., Guo, S., Yu, Z.T.F., Takahashi, H., Gschweng, E., Suh, C., Wang, X., Tang, J., McLaughlin, J., Witte, O.N. *et al.*, An integrated microfluidic culture device for quantitative analysis of human embryonic stem cells. *Lab Chip*, 2009, **9**(4): p. 555–563.
85. Kim, L., Vahey, M.D., Lee, H-Y., Voldman, J., Microfluidic arrays for logarithmically perfused embryonic stem cell culture. *Lab Chip*, 2006, **6**(3): p. 394–406.
86. Kamei, K-I., Ohashi, M., Gschweng, E., Ho, Q., Suh, J., Tang, J., For Yu, Z.T., Clark, A.T., Pyle, A.D., Teitell, M.A. *et al.*, Microfluidic image cytometry for quantitative single-cell profiling of human pluripotent stem cells in chemically defined conditions. *Lab Chip*, 2010, **10**(9): p. 1113–1119.
87. Khademhosseini, A., Ferreira, L., Blumling, J. III, Yeh, J., Karp, J.M., Fukuda, J., Langer, R., Co-culture of human embryonic stem cells with murine embryonic fibroblasts on microwell-patterned substrates. *Biomaterials*, 2006, **27**(36): p. 5968–5977.
88. Moeller, H-C., Mian, M.K., Shrivastava, S., Chung, B.G., Khademhosseini, A., A microwell array system for stem cell culture. *Biomaterials*, 2008, **29**(6): p. 752–763.
89. Kang, E., Choi, Y.Y., Jun, Y., Chung, B.G., Lee, S-H., Development of a multi-layer microfluidic array chip to culture and replate uniform-sized embryoid bodies without manual cell retrieval. *Lab Chip*, 2010, **10**(20): p. 2651–2654.
90. Park, J., Cho, C.H., Parashurama, N., Li, Y., Berthiaume, F., Toner, M., Tilles, A.W., Yarmush, M.L., Microfabrication-based modulation of embryonic stem cell differentiation. *Lab Chip*, 2007, **7**(8): p. 1018–1028.
91. Fung, W-T., Beyzavi, A., Abgrall, P., Nguyen, N-T., Li, H-Y., Microfluidic platform for controlling the differentiation of embryoid bodies. *Lab Chip*, 2009, **9**(17): p. 2591–2595.
92. Ellison, D., Munden, A., Levchenko, A., Computational model and microfluidic platform for the investigation of paracrine and autocrine signaling in mouse embryonic stem cells. *Mol Biosyst*, 2009, **5**(9): p. 1004–1012.
93. Cochran, D.M., Fukumura, D., Ancukiewicz, M., Carmeliet, P., Jain, R.K., Evolution of oxygen and glucose concentration profiles in a tissue-mimetic culture system of embryonic stem cells. *Ann Biomed Eng*, 2006, **34**(8): p. 1247–1258.
94. Flaim, C.J., Chien, S., Bhatia, S.N., An extracellular matrix microarray for probing cellular differentiation. *Nat Methods*, 2005, **2**(2): p. 119–125.
95. Flaim, C.J., Teng, D., Chien, S., Bhatia, S.N., Combinatorial signaling microenvironments for studying stem cell fate. *Stem Cells Dev*, 2008, **17**(1): p. 29–40.

96. Underhill, G.H., Bhatia, S.N., High-throughput analysis of signals regulating stem cell fate and function. *Curr Opin Chem Biol*, 2007, **11**(4): p. 357–366.
97. Soen, Y., Mori, A., Palmer, T.D., Brown, P.O., Exploring the regulation of human neural precursor cell differentiation using arrays of signaling microenvironments. *Mol Syst Biol*, 2006, **2**: p. 37.
98. Zhang, J.-Z., Gao, W., Yang, H.-B., Zhang, B., Zhu, Z-Y., Xue, Y-F., Screening for genes essential for mouse embryonic stem cell self-renewal using a subtractive RNA interference library. *Stem Cells*, 2006, **24**(12): p. 2661–2668.
99. Godara, P., McFarland, C.D., Nordon, R.E., Design of bioreactors for mesenchymal stem cell tissue engineering. *J Chem Technol Biotechnol*, 2008, **83**(4): p. 408–420.
100. Gomez-Sjoberg, R., Leyrat, A.A., Pirone, D.M., Chen, C.S., Quake, S.R., Versatile, fully automated, microfluidic cell culture system. *Anal Chem*, 2007, **79**(22): p. 8557–8563.
101. Ju, X., Li, D., Gao, N., Shi, Q., Hou, H., Hepatogenic differentiation of mesenchymal stem cells using microfluidic chips. *Biotechnol J*, 2008, **3**(3): p. 383–391.
102. Tenstad, E., Tourovskaia, A., Folch, A., Myklebost, O., Rian, E., Extensive adipogenic and osteogenic differentiation of patterned human mesenchymal stem cells in a microfluidic device. *Lab Chip*, 2010, **10**(11): p. 1401–1409.
103. Park, J.Y., Hwang, C.M., Lee, S.H., Lee, S-H., Gradient generation by an osmotic pump and the behavior of human mesenchymal stem cells under the fetal bovine serum concentration gradient. *Lab Chip*, 2007, **7**(12): p. 1673–1680.
104. Chin, V.I., Taupin, P., Sanga, S., Scheel, J., Gage, F.H., Bhatia, S.N., Microfabricated platform for studying stem cell fates. *Biotechnol Bioeng*, 2004, **88**(3): p. 399–415.
105. Chung, B.G., Flanagan, L.A., Rhee, S.W., Schwartz, P.H., Lee, A.P., Monuki, E.S., Jeon, N.L., Human neural stem cell growth and differentiation in a gradient-generating microfluidic device. *Lab Chip*, 2005, **5**(4): p. 401–406.
106. Park, J.Y., Kim, S-K., Woo, D-H., Lee, E-J., Kim, J-H., Lee, S-H., Differentiation of neural progenitor cells in a microfluidic chip-generated cytokine gradient. *Stem Cells*, 2009, **27**(11): p. 2646–2654.
107. Falconnet, D., Csucs, G., Grandin, H.M., Textor, M., Surface engineering approaches to micropattern surfaces for cell-based assays. *Biomaterials*, 2006, **27**(16): p. 3044–3063.
108. Manbachi, A., Shrivastava, S., Cioffi, M., Chung, B.G., Moretti, M., Demirci, U., Yliperttula, M., Khademhosseini, A., Microcirculation within grooved substrates regulates cell positioning and cell docking inside microfluidic channels. *Lab Chip*, 2008, **8**(5): p. 747–754.
109. Khademhosseini, A., Yeh, J., Eng, G., Karp, J., Kaji, H., Borenstein, J., Farokhzad, O.C., Langer, R., Cell docking inside microwells within reversibly sealed microfluidic channels for fabricating multiphenotype cell arrays. *Lab Chip*, 2005, **5**(12): p. 1380–1386.
110. Bhatia, S.N., Yarmush, M.L., Toner, M., Controlling cell interactions by micropatterning in co-cultures: Hepatocytes and 3T3 fibroblasts. *J Biomed Mater Res*, 1997, **34**(2): p. 189–199.
111. Hui, E.E., Bhatia, S.N., Micromechanical control of cell-cell interactions. *Proc Natl Acad Sci*, 2007, **104**(14): p. 5722–5726.
112. Wright, D., Rajalingam, B., Selvarasah, S., Dokmeci, M.R., Khademhosseini, A., Generation of static and dynamic patterned co-cultures using microfabricated parylene-C stencils. *Lab Chip*, 2007, **7**: p. 1272–1279.

113. Khademhosseini, A., Suh, K.Y., Yang, J.M., Eng, G., Yeh, J., Levenberg, S., Langer, R., Layer-by-layer deposition of hyaluronic acid and poly-L-lysine for patterned cell co-cultures. *Biomaterials*, 2004, **25**(17): p. 3583–3592.
114. Kobel, S., Limacher, M., Gobaa, S., Laroche, T., Lutolf, M.P., Micropatterning of hydrogels by soft embossing. *Langmuir*, 2009, **25**(15): p. 8774–8779.
115. Ruiz, S.A., Chen, C.S., Microcontact printing: A tool to pattern. *Soft Matter*, 2006, **3**: p. 1–11.
116. Ruiz, S.A., Chen, C.S., Emergence of patterned stem cell differentiation within multicellular structures. *Stem Cells*, 2008, **26**(11): p. 2921–2927.
117. Fukuda, J., Khademhosseini, A., Yeo, Y., Yang, X., Yeh, J., Eng, G., Blumling, J., Wang, C.F., Kohane, D.S., Langer, R., Micromolding of photocrosslinkable chitosan hydrogel for spheroid microarray and co-cultures. *Biomaterials*, 2006, **27**(30): p. 5259–5267.
118. Khademhosseini, A., Eng, G., Yeh, J., Fukuda, J., Blumling, J., Langer, R., Burdick, J.A., Micromolding of photocrosslinkable hyaluronic acid for cell encapsulation and entrapment. *J Biomed Mater Res A*, 2006, **79A**(3): p. 522–532.
119. Koh, W.-G., Itle, L.J., Pishko, M.V., Molding of hydrogel microstructures to create multiphenotype cell microarrays. *Anal Chem*, 2003, **75**(21): p. 5783–5789.
120. Leclerc, E., Furukawa, K.S., Miyata, F., Sakai, Y., Ushida, T., Fujii, T., Fabrication of microstructures in photosensitive biodegradable polymers for tissue engineering applications. *Biomaterials*, 2004, **25**(19): p. 4683–4690.
121. Khademhosseini, A., Eng, G., Yeh, J., Kucharczyk, P., Langer, R., Vunjak-Novakovic, G., Radisic, M., Microfluidic patterning for fabrication of contractile cardiac organoids. *Biomed Microdevices*, 2007, **9**(2): p. 149–157.
122. Albrecht, D.R., Underhill, G.H., Wassermann, T.B., Sah, R.L., Bhatia, S.N., Probing the role of multicellular organization in three-dimensional microenvironments. *Nat Methods*, 2006, **3**(5): p. 369–375.
123. Carlberg, B., Axell, M.Z., Nannmark, U., Liu, J., Kuhn, H.G., Electrospun polyurethane scaffolds for proliferation and neuronal differentiation of human embryonic stem cells. *Biomed Mater*, 2009, **4**: p. 045004.
124. Christopherson, G.T., Song, H., Mao, H-Q., The influence of fiber diameter of electrospun substrates on neural stem cell differentiation and proliferation. *Biomaterials*, 2009, **30**: p. 556–564.
125. Chua, K-N., Chai, C., Lee, P-C., Tang, Y-N., Ramakrishna, S., Leong, K.W., Mao, H-Q., Surface-aminated electrospun nanofibers enhance adhesion and expansion of human umbilical cord blood hematopoietic stem/progenitor cells. *Biomaterials*, 2006, **27**(36): p. 6043–6051.
126. Shih, Y-R., Chen, C-N., Tsai, S-W., Wang, Y.J., Lee, O.K., Growth of mesenchymal stem cells on electrospun type I collagen nanofibers. *Stem Cells*, 2006, **24**(11): p. 2391–2397.
127. Gallego-Perez, D., Higuita-Castro, N., Sharma, S., Reen, R.K., Palmer, A.F., Gooch, K.J., Lee, L.J., Lannutti, J.J., Hansford, D.J., High throughput assembly of spatially controlled 3D cell clusters on a micro/nanoplatform. *Lab Chip*, 2010, **10**(6): p. 775–782.
128. Hwang, C.M., Khademhosseini, A., Park, Y., Sun, K., Lee, S-H., Microfluidic chip-based fabrication of PLGA microfiber scaffolds for tissue engineering. *Langmuir*, 2008, **24**(13): p. 6845–6851.

129. Lee, K.H., Shin, S.J., Kim, C-B., Kim, J.K., Cho, Y.W., Chung, B.G., Lee, S-H., Microfluidic synthesis of pure chitosan microfibers for bio-artificial liver chip. *Lab Chip*, 2010, **10**(10): p. 1328–1334.
130. Upadhyaya, S., Selvaganapathy, P.R., Microfluidic devices for cell based high throughput screening. *Lab Chip*, 2010, **10**(3): p. 341–348.
131. Fernandes, T.G., Diogo, M.M., Clark, D.S., Dordick, J.S., Cabral, J.M.S., High-throughput cellular microarray platforms: Applications in drug discovery, toxicology and stem cell research. *Trends Biotechnol*, 2009, **27**(6): p. 342–349.
132. Bennett, M.R., Hasty, J., Microfluidic devices for measuring gene network dynamics in single cells. *Nat Rev Genet*, 2009, **10**(9): p. 628–638.
133. Theberge, A.B., Courtois, F., Schaerli, Y., Fischlechner, M., Abell, C., Hollfelder, F., Huck, W.T.S., Microdroplets in microfluidics: An evolving platform for discoveries in chemistry and biology. *Angewandte Chemie Int Edn*, 2010, **49**(34): p. 5846–5868.
134. Teh, S-Y., Lin, R., Hung, L-H., Lee, A.P., Droplet microfluidics. *Lab Chip*, 2008, **8**(2): p. 198–220.
135. Huang, W-H., Ai, F., Wang, Z-L., Cheng, J-K., Recent advances in single-cell analysis using capillary electrophoresis and microfluidic devices. *J Chromat B*, 2008, **866**(1–2): p. 104–122.
136. Gao, J., Yin, X-F., Fang, Z-L., Integration of single cell injection, cell lysis, separation and detection of intracellular constituents on a microfluidic chip. *Lab Chip*, 2004, **4**(1): p. 47–52.
137. Sohn, L.L., Saleh, O.A., Facer, G.R., Beavis, A.J., Allan, R.S., Notterman, D.A., Capacitance cytometry: Measuring biological cells one by one. *Proc Natl Acad Sci U S A*, 2000, **97**(20): p. 10687–10690.
138. King, K.R., Wang, S., Irimia, D., Jayaraman, A., Toner, M., Yarmush, M.L., A high-throughput microfluidic real-time gene expression living cell array. *Lab Chip*, 2007, **7**(1): p. 77–85.
139. Valero, A., Post, J.N., van Nieuwkasteele, J.W., ter Braak, P.M., Kruijer, W., van den Berg, A., Gene transfer and protein dynamics in stem cells using single cell electroporation in a microfluidic device. *Lab Chip*, 2008, **8**(1): p. 62–67.
140. Lee, W.G., Demirci, U., Khademhosseini, A., Microscale electroporation: Challenges and perspectives for clinical applications. *Integr Biol*, 2009, **1**(3): p. 242–251.
141. Vogelstein, B., Kinzler, K.W., Digital PCR. *Proc Natl Acad Sci* 1999, **96**: p. 9236–9241.
142. Warren, L., Bryder, D., Weissman, I.L., Quake, S.R., Transcription factor profiling in individual hematopoietic progenitors by digital RT-PCR. *Proc Natl Acad Sci* 2006, **103** (47): p. 17807–17812.
143. Diehn, M., Cho, R.W., Lobo, N.A., Kalisky, T., Dorie, M.J., Kulp, A.N., Qian, D., Lam, J.S., Ailles, L.F, Wong, M. *et al.*, Association of reactive oxygen species levels and radioresistance in cancer stem cells. *Nature*, 2009, **458**(7239): p. 780–783.
144. Faley, S.L., Copland, M., Wlodkowic, D., Kolch, W., Seale, K.T., Wikswo, J.P., Cooper, J.M., Microfluidic single cell arrays to interrogate signalling dynamics of individual, patient-derived hematopoietic stem cells. *Lab Chip*, 2009, **9**(18): p. 2659–2664.
145. Zhong, J.F., Chen, Y., Marcus, J.S., Scherer, A., Quake, S.R., Taylor, C.R., Weiner, L.P., A microfluidic processor for gene expression profiling of single human embryonic stem cells. *Lab Chip*, 2008, **8**(1): p. 68–74.
146. Lee, J., Cuddihy, M.J., Kotov, N.A., Three-dimensional cell culture matrices: State of the art. *Tissue Eng Part B Rev*, 2008, **14**(1): p. 61–86.

147. Lee, K.Y., Mooney, D.J., Hydrogels for tissue engineering. *Chem Rev*, 2001, **101**(7): p. 1869–1880.
148. Geckil, H., Xu, F., Zhang, X., Moon, S., Demirci, U., Engineering hydrogels as extracellular matrix mimics. *Nanomedicine*, **5**(3): p. 469–484.
149. Fischer, S.E., Liu, X., Mao, H.Q., Harden, J.L., Controlling cell adhesion to surfaces via associating bioactive triblock proteins. *Biomaterials*, 2007, **28**(22): p. 3325–3337.
150. Silva, G.A., Czeisler, C., Niece, K.L., Beniash, E., Harrington, D.A., Kessler, J.A., Stupp, S.I., Selective differentiation of neural progenitor cells by high-epitope density nanofibers. *Science*, 2004, **303**(5662): p. 1352–1355.
151. Hartgerink, J.D., Beniash, E., Stupp, S.I., Peptide-amphiphile nanofibers: A versatile scaffold for the preparation of self-assembling materials. *Proc Natl Acad Sci U S A*, 2002, **99**(8): p. 5133–5138.
152. Tysseling, V.M., Sahni, V., Pashuck, E.T., Birch, D., Hebert, A., Czeisler, C., Stupp, S.I., Kessler, J.A., Self-assembling peptide amphiphile promotes plasticity of serotonergic fibers following spinal cord injury. *J Neurosci Res*, 2010, **88**(14): p. 3161–3170.
153. Khetan, S., Burdick, J.A., Patterning network structure to spatially control cellular remodeling and stem cell fate within 3-dimensional hydrogels. *Biomaterials*, 2010, **31**(32): p. 8228–8234.
154. Kloxin, A.M., Kasko, A.M., Salinas, C.N., Anseth, K.S., Photodegradable hydrogels for dynamic tuning of physical and chemical properties. *Science*, 2009, **324**(5923): p. 59–63.
155. Huang, C.P., Lu, J., Seon, H., Lee, A.P., Flanagan, L.A., Kim, H.-Y., Putnam, A.J., Jeon, N.L., Engineering microscale cellular niches for three-dimensional multicellular co-cultures. *Lab Chip*, 2009, **9**(12): p. 1740–1748.
156. Takayama, S., McDonald, J.C., Ostuni, E., Liang, M.N., Kenis, P.J.A., Ismagilov, R.F., Whitesides, G.M., Patterning cells and their environments using multiple laminar fluid flows in capillary networks. *Proc Natl Acad Sci U S A*, 1999, **96**(10): p. 5545–5548.
157. Wong, A.P., Perez-Castillejos, R., Christopher Love, J., Whitesides, G.M., Partitioning microfluidic channels with hydrogel to construct tunable 3-D cellular microenvironments. *Biomaterials*, 2008, **29**(12): p. 1853–1861.
158. Frisk, T., Rydholm, S., Andersson, H., Stemme, G., Brismar, H., A concept for miniaturized 3-D cell culture using an extracellular matrix gel. *Electrophoresis*, 2005, **26**(24): p. 4751–4758.
159. Burdick, J.A., Khademhosseini, A., Langer, R., Fabrication of gradient hydrogels using a microfluidics/photopolymerization process. *Langmuir*, 2004, **20**(13): p. 5153–5156.
160. Chung, S., Sudo, R., Vickerman, V., Zervantonakis, I., Kamm, R., Microfluidic platforms for studies of angiogenesis, cell migration and cell–cell interactions. *Ann Biomed Eng*, 2010, **38**(3): p. 1164–1177.
161. Chung, S., Sudo, R., Mack, P.J., Wan, C-R., Vickerman, V., Kamm, R.D., Cell migration into scaffolds under co-culture conditions in a microfluidic platform. *Lab Chip*, 2009, **9**(2): p. 269–275.
162. Kunze, A., Giugliano, M., Valero, A., Renaud, P., Micropatterning neural cell cultures in 3D with a multi-layered scaffold. *Biomaterials*, 2011, **32**(8): p. 2088–2098.
163. Kim, M., Yeon, J., Park, J-K., A microfluidic platform for 3-dimensional cell culture and cell-based assays. *Biomed Microdevices*, 2007, **9**(1): p. 25–34.
164. Toh, Y.C., Zhang, C., Zhang, J., Khong, Y.M., Chang, S., Samper, V.D., van Noort, D., Hutmacher, D.W., Yu, H., A novel 3D mammalian cell perfusion-culture system in microfluidic channels. *Lab Chip*, 2007, **7**(3): p. 302–309.

165. Ong, S.-M., Zhang, C., Toh, Y.-C., Kim, S.H., Foo, H.L., Tan, C.H., van Noort, D., Park, S., Yu, H., A gel-free 3D microfluidic cell culture system. *Biomaterials*, 2008, **29**(22): p. 3237–3244.
166. Markx, G.H., Carney, L., Littlefair, M., Sebastian, A., Buckle, A-M., Recreating the hematon: Microfabrication of artificial haematopoietic stem cell microniches *in vitro* using dielectrophoresis. *Biomed Microdevices*, 2009, **11**(1): p. 143–150.
167. Kannan, R.Y., Salacinski, H.J., Sales, K., Butler, P., Seifalian, A.M., The roles of tissue engineering and vascularisation in the development of micro-vascular networks: A review. *Biomaterials*, 2005, **26**(14): p. 1857–1875.
168. Borenstein, J.T., Terai, H., King, K.R., Weinberg, E.J., Kaazempur-Mofrad, M.R., Vacanti, J.P., Microfabrication technology for vascularized tissue engineering. *Biomed Microdevices*, 2002, **4**(3): p. 167–175.
169. Fidkowski, C., Kaazempur-Mofrad, M.R., Borenstein, J., Vacanti, J.P., Langer, R., Wang, Y., Endothelialized microvasculature based on a biodegradable elastomer. *Tissue Eng*, 2005, **11**(1–2): p. 302–309.
170. Bettinger, C.J., Weinberg, E.J., Kulig, K.M., Vacanti, J.P., Wang, Y., Borenstein, J.T., Langer, R., Three-dimensional microfluidic tissue-engineering scaffolds using a flexible biodegradable polymer. *Adv Mater*, 2006, **18**(2): p. 165–169.
171. Cabodi, M., Choi, N.W., Gleghorn, J.P., Lee, C.S.D., Bonassar, L.J., Stroock, A.D., A microfluidic biomaterial. *J Am Chem Soc*, 2005, **127**(40): p. 13788–13789.
172. Choi, N.W., Cabodi, M., Held, B., Gleghorn, J.P., Bonassar, L.J., Stroock, A.D., Microfluidic scaffolds for tissue engineering. *Nat Mater*, 2007, **6**(11): p. 908–15.
173. Ling, Y., Rubin, J., Deng, Y., Huang, C., Demirci, U., Karp, J.M., Khademhosseini, A., A cell-laden microfluidic hydrogel. *Lab Chip*, 2007, **7**(6): p. 756–762.
174. Song, Y., Lin, R., Montesano, G., Durmus, N., Lee, G., Yoo, S-S., Kayaalp, E., Hæggström, E., Khademhosseini, A., Demirci, U., Engineered 3D tissue models for cell-laden microfluidic channels. *Anal Bioanal Chem*, 2009, **395**(1): p. 185–193.
175. Stroock, A.D., Fischbach, C., Microfluidic culture models of tumor angiogenesis. *Tissue Eng A*, 2010, **16**(7): p. 2143–2146.
176. Hoganson, D.M., Pryor, H.I., Spool, I.D., Burns, O.H., Gilmore, J.R., Vacanti, J.P., Principles of biomimetic vascular network design applied to a tissue-engineered liver scaffold. *Tissue Eng A*, 2010, **16**(5): p. 1469–1477.
177. Park, J.H., Chung, B.G., Lee, W.G., Kim, J., Brigham, M.D., Shim, J., Lee, S., Hwang, C.M., Durmus, N.G., Demirci, U. et al., Microporous cell-laden hydrogels for engineered tissue constructs. *Biotechnol Bioeng*, 2010, **106**(1): p. 138–148.
178. Bick, A., Gomez, E., Shin, H., Brigham, M., Vu, M., Khademhosseini, A., Fabrication of microchannels in methacrylated hyaluronic acid hydrogels. *IEEE 35th Annu Northeast Bioeng Conf*, 2009.
179. Paguirigan, A., Beebe, D.J., Gelatin based microfluidic devices for cell culture. *Lab Chip*, 2006, **6**(3): p. 407–413.
180. Golden, A.P., Tien, J., Fabrication of microfluidic hydrogels using molded gelatin as a sacrificial element. *Lab Chip*, 2007, **7**(6): p. 720–725.
181. Gillette, B.M., Jensen, J.A., Tang, B., Yang, G.J., Bazargan-Lari, A., Zhong, M., Sia, S.K., *In situ* collagen assembly for integrating microfabricated three-dimensional cell-seeded matrices. *Nat Mater*, 2008, **7**(8): p. 636–640.
182. Carraro, A., Hsu, W.M., Kulig, K.M., Cheung, W.S., Miller, M.L., Weinberg, E.J., Swart, E.F., Kaazempur-Mofrad, M., Borenstein, J.T., Vacanti, J.P. et al., *In vitro* analysis of a

hepatic device with intrinsic microvascular-based channels. *Biomed Microdevices*, 2008, **10**(6): p. 795–805.
183. Borenstein, J.T., Weinberg, E.J., Orrick, B.K., Sundback, C., Kaazempur-Mofrad, M.R., Vacanti, J.P., Microfabrication of three-dimensional engineered scaffolds. *Tissue Eng*, 2007, **13**(8): p. 1837–1844.
184. Hsu, W.-M., Carraro, A., Kulig, K.M., Miller, M.L., Kaazempur-Mofrad, M., Weinberg, E., Entabi, F., Albadawi, H., Watkins, M.T., Borenstein, J.T. *et al.*, Liver-assist device with a microfluidics-based vascular bed in an animal model. *Ann Surg*, 2010, **252**(2): p. 351–357.
185. Barkefors, I., Thorslund, S., Nikolajeff, F., Kreuger, J., A fluidic device to study directional angiogenesis in complex tissue and organ culture models. *Lab Chip*, 2009, **9**(4): p. 529–535.
186. Vickerman, V., Blundo, J., Chung, S., Kamm, R., Design, fabrication and implementation of a novel multi-parameter control microfluidic platform for three-dimensional cell culture and real-time imaging. *Lab Chip*, 2008, **8**(9): p. 1468–1477.
187. Khademhosseini, A., Langer, R., Microengineered hydrogels for tissue engineering. *Biomaterials*, 2007, **28**(34): p. 5087–5092.
188. McGuigan, A.P., Sefton, M.V., Vascularized organoid engineered by modular assembly enables blood perfusion. *Proc Natl Acad Sci U S A*, 2006, **103**(31): p. 11461–11466.
189. Du, Y., Lo, E., Ali, S., Khademhosseini, A., Directed assembly of cell-laden microgels for fabrication of 3D tissue constructs. *Proc Natl Acad Sci U S A*, 2008, **105**(28): p. 9522–9527.
190. Leung, B.M., Sefton, M.V., A modular approach to cardiac tissue engineering. *Tissue Eng Part A*, 2010, **16**(10): p. 3207–3218.
191. Dendukuri, D., Pregibon, D.C., Collins, J., Hatton, T.A., Doyle, P.S., Continuous-flow lithography for high-throughput microparticle synthesis. *Nat Mater*, 2006, **5**(5): p. 365–369.
192. Bruzewicz, D.A., McGuigan, A.P., Whitesides, G.M., Fabrication of a modular tissue construct in a microfluidic chip. *Lab Chip*, 2008, **8**(5): p. 663–671.
193. Chung, S.E., Park, W., Shin, S., Lee, S.A., Kwon, S., Guided and fluidic self-assembly of microstructures using railed microfluidic channels. *Nat Mater*, 2008, **7**(7): p. 581–587.
194. Moon, S., Hasan, S.K., Song, Y.S., Xu, F., Keles, H.O., Manzur, F., Mikkilineni, S., Hong, J.W., Nagatomi, J., Haeggstrom, E. *et al.*, Layer by layer three-dimensional tissue epitaxy by cell-laden hydrogel droplets. *Tissue Eng Part C Methods*, 2010, **16**(1): p. 157–166.
195. Cheung, Y.K., Gillette, B.M., Zhong, M., Ramcharan, S., Sia, S.K., Direct patterning of composite biocompatible microstructures using microfluidics. *Lab Chip*, 2007, **7**(5): p. 574–579.
196. Zguris, J.C., Itle, L.J., Koh, W-G., Pishko, M.V., A novel single-step fabrication technique to create heterogeneous poly(ethylene glycol) hydrogel microstructures containing multiple phenotypes of mammalian cells. *Langmuir*, 2005, **21**(9): p. 4168–4174.
197. Tan, W., Desai, T.A., Layer-by-layer microfluidics for biomimetic three-dimensional structures. *Biomaterials*, 2004, **25**(7–8): p. 1355–1364.
198. Phillippi, J.A., Miller, E., Weiss, L., Huard, J., Waggoner, A., Campbell, P., Microenvironments engineered by inkjet bioprinting spatially direct adult stem cells toward muscle- and bone-like subpopulations. *Stem Cells*, 2008, **26**(1): p. 127–134.
199. Ringeisen, B.R., Othon, C.M., Barron, J.A., Young, D., Spargo, B.J., Jet-based methods to print living cells. *Biotechnol J*, 2006, **1**(9): p. 930–948.

200. Demirci, U., Montesano, G., Single cell epitaxy by acoustic picolitre droplets. *Lab Chip*, 2007, **7**(9): p. 1139–1145.
201. Mironov, V., Boland, T., Trusk, T., Forgacs, G., Markwald, R.R., Organ printing: Computer-aided jet-based 3D tissue engineering. *Trends Biotechnol*, 2003, **21**(4): p. 157–161.
202. Mironov, V., Visconti, R.P., Kasyanov, V., Forgacs, G., Drake, C.J., Markwald, R.R., Organ printing: Tissue spheroids as building blocks. *Biomaterials*, 2009, **30**(12): p. 2164–2174.
203. Griffith, L.G., Swartz, M.A., Capturing complex 3D tissue physiology *in vitro*. *Nat Rev Mol Cell Biol*, 2006, **7**(3): p. 211–224.
204. Khetani, S.R., Bhatia, S.N., Engineering tissues for *in vitro* applications. *Curr Opin Biotechnol*, 2006, **17**(5): p. 524–531.
205. Khetani, S.R. Bhatia, S.N., Microscale culture of human liver cells for drug development. *Nat Biotechnol*, 2008, **26**(1): p. 120–126.
206. Huh, D., Matthews, B.D., Mammoto, A., Montoya-Zavala, M., Hsin, H.Y., Ingber, D.E., Reconstituting organ-level lung functions on a chip. *Science*, 2010, **328**(5986): p. 1662–1668.
207. Huh, D., Fujioka, H., Tung, Y-C., Futai, N., Paine, R., Grotberg, J.B., Takayama, S., Acoustically detectable cellular-level lung injury induced by fluid mechanical stresses in microfluidic airway systems. *Proc Natl Acad Sci*, 2007, **104**(48): p. 18886–18891.
208. Spangrude, G.J., Heimfeld, S., Weissman, I.L., Purification and characterization of mouse hematopoietic stem cells. *Science*, 1988, p. 58–62.
209. Baum, C.M., Weissman, I.L., Tsukamoto, A.S., Buckle, A.M., Peault, B., Isolation of a candidate human hematopoietic stem-cell population. *Proc Nalt Acad Sci*, 1992, p. 2804–2808.
210. Spence, J.R., Mayhew, C.N., Rankin, S.A., Kuhar, M.F., Vallance, J.E., Tolle, K., Hoskins, E.E., Kalinichenko, V.V., Wells, S.I., Zorn, A.M. *et al.*, Directed differentiation of human pluripotent stem cells into intestinal tissue *in vitro*. *Nature*, 2010, **470**(7332): p. 105–109.

Chapter 12

Microfluidic "On-the-Fly" Fabrication of Microstructures for Biomedical Applications

Edward Kang, Sau Fung Wong, and Sang-Hoon Lee*

Department of Biomedical Engineering
College of Health Science, Korea University
Jeongneung-dong, Seongbuk-gu
Seoul, 136-703, Republic of Korea

Monodispersed microscale gel-based particles, fibers, and tubes are rapidly being adapted to several biomedical applications, including the delivery of drugs or cells, the fabrication of photonic crystals, and tissue engineering.[1–3] In the pharmaceutical field, the encapsulation of drugs in microgels may enable stimulus-triggered drug release and targeted, site-specific drug delivery. Recent hot issues in cell therapy are the stable delivery of cells to specific regions, provision of a stable microenvironment, and providing immune protection and encapsulation of cells in a microgel may address the problem of these issues. Microgel structures have been used to model the dynamic behavior of cells in microenvironments, including guided cell culture and cell fate in three-dimensional (3D) environments.[4] To date, several methods for producing microscale gel-based microstructures have been developed. Particle fabrication was accomplished using spray drying, inverse emulsification, coacervation, and shear break-up techniques. However, these techniques produce particles that are not uniform in shape or size. Such polydispersity prevents the precise control over the volume of loaded drugs or

---

*Corresponding author. Email: dbiomed@korea.ac.kr

cells delivered to the patient. Fibrous or tubular structures have been produced using conventional wet-spinning systems or electrospinning devices. These systems display a limited range of size control and the encapsulation of biologically active species, including enzymes, proteins, drugs, and live cells, without inducing damage, remains a challenge.

Over the past decade, microfluidic technologies, which control small volumes of fluid in channels, have rapidly developed and have been widely applied in the chemical, biological, and medical fields. The continuous microfluidic synthesis and assembly of microgel structures has recently seen a steep rise in interest due to the simplicity of production and the diversity in achievable shapes and sizes. The continuous synthesis of polymeric materials in a moving flow began in the early 1900s in the fiber industry. The initial method was complicated, and the fine tuning of shapes and sizes of synthesized structures were challengeable due to difficulties associated with the precision of flow control. Flow control and *in situ* polymerization of liquid gels using microfluidic devices not only provides low-end assembly-line benefits, but also has value for high-margin products. Recently, particles, fibers, and tubes with a diversity of shapes and sizes were demonstrated to be continuously producible.[5] Janus microstructures encoded with heterogeneous materials could additionally be generated. This chapter describes the recent progress in producing microgel structures using microfluidic methods. The methods may be categorized by polymerizing methods, microgel structure formation methods, and flow focusing methods, and will be described following these categories.

Polymerizing Methods

The main advantages of microfluidic fabrication techniques are the precisely tunable flow control and the fast mixing and reaction times. Microfluidic focusing permits narrowing of a core (sample) flow to a few hundred microns in diameter. The size reduction permits ultra-fast polymerization of moving emulsions or streams in the microchannel by scale effect, which is impossible in macro-scale flow. The polymerization method can be further broken down into three methods according to the reaction source — photo-, chemical, and thermal polymerization.

Photopolymerization

Figure 1 illustrates the formation of droplets containing monomers or reactive polymers, followed by ultraviolet (UV) irradiation. The photopolymerization method may be applied to rapidly photocurable materials, such as poly(ethylene

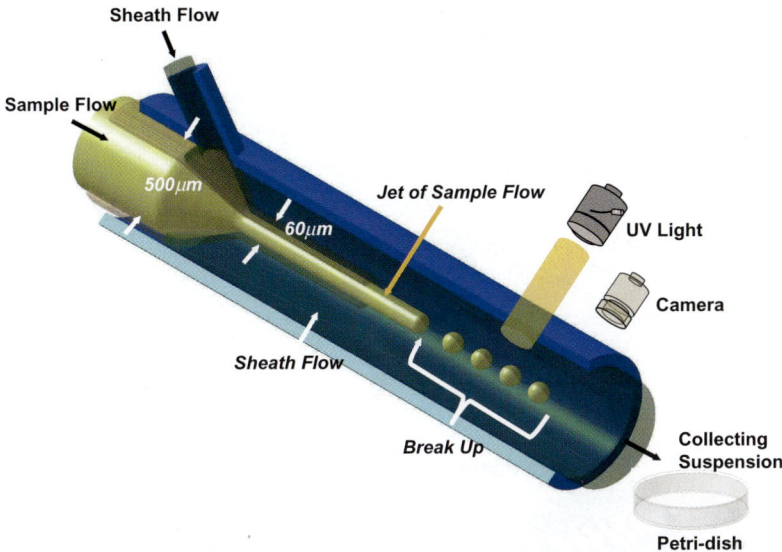

Fig. 1. Schematic of flow focusing and photopolymerization in capillary channel. (Reprinted with permission from Ref. 12. Copyright 2009, American Chemical Society.)

glycol) diacrylate (PEG-DA) or 4-hydroxybutyl acrylate (4-HBA). These backbone materials are mixed with a photoinitiator, which is a compound that decomposes into free radicals upon exposure to light. Typically, 365-nm UV light is radiated to initiate photopolymerization, and the reaction terminates before the flow has traveled through the microchannel and away from the region of irradiation. The polymerization time of a very small amount of oligomer is usually less than 200–300 ms. One disadvantage of UV irradiation is that the UV light may be harmful to bioactive species encapsulated within a microgel. By employing the digitally controllable UV radiation method, diverse particle shapes may be generated. Doyle *et al.* have reported the preparation of diverse flat structures using a mask-based photopolymerization process.[7] As shown in Fig. 2, an acrylate oligomer stream (typically PEG-DA) containing a photosensitive initiator was passed through a rectangular polydimethylsiloxane (PDMS) microchannel. Then, particle arrays of mask-defined shapes formed upon exposure of the oligomer flow to controlled pulses of UV light using an inverted microscope and the solidified particles were collected in a device reservoir. The shape of the particles in the *x–y* plane was determined by the shape of the feature on the transparency mask. Kwon *et al.* have used optofluidic maskless lithography instead of transparency masks.[8] As shown in Fig. 2, the top-view shape of the polymeric particles was dynamically controlled by patterning on a digital micromirror device. This platform could be used to easily tune the shape of particles via a computer-controlled system, and particles with wider applications and diverse 3D shapes may be produced.

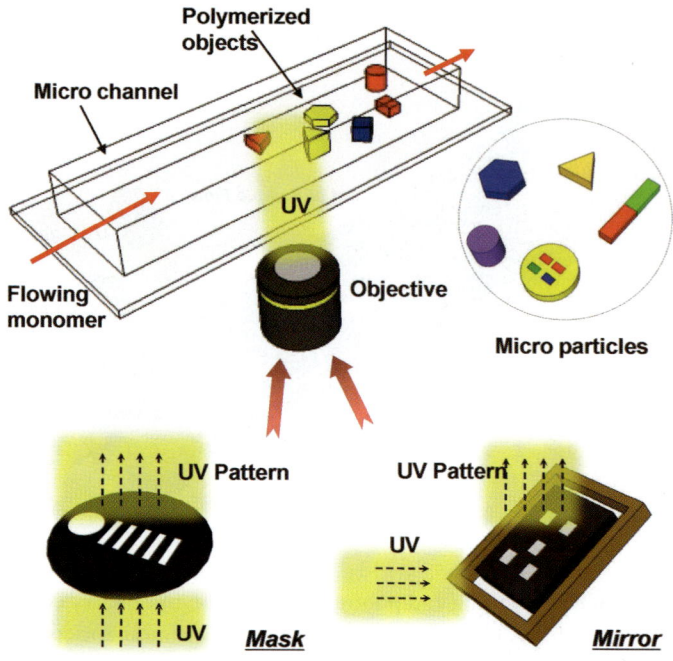

Fig. 2. Schematic of optofluidic mask- and micromirror-based lithography.

Chemical polymerization

The introduction of multiple fluids into a microchannel can produce a multiphase flow due to scale effects. The fluids can form spherical emulsions and jet streams. Figure 3 shows the production of alginate fibers generated by two-phase streams (sodium alginate and $CaCl_2$) in an outlet channel. At the interface between the two liquids, the diffusion of Ca^{2+} ions into the sodium alginate stream occurs, and the flowing stream is solidified to generate continuous fibers. The gel formation process due to Ca^{2+} ion chelation may be approximated by the following equation:[9]

$$\frac{\partial C}{\partial t} = D\frac{\partial^2 C}{\partial x^2} - \frac{\partial S}{\partial t}, \tag{1}$$

where D is the diffusion coefficient of Ca^{2+}, C is the concentration of Ca^{2+} ions in the $CaCl_2$ solution, and S is the concentration of cross-linked Ca^{2+} ions. Alginate is widely used for the chemical polymerization process due to several advantages over other biological hydrogels such as hyaluronic acid, fibrin, agar, gelatin, and chitosan. Alginate is a partially proven material

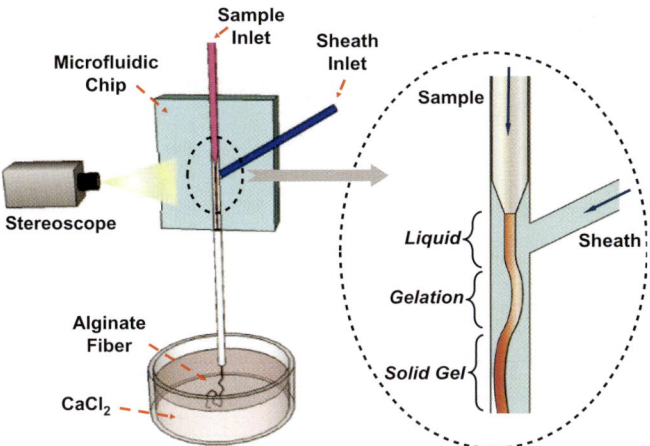

Fig. 3. Schematic of chemical polymerization of alginate fiber. During liquid alginate passing through the channel, the alginate liquid is solidified by Ca^{2+} ion diffusion.

approved by the FDA, and many studies have investigated for the use of alginate in therapy, pharmacology, and medicine. However, the practical applications in life sciences are limited by the difficulties associated with shape formation. The use of microfluidic chips for the construction of more complex structures may resolve these problems. Several hydrogels, such as chitosan, poly(lactic-*co*-glycolic acid) (PLGA), and complexes formed from alginate and chitosan, may also be used. The chemical polymerization method is especially useful for the production of biocompatible and biodegradable microscale scaffolding structures for 3D tissue engineering.

Microfluidic platforms for shaped flow generation

Microfluidic channels can generate a variety of shapes by tuning the channel design and by changing the identity of the fluid material. To this end, diverse microfluidic platforms have been developed: (1) PDMS-based planar channels, (2) capillary channels, and (3) fusion of PDMS-based planar and cylindrical channels.

PDMS-based planar channels

As illustrated by Whitesides' *Science* cover in 1999, a fluid injected into a planar microchannel assumes laminar flow, and mixing between individual streams flowing side by side proceeds only by diffusion at the interface.[10]

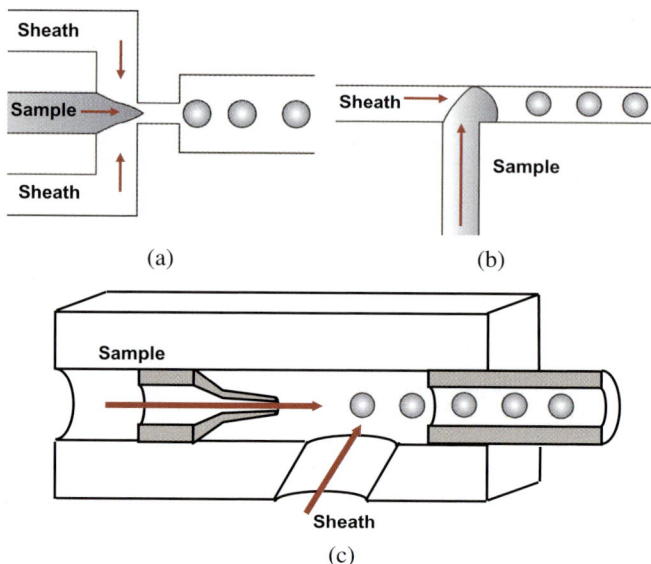

Fig. 4. Schematic of microparticle production by the (a) flow focusing method, (b) T-channel method, and (c) capillary channel method.

These properties enable the generation of focused flow, the size of which may be easily controlled by changing of the flow rate. Channels are readily fabricated using conventional soft lithographic processes, and they have been widely used for the continuous generation of microgel particles. Continuous generation begins with the emulsification of an aqueous solution containing a monomer, an oligomer, or a polymer in an immiscible nonpolar liquid. Microfluidic emulsification is usually accomplished by flow focusing and the T-junction method (Figs. 4(a) and 4(b)). The size of emulsion droplets can be adjusted according to the channel size and flow rate. In the flow focusing method, an aqueous phase and a nonpolar liquid are introduced simultaneously and meet at the junction of both flows' merging locations. A stream of aqueous droplet phases is focused by the shearing force exerted by the stream of the nonpolar liquid, and the periodic break-up of the aqueous stream yields monodisperse droplets. In the T-junction method, the continuous phase forms a thin film layer between the dispersed phase and the walls of the device. The increased pressure applied to the thin film layer squeezes the disperse phase to form monodisperse droplets. Once the droplets have been generated by both methods, polymerization may be accomplished using UV radiation or chemical reactions. A limitation of this system is that the products are mostly limited to particle structures.

Capillary channels

The combination of a pulled glass capillary and a PDMS channel enables the device to fabricate spherical and fibrous structures.[11,12] Within pre-perforated PDMS base holes, two concentric capillaries (one is pulled and the other is a regular capillary) are inserted to permit self-alignment (Fig. 4(c)). The polymerizable solution is injected into the central channel, while the sheath fluid is introduced into the sheath channel. Focused coaxial flow is generated at the position at which the two fluids merge. Due to the phenomena that dominate at the microscale (e.g., laminar flow and diffusion), a 3D coaxial sheath flow stream around the sample flow maintains its shape upon exiting the outlet channel. An advantage of the capillary channel is that a polymerizable sample flow is formed within the center of an encasing outlet flow. This flow configuration effectively prevents the clotting of polymerized structures in the channel and enables the easy extrusion of continuously produced polymerized structures and diversely shaped microstructures, including fibers, capsules, tubes, and spheres. However, a disadvantage of this technique is that the fabrication of pulled glass capillaries and the skilful treatment of the glass tips are difficult. The preparation of pulled glass microcapillaries is labor-intensive and requires skill and specialized tools, which prevents the wider adoption of the technique. Some researchers have used syringes with small needles. However, commercially available needles are limited in the extent to which diameter may be reduced. Size reduction in microfluidic device fabrication is an important challenge, and this feature is useful for forming small droplets (1–5 μm in diameter) because the size of the tip may be reduced to a few microns.[6]

Fusion of PDMS-based planar and cylindrical channels

Planar and capillary channels have their own advantages and disadvantages. For the fabrication of diverse-shaped particles, planar channels are advantageous. Capillary channels are useful for the generation of small particles and fibrous structures. Recently, a new system was reported which employed the advantages of both planar and capillary channels.[13] This work developed a method for fabricating PDMS-only cylindrical channels that generate coaxial flow using thin PDMS membranes. Figure 5(a) illustrates the method of fabricating round channels to generate 3D coaxial flows. The fusion of PDMS-based planar and cylindrical channels was formed by engraving both semi-cylindrical and conventional planar channels on a single PDMS layer, and by bonding two engraved layers (Fig. 5(b)). The fabrication method is much simpler than the fabrication of capillary channels, and several established applications of planar

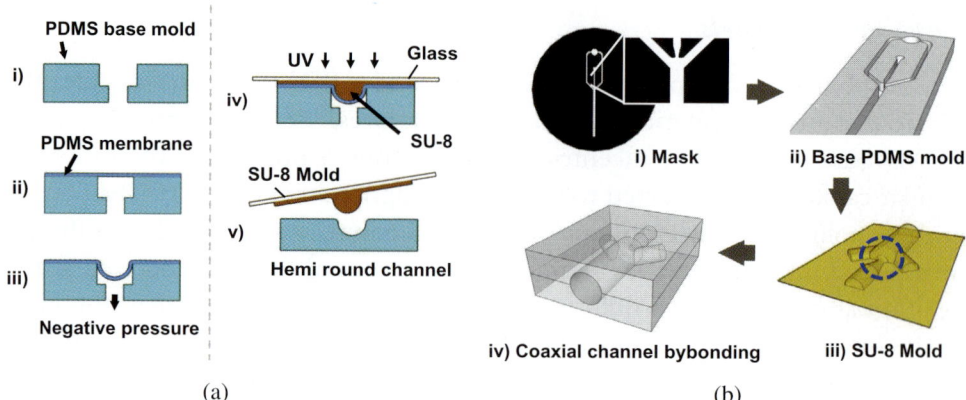

Fig. 5. Processes (a) to fabricate round channel and (b) to fabricate cylindrical channel by bonding two hemicylindrical channels. (Reprinted with permission from Ref. 13. Copyright 2010, The Royal Society of Chemistry.)

channels have been developed. Based on this technology, the integrated channel of a gradient generation channel employing 12 mixers and five cylindrical channels was developed to continuously produce five fibers having different chemical contents. Using this method, diverse microstructures can be produced easily. This method may be flexibly applied to a variety of applications because the already developed planar chip technology can be used.

Generated microstructures

The microfluidic platforms for shaped flow generation or polymerization have been used to generate diverse microscale structures. The modification of the process and channel shape increases the flexibility for tuning the shape and function of the product. In this section, we classify the synthesized microstructures as spherical, fibrous, and complicated structures.

Spherical structures

Spherical microgel particles are the most popular structures because they are easily fabricated using any microfluidic platform or polymerization method. The microfluidic generation of emulsions has been broadly studied. However, achieving fast polymerization of moving emulsions has been a challenge. Through the appropriate use of UV-curable materials (e.g., PEG-DA, 4-HBA) and UV radiation, emulsions can be rapidly polymerized. More recently, spherical alginate microgels have been produced by several methods. The production of biocompatible and biodegradable hydrogel particles is important for biomedical applications in that the

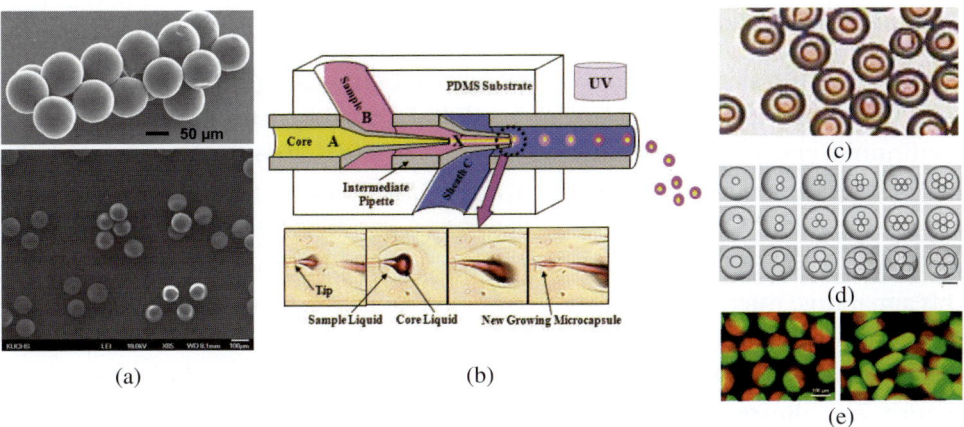

Fig. 6. (a) Top: PEG-DA microspheres, bottom: alginate microspheres, (b) Top: capillary chip to generate microcapsule containing liquid, bottom: capsule formation at the tip and core liquid in encapsulated by PEG-DA solution (adapted from Ref. 17. Copyright 2006, IOP Publishing Ltd.), (c) PEG-DA microcapsules containing dyed water, (d) single capsules trapping multiple sub-emulsions (Reprinted with permission from Ref. 14. Copyright 2007, John Wiley and Sons), (e) Janus particles stained with different fluorescence (Reprinted with permission from Ref. 15. Copyright 2006, American Chemical Society.).

fabrication process itself is biocompatible. Biodegradable particles may be used as drug or cell carriers, which is a new and important trend in medicine. Through slight modifications in the fabrication method, diverse structural shapes may be generated. The solid spherical particles are representative products fabricated by a simple process. Figure 6(a) shows PEG-DA (top) and alginate (bottom) solid particles produced via photopolymerization and chemical polymerization, respectively. The surface tension at an interface between immiscible fluids permits the formation of spherical structures. Capsular structures can be produced by the sequential connection of emulsion-generating channels, and both rectangular and capillary microchannels may be used. Figure 6(b; top) illustrates the capillary channel designed to produce capsular particles and Fig. 6(b; bottom) shows the growing capsular structure at the glass tip. Ultra-small quantities of liquid or cell suspension media were entrapped within the capsules (Fig. 6(c)). In the monodisperse double emulsions, a controlled number of monodisperse single emulsions could be trapped (Fig. 6(d)).[14] The single emulsion was subsequently emulsified in the collection tube by coaxial flow of the outermost fluid, which was injected into the outer stream through a square capillary. The size and number of both the outer and inner droplets could be precisely tuned by adjusting the dimensions of the orifices or the flow rates of the three fluids. Janus particles are particles that can be thought of as presenting two distinct hemispheres. Janus particles were produced by the simultaneous break-up of two co-flowing streams (Fig. 6(e)).[15]

Fibrous and tubular structures

Fibrous and tubular structures can be fabricated by using capillary microchannels, although these channels require skilful control of the flow rate for the continuous extrusion of polymerized fibers. In 2004, Jeong *et al.* developed a microfluidic apparatus that created a continuous process for the production of microscale cylindrical polymeric structures (e.g., fibers, tubes) by employing a 3D multiple-stream laminar flow[12] and on-the-fly photopolymerization. Using this apparatus, microstructures with diverse shapes, such as fibers or tubes, were generated by changing the channel configuration. Their dimensions can be adjusted by controlling the relative flow rates without re-tooling the device. Figure 7(a) illustrates the schematics of an apparatus for generating fibrous structures. Both the photo- and chemical polymerization methods can be employed. The curable monomer, oligomer, or polymer was introduced as a sample fluid, and the curing agent or miscible liquid was introduced as a sheath flow. The coaxial 3D continuous stream then formed in the outlet channel. During passage of the stream through the outlet channel, the central stream polymerized via UV radiation and chemical reactions at the interface of the coaxial flow. The UV-curable materials, 4-HBA or PEG-DA, have been widely used as backbone material. In contrast, alginate, PLGA, and chitosan have generally been used as chemically curable materials. A tubular structure can be produced by inserting an additional pulled micropipette.[16] Figure 7(b) shows a schematic for generating an alginate microtube. Three separate inlets permit the

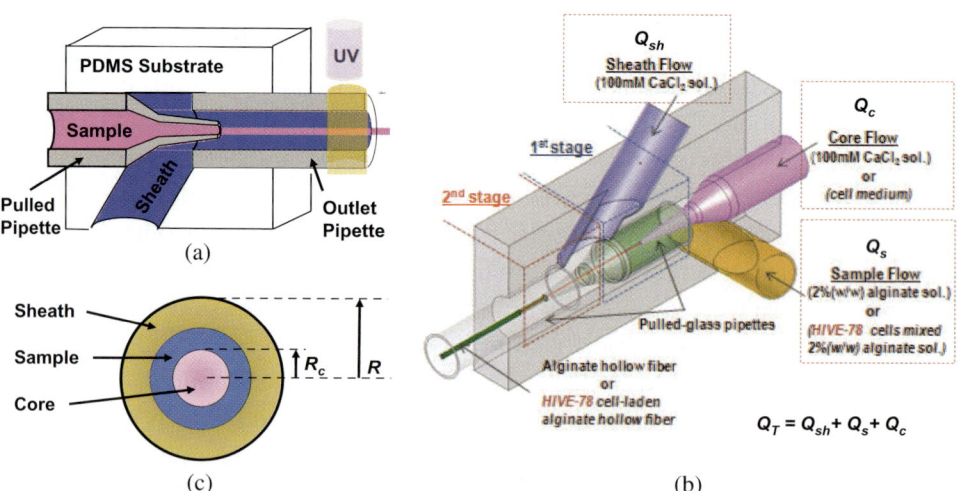

Fig. 7. (a) Schematic of capillary channel to produce microfibers (adapted with permission from Ref. 12. Copyright 2004, The Royal Society of Chemistry), (b) schematic of a capillary channel to produce alginate microtubular structures (adapted with permission from Ref. 16. Copyright 2009, John Wiley and Sons), and (c) a cross-sectional flow image.

simultaneous introduction of the core fluid (100 mM $CaCl_2$), the sample fluid (2% w/w sodium alginate, Sigma, St. Louis, MO, USA), and the sheath fluid (100 mM $CaCl_2$). During the first stage, coaxial flow (consisting of the core and sample fluid) was generated, and the solidification process began at the interface between the core and the sample flow, which formed along the inner wall of the hollow fiber (Fig. 7(c)). When this coaxial flow met the sheath flow, the outer interface of the sample polymerized within the tubular structures. The diameter of the alginate hollow fiber could be estimated according to

$$D = R_g \left(1 - \left(\frac{Q_s + Q_{sh} - Q_c}{Q_T} \right)^{\frac{1}{2}} \right)^{\frac{1}{2}}, \qquad (2)$$

where D is the diameter of the alginate hollow fiber, R_g is the inner radius of the glass pipette, Q_T is the total flow rate, Q_s is the sample flow rate, Q_{sh} is the sheath flow rate, and Q_c is the core flow rate. As with cylindrical structures, Janus fibers consisting of heterogeneous materials are produced by introducing multiple materials simultaneously into the channel.

Complicated structures and parallel production

A UV mask alone can generate complicated flat particles. As illustrated in Fig. 2, flat polygonal structures, colloidal cuboid, high-aspect ratio structures with different cross-sections, and curved particles can be continuously and massively synthesized. By varying the oligomer composition and the mask-based polymerization method, bifunctional Janus particles can also fabricated. This method enables particle synthesis, encoding, and probe incorporation into a single process to produce multifunctional particles bearing over a million unique codes. The computer-controlled micromirror-based UV exposure system has been used to synthesize multifunctional flat particles. This method enables the *in situ* alignment of the particles generated and the complex structures composed of more than 50 microstructures. Despite the many advantages, this method has disadvantages in that the material is limited to UV-curable monomers or oligomers. In addition, the fabrication platforms are complex compared to other methods. These limitations restrict widespread applicability, particularly in the biomedical and tissue engineering fields.

Biomedical applications

To date, diverse methods for fabricating microstructures have been proposed. Although researchers have suggested several potential applications, their practical use remains limited. However, applications in the biomedical field continue to be reported, and it is anticipated that applications will continue to be discovered. Here, we have classified these applications into four topics — drug and cell carriers, tissue engineering, microactuators, and bio-sensors and microdelivery systems. The cell or drug carrier and tissue engineering will be the most important of the potential applications.

Cell and drug encapsulation

Live cell encapsulation using microcapsules formed from bulk emulsifications is difficult due to the ease with which cells may be damaged during the fabrication processes, for example, during the chemical reaction, washing, centrifugation, and redispersion. However, the microfluidic system addresses these problems of stable cell encapsulation. Cell encapsulation can be achieved via both optical and chemical polymerization. Oh *et al.* described a yeast cell suspension encapsulation within microcapsules fabricated via photopolymerization.[17] They used PEG-DA as the encapsulation material. This example may constitute the first microfluidic encapsulation preparation of live cells suspended in culture media. After two days of culture, the cells were found to have proliferated and showed excellent viability. However, PEG-DA is not biodegradable and its biocompatibility has not been fully proven. This method is limited in its use as a cell or drug carrier. Alginate, which is a proven material for biomedical applications, is a good candidate for cell and drug encapsulation. The microfluidic encapsulation of cells within solid alginate particles and fibers has been reported. Shin *et al.* developed a stable encapsulation method for fibroblast (L929) cells within solid alginate fibers.[9] Takeuchi *et al.* injected a solution containing sodium alginate, $CaCO_3$, and Jurkat cells in a T-junction channel for the production of cell-encapsulating alginate particles.[18] They investigated cell viability as a function of $CaCO_3$ concentration. The emulsion was formed using corn oil with lecithin, and acetic acid was used to polymerize the alginate solution. Similar alginate-based cell encapsulation methods have also been reported by several groups. The loading of biomolecules into capsules and fibers has been demonstrated to be feasible as drug delivery carriers. As the hydrogel swells, the size of the loaded particles can affect the kinetics of cell or drug delivery. Small particles tend to rapidly secrete their encapsulated material. Bovine serum albumin–fluorescein isothiocyanate (BSA–FITC) has been broadly used as a test molecule because it displays stable loading within spherical fibrous structures. Through the appropriate selection of material, the drug release rate may be easily controlled.

Tissue engineering

Tissue engineering is a representative interdisciplinary area that applies the principles of engineering and life sciences to the development of biological substitutes that restore, maintain, or improve tissue function or a whole organ. Recently, the scope of tissue engineering has been broadened to include an understanding of the principles of tissue growth as well as control over cell behavior by providing a specific microenvironment. The final goal of tissue engineering may be the engineering of organs and their replacement. However, these goals continue to pose challenges in terms of maintaining the functional and biomechanical stabilities of laboratory-grown tissues destined for transplantation. A significant challenge is vascularization in engineered tissues. Generally, 200–300 μm is the limit for nutrient and oxygen delivery by diffusion, and microvessels must be uniformly distributed within the engineered tissue.

Hollow alginate microtubes have been continuously produced using a microfluidic chip and were aligned two-dimensionally on a flat spool.[16] Based on this technology, the feasibility of constructing 3D microvascularized structures in hydrogels using alginate hollow fibers was investigated. Vascularized tissue was formed by embedding endothelial cells within the hollow fibers, and introducing these fibers to a smooth muscle cell-laden agar–gelatin–fibronectin (AGF) environment. Cells within the hollow fibers and the hydrogel were co-cultured and remained alive with intact vascular structures for seven days. Diverse material-based fibers, including gelatin–hydroxyphenylpropionic acid (Gtn–HPA), poly-(N-isopropyl acrylamide) (poly (NIPAAM)), and polysulfone, were generated.[19] Madin–Darby canine kidney (MDCK) cells were uniformly encapsulated in the Gtn–HPA fibers. The cell-embedded tubular structures can be used to generate small vascular grafts. Cell encapsulation in the solid fibers (e.g., alginate fibers) was reported, with the achievement of cell density regulation and a uniform distribution of cells, both of which are important challenges for tissue engineering. Cell seeding and culturing on the fibers provide another important application. Cells seeded on PLGA-based fibrous structures showed behavior that depended on the curvature (i.e., on the diameter) of the fibers.[4] In the body, cells are exposed to structures that are curved in 3D, and their proliferation and shape are usually determined by the 3D extracellular matrix (ECM). Therefore, rectangular grooved structures provide only a limited understanding of cellular behavior on the 3D ECM. Microfluidics-based fibers provide excellent microenvironments for investigating the topographical effects (especially the effects of 3D curved structures) on cellular behavior. Fibroblast cells seeded on PLGA fibers demonstrated that these cells were aligned for fibers of diameters less than 20 μm. As the diameter of the fiber increased, the cells lose their orientations, and their shapes resemble those of cells cultured in Petri dishes. The guided culturing of neuron cells may be achieved along the length of the fiber, and one may envision that a bundle of fibers may enable the preparation of neuron bundles for the regeneration of nervous system components. Chitosan solid fibers were fabricated using microfluidic channels, although chitosan is mechanically weak.[20] Interestingly, the seeding of HepG2 cells (human hepatocellular carcinoma cells, which are a typical *in vitro* model for human hepatocytes) on fiber surfaces, followed by culturing for five days, produced spheroid-shaped cells aggregates on the fiber. Chitosan is a suitable scaffold material for mimicking liver tissue because its structure is similar to that of glycosaminoglycans (GAGs), which form the natural components of the liver ECM. Accordingly, the seeded cells adhered to the fiber surfaces, demonstrating the formation of spheroids. The functions of HepG2 cells grown on fibers were tested by measuring albumin secretion and urea synthesis. Recently, 3D coaxial flow by means of a hillock structure in a microfluidic focusing channel was used to produce core-shell alginate microcapsules.[21] This method was used to effectively encapsulate P19 embryonic carcinoma cells, with the formation of uniform massive embryoid bodies (EBs) for stem cell differentiation studies. This method provides an appropriate microenvironment for the formation of uniformly sized EBs, and may effectively replace the conventional hanging drop method as it is a simpler process.

Biosensing and assays

The particles produced can be used as biosensing elements by immobilizing chemicals or biological materials within the fibers. Glucose-sensing particles were fabricated using a mixture of 4-HBA, glucose oxidase, and horseradish peroxidase (HRP).[22] Exposure of the particles to glucose converted the glucose to gluconic acid and H_2O_2 by a glucose oxidase-catalyzed reaction. Fluorescent resorufin was formed via an HRP-catalyzed reaction between H_2O_2 and amplex red. Multifunctional particles that were encoded with distinct regions for analyte and target capture were fabricated based on continuous-flow lithography, which combines particle synthesis, encoding, and probe incorporation into a single process.[23] Each particle assumed an extruded two-dimensional shape, the morphology of which was determined by the photomask structures and the chemistry of which was determined by the content of the co-flowing monomer streams. PEG was used as the backbone material and eliminated the need to "block" the surfaces after probe conjugation. PEG also acted as a transparent material to allow the transmission of fluorescent signals from both particle faces. This method has the advantage of simplicity, and particles with both a fluorescent lithographically encoded region and a probe-loaded region may be synthesized in a single step. Although the application of Janus particles to sensing is not yet popular due to limitations in sensitivity and specificity, the technique is advantageous in that the encoding of arrayed biomolecules is straightforward. For more information on microfluidic glucose sensors, see Chapter 3.

Stimuli-responsive microactuators

Recently, new delivery devices that mimic the actuating processes of living systems, which respond to environmental changes, such as temperature, pH, light, electric fields, chemicals, glucose, antigens, and ionic strength, have been developed for biological and medical applications. 4-HBA microparticles were incorporated into microchannels for the fabrication of a pH-responsive delivery system (Fig. 8).[24] This approach has two advantages over conventional *in situ* polymerization methods: the fabrication process is simplified, and the response time is shortened by sensitivity to ionic diffusion in

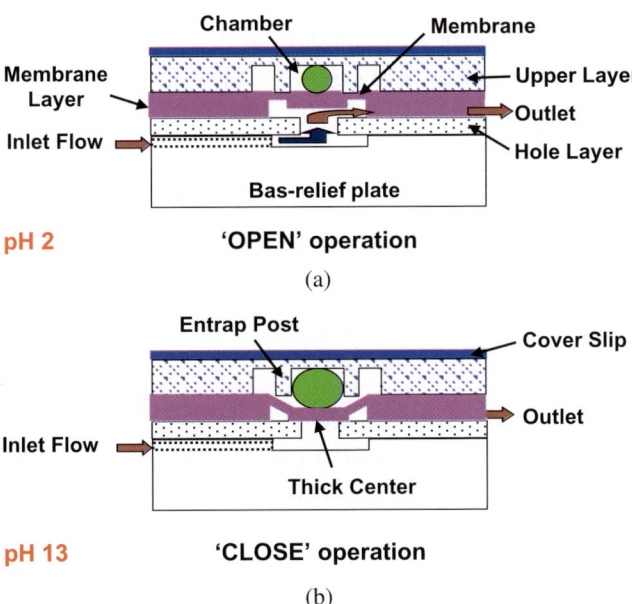

Fig. 8. Schematic of a microvalve employing pH-responsive microspheres. (a) Opening of the valve by acid (pH 2)-based shrinking of microspheres and (b) closing of the valve by base (pH 13)-based swelling of microspheres. (Adapted from Ref. 17. Copyright 2006, IOP Publishing Ltd.)

all directions. The parallel positioning of microvalve arrays that incorporate pH-responsive microspheres and the use of new multiplexing technologies that rely on multiphase laminar flow permits control over multiple microvalves without using complicated peripheral electronic devices.[25] These stimuli-responsive hydrogel-based chips may be used to deliver small amounts of drugs or molecules in response to the body conditions. This approach provides a useful tool for developing implantable drug delivery systems responding to body conditions.

Conclusions

This chapter introduced the microfluidic fabrication of diverse structures, including microscale particles, fibers, tubes, and capsules. Using the simple system and process, diverse microscale structures could be generated for use in a wide range of potential applications. Arrayed channel fabrication permits the mass production of diverse structures for realizing a factory on a table. The microstructures can be functionalized by immobilizing biological or chemical materials either within the structures or on the surfaces, and these functionalized structures could be used as scaffolds for cell positioning or for biosensors that detect multiple targets. Due to the large potential for such applications, the technology described in this chapter will likely be widely applied in a range of scientific and engineering fields.

References

1. Shah, R.K., Shum, H.C., Rowat, A.C., Lee, D., Agresti, J.J., Utada, A.S., Chu, L-Y., Kim, J.W., Fernandez-Nieves, A., Martinez, C.J., Weitz, D.A., Designer emulsions using microfluidics. *Mater Today*, 2008, **11**: p. 18–27.
2. Seiffert, S., Weitz, D.A., Microfluidic fabrication of smart microgels from macromolecular precursors. *Polymer*, 2010, **51**: p. 5883–5889.
3. Tumarkin, E., Kumacheva, E., Microfluidic generation of microgels from synthetic and natural polymers. *Chem Soc Rev*, 2009, **38**: p. 2161–2168.
4. Hwang, C.M., Park, Y., Park, J.Y., Sun, K., Khademhosseini, A., Lee, S.H., Controlled cellular orientation on PLGA microfibers with defined diameters. *Biomed Microdevices*, 2009, **11**: p. 739–746.
5. Steinbacher, J.L., TylerMcquad, D., Polymer chemistry in flow: New polymers, beads, capsules, and fibers. *J Polym Sci Part A: Polym Chem*, 2006, **44**: p. 6505–6533.
6. Shin, S.J., Hong, J.S., Lee, K.-H., Lee, S.-H., Oil-free generation of small polymeric particles using a coaxial microfluidic channel, *Langmuir*, 2009, **25**(20): p. 12361–12366.
7. Dendukuri, D., Pregibon, D.C., Collins, J., Alan Hatton, T., Doyle, P.S., Continuous-flow lithography for high-throughput microparticle synthesis. *Nat Mater*, 2006, **5**: p. 365–369.

8. Chung, S.E., Park, W., Shin, S., Lee, S.A., Kwon, S., Guided and fluidic self-assembly of microstructures using railed microfluidic channels. *Nat Mater* 2008, 7: p. 581–587.
9. Shin, S.J., Park, J.Y., Lee, J.Y., Park, H., Park, Y.D., Lee, K.B., Hwang, C.M., Lee, S.H., 'On the fly' continuous generation of alginate fibers using the microfluidic device, *Langmuir*, 2007, **23**: p. 9104–9108.
10. Kenis, J.P.A., Ismagilov, R.F., Whitesides, G.M., Microfabrication inside capillaries using multiphase laminar flow patterning. *Science*, 1999, **285**: p. 83–85.
11. Utada, A.S., Lorenceau, E., Link, D.R., Kaplan, P.D., Stone, H.A., Weitz, D.A., Monodisperse double emulsions generated from a microcapillary device. *Science*, 2005, **308**: p. 537–541.
12. Jeong, W.J, Kim, J.Y., Kim, S.J., Lee, S.H., Mensing, G., Beebe, D.J., Hydrodynamic microfabrication via "on the fly" photopolymerization of micro scale fibers and tubes, *Lab Chip*, 2004, **4**: p. 576–580.
13. Kang, E., Shin, S-J., Lee, S-H., Novel PDMS cylindrical channels that generate coaxial flow for the fabrication of microfibers and microparticles, *Lab Chip*, 2010, **10**: p. 1856–1861.
14. Chu, L-Y., Utada, A.S., Shah, R.K., Kim, J-W., Weitz, D.A., Controllable monodisperse multiple emulsions, *Angew Chem Int Edn*, 2007, **46**: 8970–8974.
15. Shepherd, R.F., Conrad, J.C., Rhodes, S.K., Link, D.R., Marquez, M., Weitz, D.A., Lewis, J.A., Microfludic assembly of homogeneous and Janus colloid-filled hydrogel. *Langmuir*, 2006, **22**: p. 8618–8622.
16. Lee, K.H., Shin, S.J., Park, Y., Lee, S-H., Synthesis of cell-laden alginate hollow fibers using microfluidic chips and microvascularized tissue-engineering applications, *Small*, 2009, **5**(11): p. 1264–1268.
17. Oh, H.J., Kim, S.H., Baek, J.Y., Seong, G.H., Lee, S.H., Hydrodynamic micro-encapsulation of aqueous fluids and cells via 'on the fly' photopolymerization. *J Micromech Microeng*, 2006, **16**: p. 285–291.
18. Tan, W-H., Takeuchi, S., Monodisperse alginate hydrogel microbeads for cell encapsulation. *Adv Mater*, 2007. **19**: p. 2696–2701.
19. Hu, M., Deng, R., Schumacher, K.M., Kurisawa, M., Ye, H., Purnamawati, K., Ying, J.Y., Hydrodynamic spinning of hydrogel fibers. *Biomaterials*, 2010, **31**: p. 863–869.
20. Lee, K.H., Shin, S.J., Kim, C-B., Kim, J.K., Cho, Y.W., Chung, B.G., Lee, S-H., Microfluidic synthesis of pure chitosan microfibers for bioartificial liver chip, *Lab Chip*, 2010, **10**: p. 1328–1334.
21. Kim, C., Lee, K.S., Bang, J.H., Kim, Y.E., Kim, M-C., Oh, K.W., Lee, S.H., Kang, J.Y., 3-Dimensional cell culture for on-chip differentiation of stem cells in embryoid body, [online].
22. Jeong, W.J, Kim, J.Y., Choo, J.B., Lee, E.K., Beebe, D.J., Seong, G.H., Lee, S.H., Hydrodynamic fabrication of microparticles using photo-polymerization and immiscible liquids in microfluidic systems. *Langmuir*, 2005, **21**: p. 3738–3741.
23. Pregibon, D.C., Toner, M., Doyle, P.S., Multifunctional encoded particles for high-throughput biomolecule analysis. *Science*, 2007, **315**: p. 1393–1396.
24. Park, J.Y., Oh, H.J., Kim, D.J., Baek, J.Y., Lee, S.H., A polymeric microfluidic valve employing a pH-responsive hydrogel microsphere as an actuating source, *J Micromech Microeng*, 2006, **16**: p. 656–663.
25. Liu, C., Park, J.Y., Xu, Y., Lee, S.H., Arrayed pH-responsive microvalves controlled by multiphase laminar flow, *J Micromech Microeng*, 2007, **17**: p. 1985–1991.

Chapter 13

Microfluidics as a Promising Tool Toward Distributed Viral Detection

*Elodie Sollier and Dino Di Carlo**

Department of Bioengineering
Henry Samueli School of Engineering and Applied Science
University of California, Los Angeles
Los Angeles, CA 90095, USA

Introduction

Viruses are ubiquitous, occuring naturally in aquatic environments and in the air, but also introduced through human activities. Viruses can be transmitted via various routes, such as sneezing, direct or sexual contact, and contaminated food or water, and consequently, cause worldwide outbreaks. Some of the most widely known historic pandemics caused by viruses include smallpox in Native American populations, HIV, viral hemorrhagic fevers like Ebola, and more recently avian influenza and SARS. According to the Global Health Program of the Bill & Melinda Gates Foundation,[1] enteric viruses, HIV, viruses causing influenza and pneumonia, and other neglected tropical viruses such as papillomavirus or viruses causing dengue and yellow fever are considered as the leading causes of childhood deaths, especially in the developing world. These viruses should now be considered priority areas of focus[2] for the global health community, with sustained attention required to tackle the lack of effective, fast, and cheap diagnosis. Indeed, new tools to diagnose these viral infections are critically needed, first to help health care providers to

---

*Corresponding author. Email: dicarlo@seas.ucla.edu

prescribe more effective and appropriate treatments the and second, to allow the early detection of potential infections and prevent pandemics.[3,4]

Viruses are also an established cause of cancer. For example, the human papillomavirus is known to cause cancers of the cervix, skin, anus, and penis, whereas the Epstein–Barr virus[5] and hepatitis B and C viruses[6] may cause lymphoma and liver cancer, respectively. Recent studies have also investigated the role of viruses as the causative agent in some neurological diseases, such as multiple sclerosis and chronic fatigue syndrome. Finally, the ability of viruses to cause pandemics and the technical possibility to recreate them in laboratories progressively lead to the fear that viruses could be weaponized for biological warfare.[7] Two of the six high-priority pathogens defined by the Center for Disease Control and Prevention are viruses, i.e., smallpox and the viruses causing hemorrhagic fevers.[8]

In addition to their obvious importance as causes of diseases, viruses are also extremely useful as research tools in various engineering processes.[9] As vectors for introducing genes into cells, viruses are used in medicine to make cells produce a foreign substance or in gene therapy or also in pharmaceutics as an alternative to antibiotics.[10–12] Due to the well-defined functional groups on their surface, viruses can be used as new-generation organic nanoparticles for surface modification.[13–15]

Thus, viral diseases are a worldwide health issue, and viral detection in clinical and environmental samples is of great interest not only in medicine and public health, but also in bioindustry and biodefense.

Viruses, Viral Samples, and Challenges of Viral Detection

Viral particles and their components

A virus is a small infectious agent that can replicate only inside other host organisms, from animal and plants to bacteria.[16] A complete virus particle, known as a virion, consists of genes, single- or double-stranded depending on the virus type, surrounded by a protective coat called a capsid. The capsid is formed from protein subunits called capsomers, all of them encoded by the viral genome. A great variety of genomic structures can be seen among viral species (Fig. 1). Briefly, DNA viruses, which have a DNA genome, replicate in the host cell's nucleus and depend on the cell's RNA processing machinery. RNA viruses, which have a genome already encoded in RNA, replicate in the host cell's cytoplasm. These reverse-transcribing viruses, which have RNA genomes, use a DNA intermediate to replicate. The capsid containing these enclosed nucleic acids defines the nucleocapsid. In addition, viruses can have

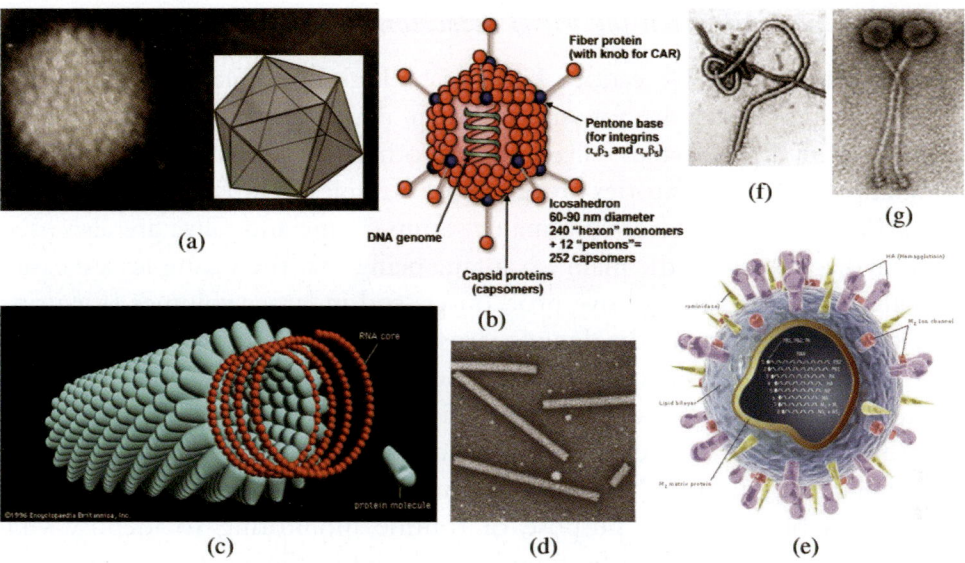

Fig. 1. Examples of viruses, with different shapes and genomic contents. (a) and (b) Adenovirus is an icosahedral DNA virus. (c) and (d) Tobacco mosaic virus has a helical nucleocapsid for a rodlike shape. Full length is 17 × 275 nm. (e) Influenza is a spherical RNA virus, surrounded by a lipid envelope containing membrane proteins. Ebola (f) and bacteriophage (g) have more exotic shapes. Ebola is a filamentous virus. (b) is extracted from Ref. 18; (c) is adapted from Encyclopaedia Brittanica, Inc.; (d) is extracted from Ref. 19; (e) is extracted from Ref. 17; and (f) is extracted from Ref. 20.

an external membrane or envelope, consisting of a phospholipid bilayer derived from the host cell membrane, but also including viral glycoproteins (Fig. 1(e)).[17]

The nucleocapsid arrangement is the basis for morphological distinctions and gives virions widely varying shapes. Nucleocapsids are mainly icosahedral[18] with at least 20 identical equilateral triangles faces, for quasi-spherical virions (Figs. 1(a) and 1(b)) or helical, for rod-like virions (Figs. 1(c) and 1(d)).[19] More unusual and complex morphologies are also possible (Fig. 1(f) and 1(g)), like the Ebola virus[20] or the bacteriophage, which is characterized by an icosahedral head attached to a rodlike tail. Virions have different surface charge and their size usually ranges from 20 to 400 nm. As a reference, bacteria are usually sized from 300 nm to 10 μm, whereas eukaryote cells are typically 10 to 100 μm across.

Thus, viral particles have various shapes, dimensions, and charges, but also different genomic and protein contents. These specificities make viral detection a challenging process without any universal solution, but often is required to be adapted and personalized for each virus considered.

Samples classically used for viral detection

Viral diagnostics involve various kinds of biological samples. Blood is obviously the sample of choice for viral pathogen detection, for example for HIV diagnosis[21] or Epstein–Barr virus[5] among others. To avoid blood cell contamination, many laboratories process blood samples to directly detect viruses like HIV[22] or dengue[23] from plasma or serum. Urine and saliva are also used for viral diagnosis with the main advantage being that these samples are easier to collect, without any invasive procedures, and in larger volumes (for urine samples). Viruses such as the dengue virus[24] and some members of the herpes virus family (cytomegalovirus (CMV), chicken pox, mononucleosis, and genital herpes) have been successfuly isolated from these samples.[25,26] Detection of viruses in urine can also reveal kidney diseases and renal injury.

Viral detection for public health also involves the analysis of environmental samples but with the purpose of routine monitoring to identify viral contamination. Virions can be found suspended in the air, for example, the influenza A viruses H5N1 and H1N1, which are highly contagious airborne diseases. This is the case particularly in confined areas with high occupant density, such as public transportation systems, airports, hospitals, and nurseries. Drinking water is also routinely monitored for viruses, especially for enteric viruses (like enteroviruses, rotaviruses, hepatitis A virus, and noroviruses) that infect the gastrointestinal system and are capable of causing a wide range of illnesses (gastroenteritis, paralysis, and meningitis). Because of their potential public health impact, these enteric viruses are on the U.S. Environmental Protection Agency List 2 for regulatory consideration for drinking water.[27] For the same reasons, sea, swimming pool, and river water are also objects of analysis for the detection of enteroviruses and polioviruses in order to prevent such outbreaks.[28,29]

In conclusion, viral detection for health diagnosis involves many different samples, including not only different bodily fluids, but also environmental samples.

Challenges of the sample matrix

Whatever the nature of the considered viral sample, sample preparation problems exist and make viral detection a challenging process. First, the sample matrix is highly variable. For clinical samples, cellular and viral composition and concentration vary with the patient, the sex, and the pathology. For environmental samples, composition, pH, sediment level, and salinity vary with the site, the type of water, and the weather. For airborne samples, an additional step is required for the long-duration collection of air particles and their transfer into a liquid sample, generally processed through electrohydrodynamic-collecting devices.[30,31] Second, the sample matrix is complex. Many larger components mask the targeted smaller viruses, like in blood, where viruses are among numerous cells and bacteria, or waterborne viruses among solid particles and parasites. Many potential contaminants for viral detection by molecular methods are also present, like humic and other organic substances in waters, known to inhibit RT-PCR enzymes, or hemoglobin in blood, known to contaminate PCR. Third, the sample has a low viral concentration (for example, 10^2–10^6 HIV virions per mL of plasma).[22] For all these reasons, viral detection is a challenge in terms of sensitivity and specificity.

To tackle these difficulties of viral samples, typical viral detection involves large volumes of sample and several steps of concentration to prepare the sample before an effective detection. The classical steps of viral sample preparation are summarized in the schematic of Fig. 2. After sampling, a first concentration step allows the pretreatment of the crude sample, generally prefiltration to remove particles larger than 300 μm. A second step of virion concentration purifies the targets but not necessarily specifically. An elution step allows the suspension of the viruses in a smaller volume, with optional stabilization procedures, until their final detection.

Thus, whatever the nature of the considered viral sample, viral detection is a multi-step and tailored process, consequently making the development of low-cost, point-of-care viral assays challenging.

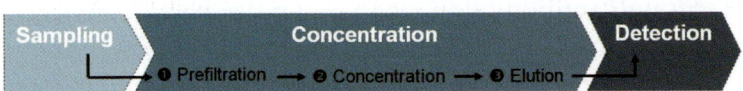

Fig. 2. Schematic of the main preparation steps for viral samples, from sampling to viral particle detection.

Macroscale Process

Methods of concentration

The concentration of viruses in biological samples generally consists of successive ultrafiltrations and removal of all cells and contaminants by centrifugation. For environmental samples, such as air and especially water, sample volume is more considerable and requires a higher concentration ratio. Ultrafiltration methods function by viral adsorption on charged membranes followed by elution.[27,32] Indeed, viruses are negatively charged in natural water environments and can be consequently adsorbed on numerous matrices by electrostatic and hydrophobic interactions.[27] Considering the example of contaminated water, filter membranes can be negatively (like HA Millipore) or positively (like NanoCeram) charged, more adapted to sea water or drinking water, respectively. These charged membranes seem efficient, but are expensive and require an adjustment of the protocol to each targeted virus, such as the adsorption time, the pH and temperature, or even the elution protocol. Besides, such ultrafiltration membranes require two steps of treatment (adsorption of viruses and elution in smaller volume), long elution periods, and preliminary prefiltration steps.[27] Vortex flow filtration and tangential flow filtration are promising alternatives to these adsorption–elution techniques, based on a flow pattern forcing water through a cylindrical filter with pressure, while keeping and retaining particles from filter pores to avoid clogging. These approaches are cheaper and less time-intensive, allow minimal manipulation of sample, under natural pH, and elution steps are not necessary. Thus, macrotechniques of concentration are efficient but time-consuming, require multiple steps, and above all are virus- and sample-dependent.

Methods of detection

Direct detection of viral particles

The traditional gold standard methods for pathogenic viral detection are based on the direct identification of viral particles in cell cultures.[33] For example, the hemagglutination assay consists of the agglutination of viruses with themselves through antibody bridges that increase particulate size and allow direct detection. The plaque assay allows the quantification of the number of infectious viral particles by culturing a dilute sample of viral particles on a plate covered with host cells and then counting the number of local lesions developed, called plaques.[16] Cell culture-based techniques, traditional and efficient, were the most widely used methods before the development of more precise and efficient molecular methods.[34–36] Indeed, cell culture-based assays are labor-intensive, time-consuming, expensive, and require considerable technical expertise. These approaches are also impractical for in-field monitoring, because the techniques require large sample volumes and days to weeks for results. Finally, such assays cannot be applicable to some virus types, such as noroviruses, which do not grow in tissue culture. However, numerous optimizations of cell culture formats and technologies have improved these approaches,[35,37] especially in reducing the average time for virus detection from five to 10 days to only 24 to 48 h for many viruses. Also, protocols are being developed to make them more efficient and less technically demanding.[36] Consequently, cell cuture-based techniques still remain the method to which all others are compared.[38]

Detection of viral molecular fingerprints

Nowadays, the most common laboratory approach for viral assays consists of the detection of a virus' molecular fingerprints, including viral proteins or nucleic acids. These various methods make possible more rapid viral detection directly on the sample, not requiring the cultivation of the virus on host cells, and enabling detection even for viruses that are no longer infectious.

Detection of anti-viral antibodies

A common strategy is to analyze the host organism's response to the virus, especially the virus-specific antibodies produced by the immune system after contact with the virus. Such indirect techniques, named antibody serology, have become simple and mature, commonly used to diagnose viral infection. Antibodies are detected via their binding to antigens grafted on surfaces, with readouts of such antibody/antigen complexes by fluorescence assays, enzyme-linked assays, or Western blotting.[7,39,40] Due to the speed of the antibody/antigen recognition event, such assays are fast (usually requiring less than one day), sensitive, and specific, but because of the time required after exposure for the body to secrete antibodies directed against the viral antigens, this indirect approach is not adapted for acute viral infections. For such reasons, serology is also complicated by difficulties in distinguishing recent from older infections. However, the detection of virus-specific antibodies allows the diagnosis from a single specimen, and is routinely used to detect CMV and hepatitis A and B viruses among others, or for certain chronic infections such as HIV,[40,41] in which the presence of any antiviral antibodies is always indicative of the current infection.

Detection of viral antigens

Viral antigen detection corresponds to the detection of the specific antigen present on viral surfaces. Methods are approximatively identical to those used for antibody serology, but also electrochemiluminescence and radioimmunoassays are employed for readout. Detecting viral protein antigens generally requires an additional step to lyse intact virions or host cells to extract the target protein they contain. Viral antigen detection suffers from a lack of sensitivity and low detection limits,[36,42] but allows quick diagnosis (minutes to hours) and requires less technical expertise, making it a solution of choice for biological warfare agent detection.[7,43,44] In addition, antigen detection is useful for viruses that do not proliferate in standard cell cultures. Examples of common targets for such approaches are influenza and adenoviruses.[43–46]

Detection of viral genomes

Diagnosis based on viral DNA and RNA amplification (by PCR and RT-PCR) and detection are now the best alternatives to classical cell culture and currently the most sensitive, fast, and specific approaches for viral detection and differentiation of virus subtypes.[33,35,47] However, these approaches generally require high levels of purification of the virus before RNA extraction to avoid any contamination, and obviously virus lysis steps, making any integration and automation of such processes challenging. Molecular methods also need high-end laboratory equipment, skilled and dedicated technicians, and cumbersome infrastructure, leading to higher operating costs. Many laboratories have developed their own "home-made" PCR assays as few assays are calibrated and officially licensed to be used in routine diagnosis (such as the Amplicor HIV-1 Monitor assay for HIV RNA from Roche Molecular Systems, or the capture assays for CMV and human papillomavirus from Digene Corporation). With recent optimizations, detection is possible in parallel for different viruses with multiplex techniques.[48] With "real-time PCR," a fluorescent signal is generated as PCR occurs and detected by automated optical systems. This combination and automation of nucleic acid amplification and signal detection significantly reduces the time required for detection (30 min for the RT-PCR instrument, Light Cycler, Roche Diagnostics), decreasing the probability of contamination.[5,49,50]

Thus, multiple methods are used for viral diagnosis, not only varying by their sensitivities and specificities, but also by their cost, the processing time, and the technical expertise they require. It can be concluded that no single approach is optimal or even sufficient for detecting all viruses in all samples, and specific combined approaches are suited for each given viral pathogen.

Toward viruses-on-a-chip

The quality and performance of these macroassays vary widely depending on the lab and protocols used, hence the standardization of these assays is needed. To tackle this variability issue, many different automated systems recently aim to integrate the entire process, such as the GeneXpert from Cepheid or the Viper from Becton Dickinson Diagnostics, which brings together nucleic acid isolation, amplification, and detection. However, the need for technical skill remains, both in specimen manipulation and in the monitoring of instrument performance, which despite continuous technical improvements, still limits the broad surveillance applications of viral assays. The diagnosis of viral diseases remains cumbersome. Thus, there is an obvious need for "viruses-on-a-chip," a miniaturized system that would bring viral detection from the centralized laboratory to the field, greatly enabling homeland and public health surveillance, and also the detection of biological warfare pathogens. Such a device should be usable by not only non-highly skilled technicians, but also physicians, nurses, or caregivers at the point of care; they should operate with low reagent consumption, cost-effectively, and in a small transportable platform to permit viral detection in mobile and rural clinics.

In parallel, progresses in micro- and nano-technologies make conceivable the whole integration of viral diagnosis on chip. First, the microfluidics community has progressively increased its fundamental understanding of processing fluids and particles down to the nanoscale,[52–55] whereas methods of microseparation and detection have increased in sensitivity and decreased in the size of targets.[52,56,57] Second, techniques of fabrication have become more precise, allowing for easier scaling down to the nanoscale.[58] The goal here is to assess these promising existing techniques toward viral detection on chip.

Viral Detection on Chip

Molecular detection of viruses on chip

Schulze et al.[57] and Cheng et al.[59] have presented all recent integrated devices for molecular viral detection, especially sample-to-answer systems.

Detection of anti-viral antibodies

Integrated serological tests suffer from the same drawbacks of macroscale indirect methods, i.e., the difficulties in detecting recent infections, limiting their application to some well-defined applications. Nevertheless, several microdevices have been proposed, taking advantage of the progress in immunofluorescent and enzyme-linked assays, to allow antibody detection with an improvement of the detection limit and sensitivity, especially amperometric or piezoelectric immunosensors.[60,61] As explained by Cheng et al.,[59] membrane strip assays have also become popular for fast, simple, and sample-to-answer diagnosis (Figs. 3(a) and 3(b)). Consequently, many commercial kits are available for various viral infections such as hepatitis B,[62] dengue,[63,64] or HIV,

320 *Microfluidic Technologies for Human Health*

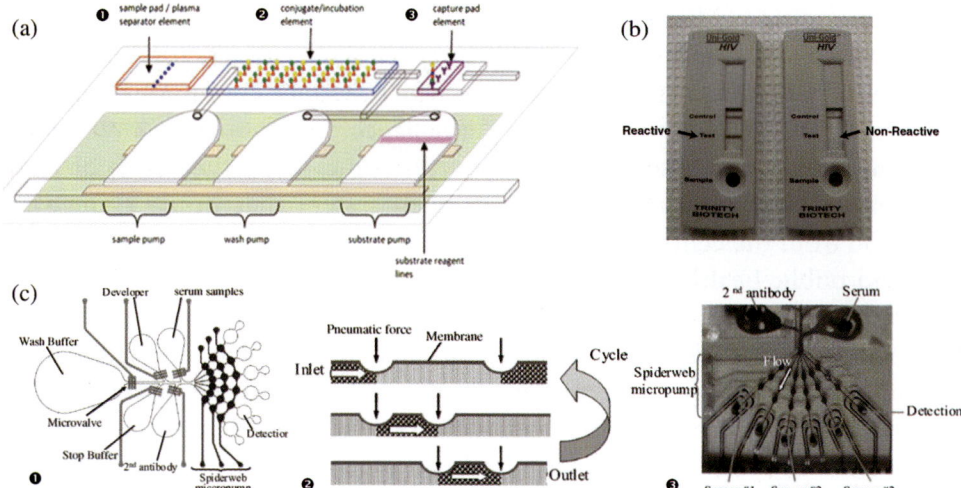

Fig. 3. Microsystems for viral molecular detection by serology. (a) Membrane strip immunoassay and lateral diffusion of target antibodies. Schematics of the Fluidics-on-Flex, from Epocal Corporation. ❶ The sample is deposited on the sample pad and moves forward by capillary force. ❷ The target antibodies present in the sample bind to the detection molecule and such complexes are carried towards the capture pad, composed of lines or spots coated with antigens directed against the target. ❸ Binding of antibodies + detection molecules on the capture pad leads to a visible color change. (b) Example of Uni-Gold Recombigen™ HIV test.[65] (c) To replace capillary strips or immunoassays with macroscale pumps, Wang *et al.* proposed a serological test regulated by automatic micropumps.[67]

including the OraQuick HIV and Uni-Gold Recombingen™ HIV tests.[64,65] The same principle can also be applied in microsystems, as in the study by Wang *et al.*,[67] where antigens against hepatitis C and syphilis were immobilized in microfluidic channels instead of a strip, with a flow control managed by pneumatic functions (valves and pump) (Fig. 3(c)). Such device allows for fast (20 min) and automated detection with multiplexing capabilities from a small amount of sample. Compared to strip approaches, this microfluidic device is (1) more sensitive, exhibiting a signal similar to conventional ELISA, (2) capable of quantitatively detecting the number of pathogens contrary to the YES/NO result from strips, and finally, (3) test reagents are driven into the chip just before the sample itself, avoiding reagent degradation and simplifying storage.

Detection of viral antigens

As for macroscale methods, miniaturized systems developed for antigen detection are generally similar to those proposed for antibody serology. Many microdevices operate using antigen detection on solid supports, like lateral diffusion strips with many commercialized examples available. For example, Smit et al.[67] compared the Binax NOW Flu assay and the Becton Dickinson Directigen Flu assay with macroscale immunofluorescence and viral culture, showing sensitivities from 53% to 59% for influenza A detection compared to 80% for reference methods. Similar conclusions are presented by Weinberg et al.[69] and De Lima et al.,[70] which confirmed that although viral antigen detection suffers from a lack of sensitivity, these assays are a solution of choice for the screening of outbreaks. Indeed, their fast, simple, and automated process may be an advantage in some settings as it requires no technical expertise. Examples of common targets for such approaches have been influenza and adenoviruses. In parallel with lateral diffusion solutions, more and more approaches integrate the use of magnetic beads coated with antiviral antibodies. The Abbott ARCHITECT, presented by Leary et al.,[71] illustrates such magnetic bead-based sandwich assays, where the target antigen (here associated with viruses transmitted by transfusion) is captured by magnetic beads coated with antiviral antibodies, bound to secondary chemiluminescent antibodies and detected after concentration with a magnetic field gradient. This device allows for the high-throughput and fully automated detection of viral antigens from serum or plasma, with a sensitivity of 99%, which is almost equivalent to nucleic acid testing.

Detection of viral genomes

As discussed previously, PCR amplification of portions of the viral genome is considered as the gold standard and the most sensitive approach for the diagnosis of many viral infections, but requires high-end and costly equipment and skilled technicians. There have been efforts to miniaturize and integrate RT-PCR amplification and detection on microchips, but to date, the lack of reliable microscale methods for extracting nucleic acid from virions remains an obstacle to viable point-of-care complete viral assays. Integrated methods of virion concentration, generally based on magnetic beads[72,73] or microfilters,[74] will be further presented later. Different approaches have been reported for RNA extraction: (1) after a preliminary lysis of virions by laser irradiation,[72] chemical lysis,[74] or with a high-electric field generated by nano-electrode arrays,[75] and (2) RNA targets are captured, concentrated, and isolated from contaminants by binding on silica surfaces[74] or on magnetic beads conjugated with DNA strands. Numerous chips have been proposed to integrate PCR on chip,[47,57,59,76] with droplets or continuous flow, modules of thermal cycling consisting in thermoelectric modules or infrared lamps. The detection of amplified DNA is commonly realized by capillary electrophoresis (CE) such as in Kaigala et al.[77] Other techniques are based on sandwich assays, where detection tags can not only use traditional fluorescent and enzymatic molecules, but also quantum dots and nanoparticles. Many innovative techniques for amplicon detection are presented and assessed in the review by Cheng et al.[59] There are only a few complete systems, including viral sample preparation, RT-PCR, target capturing, and detection on a single device. For example, Lien et al.[73] integrated in the same chip not only sample preparation and RT-PCR for detection of dengue and enteroviruses, but with external traditional detection. Lee et al.[78] combined RT-PCR and HIV enzymatic detection in less than one hour, but with sample preparation not included in this process, whereas Kaigala et al.[77] coupled PCR and CE for polyoma virus detection from urine. Some rare designs integrate all steps, such as the droplet device proposed by Pipper et al.[79] for avian influenza H5N1 detection, or the all-automated device by Pal et al.[80] Recent integrated systems like the GeneXpert (Cepheid) represent promising solutions toward the accessibility and utility of miniaturized viral molecular diagnostics. However, the use of cartridges pre-loaded with chemical reagents brings forward the problem of denaturation and shelf-life during transport and storage. Thus, in spite of continuous technical improvements in terms of speed, detection limit, accuracy, and cost-effectiveness, existing molecular assays require multi-step sample preparation, multiple temperature-sensitive reagents for amplification and detection, and/or highly skilled laboratory technicians, which limit the broad deployment of viral assays. For these reasons, a recent trend for viral integration involves exploiting direct virion detection.

Direct detection of viruses on chip

New methods are currently under development for the direct sensing of whole viral particles on chip. Due to their nanoscale size, most viral particles are too small to image with light microscopy, so other methods such as scanning and transmission electron microscopies are commonly used to visualize virion morphology.[81] Innovative optical methods developed to tackle these issues are assessed in the review by Myers et al.[82] Among them, some studies focused on chemiluminescence viral detection, where virus binding through their interaction with antibodies grafted on a substrate induces photochemical emission, directly or with enzymatic labeling. In the work from Yacoub-George et al.,[83] a chemiluminescent capillary enzyme immunoassay sensor is validated for the detection of bacteriophage M13 within 29 min, with no equipment required for excitation and only sensitive photomultiplier tubes (PMTs) as luminescence readers. Another attractive and highly sensitive

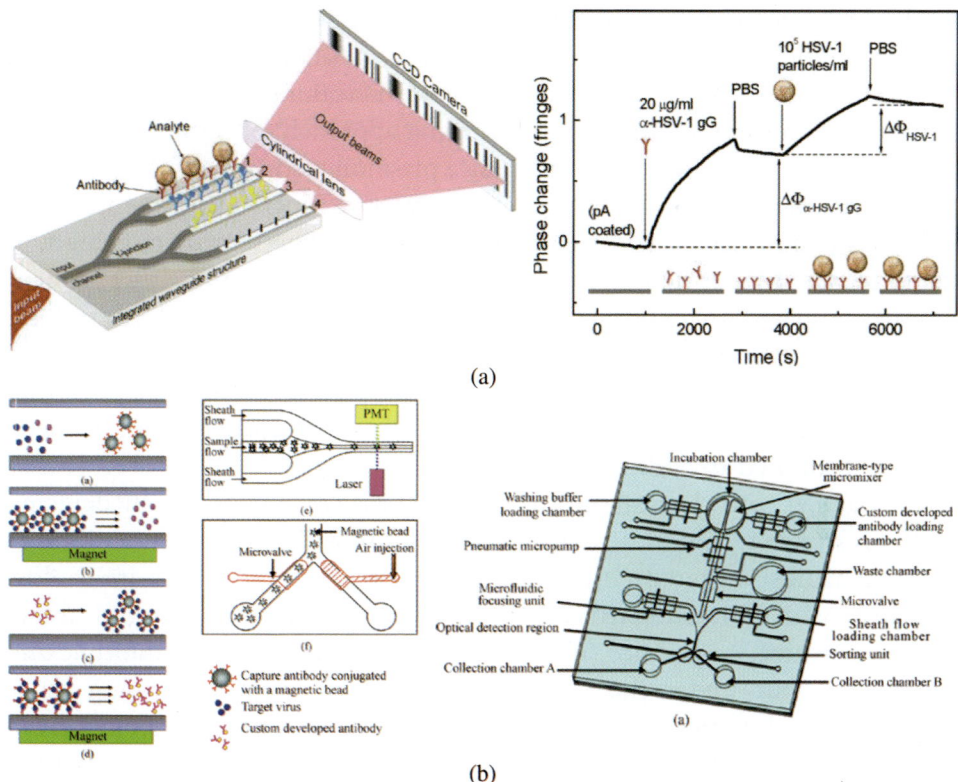

Fig. 4. Optical methods for direct virion detection. (a) Representation of the interferometric sensor proposed by Ymeti et al.[84] (left). The integrated optical sensor contains four channels; 1, 2, and 3 are the functionalized measuring channels, and 4 is the reference non-functionalized channel. When a virion binds to one channel, a phase change is measured (right). (b) Microflow cytometer using magnetic functionalized beads for viral detection on chip from Yang et al.[98] Schematic illustration of the microfluidic chip, integrating different functional components on a single chip. The incubation module involves a) mixing of viral samples with magnetic beads, b) the purification and concentration of the target viruses through a magnetic washing process, c) the staining of viruses/bead complexes with custom developed antibodies, and d) the washing of fluorescent unbound antibodies. The cytometry module allows e) the fluorescent detection of the bead/viruses passing through an optical detection region (laser and PMT) and f) the sorting of these complexes utilizing microvalves of the pneumatic flow sorting module.

approach for viral detection on chip is interferometric detection, where viral binding on a substrate causes a refractive index change along the optical path, resulting in a phase shift compared to the reference path. Such devices are composed of several optical waveguides, one non-functionalized waveguide being the reference, whereas three other waveguides are coated with different anti-viral antibodies to detect three different viral targets (Fig. 4(a)). This method has been successfully exploited[84,85] for the detection of herpes simplex and avian influenza viruses, respectively, with a sensitivity approaching unique virion, but signicant time is needed to adjust microfluidic channels and waveguides. Last but not least, another promising approach for optical viral detection on chip is detection based on surface plasmon resonance (SPR).[86] SPR is a label-free method for the detection of refractive index changes on a metal surface functionalized against targeted viruses.[87,88] The sensitivity is high, as demonstrated by Lei et al.[87] with an increase of more than one order of magnitude compared to conventional ELISA, for a real-time and continuous detection, but under the condition of complex and expensive equipment. Localized SPR (LSPR), based on the collective plasmon resonance occurring at the surface of metal nanoparticles instead of a metal-coated prism, has been recently proposed to address these issues.[89]

In parallel with classical optical methods, viruses are commonly detected by fluorescence techniques after specific staining with fluorescent antibodies directed against targeted viruses.[90] Here, fluorescent antibodies are in flow and help with virion identification. Progress in such technologies allows highly sensitive,[91] specific, and quantitative detection and multiplexing capabilities, but under the condition of purified samples. Among those fluorescent techniques recently reviewed by Schulze et al.,[57] we point out in particular laser-induced fluorescence (LIF), evanescent wave excitation, fluorescence resonance energy transfer (FRET), and fluorescence cross-correlation spectroscopy (FCCS). In view of the total integration of fluorescence detection, LEDs can be integrated on chip to induce fluorescence with the detection of fluorescent emission by CMOS sensors or silicon photodetector diodes.[44,92] To increase the fluorescence signal, the antiviral antibodies tend to be grafted on quantum dots[93,94] or fluorescent nano-/microbeads.[56,95,96] After their specific staining, fluorescent virions can be selectively detected by flow cytometry (FCM), where each viral passage event is counted. For example, Ferris et al. proposed their own integrated cytometer and validated it for the detection of adenovirus, influenza A, and respiratory syncytial virus, within one hour and with viral concentrations ranging from 5×10^6 to 5×10^9/mL.[97] More recently, Yang et al.[98] presented a miniature FCM integrating virion purification and concentration with magnetic fluorescent beads, and subsequent detection with micro-FCM and optical detection modules (Fig. 4(b)). Such a device was validated for dengue virus diagnosis within 40 min. Hence, integrated FCM allows the continuous and quantitative high-throughput detection of virions, but requires several steps of sample preparation target staining and also cumbersome optical equipment.

Alternative approaches for direct viral particle detection that enable accurate viral detection are electromechanical and electrical methods. Among electromechanical techniques, quartz crystal microbalance (QCM) techniques use piezoelectric materials, such as disks of crystalline quartz. The application of an external electrical potential to such a material induces an acoustic wave throughout the crystal, directed perpendicularly to the plate surfaces. Most studies use QCM as a mass detector. The quartz crystal surface, being coated with antibodies, viral binding on the piezoelectric material will result in an increase in mass that lowers the chip resonant frequency.[99] In Zuo et al.,[100] QCM allows the detection of SARS coronavirus in gas phase. In studies from Peduru Hewa et al.,[101] a QCM device is evaluated for the detection of influenza A and B viral. The coupling of viruses with nanoparticles results in a 10-fold increase in mass sensitivity for a detection limit comparable in sensitivity and specificity to macroscale methods. An original QCM mode presented in Cooper et al.,[102] and named rupture event scanning (REVS), involves increasing the voltage until the virions detach from the surface (Fig. 5(a)). The magnitude of the critical surface acceleration has been proven to depend on the mass of the particle and the properties of viral binding to the surface. Validated for the detection of herpes simplex virus, this method is rapid and quantitative, with a sensitivity approaching detection of a single virus particle, for a detection limit of 1,000/mL.

Microcantilevers work according to a similar principle. Based on the operation mechanisms and materials, microcantilevers can be classified into silicon-based, piezoresistive, capacitive, magnetoresistive, and piezoelectric microcantilevers. In static (or strain) mode, microcantilevers are coated with a gold film and antiviral antibodies and viral binding generates a strain on the microcantilever, the deflection of which is measured by piezoresistive or optical methods.[103] In vibrational (or resonant) mode, the microcantilever is actuated at a given resonant frequency and viral binding on the microcantilever decreases the resonant frequency. In Johnson et al.,[104] such microcantilevers are used for the detection of the vaccinia virus, with the future prospect of a direct and rapid detection of airborne viruses (Fig. 5(b)). A main drawback of this technique is generally its reduced efficiency in liquids, because viscosity degrades the quality factor of the resonator.[105,106] To address this challenge, Manalis et al. proposed an original approach, injecting the suspension inside a hollow resonator surrounded by vacuum, i.e., using microchannels embedded in cantilevers.[107,108] This alternative allows weighing single nanoparticles in water with sub-femtogram resolution.[106]

Electrical conductance methods[109] detect viral binding by mass or charge sensing. Semi-conducting nanowires or carbon nanotubes can be designed

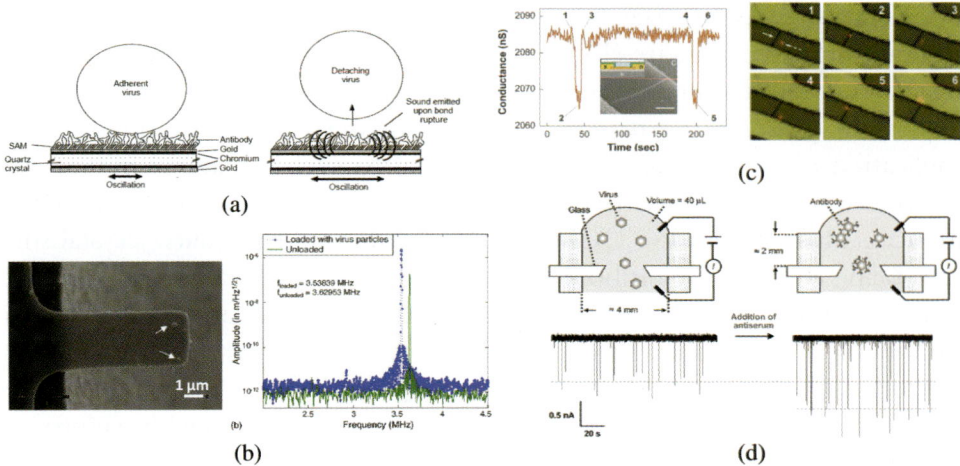

Fig. 5. Electromechanical and electrical methods for direct virion detection. (a) Quartz crystal microbalance (QCM) and rupture event scanning (REVS) mode.[102] QCM is composed of a disk coated with chromium, gold, a self-assembled monolayer (SAM), and antiviral antibodies. Applied voltage induces transverse oscillations, in which amplitude is increased leading to final surface/virus bond breakage. As demonstrated by Cooper et al.,[102] acoustic energy resulting from this viral detachment propagates on the disk surface and is detected by the QCM. (b) Microcantilevers driven by a piezoelectric ceramic from Johnson et al.[104] Scanning electron micrographs of two viruses on a cantilever and resonant frequency modification after viral binding. (c) Selective detection of single influenza A viruses on nanowires[109] by conductance changes associated with virion binding from Patolsky et al. In images 1–6, also indicated in the conductance data, the virion appears as a red dot, its position as a dashed arrow, and the position of the nanowire as a plain white arrow. (d) Resistive-pulse technique for unique virion detection as demonstrated by Uram et al.[111] Single virions passing through the pore cause a transient reduction in current (named resistive pulse or current spike). Binding of antibodies to virions increases their volume, leading to an increase in the peak amplitude detected.

as field-effect transistors, where charged virion binding produces a depletion or gain of nanowire charge (depending on the viral charge), measured as a conductance change (Fig. 5(c)). In Patolsky,[109] such a technique allows the direct and real-time electrical detection of influenza A with high selectivity. In Coulter methods, viruses are continuously injected in a narrow channel — or orifice — and increase the resistance measured through this orifice by their passage, the resistance change being directly proportional to the virion volume. This macroscale technique[110] has been recently miniaturized by Uram et al.[111] and An et al.,[112] where the passage of virions through submicrometric pores allows the detection of a 190-nm–large chlorella virus, and also the determination of virus size and the number of antibodies bound to the virus over time (Fig. 5(d)). Rapid, simple, continuous, and label-free, this highly attractive technique requires no immobilization of the virion and operates efficiently for viral concentration from 4.4×10^7 to 2.5×10^9/mL. However, effective purification steps are needed to isolate pure viral populations from the original sample.

Thus, many promising techniques — efficient and sensitive — have emerged for integrated and direct virion detection, but operate reliably only with pure and not real viral samples, i.e., without any other particles or chemicals as contaminants. Therefore, matching purification steps that can isolate pure viral populations from the original sample are critically needed toward complete integration.

Microfluidic Methods of Virion Concentration

Integrated ultrafiltration

Several approaches have been proposed to address the need for pure viral particles. Particularly, many recent studies tend to integrate on chip the principles of macroscale ultrafiltration, with various methods such as electrophoresis,[90] porous membranes, filtering pillars, or variable height channel filters combined with microfluidics. As an example of microfiltration, Yobas *et al.*[74] proposed a chip to filter viruses from whole blood and carry out the viral RNA extraction and nasba amplification on the same chip (Fig. 6(a)).[113] Based on silicon filters smaller than 1 μm, this chip seems efficient and advantageously presents an integrated protocol. Still, few details are given about the device. In a previous study by Zhang *et al.*,[91] an array of nanoscale fluidic chambers allows the concentration of dengue viruses in a femtoliter volume and helps to improve their fluorescent detection down to the nanomolar range. Recent work, taking advantage of micro-/nanofabrication advances, proposed original geometries[114,115] (Fig. 6(b)). Reichmuth *et al.*[116] also illustrated an example of a nanoporous membrane integrated on chip, in which size-exclusion properties are used for virion concentration (Fig. 6(c)). A UV-polymerizable polyacrylamide membrane, located in the loading channel, contains pores large enough to allow liquid to pass through, but is smaller than influenza virions. After virion concentration, the concentrated plug of virions is driven off the membrane by changing the voltage and sent toward another channel for further analysis. In parallel to passive microfiltration, integrated dielectrophoresis (DEP) is also an attractive approach as it may be able to differentiate viruses with different charges, although volume throughput may be an issue. The reader interested in learning more about DEP is directed to the comprehensive reviews of Gascoyne *et al.*[117] and

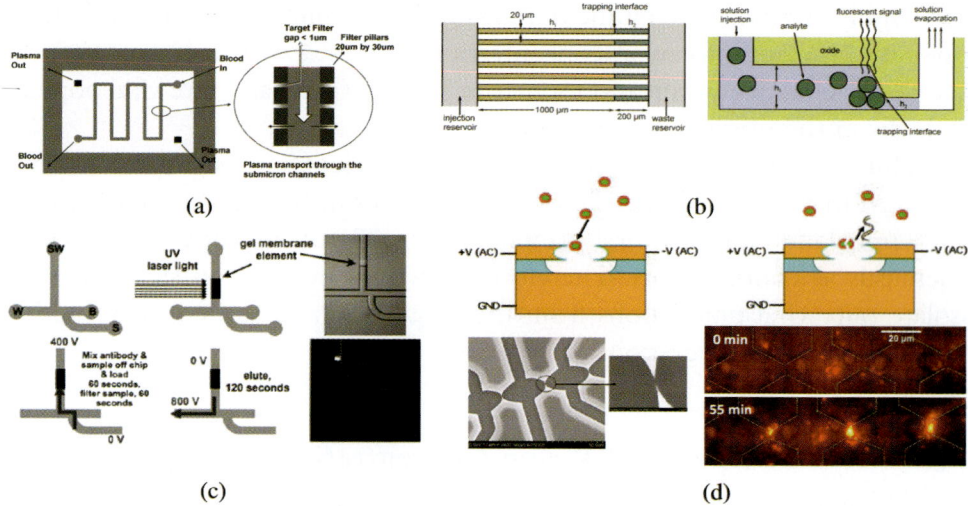

Fig. 6. Integrated ultrafiltration for direct virion detection. (a) Microfilter to separate plasma and viral particles from blood cells by Yobas et al.[74] (b) Schematic of the trapping device proposed by Hamblin et al.[115] Viral sample is injected through the filtering parts by capillary action, virions being trapped at the entrance of second thinner section. (c) Viral concentration *in situ* by porous membrane, as proposed by Reichmuth et al.[116] A polyacrylamide plug is formed in a glass channel by projecting a UV laser beam onto the acrylamide-containing channel, whereas the viral sample is injected by electrophoresis. (d) Capture and lysis of the vaccinia virus by silicon nanoscale probe array as demonstrated by Lien et al.[73] Schematic principle, SEM images, and progression of viral capture by ethidium bromide staining.

Hughes et al.[118] As an example, Park et al. integrated in SOI (silicon on insulator) wafers a nanoscale probe array for the DEP capture of vaccinia viruses and, with higher voltage and without any chemicals, the subsequent lysis of captured viral particles (Fig. 6(d)).[75]

Integrated ultrafiltration is an attractive technique for viral concentration, but these filters, with or without electrophoretic forces, suffer from the same downsides as macroscale filtration systems, especially due to the requirement of a difficult elution step to remove clogged viral particles from the membrane pores for further analysis.

Microfluidics and functionalized beads

Recently, functionalized magnetic bead-based separations have been used for viral concentration from bodily fluids and several successful attempts of miniaturization have been reported. Magnetic antibody-coated beads are micrometer-large beads with paramagnetic cores or nanometer-large beads with superparamagnetic cores, both coated with silica or polymers that are functionalized with antibodies to specifically target the viral antigens. Due to the specificity of the antibody/antigen interaction, this method is widely used and characterized by high separation efficiency, a high throughput, and a significant sample concentration, such as for dengue virus concentration.[73] Also utilized for HIV detection by Chen et al.,[22] magnetic nanoparticles allow an 80-fold concentration of virions at 10 µL/min. Viral binding to beads coated with antiviral antibodies is also a new preferential solution to push the limits of viral detection methods.[56,98,119] Magnetic PEI (polyethyleneimine) beads are another approach, commonly used to efficiently adsorb many model viruses.[120,122] However, several small viruses like hepatitis B cannot be concentrated by such a method, and the mechanism of viral adsorption by PEI beads, which is less specific and pH-dependent, is not elucidated yet.

Such bead techniques involve several steps, including the mixing of beads and the sample for target binding on the beads, the magnetic retention of the beads plus viral complexes, the final rinsing of the beads to wash off unbound components, and the elution into a smaller volume, making integration of all these steps in one chip challenging. This technique is not continuous and is hardly adaptable for the screening of larger fluid volumes, which are necessary for public health surveillance. Coupling of bead protocols with methods of continuous and integrated buffer exchange should solve that problem, like magnetic beads crossing successive laminar flows by Peyman et al.[122] or acoustic liquid exchange by Augustsson et al.[123] Furthermore, due to the antibodies used in these systems, which degrade rapidly if not refrigerated, magnetic bead-based techniques remain a short shelf-life technique. However, these techniques are still one of the sole approaches that allow a complete protocol, from virus concentration to detection, on an integrated chip. A recent alternative consists of the functionalization of the microfluidic channels on a fixed support.[21,124] Specific and efficient, this promising approach would allow viral extraction from of the initial sample matrix, with no need for initial concentration steps and a simplified protocol.

Microfluidics and continuous virion separation

From micro- to nanoscale bioparticles

Due to the small size of viruses, some techniques commonly used for the continuous separation of cells or bacteria on chip[125–127] are difficult to apply to viruses, whereas others that can be tuned over a range of sizes may be successfully applied on viruses in the future.

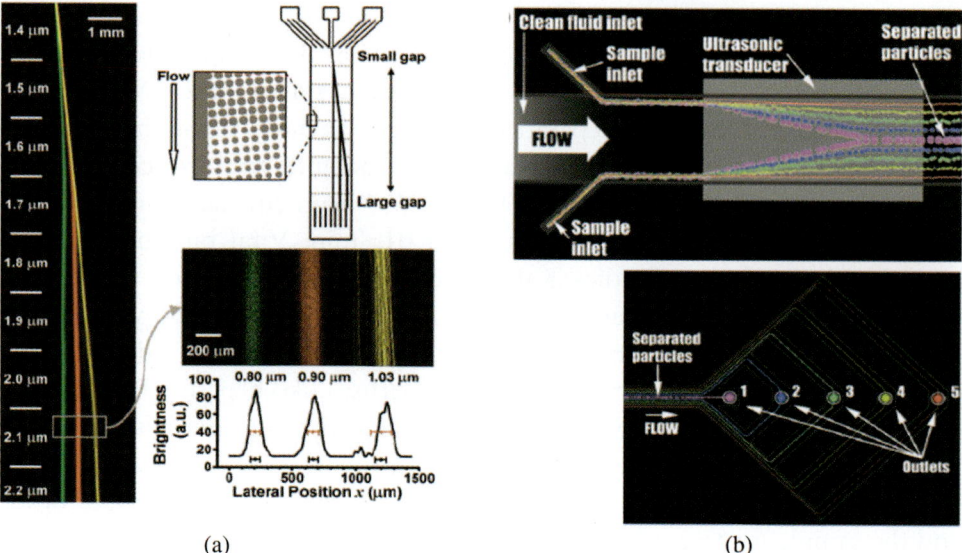

Fig. 7. Microfluidics and continuous virion separation. (a) Deterministic lateral displacement (DLD), by Huang et al.[131] High-resolution separation of 0.80 μm (green), 0.90 μm (red), and 1.03 μm (yellow) beads, with a design of varying gap size. (b) Free-flow acoustophoresis (FFA) by Peterson et al.[135] Illustration of size separation of particles in different pressure nodes and fractionation in different outlets.

For example, deterministic lateral displacement (DLD) uses flow through an array of microposts to modify particle trajectory and sort them by their size (Fig. 7(a)).[128–131] Especially, Huang et al.[131] separated microspheres with diameters of 0.80, 0.90, and 1.03 μm, with 10-nm size resolution, giving evidence that DLD may be soon applied for viral particle separation and in any case for crude viral sample and contaminant prefiltration.

Another novel method, named free-flow acoustophoresis (FFA), uses acoustic forces generated by high-frequency acoustic resonators to separate particles or cells based on their size and density (Fig. 7(b)).[132–134] The continuous separation in four different outlets of 2-, 5-, 8-, and 10-μm particles have been presented by Petersson et al.[135] in a 350-μm–large channel. Such sorting is fast, efficient, continuous, and represents an appealing solution for future viral separation.[129] The manipulation of medium density can also be combined to FFA to make contaminant separation more efficient, like for the separation of red cells, platelets, and leukocytes.[135]

In parallel, Di Carlo and others have recently shown that inertial focusing can separate and concentrate microparticles and cells in microchannels.[136–138] Briefly, this particle focusing is based on the superposition of (1) a shear gradient lift force, due to the shear rate of a parabolic flow that is directed toward

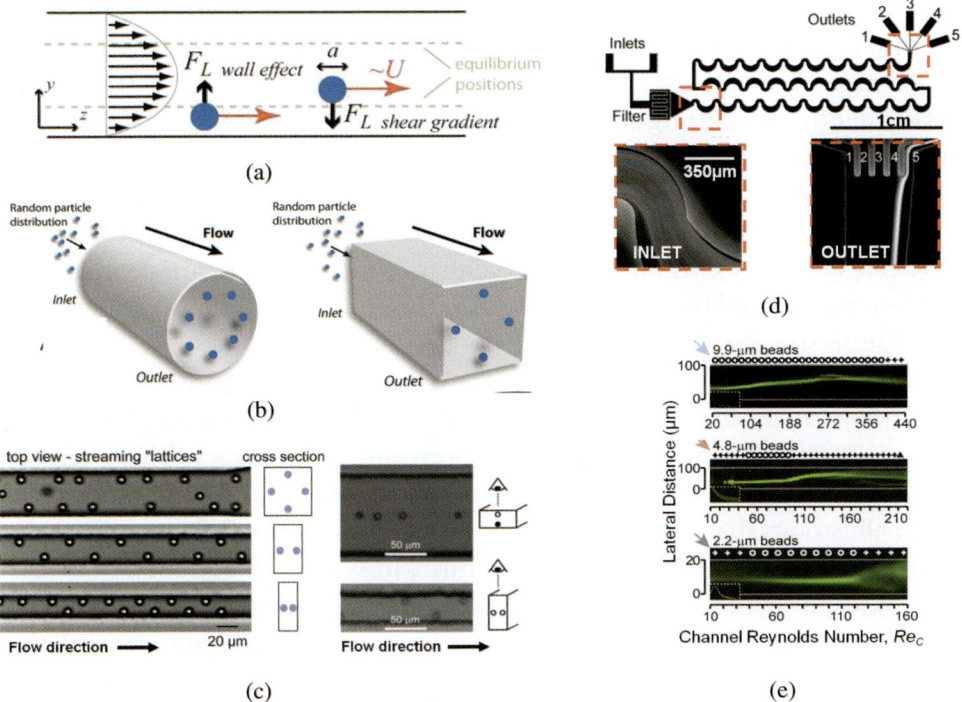

Fig. 8. Inertial microfluidics and continuous virion separation.[136,138] (a) Two lift forces perpendicular to the flow direction act on the particle to create its equilibrium position; a "wall effect" and a "shear gradient." (b) In a cylindrical or square channel, randomly distributed particles focus respectively to an annulus or to four equilibrium regions centered at the faces of the channel. (c) Aspect ratio effects on inertial focusing of 10-μm polystyrene particles.[140] (d) Curving focusing and concentration device consists of inlets, a coarse filter with a 50-μm cutoff, 62 asymmetric separation turns, and 5 outlets to collect the filtrate.[137] Fluorescent images are shown for 9-μm beads at the intlet and outlet. (e) Reynolds number–space map of particles focusing for 9.9-, 4.8-, and 2.2-μm polystyrene particles. Slices of fluorescent streak images are assembled over many flow rates to visualize particle focusing accuracy (o) as a function of fluid dynamic conditions.[140]

the channel walls, and (2) a wall effect lift force, due to reflection of a particle's wake that is directed away from channels walls (Fig. 8(a)). Consequently, dilute suspensions of particles in confined channel flows with finite inertia will migrate across streamlines to occupy positions of dynamic equilibrium located approximately halfway between the channel centerline and walls. In cylindrical channels, particles focus to an annulus within the channel cross-section.[139] In square or rectangular channels, inertial lift forces also focus particles, reducing these positions from an annulus to four precise locations, at a rate that depends on the particle and channel diameter (Fig. 8(b)).[136] By changing the symmetry of the system (i.e., for a very wide or very tall channel), focusing to single points in channels can be achieved (Fig. 7(c)) and has been successfully applied to blood cells for high-throughput cytometry.[136] Subsequently, by splitting the outlet channel in correspondence to the streamlines, focused particles can be collected in a smaller volume and be significantly concentrated (Fig. 8(d)).[136] By using channels with curvature, an additional drag force named Dean force balances these inertial lift forces and has been shown by Gossett et al.[141] to enhance the speed of particle migration to more stable equilibrium positions, and consequently to allow a faster focusing than in straight channels (Fig. 8(d)).

Inertial focusing has been demonstrated predominantly on 10- to 20-μm–large bioparticles and cells, but recent studies have elucidated particle size effects.[136,140,142,143] Figure 8(e) shows fluorescent streak images, which demonstrate how focusing evolves for three particle sizes. Particularly, the bottom image illustrates that accurate focusing is observed (o) for 2.2-μm particles in a 20-μm–wide channel into single streams and with high accuracy.[141] Bhagat et al. also presented the successful separation of 590-nm nanoparticles from 1.9-μm particles.[142] Thus, we and others experimentally demonstrated that inertial focusing can achieve particle separation over an order of magnitude of sizes, suggesting that inertial focusing could be scaled down further for nanosized particles. Furthermore, these results also confirmed that inertial focusing could be useful for contaminant filtration prior to entering a nanofluidic channel.

Inertial effects at the micro-/nanoscale appear extremely promising in addressing the viral concentration challenge, potentially enabling an approach that is (1) simple and requires no complex fabrication, (2) low-cost so that it can be widely deployed, with (3) high filtration efficiency (no contamination), (4) high throughput, resulting in significant viral concentration levels, and finally, (5) able to operate on large volumes of sample. For all these reasons, we anticipate that inertial focusing would critically address the current challenges for viral concentration from large sample volumes like water screening, but may be less useful for blood screening from small finger-prick volumes.

Therefore, techniques commonly used for continuous microparticle separation may be promising for viral concentration, and reducing the operational range to lower size limits of such methods is in progress (Table 1). However, by reducing the size of particles and channels, mechanisms governing these microscale techniques are expected to shift. This makes it difficult to predict the future operating space of each technique. In particular, the significance of Brownian motion in concentration accuracy is expected to increase. Another conceivable way of concentrating

Table 1. Viral detection, challenges, constraints, and microfludic approaches.

Challenges of viral detection	Consequences for macro-scale systems	Needs and constraints on micro-scale systems	Technologies
✓ virus variability - size - morphology - charge - content	✓ cumbersome equipments ✓ expensive equipments ✓ skilled technicians ✓ time-consuming process ✓ multi-step process ✓ sample-dependant	✓ transportable ✓ cheap ✓ usable by non-skilled persons ✓ less-time effective ✓ long shelf-life	❶ **Virion concentration** ✓ Microfiltration ✓ Dielectrophoresis ✓ Magnetic beads ✓ Deterministic lateral displacement ? ✓ Acoustophoresis ? ✓ Inertial microfluidics ?
✓ sample variability - pH - salinity - concentration - composition			❷ **Direct virion detection** ✓ Chemiluminescence ✓ Fluorescence ✓ Interferometry ✓ SPR
✓ low viral concentration ✓ many contaminants		✓ selectivity ✓ sensitivity ✓ throughput ✓ target stabilization	✓ µFlow cytometry ✓ Electromechanical ✓ Electrical ✓ Detection of anti-viral antibodies ✓ Detection of viral antigens ✓ RT-PCR and genome detection

virions, inversely to going progressively from macro to micro and nano, could be going from the physics investigated at the nanoscale and increasing operational limits up to the microscale.

Conclusion

Effective detection of viruses in clinical and environmental samples is of great interest in public health, bioindustry, and biodefense, but the diagnosis of most viral diseases still remains cumbersome, multi-step, multi-reagent, technically demanding, and expensive. Thus, highly distributed systems critical for the early detection of viruses are still limited and there is a crucial need for a low reagent consumption and cost-effective viral assay.

New miniaturized methods are being developed based on advances in microtechnologies and micro-optics, which enable more accurate viral detection, including SPR imaging, mechanical resonating cantilevers, or electrical conductance methods among others. These microdevice approaches are faster, cheaper, and more sensitive, but operate reliably only with pure viral samples. In parallel, some techniques have been integrated on chip for viral particle concentration, such as functionalized magnetic beads, functionalized channels, and DEP microfiltration. However, many promising and simpler techniques in which size limitations have still to be investigated exist.

The next step toward the whole integration of viral diagnostics on chip and especially the commercial stage of such a device will be the coupling of these two steps in a unique system. Indeed, future works will investigate micro–nano approaches for concentrating and purifying viruses from various complex samples and how they can be coupled with subsequent direct viral detection. Such a synergistic system would propose solutions to early detection and screening applications from various samples such as water, blood, and air. For example, the on-chip detection and determination of viral load for HIV and dengue, prevalent in impoverished settings and that infect millions of people worldwide, would have a wide-reaching impact on healthcare. Other nonweaponized viruses such as influenza can be considered national threats and the quick, inexpensive viral identification from a larger sample population would allow earlier flu pandemic warnings. Another application may involve routine viral monitoring of drinking water, especially with the detection of enteric noroviruses capable of causing a wide range of illnesses.

References

1. Bill & Melinda Gates Foundation, Global Health Program, Retrieved September 1, 2010, from http://www.gatesfoundation.org/global-health.
2. Nichol, S.T. et al., Emerging viral diseases. *Proc Natl Acad Sci*, 2000, **97**(23): p. 12411–12412.
3. Bill & Melinda Gates Foundation, Pneumonia, Strategy Overview, Global Health Program, November 2009, Retrieved September 1, 2010, from http://www.gatesfoundation.org.
4. Bill & Melinda Gates Foundation, Enteric and diarrheal diseases, Strategy Overview, Neglected and other infectious diseases, Strategy Overview, Retrieved September 1, 2010, from http://www.gatesfoundation.org.
5. Dehee, A. et al., Quantification of Epstein-Barr virus load in peripheral blood of human immunodeficiency virus-infected patients using real-time PCR. *J Med Virol*, 2001, **65**(3): p. 543–552.
6. Chao, J. et al., Hepatitis B and liver cancer knowledge and practices among healthcare and public health professionals in China: A cross-sectional study. *BMC Public Health*, 2010, **10**: p. 98.
7. Peruski, A.H. et al.,, Immunological methods for detection and identification of infectious disease and biological warfare agents. *Clin Diagnostic Lab Immun*, 2003, **10**(4): p. 506–513.
8. Center for Disease Control and Prevention, List of high-priority pathogens, Retrieved September 1, 2010, from http://www.bt.cdc.gov/agent/agentlist-category.org.
9. Mao, C. et al., Virus-based chemical and biological sensing. *Angew Chem Int Ed*, 2009, **48**: p. 6790–6810.
10. Carrera, M.R.A. et al., Treating cocaine addiction with viruses. *Proc Natl Acad Sci*, 2004, **101**(28): p. 10416–10421.
11. Kovacs, E.W. et al., Dual-surface-modified bacteriophage MS2 as an ideal scaffold for a viral capsid-based drug delivery system. *Bioconjugate Chem*, 2007, **18**(4): p. 1140–1147.

12. Kwak, M. et al., Virus-like particles templated by DNA micelles: A general method for loading virus nanocarriers. *J Am Chem Soc*, 2010, **132**: p. 7834–7835.
13. Mao, C. et al., Viral assembly of oriented quantum dot nanowires. *Proc Natl Acad Sci*, 2003, **100**(12): p. 6946–6951.
14. Mao, C. et al., Virus-based toolkit for the directed synthesis of magnetic and semiconducting nanowires. *Science*, 2004, **303**(5655): p. 213–217.
15. Balci, S. et al., Self-assembly of metal-virus nanodumbbells. *Angew Chem Int Ed*, 2007, **46**(17): p. 3149–3151.
16. Lodish, H. et al., *Molecular Cell Biology*, 5th edn. WH Freeman and Company, New York, 2003.
17. Kalser, J. A one-site-fits-all fluvaccine? *Science*, 2006, 312: p. 380–382.
18. Khurana, V.G., Meyer, F.B., Translational paradigms in cerebrovascular gene transfer. *J Cerb Blood Flow Metab*, 2003, **23**: p. 1251–1262.
19. Brookhaven Instruments, What is a particle, Retrieved September 1, 2010, from http://www.brookhaveninstruments.com/literature/what%20is/lit_what_is_a_particle.html.
20. Murphy, F., Transmission electron micrograph of the Ebola virus. Hemorrhagic fever. RNA virus, Center for Disease Control and Prevention, Identification No. 1833, 1976.
21. Kim, Y.G. et al., Quantum dot-based HIV capture and imaging in a microfluidic channel. *Biosensors Bioelectron*, 2009, **25**: p. 253–258.
22. Chen, G.D. et al., Concentration and purification of human immunodeficiency virus type 1 virions by microfluidic separation of superparamagnetic nanoparticles. *Anal Chem*, 2010, **82**: p. 723–728.
23. Dussart, P. et al., Evaluation of an enzyme immunoassay for detection of dengue virus NS1 antigen in human serum. *Clin Vaccine Immunol*, 2006, **13**(11): p. 1185–1189.
24. Poloni, T.R. et al., Detection of dengue virus in saliva and urine by real-time RT-PCR. *Viro J*, 2010, 7(22).
25. Yamamoto, A.Y. et al., Is saliva as reliable as urine for detection of cytomegalovirus DNA for neonatal screening of congenital CMV infection? *J Clin Virol*, 2006, **36**(3): p. 228–230.
26. Testereci, H. et al., The determination of acyclovir in sheep serum, human serum, saliva and urine by HPLC. *Eastern J Med*, 1998, **3**(2): p. 62–66.
27. Karim, M.R. et al., New electropositive filter for concentrating enteroviruses and noroviruses from large volumes of water. *Appl Env Microbiol*, 2009. **75**(8): p. 2393–2399.
28. Haramoto, E. et al., Recovery of naked viral genomes in water by virus concentration methods. *J Virol Methods*, 2007, **142**: p. 169–173.
29. Fuhram, J.A. et al., Rapid detection of enteroviruses in small volumes of natural waters by real-time quantitative reverse transcriptase PCR. *Appl Env Microbiol*, 2005, **71**(8): p. 4523–4530.
30. Sandström, N. et al., Electrohydrodynamic enhanced transport and trapping of airborne particles to a microfluidic air-liquid interface. *IEEE 21st Int Conf on Micro Electro Mechanical Systems* 2008, p. 595–598.
31. Frisk, T. et al., A micromachined interface for airborne sample-to-liquid transfer and its application in a biosensor system. *Lab Chip*, 2006, **6**: p. 1504–1509.
32. Fong, T.T. et al., Enteric viruses of humans and animals in aquatic envirnnoments: Health risks, detection and potential water quality assessment tools. *Microbiol Mol Biol Rev*, 2005, **69**(2): p. 357–371.
33. Storch, G.A., Diagnostic virology. *Clin Infect Dis*, 2000, **31**: p. 739–751.

34. Schmidt, N.J., Rapid viral diagnosis. *Med Clin North Am*, 1983, **67**(5): p. 953–972.
35. Mann, C.M. et al., Rapid diagnosis of viral pathogens. *Clin Lab Med*, 1995, **15**(2): p. 389–405.
36. Leland, D.S. et al., Role of cell culture for virus detection in the age of technology. *Clin Microbiol Rev*, 2007, **20**(1): p. 49–78.
37. Buck, G.E. et al., Comparison of mixed cell culture containing genetically engineered BGMK and Caco-2 cells (Super E-Mix) with RT-PCR and conventional cell culture for the diagnosis of enteroviruses meningitis. *J Clin Virol*, 2002, **24**: p. S13–S18.
38. Hsiung, G.D. Diagnostic virology: From animals to automation. *Yale J Biol Med*, 1994, **57**: p. 727–733.
39. Koivunen, M.E. et al., Principles of immunochemical techniques used in clinical laboratories. *Labmedecine*, 2006, **37**(8): p. 490–497.
40. Hodinka, R.L. et al., Detection of human immunodeficiency virus antibodies in oral fluids. *Clin Diag Lab Immun*, 1998, **5**(4): p. 419–426.
41. Granade, T.C. et al., Detection of antibodies to human immunodeficiency virus type 1 in oral fluids: A large scale evaluation of immunoassay performance. *Clin Diag Lab Immun*, 1998, **5**(2): p. 171–175.
42. Landry, M.L. et al., Detection of herpes simplex virus in clinical specimens by cytospin-enhanced direct immunofluorescence. *J Clin Microbiol*, 1997, **35**: p. 302–304.
43. Fedorko, D.P. et al., Performance of the rapid tests for detection of avian influenza A virus types H5N1 and H9N2, *J Clin Microbiol*, 2006, **44**: p. 1596–1597.
44. Schmidt, O. et al., Fluorescence spectrometer-on-a-fluidic-chip. *Lab Chip*, 2007, **7**: p. 626–629.
45. Cazacu, A.C. et al., Comparison of a new lateral-flow chromatographic membrane immunoassay to viral culture for rapid detection and differentiation of influenza A and B viruses in respiratory specimens. *J Clin Microbiol*, 2004, **42**: p. 3661–3664.
46. Cruz, A.T. et al., Performance characteristics of a rapid immunochromatographic assay for detection of influenza virus in children during the 2003 to 2004 influenza season. *Ann Emergency Med*, 2006, **47**: p. 250–254.
47. Yang, S. et al., PCR-based diagnostics for infectious diseases: Uses, limitations, and future applications in acute-care settings. *Lancet Infect Dis*, 2004, **4**(6): p. 337–348.
48. Roberts, T.C. et al., Multiplex polymerase chain reaction for diagnosis of AIDS-related central nervous system lymphoma and toxoplasmosis. *J Clin Microbiol*, 1997, **35**: p. 268–269.
49. Caliendo, A.M. et al., Comparison of qualitative and quantitative PCR assays for cytomegalovirus DNA in plasma. *J Clin Microbiol*, 2001, **39**: p. 1334–1338.
50. Espy, M.J. et al., Real-time PCR in clinical microbiology: Applications for routine laboratory testing. *Clin Microbiol Rev*, 2006, **19**: p. 165–256.
51. Karniadakis, G. et al., *Microflows and Nanoflows: Fundamentals and Simulation*, Springer, New York, 2005.
52. Eijkel, J.C.T. et al., Nanofluidics: What is it and what we can expect from it? *Microfluid Nanofluid*, 2005, **1**: p. 249–267.
53. Pennathur, S. et al., Electrokinetic transport in nanochannels. 1. Theory. *Anal Chem*, 2005, **77**: p. 6772–6781.
54. Pennathur, S. et al., Electrokinetic transport in nanochannels. 2. Experiments. *Anal Chem*, 2005, **77**: p. 6782–6789.

55. Napoli, M. et al., Nanofluidic technology for biomolecule applications: A critical review. *Lab Chip*, 2010, **10**: p. 957–985.
56. Wittenberg, N.J. et al., Using nanoparticles to push the limits of detection. *Wiley Interdisciplinary Rev: Nanomed Nanobiotechnol*, 2009, **1**(2): p. 237–254.
57. Schulze, H. et al., Multiplexed optical pathogen detection with lab-on-chip devices. *J Biophoton*, 2009, **2**(4): p. 199–211.
58. Mijatovic, D. et al., Technologies for nanofluidic systems: Top-down *vs.* bottom-up a review. *Lab Chip*, 2005, **5**: p. 492–500.
59. Cheng, X. et al., Micro and nanotechnology for viral detection. *Anal Bioanal Chem*, 2009, **393**: p. 487–501.
60. Ionescu, R.E. et al., Amperometric immunosensors for the detection of anti-West Nile virus IgG. *Anal Chem*, 2007, **79**: p. 8662–8668.
61. Dai, Z. et al., Novel amperometric immunosensors for rapid separation-free immunoassay of carcinoembryonic antigen. *J Immunol Methods*, 2004, **287**: p. 13–20.
62. Lin, Y.H. et al., Evaluation of a new hepatitis B virus surface antigen rapid test with improved sensitivity. *J Clin Microbiol*, 2008, **46**(10): p. 3319–3324.
63. Dussard, P. et al., Evaluation of an enzyme immunoassay for detection of dengue virus NS1 antigen in human serum. *Clin Vaccine Immunol*, 2006, **13**(11): p. 1185–1189.
64. Shu, P.Y. et al., Current advances in dengue diagnosis. *Clin Diagnostic Lab Immunol*, 2004, **11**(4): p. 642–650.
65. Greenwald, J.L. et al., A rapid review of rapid HIV antibody tests. *Curr Infect Dis Rep*, 2006, **8**: p. 125–131.
66. Greer, L. et al., Rapid diagnostic methods in sexually transmitted infections. *Infect Dis Clin North Am*, 2008, **22**: p. 601–617.
67. Wang, L.H. et al., Automatic bio-sampling chips integrated with micro-pumps and micro-valves for disease detection. *Biosensors Bioelectron*, 2005, **21**: p. 419–425.
68. Smit, M. et al., Comparison of the NOW Influenza A & B, NOW Flu A, NOW Flu B, and Directigen Flu A+B assays, and immunofluorescence with viral culture for the detection of Influenza A and B viruses. *Diagn Microbiol Infect Dis*, 2007, **57**: p. 67–70.
69. Weinberg, A. et al., Evaluation of three immunoassay kits for rapid detection of Influenza Virus A and B. *Clin Diagnostic Lab Immunol*, 2005, **12**(3): p. 367–370.
70. De Lima, C. et al., Utility of the BD Directigen Flu A+B rapid antigen detection assay as an influenza outbreak detection tool. *Int J Infect Dis*, 2009, **13**: p. 327–328.
71. Leary, T.P. et al., A chemiluminescent, magnetic particle-based immunoassay for the detection of hepatitis C virus core antigen in human serum or plasma. *J Med Virol*, 2006, **78**: p. 1436–1440.
72. Cho, Y.K. et al., One-step pathogen specific DNA extraction from whole blood on a centrifugal microfluidic device. *Lab Chip*, 2007, **7**: p. 565–573.
73. Lien, K.Y. et al., Purification and enrichment of virus samples utilizing magnetic beads on a microfluidic system. *Lab Chip*, 2007, **7**: p. 868–875.
74. Yobas, L. et al., Nucleic adic extraction, amplification, and detection on Si-based microfluidic platforms. *IEEE J Sol-State Circuit*, 2007, **42**(8): p. 1803–1813.
75. Park, K. et al., Electrical capture and lysis of vaccinia virus particles using silicon nanoscale probe array. *Biomed Microdevices*, 2007, **9**: p. 877–883.
76. Cho, Y.K. et al., Clinical evaluation of micro-scale chip-based PCR system for rapid detection of hepatitis B virus. *Biosens Bioelectron*, 2006, **21**(11): p. 2161–2169.

77. Kaigala, G.V. et al., Automated screening using microfluidic chip-based PCR and product detection to assess risk of BK virus-associated nephropathy in renal transplant recipients. *Electrophoresis*, 2006, **27**: p. 3753–3763.
78. Lee, J. et al., Real-time detection of airborne viruses on a mass-sensitive device. *Appl Phys Lett*, 2008, **93**: p. 013901.
79. Pipper, J. et al., Catching bird flu in a droplet. *Nat Med*, 2007, **13**: p. 1259–1263.
80. Pal, R. et al., An integrated microfluidic device for influenza and other genetic analyses. *Lab Chip*, 2005, **5**: p. 1024–1032.
81. De Zoeten, G.A. Application of scanning microscopy in the study of virus transmission of Aphids1. *J Virol* 1968, p. 745–751.
82. Myers, F.B. et al., Innovations in optical microfluidic technologies for point-of-care diagnostics. *Lab Chip*, 2008, **8**(12): p. 215–231.
83. Yacoub-George, E. et al., Automated 10-channel capillary chip immunodetector for biological agents detection. *Biosens Bioelectron*, 2007, **22**: p. 1368–1375.
84. Ymeti, A. et al., Fast, ultrasensitive virus detection using a Young interferometer sensor. *Nano Lett*, 2007, **7**(2): p. 394–397.
85. Xu, J. et al., Detection of avian influenza virus using an interferometric biosensor. *Anal Bioanal Chem*, 2007, **389**: p. 1193–1199.
86. Homola, J. Surface Plasmon resonance sensors for detection of chemical and biological species. *Chem Rev*, 2008, **108**: p. 462–493.
87. Lei, Y. et al., Electroless-plated gold films for sensitive surface plasma resonance detection of white spot syndrome virus. *Biosens Bioelectron*, 2008, **23**: p. 1200–1207.
88. Chinowsky, T.M. et al., Portable 24-analyte surface plasma resonance instruments for rapid, versatile biodetection. *Biosens Biolectron*, 2007, **22**: p. 2268–2275.
89. Stewart, M.E. et al., Nanostructured plasmonic sensors. *Chem Rev*, **108**: p. 494–521.
90. Akin, D. et al., Real-time virus trapping and fluorescent imaging in microfluidic devices. *Nano Lett*, 2004, **4**(2): p. 257–259.
91. Zhang, Y. et al., Toward the detection of single virus particle in serum. *Anal Biochem*, 2006, **356**: p. 161–170.
92. Balslev, S. et al., Lab-on-a-chip with integrated optical transducers. *Lab Chip*, 2006, **6**: p. 213–217.
93. Joo, K.J. et al., Imaging and tracking of viruses using quantum dots. In: Kumar, C.S.S.R. ed., *Nanomaterials for the Life Science: Semiconductor Nanomaterials*, Vol. 6. Wiley–VCH, 2010.
94. Kim, Y.J. et al., Quantum dot-based HIV capture and imaging in a microfluidic channel. *Biosens Bioelectron*, 2009, **25**: p. 253–258.
95. Agrawal, A. et al., Counting single native biomolecules and intact viruses with color-coded nanoparticles. *Anal Chem*, 2006, **78**(4): p. 1061–1070.
96. Agrawal, A. et al., Real-time detection of virus particles and viral protein expression with two-color nanoparticles probes. *J Virol*, 2005, **79**(13): p. 8625–8628.
97. Ferris, M.M. et al., Rapid enumeration of respiratory viruses. *Anal Chem*, 2002, **74**: p. 1849–1856.
98. Yang, S.Y. et al., Micro flow cytometry utilizing a magnetic bead-based immunoassay for rapid virus detection. *Biosen Bioelectron*, 2008, **24**: p. 885–862.
99. Lee, S.H. et al., A polymer lab-on-a-chip for reverse transcription (RT)-PCR based point-of-care clinical diagnostics. *Lab Chip*, 2008, **8**: p. 2121–2127.

100. Zuo, B. et al., Piezoelectric immunosensor for SARS-associated coronavirus in sputum. *Anal Chem*, 2004, **76**: p. 3536–3540.
101. Peduru Hewa, T.M., et al.,, The detection of influenza A and B viruses is clinical specimens using a quartz crystal microbalance. *J Virol Methods*, 2009, **162**: p. 14–21.
102. Cooper, M.A. et al., Direct and sensitive detection of a human virus by rupture event scanning. *Nat Biotechnol*, 2001, **19**: p. 833–837.
103. Gunter, R.L. et al., Viral detection using an embedded piezoresistive microcantilever sensor. *Sens Actuators A*, 2003, **107**: p. 219–224.
104. Johnson, L. et al., Characterization of vaccinia virus particles using microscale silicon cantilever resonators and atomic force microscopy. *Sens Actuators B*, 2006, **115**: p. 189–197.
105. Ekinci, K.L. et al., Ultimate limits to inertial mass sensing based upon nanoelectromechanical systems. *J Appl Phys*, 2004, **95**(5): p. 2682–2689.
106. Burg, T.P. et al., Weighing of biomolecules, single cells and single nanoparticles in fluid. *Nat Lett*, 2007, **446**: p. 1066–1069.
107. Datar, R. et al., Cantilever sensors: Nanomechanical tools for diagnostics. *MRS Bull*, 2009, **34**: p. 449–454.
108. Chunara, R. et al., Mass-based readout for agglutination assays. *Appl Phys Lett*, 2007, **91**: p. 193902.
109. Patolsky, F. et al., Electrical detection of single viruses. *Proc Natl Acad Sci U S A*, 2004, **101**: p. 14017–14022.
110. De Blois, R.W. et al., Comparative measurements of size and polydispersity of several insect viruses. *Anal Biochem*, 1978, **90**(1): p. 273–288.
111. Uram, J.D. et al., Submicrometer pore-based characterization and quantification of antibody-virus interactions. *Small*, 2006, **2**(8–9): p. 967–972.
112. An, R. et al., Ultrafast laser fabrication of submicrometer pores in borosilicate glass. *Opt Lett*, 2008, **33**(10): p. 1153–1155.
113. Hui, W.C. et al., Microfluidic systems for extracting nucleic acids for DNA and RNA analysis. *Sens Actuators A*, 2007, **133**: p. 335–339.
114. Shen, C. et al., Nano particle filtration module for lab-on-a-chip devices. *IEEE*, 2010.
115. Hamblin, M.N. et al., Selective trapping and concentration of nanoparticles and viruses in dual-height nanofluidic channels. *Lab Chip*, 2010, **10**: p. 173–178.
116. Reichmuth, D.S. et al., Rapid microchip-based electrophoretic immunoassays for the detection of swine influenza virus. *Lab Chip*, 2008, **8**: p. 1319–1324.
117. Gascoyne, P.R.C. et al., Particle separation by dielectrophoresis. *Electrophoresis*, 2002, **23**: 1973–1983.
118. Hughes, M.P. Strategies for dielectrophoretic separation in laboratory-on-a-chip systems. *Electrophoresis*, 2002, **23**: p. 2569–2582.
119. Abraham, A.M. et al., Nanotechnology: A new frontier in virus detection in clinical practice. *Indian J Med Microbiol*, 2008, **26**(4): p. 297–301.
120. Satoh, K. et al., Virus concentration using polyethyleneimine-conjugated magnetic beads for improving the sensitivity of nucleic acid amplification tests. *J Virol Methods*, 2003, **114**: p. 11–19.
121. Uchida, E. et al., Optimization of the virus concentration using polyethyleneimine-conjugated magnetic beads and its application to the detection of human hepatitis A, B and C viruses. *J Virol Methods*, 2007, **143**: p. 95–103.

122. Peyman, S.A. et al., Rapid on-chip multi-step (bio)chemical procedures in continuous flow manoeuvring particles through co-laminar reagent streams. *Chem Commun*, 2008, **10**: p. 1220–1222.
123. Augustsson, P. et al., Buffer medium exchange in continuous cell and particle streams using ultrasonic standing wave focusing. *Microchim Acta*, 2009, **164**: p. 269–277.
124. Rowe, C.A. et al., Array biosensor for simultaneous identification of bacterial, viral, and protein analytes. *Anal Chem*, 1999, **71**: p. 3846–3852.
125. Gossett, D.R. et al., Label-free separation and sorting in microfluidic systems. *Anal Bioanal Chem*, 2010, **397**(8): p. 3247–3267.
126. Tsutsui, H. et al., Cell separation by non inertial force fileds in microfluidic systems. *Mech Res Commun*, 2009, **36**(1): p. 92–103.
127. Lenshof, A. et al., Continuous separation of cells and particles in microfluidic systems. *Chem Soc Rev*, 2010, **39**: p. 1203–1217.
128. Davis, J.A. et al., Deterministic hydrodynamics: Taking blood apart. *Proc Natl Acad Sci*, 2006, **103**(40): p. 14779–14784.
129. Inglis, D.W. et al., Critical particle size for fractionation by deterministic lateral displacement. *Lab Chip*, 2006, **6**(5): p. 655–658.
130. Inglis, D.W. et al., Highly accurate deterministic lateral displacement device and its application to purification of fungal spores. *Biomicrofluidics*, 2010, **4**: p. 024109.
131. Huang, L.R. et al., Continuous particle separation through deterministic lateral displacement. *Science*, 2004, **304**: p. 987–990.
132. Laurell, T. et al., Chip integrated strategies for acoustic separation and manipulation of cells and particles. *Chem Soc Rev*, 2007, **36**: p. 492–506.
133. Nilsson, A. et al., Acoustic control of suspended particles in micro fluidic chips. *Lab Chip*, 2004, **4**: p. 131–135.
134. Petersson, F. et al., Continuous separation of lipid particles from eryhtorcytes by means of laminar flow and acoustic standing wave forces. *Lab Chip*, 2005, **5**: p. 20–22.
135. Petersson, F. et al., Free flow acoustophoresis: Microfluidic-based mode of particle and cell separation. *Anal Chem*, 2007, **79**: p. 5117–5123.
136. Di Carlo, D. et al., Continuous inertial focusing, ordering, and separation of particles in microchannels. *Proc Natl Acad Sci*, 2007, 104. **48**: p. 18892–18897.
137. Di Carlo, D. et al., Equilibrium separation and filtration of particles using differential inertial focusing. *Anal Chem*, 2008, **80**: p. 2204–2211.
138. Di Carlo, D. et al., Inertial microfluidics. *Lab Chip*, 2009, **9**: p. 3038–3046.
139. Segré, G. et al., Radial particle displacements in Poiseuille flow of suspensions. *Nature*, 1961, **189**: p. 209–210.
140. Hur, S.C. et al., Sheathless inertial cell ordering for extreme throughput flow cytometry. *Lab Chip*, 2010, **10**: p. 274–280.
141. Gossett, D.R. et al., Particle focusing mechanisms in curving confined flows. *Anal Chem*, 2009, **81**: p. 8459–8466.
142. Bhagat, A.A.S. et al., Inertial microfluidics for continuous particle filtration and extraction. *Microfluid Nanofluid*, 2009, **7**: p. 217–226.
143. Bhagat, A.A.S. et al., Inertial microfluidics for sheath-less high throughput flow cytometry. *Biomed Microdev*, 2010, **12**: p. 187–195.

Chapter 14

Electrophoresis and Dielectrophoresis for Lab-on-a-Chip (LOC) Analyses

Yağmur Demircan, Gürkan Yilmaz, and Haluk Külah*

METU-MEMS Center and
Department of Electrical and Electronics Engineering
Middle East Technical University (METU)
Ankara, Turkey

Introduction

Early and precise diagnosis plays an important role in the treatment of almost all diseases. The quality of the diagnosis directly relies on the quality of the laboratory analyses. Micro-electro-mechanical systems (MEMS) and nanotechnology introduced miniaturized fluidic and biomedical components to improve the quality of today's laboratory operations in the last decade. Lab-on-a-chip (LOC) systems developed by means of these technologies offering fast analysis with reduced cost and size, possessing portability, operating without a need for trained personnel, and requiring less amounts of biological samples and reagents for analysis. These systems find applications in medicine, biology, genetics, pharmaceuticals, the food industry, the defense industry, and environmental monitoring, either by enhancing the current analysis and detection methods by implementing state-of-the-art macro-scale counterparts or by enabling new methods unique to such miniaturized biomedical systems.

Among numerous application areas of LOC systems, the separation and detection of rare cells, bacteria, and viruses have drawn considerable attention as miniaturized systems may result in a much better performance in terms of

---

*Corresponding author: Email: kulah@metu.edu.tr

detection resolution, accuracy, and analysis time. For example, these systems may enable early diagnosis of some cancers, which requires the detection of a few cancer cells out of millions of healthy ones. This development not only allows very early diagnosis, but also constitutes an enabling technology for cell biology and medicine.

The separation of the target particles (cells, bacteria, viruses, etc.) is important for rare particle detection applications, and it can be realized by many different ways (e.g., mechanical, magnetic, antibody-based, electrical, etc.) depending on the target particle properties. Among these different strategies, electrical separation is especially important as the electrical properties of the biological particles are stable, specific, and distinctive for separation applications. Electrical separation techniques are based on two main electrical phenomena — electrophoresis and dielectrophoresis. Electrophoresis is related with the separation of charged particles (e.g., DNA and proteins), while dielectrophoresis of deals with dielectric particles (e.g., cells and bacteria). This chapter introduces these two main electrical separation techniques for clinical laboratory applications. The first section presents the theory and the operating principle for electrophoresis together with the macro- and micro-scale examples from the literature. In this section, the main focus is on DNA and protein electrophoresis. The second section summarizes the theoretical background, micro-scale implementation, various operation strategies, and cell-based examples for dielectrophoresis.

Electrophoresis

Electrophoresis is an electrokinetic phenomenon that was observed in the beginning of 19th century. Thus, it possesses a mature theory, which has been developed since then. Electrophoresis, basically, defines the movement of charged particles dispersed in a medium under a constant electric field. The force causing this movement is called electrophoretic force, and it depends on the charge of the particle (q) and the applied electric field (E) as shown below:

$$F_{EP} = qE. \qquad (1)$$

This equation indicates that particles having different charges will be subjected to different electrophoretic forces and hence have different velocities, meaning that they can be separated. The separation of charged particles by electrophoresis can be utilized in many different applications, including single nucleotide polymorphism (SNP) detection by DNA electrophoresis,[1] diagnosis of meningitis by protein electrophoresis,[2] etc. Equation (1) is a first-order approximation to explain the electrophoresis theory, and other effects (e.g., electroosmosis, double-layer generation at the fluid–particle interface, and viscous drag force) should be considered when analyzing a system aiming at electrophoretic separation. This section discusses the theoretical explanations regarding electrophoresis together with other effects. Besides, the macro- and micro-scale applications of electrophoresis are investigated.

Theory of electrophoresis

Electrophoretic force

Electrophoretic force is defined as the product of the charge and the applied electric field (Eq. (1)). The direction of the force depends on the sign of the charge as depicted in Fig. 1.

The magnitude of the force depends on the magnitude of the applied electric field and the net charge stored on the particle. Therefore, increasing the applied electric field is the only way to speed up a certain particle. The electric field can be increased by either increasing the applied potential or decreasing the distance between the electrodes. Increasing the potential is not a practical method for biological applications, since high voltages may be harmful for samples under inspection. Therefore, it is generally preferred to optimize the geometry according to the application.

Another critical parameter for electrophoretic separation is the mass-to-charge ratio of the injected particles. According to Eq. (2),

$$a = \frac{E}{m/q}, \qquad (2)$$

particles observe the same acceleration due to electrophoresis, provided that they have the same mass-to-charge ratio.

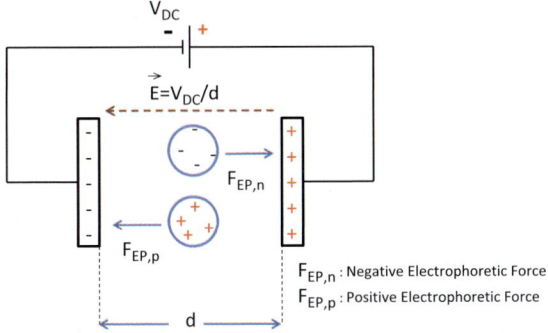

Fig. 1. Illustration of positive and negative electrophoretic forces on charged particles.

A practical example of this situation can be seen in DNA electrophoresis, where the total charge of the DNA sample is directly proportional to its base-pair number, and hence to its mass. To realize DNA electrophoresis, a gel-like medium, which practically eliminates the effect of mass-to-charge ratio, should be utilized. On the other hand, electrophoretic force is not the single force acting on the injected particles. While electrophoresis occurs, there exist other effects, such as double-layer generation, electroosmotic flow (EOF), and viscous drag force.[3,4]

Double-layer generation

When a charged particle is immersed in a solution, it disturbs the charge distribution around itself, since the surface charges attract the oppositely charged ions that are mobile in the solution (Fig. 2).[3,5] Helmholtz, for the first time in the literature, modeled the double layer as a simple capacitor, consisting of an inner plane of ions (attracted from the solution by the charged particle) and an outer plane of ions (placed as a negative response to the inner layer). Afterward, the Gouy–Chapman model introduced an exponentially decreasing electric potential away from the surface of the charged particle. Finally, the Stern model combined both models to improve the accuracy. In this model, ions next to the charged particle are considered as permanently absorbed by the surface of the particle, and the second ion layer is modeled according to the Gouy–Chapman model. The electric double layer is the combination of these two layers — Stern layer, which consists of the immobile charges and the diffuse layer, which contains mobile charges.[3,4]

Due to double-layer generation, if a colloidal particle moves in the dispersion medium, the surrounding liquid layer stays as attached to the particle. The boundary of this layer is named slipping plane, on which the zeta potential is formed.[5] In addition, as depicted in Fig. 2, the distance between the surface of the charged particle and the Stern plane is named the Stern layer, and its thickness is approximated to the Bjerrum length, which is expressed as:[6]

$$\lambda_B = \frac{q^2}{4\pi\varepsilon_r\varepsilon_0 kT}, \qquad (3)$$

where q is the elementary electron charge, ε_r is the relative permittivity of the medium, ε_0 is the vacuum permittivity, k is the Boltzmann constant, and T is the medium temperature (in Kelvin).

In order to investigate the electrostatic interaction of these ions, the Poisson–Boltzmann equation, constituting the basis of the Gouy–Chapman

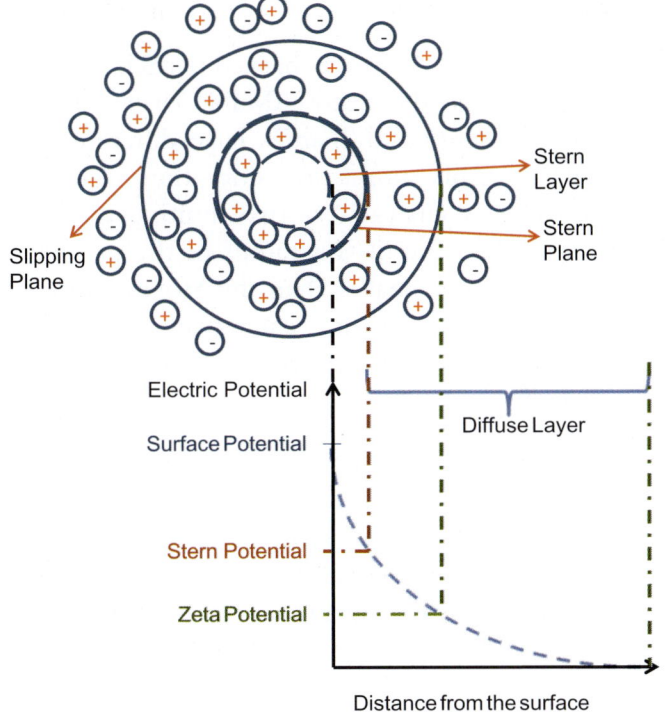

Fig. 2. Electric double-layer generation at liquid-particle interface.[5]

model, has to be solved. The Poisson–Boltzmann equation simplifies to the Helmholtz equation if the electrostatic potential is smaller than kT/q [7]:

$$\nabla^2 \varphi = k^2 \varphi, \qquad (4)$$

where φ and k are the electrical potential and the inverse of the Debye screening length (D), respectively. κ is formulated in Eq. (5)[4,8,9] and the Debye screening length can be defined as the scale length to which a charge in the plasma is shielded:[5]

$$k = D^{-1} = \sqrt{\frac{q^2 \Sigma_i \, z_i^2 C_i}{\varepsilon_r \varepsilon_0 KT}}, \qquad (5)$$

where C_i and z_i denote the concentration and the number of charges of ions of species i, respectively, q is the elementary electron charge, ε_r is the relative permittivity of the medium, ε_0 is the vacuum permittivity, k is the Boltzmann constant, and T is the medium temperature (in Kelvin).

Viscous drag force and electroosmosis

When an electric field is applied onto a charged particle in an ionic solution, both the charged particle and the electric double-layer associated with it are affected. Note that the mentioned charges have different signs; therefore, there will be a hydrodynamic interaction since they will tend to move in opposite directions. This is analyzed at two limiting cases: (i) the Debye length is much greater than the Stokes radius of the particle ($D \gg R$), and (ii) the Debye length is much smaller than the Stokes radius of the particle ($D \ll R$). If the Debye length is much greater than the Stokes radius of the particle, force balance is formed between the electrophoretic force and the viscous drag force:[10]

$$qE = 6\pi\eta v R, \qquad (6)$$

where E is the electric field, q is the elementary electron charge, η denotes the viscosity of the fluid, v denotes the velocity of the particle, and R indicates the Stokes radius of the particle. This radius can be formulated for a perfect sphere as:[11]

$$R_H = \frac{kT}{6\pi\eta C_d}, \qquad (7)$$

where k is the Boltzmann constant, T is the medium temperature (in Kelvin), η denotes the viscosity of the fluid, and C_d is the diffusion coefficient.

The velocity of the particle (v) can be expressed as the product of the electrical mobility of the particle (μ) and the applied electric field:

$$v = \mu E. \qquad (8)$$

As a result, the mobility of the particles which possess $D \gg R$, can be expressed as:

$$\mu = q/6\pi\eta R. \qquad (9)$$

For the other case, i.e., $D \ll R$, solving the Navier–Stokes equation results in mobility expression, which is independent of the particle size and shape (the Smoluchowski theory):[3]

$$\mu = \frac{\varepsilon_r \varepsilon_o \zeta}{\eta}, \qquad (10)$$

where η denotes the viscosity of the fluid, ε_r and ε_o are the relative permittivity of the medium and vacuum permittivity, respectively, and ξ denotes the zeta potential, which is described as the electric potential at the slipping plane. Considering electrophoretic separation, size-independent velocity expression possesses important consequences. For example, if the Debye length is too small, size-based electrophoretic separation cannot be realized.

The derivations listed up to now describe the motion of a charged particle with respect to the surrounding medium. However, the surrounding fluid may also be affected from the applied electric field. If the fluid is flowing through a microchannel on which immobile surface charges exist, there exists an electric double layer at the interface of the ionic solution and the channel surface. When an electric field is applied to perform electrophoresis, the net charge in the double layer tends to move, and therefore an EOF is generated. A general example presented for this case is the water flow inside a glass microchannel. If the surface of the glass microchannel is negatively charged, the formed Debye layer will be positively charged. Under an applied electric field, the bulk fluidic motion starts in the same direction with the electric field. The most important point to be remarked is that electrophoresis defines the movement of the particle with respect to the medium; however, electroosmosis defines the bulk fluidic movement, thus it is expected to significantly affect electrophoretic separations. As a result, to enhance the separation performance, the EOF should be controlled. Of course, there exist cases where electroosmosis is not significant. For instance, the separation of particles having a Debye layer much larger than their radiuses is based on their size, and since electroosmotic flow affects the particles in the same amount, the relative movements of the particles with respect to each other will be independent from the EOF.

In this section, the theoretical background of electrophoresis is introduced with an emphasis on velocity expressions under different circumstances. It is critical to understand the mobility behavior since electrophoretic separation methods mainly depend on forming different velocities for different particles. Therefore, different strategies have been developed for various applications. The proceeding section presents these applications and techniques.

Applications of electrophoresis

DNA electrophoresis

With the Human Genome Project, DNA sequencing techniques have become more important because although a great amount of human genome has been sequenced, the most important parts, which vary from person to person, are still awaiting to be decoded. These differences are caused by SNPs.[1] In order to detect SNPs and to speed up DNA sequencing analysis, new techniques have been developed. DNA electrophoresis is one of the most important examples to these techniques. DNA fingerprinting,[12] pathogen identification, and the detection of diseases are the most popular applications of DNA electrophoresis.

Since DNA structure has a phosphate backbone, which is negatively charged due to O^- (Fig. 3), it is suitable for electrophoretic separation.[13] Each base-pair of a DNA has two phosphate groups, and hence $2e^-$. Thus, the charge of a DNA is proportional to the number of base-pairs, in other words, to its mass. This means that, according to Eq. (2), the electrophoretic mobility of different DNA fragments is the same under the same electric field, regardless of their sizes. This problem is solved by using a sieving matrix in the separation zone. Using a sieving medium creates different resistances for different-sized DNA fragments, and therefore separation becomes possible as depicted in Fig. 4.

A sieving matrix has pores through which DNA fragments can pass. There are two main sieving media used in this process — agarose gel to separate large molecules and polyacrylamide gel to separate smaller molecules. Molecule size is defined by the base-pair number. The polyacrylamide gel provides better resolution due to its smaller pores. Another method to enhance resolution is to increase the applied voltage, thereby increasing the

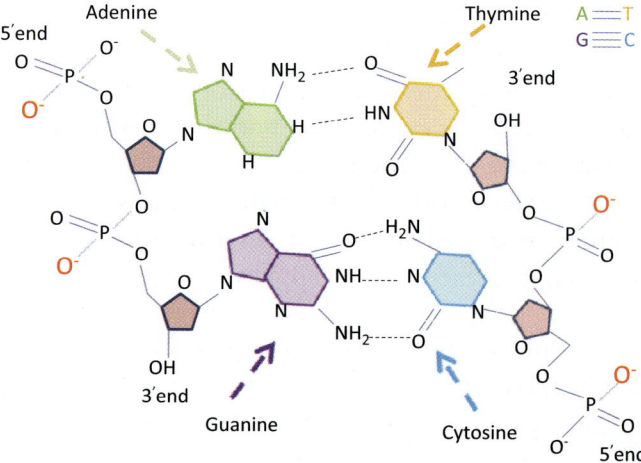

Fig. 3. Chemical structure of a DNA with two base-pairs. Each base-pair has 2 O^- and therefore introduces $2e^-$.[14]

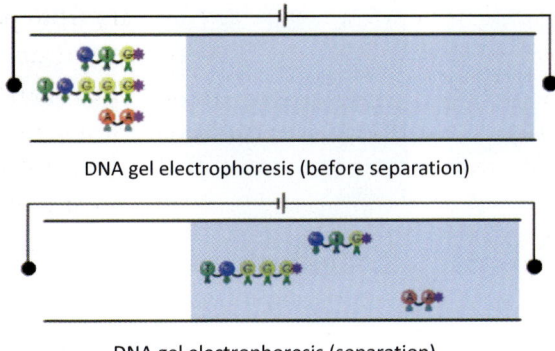

Fig. 4. Illustration of DNA gel electrophoresis.[15]

electric field and the velocity. However, Joule heating should be considered in this case.[16] When the applied voltage is larger, the effect of Joule heating increases and the samples may be degraded.

In conventional macrosystem DNA electrophoresis, there is a tank filled with the sieving matrix and the buffer solution. The dye-labeled samples are loaded to this tank and then, the electric field is applied to the system. DNA fragments travel toward the anode with the effect of the electric field. Increasing the running duration lengthens the traveling time of the DNA fragments, resulting in improved resolution. However, increasing the running time too much may cause band dispersion.[17] Therefore, separation time should be optimized according to the application (e.g., PCR, plasmid preparation, fast separation, etc.), applied voltage levels, type of the sieving matrix, and the size of the DNA fragments.

After the separation phase, formed bands on the sieving matrix are observed under the microscope. The whole electrophoretic separation process may take between 12 to 48 h, depending on the sample. Figure 5 shows two typical macroscale electrophoresis systems, one with horizontal and the other with vertical tanks.[18] In these systems, the well volume, which defines the volume of the sample to be separated, is adjusted by using a comb-like structure.

While miniaturizing conventional macrolevel electrophoresis systems, the experimental procedure is conserved. The miniaturization of electrophoresis systems brings several advantages, such as fast separation, low analyte/reagent consumption, parallel processing ability, and low cost.[19] Utilizing microsystems for electrophoresis provides higher electric field, and hence fast separation, even at very low voltage levels as the distance is much smaller. Also, less amounts of sample and sieving matrix are needed due to the overall

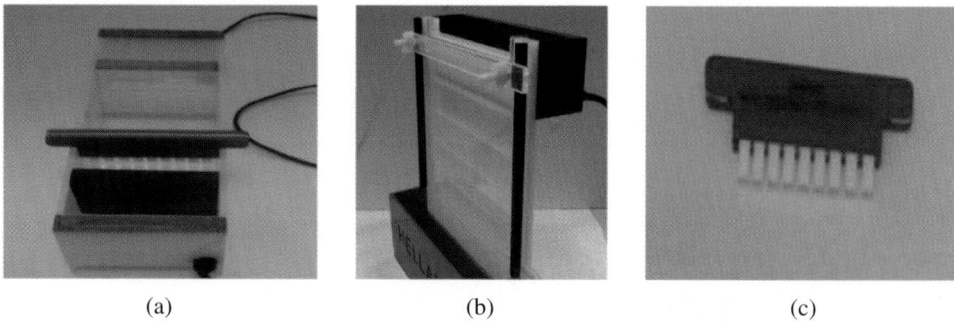

Fig. 5. Two typical electrophoresis tanks: (a) horizontal tank and (b) vertical tank. (c) A comb-like structure for defining the well volume. (Reprinted with permission from Ref. 18.)

volume, which is in the order of microliters. Another advantage in micro-scale implementation is the parallel processing ability. The throughput can be further increased by performing electrophoresis in parallel microchannels simultaneously. Moreover, micro-scale implementation brings batch fabrication and low cost, which are the major advantages of MEMS technology. On the other hand, the performance of the electrophoresis in microscale is limited by the interaction between the capillary wall and the analyte, EOF, and the high viscosity of the sieving matrix.[19] Due to the interaction between the capillary wall and the analyte, DNA fragments can be lost. Polymer wall coatings can be used to solve this problem. As it is seen in the section on the theory of electrophoresis, if DNA fragments are under the effect of EOF, they cannot be separated. To overcome this, the self-coating polymer matrices can be utilized. Finally, high-viscosity matrices need a longer loading time and higher pressure, and therefore are automated injections are unfeasible. Matrices with thermally adjustable viscosity are a possible solution for this limitation.

In microscale electrophoresis systems, there are three important issues — the polymerization of the sieving matrix, injection of the samples, and interpretation of the results.

(1) *Polymerization of the sieving matrix*: For microscale electrophoresis systems, the polymerization of the sieving matrix is critical, as it defines the separation zone. The most common technique to define the separation zone is photopolymerization (UV cross-linking). Figure 6 shows the steps for photopolymerization. In this technique, the sieving matrix is loaded in liquid form to the microchannels. Then, by using a photo-mask, only the part of the matrix located at the separation zone is subjected to UV light, and

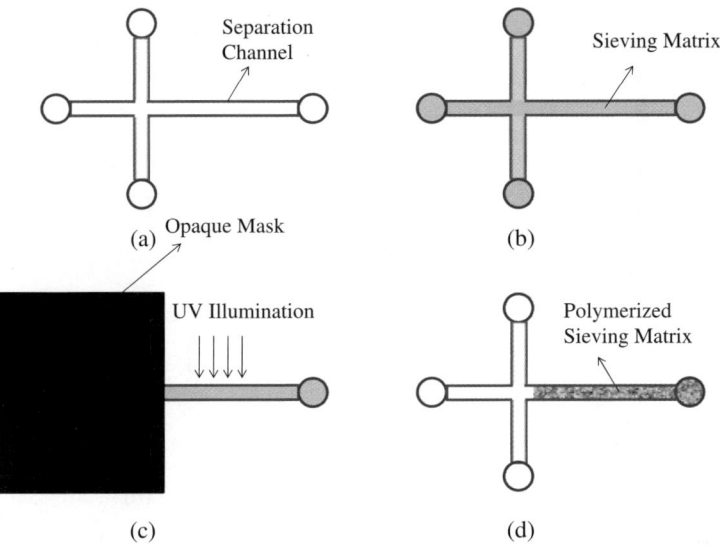

Fig. 6. Photopolymerization steps in micro-scale electrophoresis: (a) empty microchannel, (b) loading of the sieving matrix in liquid form, (c) UV illumination with an opaque mask for polymerization of the sieving matrix, (d) sieving matrix in liquid form is washed away and the separation zone is defined.[20]

hence, polymerized. In this process, polymerization time should be adjusted depending on the application (e.g., base-pair number of the samples, desired resolution, etc.).[20]

(2) *Injection of the samples*: For DNA analysis, if microsystems are used, the selection of the injection technique is critical, since sample injection should be reproducible and accurate.[19] In order to interpret the resulting band formation, the DNA sample volume directed into the gel matrix should be known. In microscale electrophoresis systems, some electrokinetic injection techniques, such as cross (Fig. 7(a)), double T (Fig. 7(b)), and triple T injectors (Fig. 7(c)) are used.[19]

In cross injectors, the volume of the sample plug is determined by the cross-sectional volume of the two microchannels. For this injector, first, a voltage is applied to the sample loading reservoir (between the bottom and top electrodes) in order to load the sample. When the sample fills the cross-section, a voltage is applied to the buffer reservoir (between the left and right electrodes) to drive the sample. While the sample is injected, leakage occurs, resulting in decreased resolution and hence, separation performance. In order to minimize this leakage effect, the pullback injection method has been developed.[21]

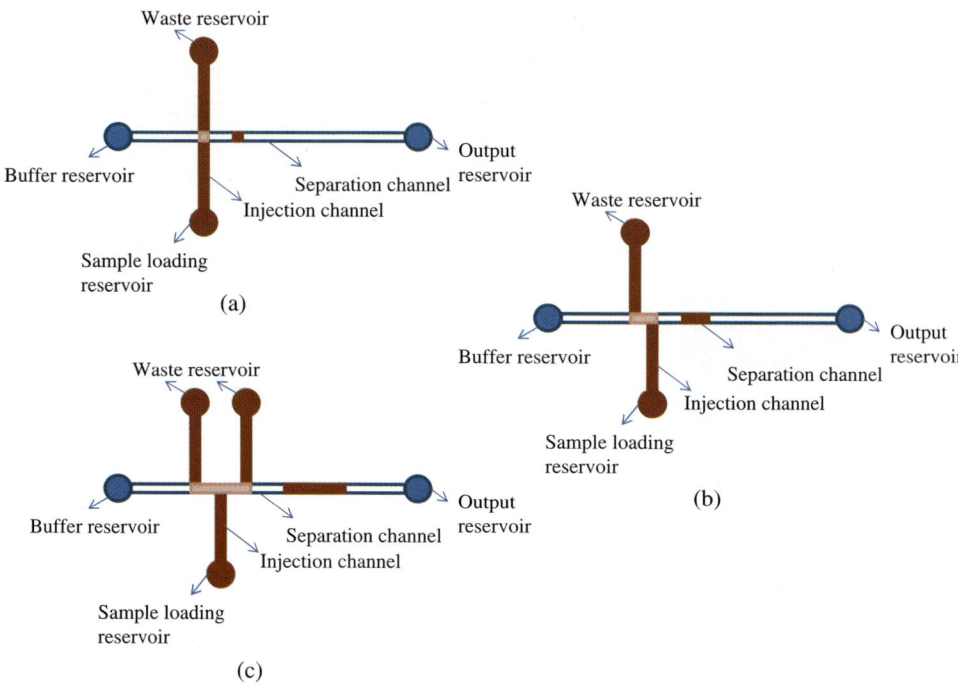

Fig. 7. Electrokinetic injection techniques for sample loading. (a) Cross injector, (b) double-T injector, and (c) triple-T injector.

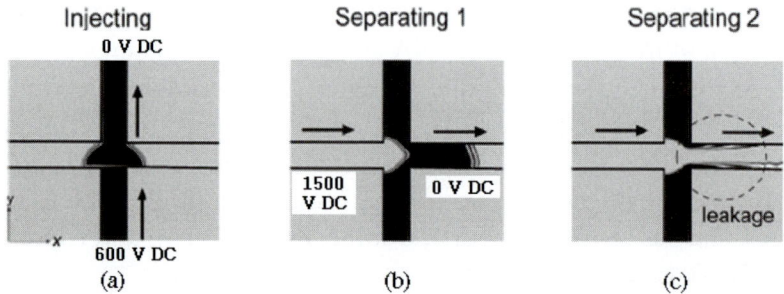

Fig. 8. Simulation results of the injection sequences in the cross injector.[22] (Copyright Wiley-VCH Verlag GmbH & Co. KGaA. Reproduced with permission.)

Figure 8 shows the simulation images of the injection sequences in the cross injector. According to these results, the leakage is 35.2%. When the pullback method is applied (Fig. 9), the leakage decreases to 2.3%.[22] With double and triple T injectors, a larger amount of the sample can be directed to the separation zone. The loading and separation steps are the same with the cross T injector.[23] Figure 10 shows these steps. According to simulations,

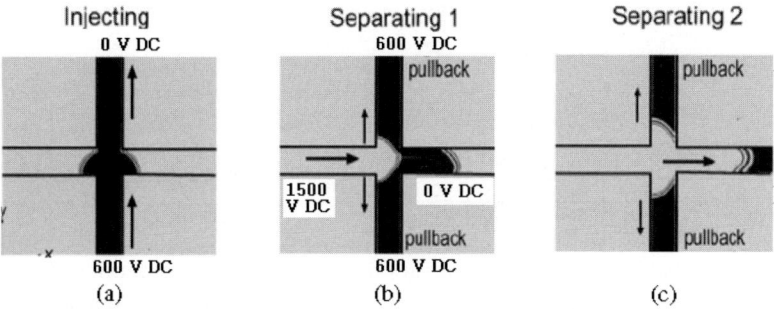

Fig. 9. Simulation results of pull-back injection for the cross injector.[22] (Copyright Wiley-VCH Verlag GmbH & Co. KGaA. Reproduced with permission.)

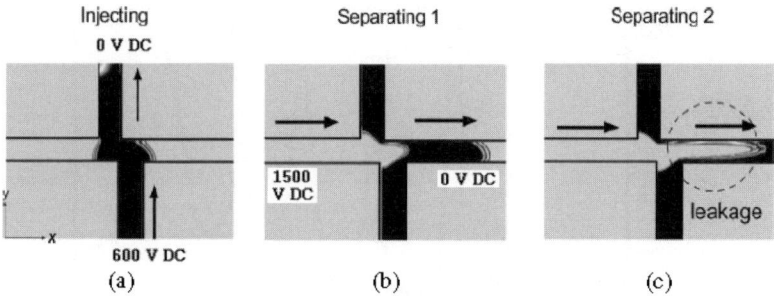

Fig. 10. Simulation results of the injection sequences in the double-T injector.[22] (Copyright Wiley-VCH Verlag GmbH & Co. KGaA. Reproduced with permission.)

the leakage is 48.5%. The authors reported that the leakage can be decreased down to 3.3%, by using pullback.[22]

(3) *The interpretation of the results*: Although experimental steps are conserved while miniaturizing conventional macrolevel electrophoresis systems, the results are interpreted with different methods. For example, if a fluorescence dye is used, the fluorescence intensity of each band is measured. According to this measurement, a histogram is prepared, which can be fluorescence intensity versus time or fragment distance from the gel interface.

Figure 11 shows the schematic diagram, applied voltages for injection and separation, and resulting histogram for a micro-scale electrophoresis system utilizing T-injection.[24] The authors reported that samples are separated in 90 s with 200 V/cm electric field strength. In this study, DNA fragments were labeled with Sytox Orange dye. Due to this dye, accurate results can be consistently obtained without being concerned about the ratio of the staining dye to base-pairs.

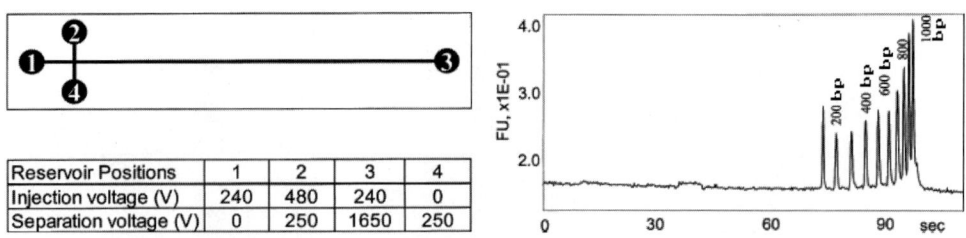

Fig. 11. Schematic diagram, applied voltages for injection and separation, and resulting histogram for a micro-scale electrophoresis system utilizing T-injection.[24] (Copyright Wiley-VCH Verlag GmbH & Co. KGaA. Reprinted with permission.)

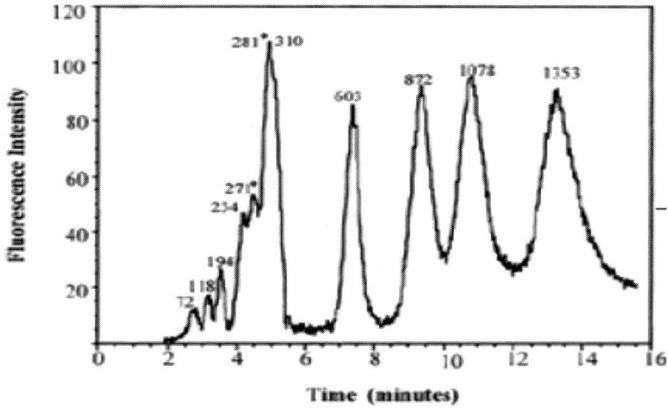

Fig. 12. Experimental results for a micro-electrophoresis system utilizing the T-injection technique.[20] (Copyright Wiley-VCH Verlag GmbH & Co. KGaA. Reprinted with permission.)

Figure 12 shows the results of another micro-electrophoresis system utilizing T-injection. The authors reported that the samples are separated in 16 min by applying 16 V/cm electric field. In this experiment, YOYO-1 was used as the labeling dye, and results were interpreted with a fluorescence intensity versus time histogram. In addition, the numbers at the peak points display base-pair numbers in DNA fragments and asterisks are used to show anomalous results.[20]

Figure 13 depicts another micro-electrophoresis system utilizing a modified cross T-injector for the separation of single-stranded DNA (ssDNA). Polyacrylamide was used as the sieving matrix and the electric field was optimized as 50 V/cm. In addition, an on-chip photodiode was used to facilitate the observation of the band structures.[25]

Figure 14 shows the design and the test results for a 96-channel radial capillary array electrophoresis chip, which was proposed by Mathies and

Electrophoresis and Dielectrophoresis for Lab-on-a-Chip Analyses 355

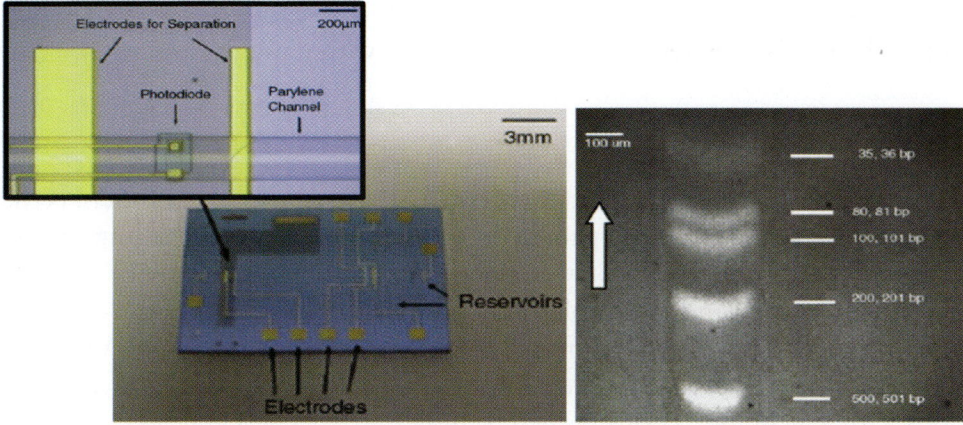

Fig. 13. A micro-electrophoresis system utilizing a modified cross T-injector with its experimental results. The device employs photodiode for better observation of the band structure.[25] (Reprinted with permission from Elsevier.)

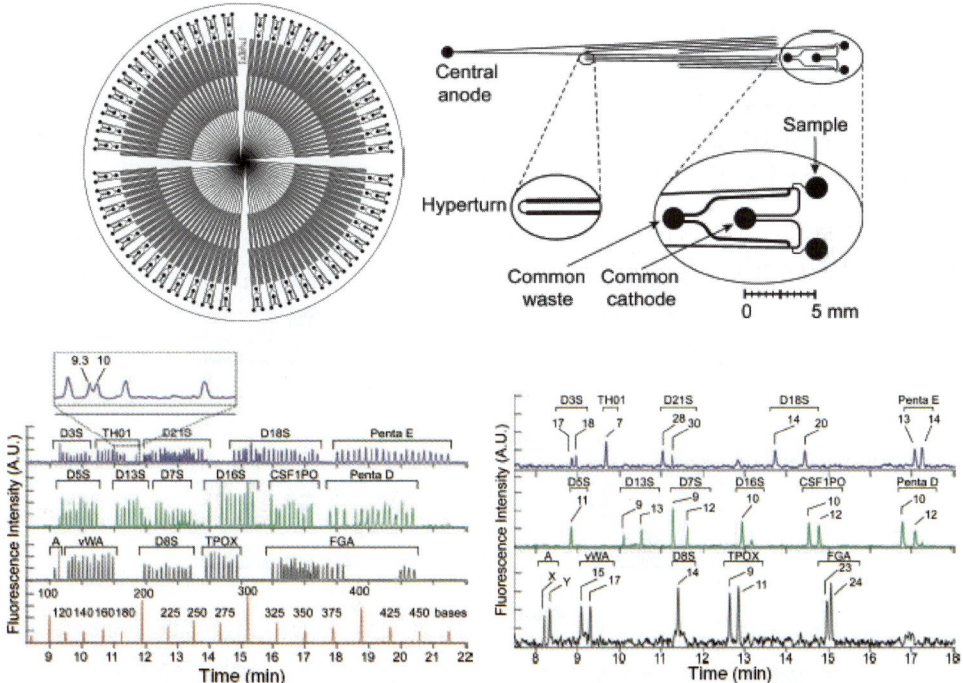

Fig. 14. The design and the test results for a 96-channel radial capillary array electrophoresis chip.[26] (Copyright Wiley. Reprinted with permission.)

coworkers for forensic applications.[26] The authors reported that 96 samples in parallel can be analyzed in less than 25 min with enough resolution.

Mutation detection

Mutations may induce different effects on a DNA sample. One of them is the deletion or addition of base-pairs to the original fragment, resulting in size change. This can be detected by comparing the mutant sample with its wild-type counterpart using electrophoresis. In addition, mismatches can occur in one or more bases, where the overall fragment size stays the same. Although the size remains unchanged, electrophoresis may help in detecting such mutations with a technique known as heteroduplex analysis (HDA). HDA is based on the separation of the double-stranded DNA (dsDNA) fragments with perfect complementaries (homoduplexes) those with mismatches (heteroduplexes) (Fig. 15). Because of the heteroduplexes, the usual confirmation of the DNA is distorted and therefore, these DNA samples have slower migration patterns in high-resolution separation matrices. Even when only one mismatch occurs, the heteroduplex DNA fragment has a different mobility compared to its homoduplex counterpart. Thus, if a DNA size ladder or an ideal homoduplex wild-type DNA and the test sample are run together, heteroduplex structures can be distinguished.[15] HDA electrophoresis has been commonly used at the macro-scale level.

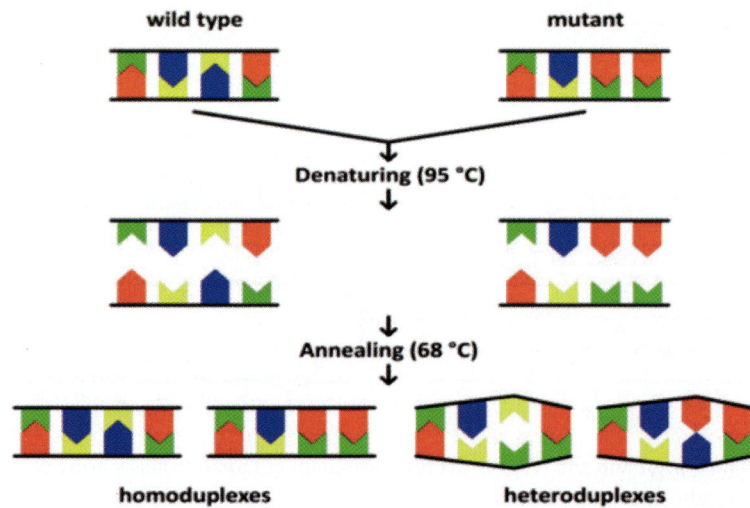

Fig. 15. Illustration of generation of homoduplexes and heteroduplexes from wild-type and mutant DNA fragments.[15]

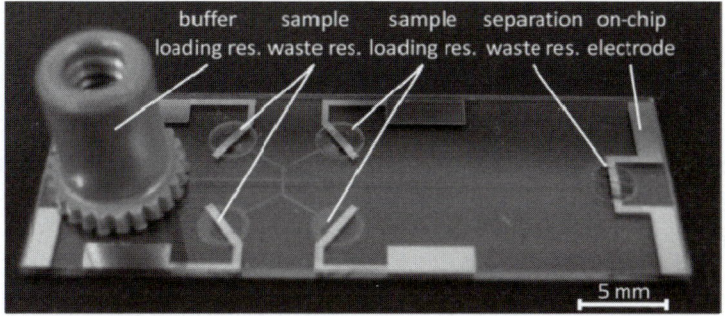

Fig. 16. A micro electrophoresis system for heteroduplex analysis (HDA).[27]

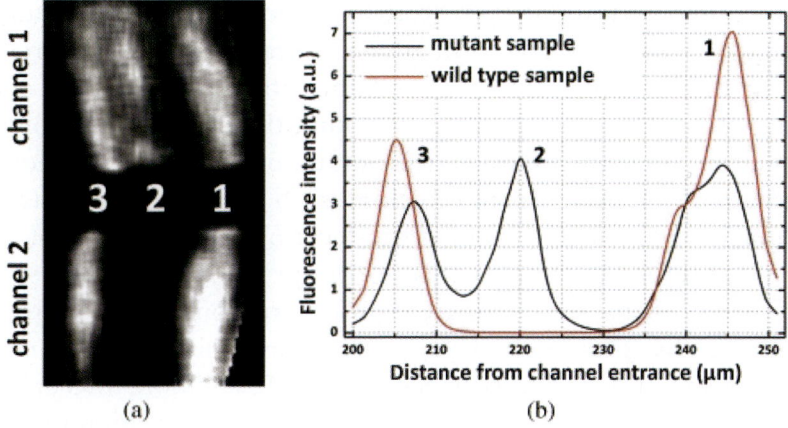

Fig. 17. Test results of the microchip. (a) Band formation and (b) histogram: (1) homoduplexes, (2) heteroduplex, and (3) gel interfaces. Homoduplexes migrate faster than heteroduplexes.[27]

A micro-scale approach with differential reading may improve the results for HDA. In this approach, two identical microchannels with independent sample loading and waste reservoirs are employed to run wild-type and mutant DNA samples (Fig. 16). It was reported that mutation detection was successfully achieved in a total separation length of 250 μm in less than 3 min for a 590-bp DNA sample harboring a 3-bp mutation causing an amino acid change (Fig. 17).[27] Parylene-C was used as the structural material for fabricating the microchannels as it provides conformal deposition, transparency, biocompatibility, and low background fluorescence without any surface treatment. The use of HDA as the mutation detection method on a cross-linked polyacrylamide medium provided accurate mutation detection in an extremely short length of time. The presence of the two channels on the microchip offers the opportunity of rapidly comparing the sample to be tested with a desired control sample, which is very critical for the accuracy and reliability of the mutation analyses, especially for clinical and research purposes.

The performance of mutation analysis is defined with the detection of SNPs at the longest chain of DNA fragment. It is evident that for longer DNA fragments, detecting SNPs will require higher resolution.

Protein electrophoresis

Separating and analyzing proteins is an important task to fully understand their functions in living organisms. Isoelectric focusing and protein electrophoresis are used for this purpose. Proteins are neutral at their specific isoelectric point (pI) — positive under the pI, and negative above it. Therefore, if a protein solution is subjected to a medium with a pH gradient, and if an electrical potential is applied on this medium, then the protein samples will be focused on different positions depending on their pI value. However, if the pI values of the proteins are too close, such as hemoglobin A (pI = 7.1) and hemoglobin C (pI=7.4), isoelectric focusing method may not be sufficient for separation. Because of this reason, protein electrophoresis can be used on such proteins, with a different procedure from DNA electrophoresis (Fig. 18). In this procedure, first, proteins are uniformly and negatively charged by using SDS (sodium dodecyl sulfate). In addition, SDS disrupts the hydrogen bonds by heating and hence, unfolding (linearization or denaturation) of the proteins occurs.[28] Next, the electrophoresis run is started, while pH is kept constant in the separation zone. The separation in this method is based on the sizes of the proteins. Alkaline or acid PAGE (polyacrylamide gel electrophoresis) can be used instead of SDS.[29]

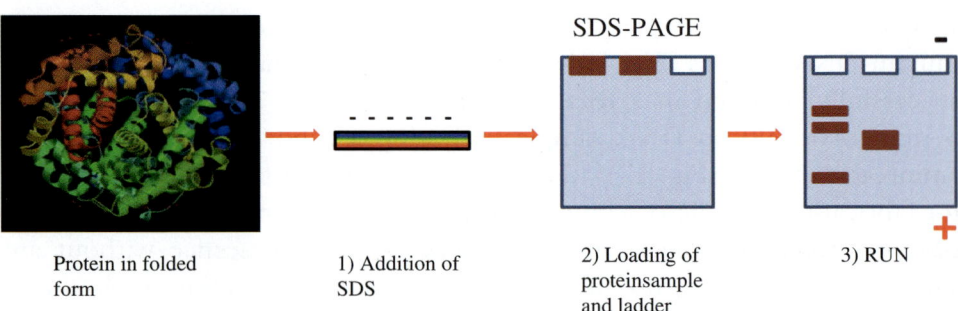

Fig. 18. Steps of SDS-PAGE protein electrophoresis.

Besides the common advantages of the micro-scale implementation, such as low cost, fast operation, portability, etc., reduced toxicity due to the decreased volumes of reagents is another advantage of miniaturized protein electrophoresis devices.[30] Up to now, some of the protein separation techniques, including capillary zone electrophoresis, capillary gel electrophoresis, and SDS-PAGE, have been applied to micro-scale devices, where the electrophoretic separation steps are the same with their macro-scale counterparts.

Yao et al. developed a micro-scale device that is capable of separating six different proteins in 30 s with 5 kV applied potential, using the SDS capillary gel electrophoresis technique.[31]

Figure 19 shows the experimental results for another device with double T-injector, which is capable of separating four different proteins (α-lactalbumin, trypsin inhibitor, carbonic anhydrase, and alcohol dehydrogenase) in 6.4 s with 1.82 kV.[32]

Another type of protein electrophoresis is native PAGE, which does not measure the molecular weight of a proteindirectly. This technique provides some useful information about the charge of the protein samples and their subunit composition.[33] One of the major advantages of this technique is that proteins preserve their activity as they are not denaturated, and therefore they can be detected in their biologically active forms by using proper labeling dyes.

Figure 20 shows a microchip utilizing both native and SDS-PAGE with 36 parallel capillaries and its test results. In this study, the electrophoresis of the native proteins (in nondenaturized form) and SDS proteins (in denaturized form) in 20 min were observed in a single chip, simultaneously. In this device, planar electrodes were used in order to provide more uniform distribution of the electric potential into 36 parallel capillaries.[34]

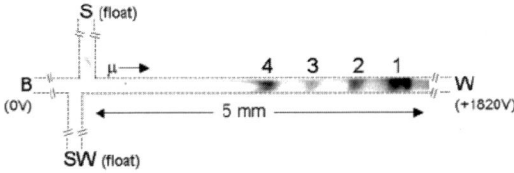

Fig. 19. Experimental results for a protein electrophoresis system utilizing double-T injector. (1. α-lactalbumin, 2. trypsin inhibitor, 3. carbonic anhydrase, 4. alcohol dehydrogenase).[32] (Copyright American Chemical Society. Reprinted with permission.)

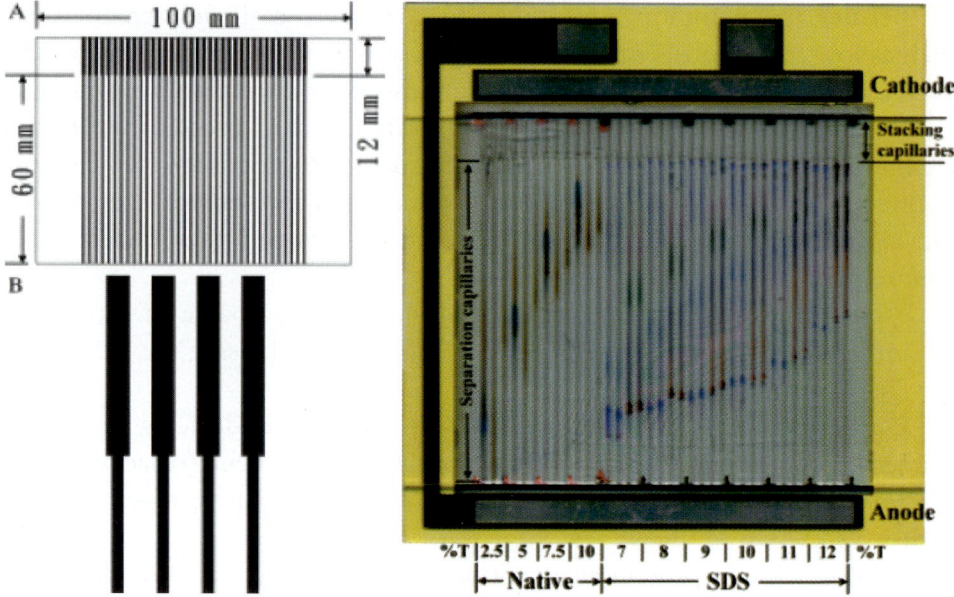

Fig. 20. A micro-chip utilizing both native and SDS-PAGE with 36 parallel capillaries.[34] (Copyright Wiley-VCH Verlag GmbH & Co. KGaA. Reproduced with permission.)

Conclusion

In this section, the theoretical explanation of electrophoresis is presented with emphasis on electric double layer, electroosmosis, and viscous drag force. Then, the most common electrophoretic separation applications, DNA and protein electrophoresis, have been introduced in macro- and micro-scale with examples from the literature.

The electrophoresis theory is significantly mature and the practical applications in macro-scale have an important place in the diagnosis of some neural diseases (multiple sclerosis, Alzheimer Disease[35]), blood diseases (mild hemolytic anemia[36]), oncologic diseases (monoclonal gammopathy[37]), cardiovascular diseases, (detection of apoliprotein E genotypes[38]), forensic medicine,[28] pharmacology,[28] and molecular biology (SNP[1]). Micro-scale implementation improves the performance of electrophoresis in these applications, while bringing more advantages such as fast separation, low reagent and sample consumption, reduced layout area, and multiple processing in a very small area.

Dielectrophoresis

Dielectrophoresis (DEP) is another electrical separation method which can be used in biological applications. In contrast to electrophoresis, DEP is defined as the movement of the dielectric particles in the presence of a nonuniform electric field, when the particle and the surrounding medium have different dielectric constants and polarizabilities. This method depends on the dielectric properties of the particle, which is directly related to its chemical, structural, and morphological characteristics. Therefore, DEP provides a more sensitive analysis compared to other separation techniques.[39] This section reviews the theoretical explanations of DEP separation techniques. Besides, a classification of DEP devices is presented according to operation strategies and electrode configurations.

Theory of dielectrophoresis

DEP is defined the movement of dielectric particles dispersed in a medium under a nonuniform electric field. The force causing this movement is called as *dielectrophoretic force*, and it can be formulated as[40]:

$$F_{DEP} = (p \cdot \nabla) E, \qquad (11)$$

where p is the dipole moment of the particle and E is the applied electric field. The dipole moment of a homogenous solid spherical particle with radius R is defined as:[40]

$$p = 4\pi\varepsilon_m \left(\frac{\varepsilon_p^* - \varepsilon_m^*}{\varepsilon_p^* + 2\varepsilon_m^*} \right) R^3 E, \qquad (12)$$

where ε_m denotes the electrical permittivity of the medium, ε_p^* and ε_m^* are the complex permittivities of the particle and the medium, respectively. Here, complex permittivity[41] is defined as:

$$\varepsilon^* = \varepsilon - j\frac{\sigma}{\omega}, \qquad (13)$$

where ω is the angular frequency of the applied electric field, and ε and σ denote the electrical permittivity and the conductivity, respectively.

Equation (14)[41] represents the time-averaged dielectrophoretic force exerted on a homogenous spherical particle with radius R:

$$\langle F_{DEP} \rangle = 2\pi\varepsilon_m R^3 Re(f_{CM}) \nabla |E|^2, \qquad (14)$$

where f_{CM} is the Clausius–Mossotti factor (Eq. (15)),[41] which indicates whether the medium or the particle is more polarizable.

Additionally,

$$f_{CM} = \frac{\varepsilon_p^* - \varepsilon_m^*}{\varepsilon_p^* + 2\varepsilon_m^*}, \qquad (15)$$

where ε_p^* and ε_m^* are complex electrical permittivities of the particle and the medium, respectively.

The equations presented above give the following outcomes:

(1) Under uniform electric field, DEP force does not exist.
(2) The polarity of the applied potential has no effect on the direction of the DEP force. This direction depends on the intensity of the electric field, the dielectric properties of the particle, and the frequency of the applied potential.
(3) The frequency of the applied potential plays an important role in the direction of the DEP force since this frequency is the only adjustable value in the complex permittivity equation. Note that the complex permittivity defines the polarizability of the particle and the medium and hence, the direction of the DEP force. If the particle and the medium have the same polarizabilities, the force becomes zero since f_{CM} is zero.

DEP can be classified as positive or negative depending on the sign of $Re(f_{CM})$, which indicates the direction of movement of the dielectric particle. If the complex permittivity of the particle is larger than that of the medium, i.e., $\varepsilon_p^* > \varepsilon_m^*$, then the particle moves to the region where the electric field is more intense. Inversely, if the complex permittivity of the particle is smaller than the medium, i.e., $\varepsilon_p^* > \varepsilon_m^*$, then the particle moves to the region where the electric field is less intense[40] (Fig. 21).

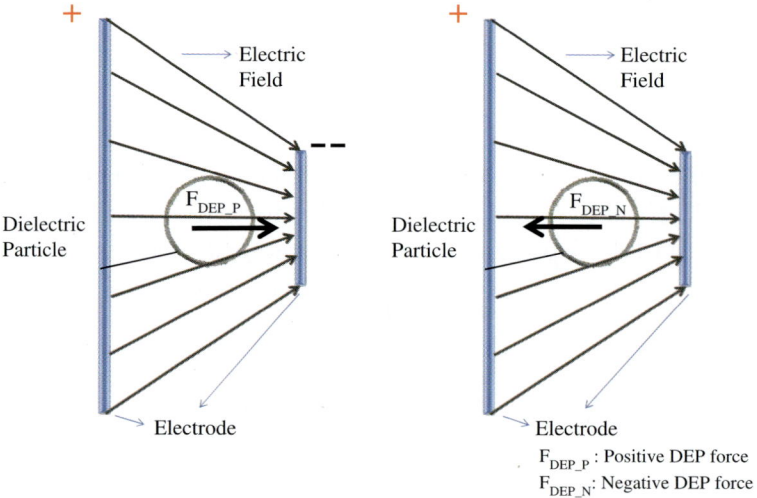

Fig. 21. Illustration of the dielectrophoretic force showing negative and positive dielectrophoretic forces.

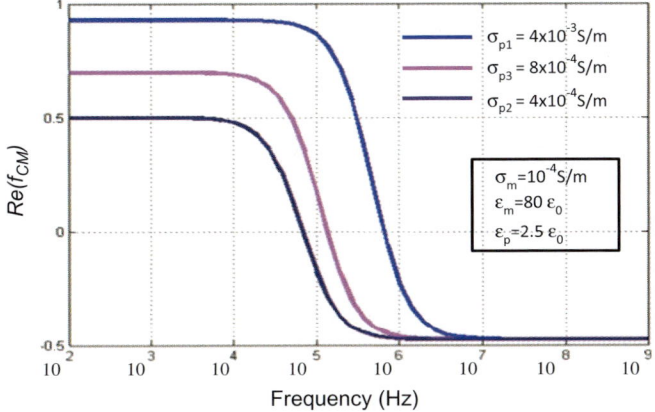

Fig. 22. Real part of the Clausius–Mossotti factor with respect to the applied voltage frequency. (σ_{p1}, σ_{p2}, and σ_{p3} are the particle conductivities and σ_m is the medium conductivity.)

Figure 22 shows the variation of the real part of the Clausius–Mossotti factor with respect to the applied voltage frequency. This plot indicates that a particle can be subjected to positive or negative DEP force depending on the frequency of the applied voltage.

Limiting cases of the Clausius–Mossotti factor for frequency spectrum can be expressed as:[42]

$$\lim_{\omega \to 0} Re(f_{CM}) = \frac{\sigma_p - \sigma_m}{\sigma_p + 2\sigma_m} \tag{16a}$$

and

$$\lim_{\omega \to \infty} Re(f_{CM}) = \frac{\varepsilon_p - \varepsilon_m}{\varepsilon_p + 2\varepsilon_m}. \tag{16b}$$

These results show that the sign of the Clausius–Mossotti factor is determined by the electrical conductivities of the particle and the medium at low frequencies, whereas it depends on the permittivity values at high frequencies. Another point is that when the frequency goes to zero (DC), and if σ_p is much larger than σ_m, the maximum value of the Clausius–Mossotti factor occurs (1). However, if σ_m is much larger than σ_p, in the same case, then the Clausius–Mossotti factor will be at its minimum value (−0.5). These outcomes are also valid for high frequencies when the permittivities are concerned.[42]

Analysis on the Clausius–Mossotti factor equation with the complex permittivity definition shows that there is a frequency at which the Clausius–Mossotti factor becomes zero and hence, the DEP force.

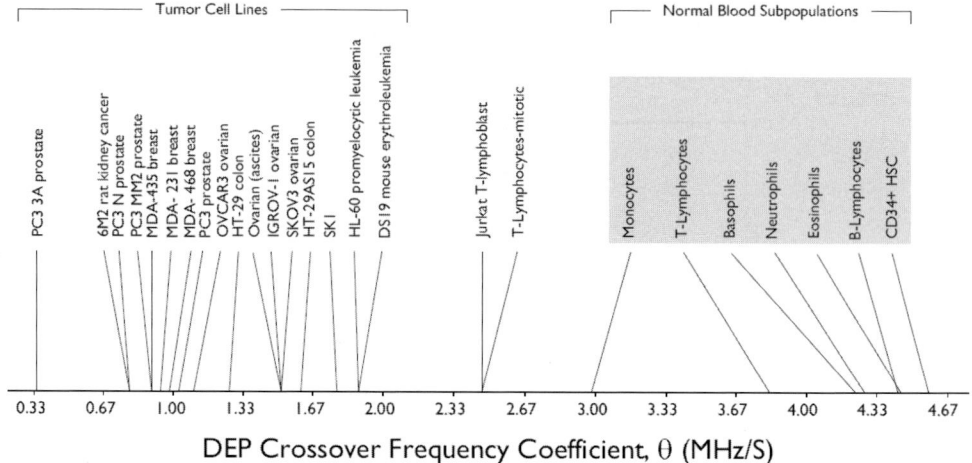

Fig. 23. The distribution of the crossover frequency for different types of tumor cells and normal peripheral blood mononucleocytes. (Courtesy of Gascoyne et al.[43])

This frequency is called DEP crossover frequency (f_{cross}), which can be formulated as:[42]

$$f_{cross} = \frac{1}{2\pi}\sqrt{\frac{(\sigma_m - \sigma_p)(\sigma_p + 2\sigma_m)}{(\varepsilon_p - \varepsilon_m)(\varepsilon_p + 2\varepsilon_m)}}. \qquad (17)$$

The separation of biological particles (e.g., cells) can be provided by utilizing the crossover frequency. The distribution of the crossover frequency for different tumor cell lines and normal peripheral blood mononucleocytes has been reported in the literature.[43] Figure 23 shows a similar distribution for DEP crossover frequency coefficients.

If the crossover frequencies of the particles are not too close, the DEP separation technique can be utilized. On the other hand, if f_{cross} of the particles are too close, DEP may not be sufficient for proper separation. Therefore, size-based separation can be a solution for particles with similar electrical properties.

According to Eq. (14), two different-sized particles with the same electrical properties and density will observe different DEP forces proportional to the cube of their radiuses (R^3). Since the masses of the particles are also proportional to R^3, DEP force will create the same acceleration for these particles (Eq. (18)). However, in a liquid medium, due to the viscous drag force balancing the DEP force (Eq. (19)),[44,45] the particles will eventually observe zero acceleration.

$$a = \frac{3}{2}\frac{\varepsilon_m}{d}Re(f_{CM})\nabla|E|^2 \qquad (18)$$

and

$$2\pi\varepsilon_m R^3 Re(f_{CM})\nabla|E|^2 = 6\pi\eta R\upsilon, \quad (19)$$

where ε_m denotes the electrical permittivity of the medium, R is the radius of the particle, f_{CM} is the Clausius–Mossotti factor, ϑ is the velocity of the particles, and η is the viscosity of the medium.

From Eq. (19), average velocity of the particles ($\langle v \rangle$) can be obtained as[44]:

$$\langle v \rangle = \frac{\langle \varepsilon_m R^2 Re(f_{CM})\nabla|E|^2 \rangle}{3\eta} \quad (20)$$

According to Eq. (20), for two different particles with different radii (R_1, $R2$) and different Clausius–Mossotti factors ($f_{CM},1$, $f_{CM},2$), the following relation between the average speeds ($\langle \vartheta_1 \rangle$, $\langle \vartheta_2 \rangle$) can be expressed as:[44]

$$\left\langle \frac{\vartheta_1}{\vartheta_2} \right\rangle = \frac{\left\langle \frac{F_{DEP,1}}{6\pi R_1 \eta} \right\rangle}{\left\langle \frac{F_{DEP,2}}{6\pi R_2 \eta} \right\rangle} = \frac{R_1^2 Re(f_{CM,1})}{R_2^2 Re(f_{CM,2})} \quad (21)$$

Therefore, the average speed, or equivalently the traveled distance for a predefined time period, can be used as the discrimination factor for these two different particles.

If the f_{CM} values are the same, Eq. (21) reduces to:

$$\left\langle \frac{\vartheta_1}{\vartheta_2} \right\rangle = \frac{\left\langle \frac{F_{DEP,1}}{6\pi R_1 \eta} \right\rangle}{\left\langle \frac{F_{DEP,2}}{6\pi R_2 \eta} \right\rangle} = \frac{R_1^2}{R_2^2}. \quad (22)$$

Equation (22) shows that size-based separation by DEP is feasible.

Classification of DEP systems according to operation strategies

Since there exists many ways to create a nonuniform electric field in a microchannel, there are various strategies reported in the literature for the manipulation and separation of particles. This section presents the unique features of some common strategies: field-flow-fraction DEP (FFF-DEP), travelling wave DEP (twDEP), multiple frequency DEP (MF-DEP), electrothermal-assisted DEP (ETA-DEP), and immune-assisted DEP (IA-DEP) strategies.[39]

Field-flow-fraction dielectrophoresis (FFF-DEP)

In this strategy, a negative DEP force is created such that it levitates the particles in the microchannel. Through this way, particles with different masses will be levitated to different heights. Since the flow inside the channel has parabolic velocity profile in vertical axis, these particles will have different velocities and therefore, separation is realized. This technique is more suitable for living cell separation as it requires lower electric fields than the other strategies. However, long electrode arrays and microchannels are required for proper separation.[39,46,47]

Travelling wave dielectrophoresis (twDEP)

In this method, a traveling electric field is generated by using electrodes biased with 90° phase-shifted AC voltages (e.g., 0°, 90°, 180°, and 270°). This creates a negative DEP force due to $Re\,(f_{CM})$ levitating the particles in the microchannel, as in the case of FFF-DEP. Besides this, the traveling electric field creates another DEP force in parallel direction with the microchannel due to $Im\,(f_{CM})$. Due to these two forces, separation of the particles is realized.[39,48]

Multiple frequency dielectrophoresis (MF-DEP)

In this technique, more than two electric fields with different frequencies are used to separate the particles. These electric fields create different DEP forces in different directions and amplitudes, and the particles move depending on the superposition of these forces. MF-DEP provides high-resolution separation for particles with very close f_{CM} values.[39,49]

Electrothermal-assisted dielectrophoresis (ETA-DEP)

This separation strategy utilizes Joule heating, which is related to the power dissipation in a liquid medium (e.g., buffer solution), and hence electrothermal heating. This heating creates electrothermal vortices inside the channel, realizing DEP movement if the medium has high conductivity. ETA-DEP can be used for the accumulation of dielectric particles rather than separation.[39,50]

Immuno-assisted dielectrophoresis (IA-DEP)

In this strategy, DEP force is used for improving the capture of target particles by a surface with immobilized antibodies. Here, the DEP force is created such that it drags only the the desired particles (e.g., rare cells) on a activated surface to improve the detection efficiency.[39,51,52]

Classification of DEP systems according to electrode configuration

In the previous section, some DEP strategies are discussed along with their features. In DEP microdevices, different electrode configurations (e.g., castellated, ratchet, concentric, quadrupole, parallel, and spiral[44]) can be utilized according to the properties of these strategies. Besides altering the electrode configuration, a non-uniform electric field can also be formed by placing obstacles inside the microchannel. This section summarizes these different approaches for creating a non-uniform electric field.

Castellated electrodes

This electrode configuration consists of electrodes that are patterned such that the distance between them periodically increases and decreases in a square wave formation (Fig. 27). The opposite arrays are stimulated with the same sinusoidal signal with a 180° phase shift. The electric field on the sharp edges of the facing electrodes is more intense than the electric field formed at the inner regions.[44,53]

Ratchet electrodes

This electrode configuration is patterned such that parallel electrodes are placed into the microchannel with a constant inclination angle with the wall of the microchannel (Fig. 25). Since the electric field is stronger at the tips of

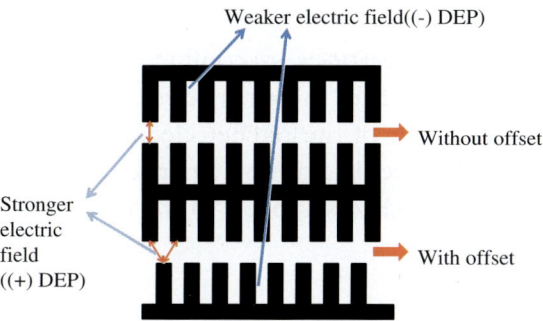

Fig. 24. Castellated electrodes are utilized to create dielectrophoretic force without and with offset, respectively.[54]

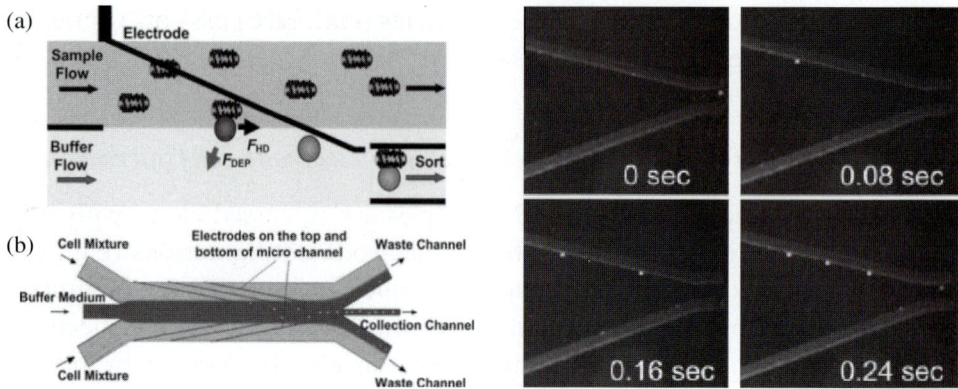

Fig. 25. An application of ratchet electrodes. Reprinted with permission from Ref. 55.

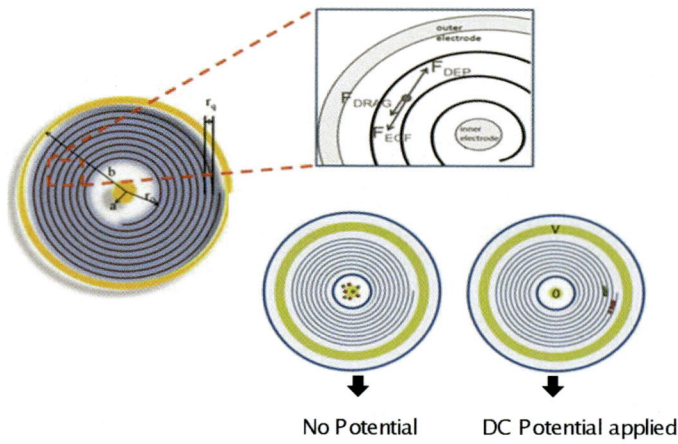

Fig. 26. Concentric electrode configuration with spiral channels and force diagram on particle. (a: the inner electrode radius, b: the outer electrode radius, r_0: the initial spiral radius, and r_0: the radius increment per turn, F_{DRAG}: viscous drag force, F_{EOF}: electroosmotic flow force).[57]

the electrodes, positive DEP particles are confined into the middle of the channel and by using a three-way exit, these particles can be collected.[55] Figure 25 shows an experimental result of the ratchet electrodes with the schematic of the microdevice and the force diagram on a particle. In this experiment, a 20 V_{pp} sinusoidal signal at 500 kHz is applied to the electrodes.[55]

Concentric electrodes

This electrode configuration provides a nonuniform electric field in a radial direction (Fig. 26). Considering the electric field equation for coaxial lines

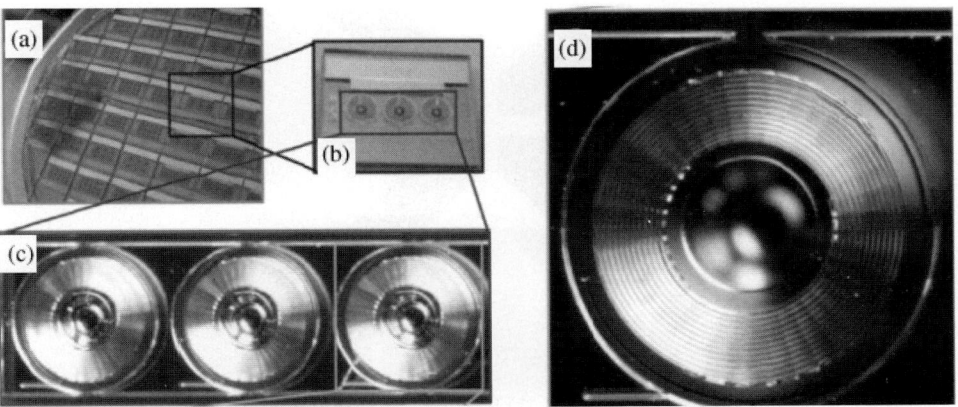

Fig. 27. Photograph of a micro dielectrophoresis chip utilizing concentric electrodes with spiral microchannels.[57]

(Eq. (23)), strong electric field is generated around the inner electrode, whereas weak electric field is observed around the outer electrode:

$$\overrightarrow{E(r)} = \frac{V}{\ln(b/a)r} a_r. \quad (23)$$

Figure 26 shows an application of this configuration with spiral microchannels and the force diagram on a particle. In this structure, a DC electric field induces a sized-based DEP separation. Figure 27 shows the fabricated devices.[56,57]

Quadrupole electrodes

Quadrupole electrode configuration consists of four orthogonal electrodes facing each other at a center point. In this configuration, electrodes next to each other are biased with 180° phase-shifted sinusoidal voltages. This results in weak electric field regions at the center and strong electric fields along the arms. Figure 28 shows a micro-DEP system utilizing quadrupole electrodes for bovine red blood cell separation. In this example, a $10 V_{pp}$ sinusoidal voltage is applied with a frequency varying from 50 kHz to 5 MHz. It has been reported that red blood cells observe (+) DEP in between 50 kHz and 5 MHz (Fig. 28(b)), whereas they observe (−) DEP under 50 kHz and above 5 MHz (Fig. 28(c)).[58]

Parallel electrodes

This electrode configuration consists of a series of parallel electrodes biased with signals having constant phase differences (e.g., 0°, 90°,180°, 270°, 0°, 90°…).

Fig. 28. A micro-DEP system utilizing quadrupole electrodes for bovine red blood cell separation.[58] (a) Quadrupole electrode structure, (b) red blood cells are located on the arms indicating that they observe (+) DEP between 50 kHz and 5 MHz, (c) red blood cells are concentrated at the center indicating that they observe (−) DEP under 50 kHz and above 5 MHz.[58] (Copyright American Institute of Physics. Reprinted with permission.)

Maturos *et al.* implemented a micro-scale twDEP system utilizing parallel electrodes, where they observed that the cells under test observed (−) DEP under a 8V, 50 KHz sinusoidal signal.[59]

Spiral electrodes

This electrode configuration is the wrapped version of very long parallel electrodes and it can be utilized by either rectangular or spherical spiral electrodes. Figure 29 shows a spiral electrode configuration and its biasing. In this figure, the experimental results for concentrating malaria-infected erythrocytes are presented.[60]

In this section, different methods for creating a nonuniform electric field by changing the electrode configuration are presented. Another way of forming a nonuniform electric field is to place insulating obstacles inside the channel, in this section which will disturb the field distribution. The insulating obstacle can be either inside the channel or on the surface of the channel wall. Figure 30 presents a study in which an insulating hurdle is constructed between the electrodes to create a nonuniform electric field to separate latex particles and yeast cells.[61]

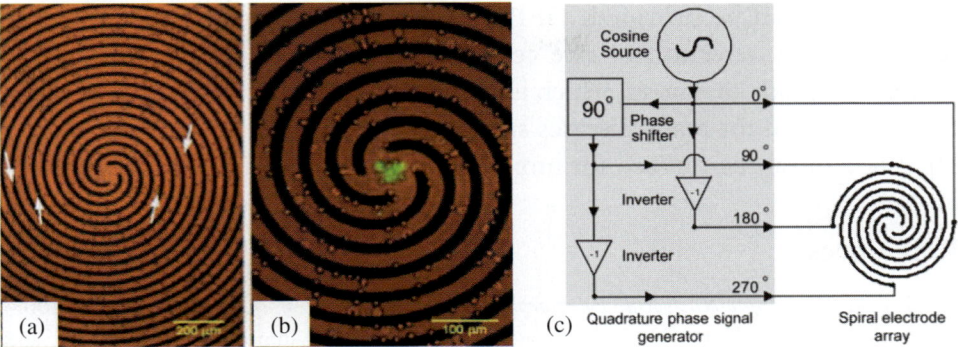

Fig. 29. (a) parasitic cells (green) are marked with arrows and (b) a sinusoidal voltage with 2 V_{pp} and 2 MHz was applied and parasitic cells were collected at the center.[60] (Reprinted with permission.)

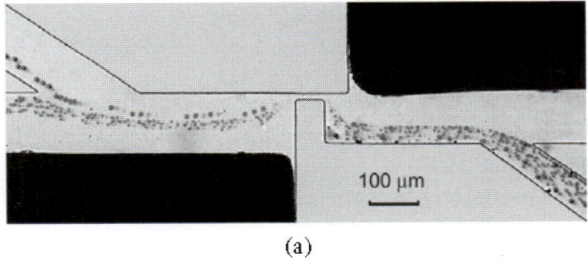

Fig. 30. Utilization of an insulating hurdle to create nonuniformity in the electric field distribution and separation of 5-um and 10-um particles at 10 V.[61] (Reprinted with permission and courtesy of Cetin.)

Conclusion

The theory of DEP is discussed in this section, with emphasis on the important parameters of DEP force. In addition, DEP separation strategies and electrode configurations are presented with practical examples.

Although the DEP theory is well defined over the years, laboratory and micro-scale applications are limited due to some practical reasons. In DEP separation, the separation performance is dependent on the secondary effects (such as medium conditions, particle–medium interactions, particle–channel surface interactions, sensitivity of the biological samples (e.g., living cells) to the medium conditions[62] and the applied field strength, electrical shielding effects,[63] etc.) other than the DEP force. Therefore, DEP separation systems require extensive efforts on optimization and experimental verification. However, if high-performance separation in accordance with the theoretical expectations can be provided, DEP may become an invaluable tool for the separation and detection of rare cells and viruses (e.g., cancer cells, HIV). Using DEP, early diagnosis, which is an important step for the curing process, can be achieved for many diseases such as cancers and AIDS. Besides, such a DEP system can be utilized for monitoring the treatment process.

References

1. Kwok, P.Y., *Single Nucleotide Polymorphisms: Methods and Protocols*. Humana Press, 2003.
2. Dirks-Go, S.I., Zanen, H.C., Latex agglutination, counter immunoelectrophoresis, and protein A co-agglutination in diagnosis of bacterial meningitis. *J Clin Pathol*, 1978, **31**: p. 1167–1171.
3. Righetti, P.G., *Electrokinetic Separation Methods*. Elsevier Science, 1979.
4. Li, D., *Electrokinetics in Microfluidics*. Academic Press, 2004.
5. Kopeliovich, D., Stabilization of colloids, Retrieved in 2011, from http://www.substech.com/dokuwiki/docu.php?id=stabilization_of_colloids.
6. Russel, W.B., Saville, D.A., Schowalter, W.R., *Colloidal Dispersions*. Cambridge University Press, New York, 1989.
7. Howe, M.S., *Acoustics of Fluid-Structure Interactions*. Cambridge University Press, 1998.
8. Goldston, R.J., Rutherford, P.H., *Introduction to Plasma Physics*. Institute of Physics Publishing, Philadelphia, 1997.
9. Lyklema, J., *Fundamentals of Interface and Colloid Science*. Academic Press, New York, 1993.
10. Jorgenson, J.W., Lukacs, K.D., Zone electrophoresis in open-tubular glass capillaries, *Anal Chem*, 1981, **53**, p. 1298–1302.
11. Strobl, G.R., *The Physics of Polymers Concepts for Understanding Their Structures and Behavior*. Springer-Verlag, 1996.
12. Vos, P. *et al.* AFLP: A new technique for DNA fingerprinting, *Nucleic Acids Res*, 1995, **23**, p. 4407–4414.
13. Watson, J.D., Crick, F.H.C., Molecular structure of nucleic acids; a structure for deoxyribose nucleic acid, *Nature*, 1953, **171**: p. 737–738.
14. Ghosh, A., Bansal, M., A glossary of DNA structures from A to Z, *Acta Crystallographica D Biol Crystallography*, 2003, **59**: p. 620–626.
15. Sukas, S., Design and fabrication of a DNA electrophoresis chip based on MEMS technology, M.Sc. Dissertation, METU, 2007.
16. Petersen, N.J., Nikolajsen, R.P.H., Mogensen, K.B., Kutter, J.P., Effect of Joule heating on efficiency and performance for microchip-based and capillary-based electrophoretic separation systems: A closer look, *Electrophoresis*, 2004, **25**: p. 253–269.
17. Lodish, H.F., *Molecular Cell Biology*. W. H. Freeman and Company, New York, 2008.
18. Hellabio Company, Retrieved in 2011, from http://www.hellabio.com/index2.htm.
19. Ashton, R., Padala, C., Kane, R.S., Microfluidic separation of DNA. *Curr Opin Biotechnol*, 2003, **14**: p. 497–504.
20. Brahmasandra, S.N., Ugaz, V.M., Burke, D.T., Mastrangelo, C.H., Burns, M.A., Electrophoresis in microfabricated devices using photopolymerized polyacrylamide gels and electrode-defined sample injection, *Electrophoresis*, 2001, **22**: p. 300–311.

21. Wu, C.H., Yang, R.J., Improvements on the electrokinetic injection technique for microfluidic chips. *Electrophoresis*, 2006, **27**: p. 4970–4981.
22. Tsai, C.H., Yang, R.J., Tai, C.H., Fu, L.M., Numerical simulation of electrokinetic injection techniques in capillary electrophoresis microchips. *Electrophoresis*, 2005, **26**: p. 674–686.
23. Fu, L.M., Yang, R.J., Lee, G.B., Liu, H.H., Electrokinetic injection techniques in microfluidic chips. *Anal Chem*, 2002, **74**: p. 5084–5091.
24. Ronai, Z., Barta, C., Sasvari-Szekely, M., Guttman, A., DNA analysis on electrophoretic microchips: Effect of operational variables. *Electrophoresis*, 2001, **22**: p. 294–299.
25. Zheng, J., Webster, J.R., Mastrangelo, C.H., Ugaz, V.M., Burns, M.A., Burke, D.T., Integrated plastic microfluidic device for ssDNA separation. *Sens Actuators B*, 2007, **125**: p. 343–351.
26. Yeung, S.H.I., Greenspoon, S.A., McGuckian, A., Crouse, C.A., Emrich, C.A., Ban, J., Mathies, R.A., Rapid and high-throughput forensic short tandem repeat typing using a 96-Lane microfabricated capillary array electrophoresis microdevice. *J Forensic Sci*, 2006, **51**: p. 740–747.
27. Sukas, S., Erson, A.E., Külah, H., Sert, C., A parylene based double-channel micro-electrophoresis system for rapid mutation detection via heteroduplex analysis. *Electrophoresis*, 2008, **29**: p. 3752–3758.
28. Westermeier, R., *Electrophoresis in Practice*. Wiley-VCH Verlag GmbH and Co., KGaA, Weinheim, FRG, 2004.
29. Garfin, D.E., Gel electrophoresis of proteins. In: Alberts, B., Bray, D., Hopkin, K. *et al.*, *Essential Cell Biology*. Oxford University Press, 2003.
30. Anema, S.G., The use of 'lab-on-a-chip' microfluidic SDS electrophoresis technology for the separation and quantification of milk proteins. *Int Dairy J*, 2009, **19**: p. 198–204.
31. Yao, S., Anex, D.S., Caldwell, W.B., Arnold, D.W., Smith, K.B., Schultz, P.G., SDS capillary gel electrophoresis of proteins in microfabricated channels. *Proc Natl Acad Sci U S A*, 1999, **96**: p. 5372.
32. Herr, A.E., Singh, A.K., Photopolymerized cross-linked polyacrylamide gels for on-chip protein sizing. *Anal Chem*, 2004, **76**: p. 4727–4733.
33. Krause, F., Detection and analysis of protein–protein interactions in organellar and prokaryotic proteomes by native gel electrophoresis: (Membrane) protein complexes and supercomplexes. *Electrophoresis*, 2006, **27**: p. 2759–2781.
34. Tsai, S.W., Loughran, M., Suzuki, H., Karube, I., Native and sodium dodecyl sulfate-capillary gel electrophoresis of proteins on a single microchip. *Electrophoresis*, 2004, **25**: p. 494–501.
35. Choe, L.H., Werner, B.G., Lee, K.H., Two-dimensional protein electrophoresis: From molecular pathway discovery to biomarker discovery in neurological disorders. *NeuroRX*, 2006, **3**: p. 327–335.
36. Steinberg, M.H., The paradox of hemoglobin SC disease. *Blood Rev*, 2003, **17**: p. 167–178.
37. Landgren, O. *et al.* Monoclonal gammopathy of undetermined significance (MGUS) consistently precedes multiple myeloma: A prospective study. *Blood*, 2009, **113**: p. 5412–5417.
38. Koay, E.S.C., Zhu, M., Wehr, T., Choong, M.L., Khaw, M.C., Sethi, S.K., Aw, T.C., Detection of apolipoprotein E genotypes by capillary electrophoresis. *Talanta*, 1998, **45**: p. 673–681.

39. Khoshmanesh, K., Novel dielectrophoretic platforms for manipulation and separation of particles, Ph.D. Dissertation, Deakin University, 2010.
40. Vykoukal, J., Vykoukal, D.M., Dielectrophoretic methods for biomedical applications. In: Yih, T., Talpasanu, I., eds. *Micro and Nano Manipulations for Biomedical Applications*, Artech House, Inc., 2008.
41. Jones, T.B., *Electromechanics of Particles*. Cambridge University Press, 1995.
42. White, C.M., Holland, L.A., Famouri, P., Application of capillary electrophoresis to predict crossover frequency of polystyrene particles in dielectrophoresis. *Electrophoresis*, 2010, **31**: p. 2664–2671.
43. Gascoyne, P.R.C., Vykoukal, J.V., Dielectrophoresis-based sample handling in general-purpose programmable diagnostic instruments. *Proc IEEE*, 2004, **92**: p. 22–42.
44. Yılmaz, G., Design and implementation of a MEMS based spiral channel dielectrophoretic separator for cytometry applications, M.Sc. Dissertation, METU, 2010.
45. Çiftlik, A.T., Design and implementation of a dielectrophoretic separator and a gravimetric detector for MEMS based cytometry applications, M.Sc. Dissertation, METU, 2008.
46. Gascoyne, P.R.C., Noshari, J., Anderson, T.J., Becker, F.F., Isolation of rare cells from cell mixtures by dielectrophoresis. *Electrophoresis*, 2009, **30**: p. 1388–1398.
47. Wang, X.B., Vykoukal, J., Becker, F.F., Gascoyne, P.R., Separation of polystyrene microbeads using dielectrophoretic/gravitational field-flow-fractionation. *Biophys J*, 1998, **74**: p. 2689–2701.
48. Choi, E., Kim, B., Park, J., High-throughput microparticle separation using gradient traveling wave dielectrophoresis. *J Micromech Microeng*, 2009, **19**: p. 125014.
49. Urdaneta, M., Smela, E., Multiple frequency dielectrophoresis. *Electrophoresis*, 2007, **28**: p. 3145–3155.
50. Grom, F., Kentsch, J., Müller, T., Schnelle, T., Stelzle, M., Accumulation and trapping of hepatitis A virus particles by electrohydrodynamic flow and dielectrophoresis. *Electrophoresis*, 2006, **27**: p. 1386–1393.
51. Yang, L., Dielectrophoresis assisted immuno-capture and detection of foodborne pathogenic bacteria in biochips. *Talanta*, 2009, **80**: p. 551–558.
52. Koo, O.K., Liu, Y., Shuaib, S., Bhattacharya, S., Ladisch, M.R., Bashir, R., Bhunia, A.K., Targeted capture of pathogenic bacteria using a mammalian cell receptor coupled with dielectrophoresis on a biochip. *Anal Chem*, 2009, **81**: p. 3094–3101.
53. Becker, F.F., Wang, X.B., Huang, Y., Pethig, R., Vykoukal, J., Gascoyne, P.R., Separation of human breast cancer cells from blood by differential dielectric affinity. *Proc Natl Acad Sci U S A*, 1995, **92**: p. 860–864.
54. Morgan, H., Hughes, M.P., Green, N.G., Separation of submicron bioparticles by dielectrophoresis. *Biophys J*, 1999, **77**: p. 516–525.
55. Hu, X., Bessette, P.H., Qian, J., Meinhart, C.D., Daugherty, P.S., Soh, H.T., Marker-specific sorting of rare cells using dielectrophoresis. *Proc Natl Acad Sci U S A*, 2005, **102**: p. 15757–15761.
56. Yilmaz, G., Ciftlik, A.T., Külah, H., A dielectrophoretic cell/particle separator fabricated by spiral channels and concentric gold electrodes. *Solid-State Sensors, Actuators and Microsystems Conference, Transducers International*, p. 73–76, 2009.
57. Yılmaz, G., Çiftlik, A.T., Külah, H., A MEMS based spiral channel dielectrophoretic chromotography system for cytometric applications. *BioTechnol J*, 2011, **6**: p. 185–194.

58. Gordon, J.E., Gagnon, Z., Chang, H.-C., Dielectrophoretic discrimination of bovine red blood cell starvation age by buffer selection and membrane cross-linking. *Biomicrofluidics*, 2007, **1**: p. 044102.
59. Maturos, T., Jaruwongrungsee, K., Bunthawin, S., Sappat, A., Wisitsora-at, A., Wanichapichart, P., Tuantranont, A., Development of traveling wave dielectrophoretic (twDEP) microfluidic system, *5th International Conference on Electrical Engineering/Electronics, Computer, Telecommunications and Information Technology*, ECTI-CON, 2008, p. 837–840.
60. Gascoyne, P., Mahidol, C., Ruchirawat, M., Satayavivad, J., Watcharasit, P., Becker, F.F., Microsample preparation by dielectrophoresis: Isolation of malaria. *Lab Chip*, 2002, **2**: p. 70–75.
61. Çetin, B., Microfluidic continuous separation of particles and cells by ac- dielectrophoresis, Ph.D. Dissertation, Vanderbilt University, 2009.
62. Grimnes, S., Rikshospitalet, O., Schwan, N.H.P., *Interface Phenomena and Dielectric Properties of Biological Tissue, Encyclopedia of Surface and Colloid Science*, p. 2643–2651, CRC Press, 2002.
63. Mukundan, V., Ponce, P., Butterfield, H.E., Pruitt, B.L., Modeling and characterization of electrostatic comb-drive actuators in conducting liquid media. *J Micromech Microeng*, 2009, **19**: p. 065008.

Chapter 15

Ultrasonic Embossing of Carbon Nanotubes for the Fabrication of Polymer Microfluidic Chips for DNA Sample Purification

Puttachat Khuntontong*, Min Gong†, and Zhiping Wang*,‡

**Microfluidics Manufacturing Programme
Singapore Institute of Manufacturing Technology
71 Nanyang Drive, Singapore 638075*

*†Genome Institute of Singapore
60 Biopolis Street, Singapore 138672*

Introduction

Microfluidics technology has been in existence for more than a decade,[1] but it has not been widely used in end applications. One reason could be due to the relative high cost of microfabrication. Polymer microfabrication techniques such as injection molding, thermoforming, and hot embossing allow the mass production of microdevices from a variety of materials.[2] Several types of polymers are available for a number of applications. However, the required investment may not be affordable to small companies. Moreover, the existing polymer microfabrication techniques demand complex equipment and a comparatively long processing time. Therefore, a fabrication technique which allows small enterprises to generate products with microfeatures in small scale and with low investment is desirable. Ultrasonic embossing of polymers is a process that appears to meet these needs as microstructures can be generated

---

‡Corresponding author. Email: zpwang@simtech.a-star.edu.sg

within a few seconds from a master tool.[3-4] Additionally, it is much aligned to the requirement of laboratory scale for rapid prototyping.

In ultrasonic embossing, polymer films are softened by heating with ultrasound and a pattern with microfeatures is molded. The required equipment is a commercially available ultrasonic welding machine and a micropatterned tool. This technique has been proven through the molding of microstructures and microchannels with cycle times of a few seconds and at a low cost.[5-8] The thermal flow sensor[9] and micromixer[10] were fabricated by ultrasonic embossing and the functions of these devices have been successfully demonstrated. The same technique has also been applied to emboss a conductive layer onto polymer films. Furthermore, transponders and flexible key pads have been fabricated on thin polymer films with rapid and low cost by ultrasonic embossing.[11]

In the ultrasonic embossing process, the horn transfers a high-frequency vibrational energy to the polymer film, which is between the horn and a master mold. The polymer is heated up locally at the protruding microstructures of the mold. The mechanism of heating can be described by the viscoelastic theory. Polymers exhibit a viscoelastic behavior; they behave as a combination of an elastic material and a viscous material. In viscoelastic materials, a part of the energy input is recovered, while the rest of the energy is dissipated. Therefore, for a viscoelastic material, the dissipated energy will result in the heating and melting of the polymer during ultrasonic embossing. The energy dissipated is a function of frequency and amplitude of vibrations, together with the loss modulus of the polymer. The average of energy dissipation per unit time and per unit volume is shown:[12]

$$\dot{Q}_{avg} = \frac{\omega E'' \varepsilon_0^2}{2}, \tag{1}$$

where E'' is the loss modulus, representing the ability of the material to convert mechanical energy into heat. ε_o is the strain amplitude and ω is the operation frequency, which is 35 kHz in this study. The dynamic loss modulus of the polymer is both frequency- and temperature-dependent.

Ultrasonic embossing can be divided into five phases:[13] pre-process, force build-up phase, transient phase, cooling phase, and de-molding phase as shown in Fig. 1. In pre-processing (Fig. 1(a)), polymer films are positioned between the horn and the metal tool. The horn moves downward and presses the polymer films between the horn and the tool. In the force build-up phase (Fig. 1(b)), pressure is applied gradually until a certain level is reached, which triggers the start of ultrasonic vibration. In the third phase (Fig. 1(c)), the vibrational energy is dissipated as heat, elevating the polymer temperature to a level

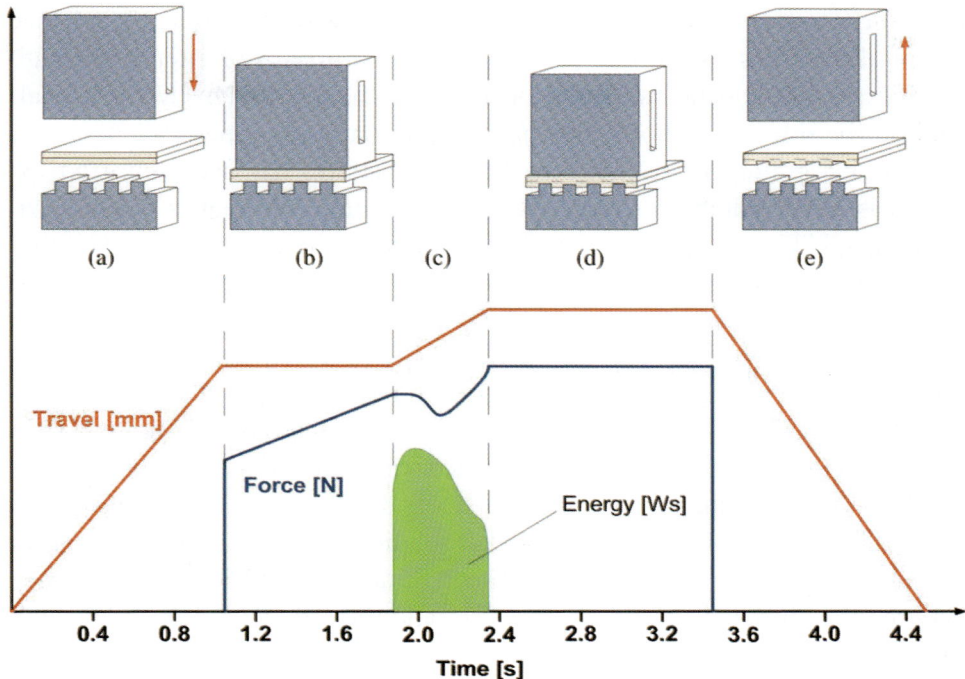

Fig. 1. Ultrasonic embossing process: (a) pre-processing phase, (b) force build-up phase, (c) transient phase, (d) cooling phase, and (e) de-molding phase.

sufficient to promote the embossing of micropatterns onto the polymer surface. The temperature can increase beyond the glass transition temperature (T_g), which depends on the vibration period, amplitude, and frequency of vibrations. Energy dissipation is concentrated at polymer parts, which are in contact with the microstructures on the tool. Therefore, a thin molten polymer layer is generated next to the protruding structures on the tool. The forces drop slightly in this phase because the polymer is softening and the melting polymer adapts to the metal tool. When the horn travels to a predefined distance, the ultrasound is stopped and cooling and solidification of the polymer begin (Fig. 1(d)). After cooling, the ultrasonic horn moves back up to its original position and the molded polymer can be easily removed from the tool (Fig. 1(e)).

Ultrasonic Embossing of Carbon Nanotubes onto Microchannels of a Polymer Device

Carbon nanotubes (CNTs) have been used in microfluidic applications[14–18] because of their advantageous properties such as excellent electrical conductivity, chemical resistance, thermal conductivity, mechanical strength, and high surface area-to-volume ratio. CNTs can be functionalized to enhance chemical reactions and to interact with molecular groups of interest through specific chemicals. The tubular nanostructure can be selectively functionalized on the side wall,[19] both ends of the tube,[20] and the inner wall.[21]

The integration of CNTs onto silicon substrates has been reported. As silicon is not a suitable material for low-cost applications, it is desirable to integrate CNTs into polymer microfluidic devices. Commonly used approaches for the integration of CNTs to polymer substrates include solution-based processing, soft lithography, casting by polydimethylsiloxane (PDMS),[22,23] and hot embossing.[24] In this study, we proposed a new approach to locally transfer a CNT layer into polymer microchannels by using ultrasonic embossing. With a rapid fabrication process, microchannels and embedded CNTs in the channels can be produced at the same step.

The carboxyl functional group on CNTs has been used for temporarily immobilizing DNA. Under high ionic salt solutions and high concentrations of hydrogen ions (H^+) at a pH lower than 6.5, an electrostatic force is formed between the phosphate group of DNA, H^+, and the carboxylic group. The high surface area-to-volume ratio in nanotubes increases the binding capacity of the DNA sample. DNA can be released from the carboxylic group when the solution has a pH higher than 6.8. Therefore, the immobilization of DNA onto CNTs can be controlled by adjusting the pH value, which is the same as in the commonly used method of silica surfaces for DNA extraction. Ultrasonic hot embossing is used to pattern microchannels and CNTs onto polymer films at the same time; the latter are deposited onto the surface of another polymer film. Figure 2 outlines the process of the ultrasonic embossing of CNTs onto polymer films.

First, a layer of CNTs needs to be deposited onto a polymer substrate, called the CNT carrier film. We used carboxylic group-functionalized single- walled carbon nanotubes (SWNTs) 90 wt%, 1–2 nm from Cheap Tubes, Inc. The SWNTs were dissolved in water with 0.5% of CNT concentration by molecular weight and were sonicated for an hour. The purification of the SWNT solution was performed by centrifugation at 8,000 rpm for 30 min and the removal of suspending particles. The deposition of CNT solution on a carrier film can be achieved by various techniques, for example, spin coating, air spray, and ink jet. We used an air gun to spray the solution onto the heated polymer film. The carrier film, made of cyclic olefin copolymer (COC) material of 100-μm thickness, was put on a hot plate at 105°C. COC was selected because its T_g of 135°C is higher than the temperature of water vapor. When the SWNT solution was sprayed on the carrier film, the water in a mixed solution evaporated, leaving only the SWNTs deposited on the carrier film. With this method, multi-layers of SWNTs were thus deposited onto the COC to achieve the desired thickness.

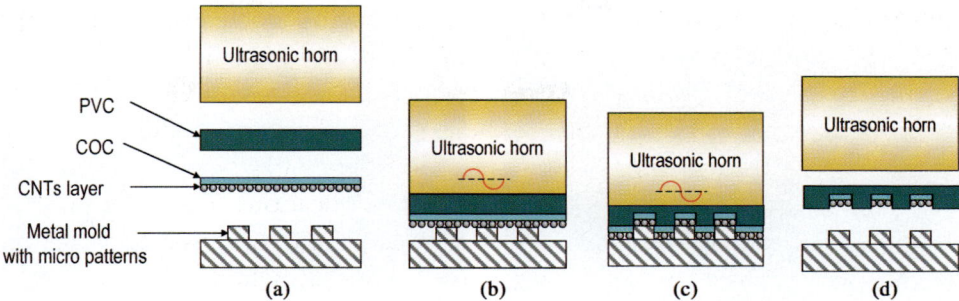

Fig. 2. Process of ultrasonic embossing of CNTs onto a polymer substrate. (a) SWNTs are deposited on the carrier cyclic olefin copolymer (COC), and placed between the master mold and the target polymer. SWNTs face the master mold. (b) Pressure from the ultrasonic machine is applied against the polymer layers and mold. At the preset pressure, ultrasound is started, (c) heat generates at the local area of the protrusion structure of the master mold, the temperature is raised above the glass transition temperature of the target polymer and the polymer is molded under the applied pressure. (d) Within a few seconds, the temperature drops to below T_g, and only the local area from the mold structures, SWNTs, and carrier layer are embedded onto the polymer substrate. Non-welded areas are removed easily.

Thereafter, a master mold is designed and fabricated for the replication of the microstructures and microchannels. Various mold-making techniques can be applied such as micromilling, electrodischarge machining (EDM), electroplating, and lithography with etching.

Finally, putting the SWNT carrier film between the master mold and polymer substrate with the SWNT side facing the master mold, nanotubular particles are embossed and welded onto the polymer substrate at the same step as the microstructures by ultrasonic embossing. Figures 2(a)–(d) depict the process of embossing SWNT layers onto the polymer substrate. Two layers of polymer films, with 180-µm thickness for each layer, are used as the polymer substrate. Ultrasonic embossing is conducted at the frequency of 35 kHz and 16.25-µm amplitude. The vibration starts once the trigger force of 250 N is reached. The embossing force is increased to 280 N as per preset values. The total vibration time is 0.3 s. During this time, the polymer is softened and molded into the master mold. The COC film with SWNTs is embossed and welded onto the polymer films at the location of the protruding structures. Non-welded areas are removed and only the embedded SWNTs are left onto the microchannels.

Scanning electron microscopy (SEM) and Raman spectroscopy were used to study the SWNTs before and after ultrasonic embossing. Figure 3(a) shows SEM images of SWNTs after deposition on the COC layer and Fig. 3(b) shows the CNTs embedded in the microchannels by ultrasonic embossing.

382   *Microfluidic Technologies for Human Health*

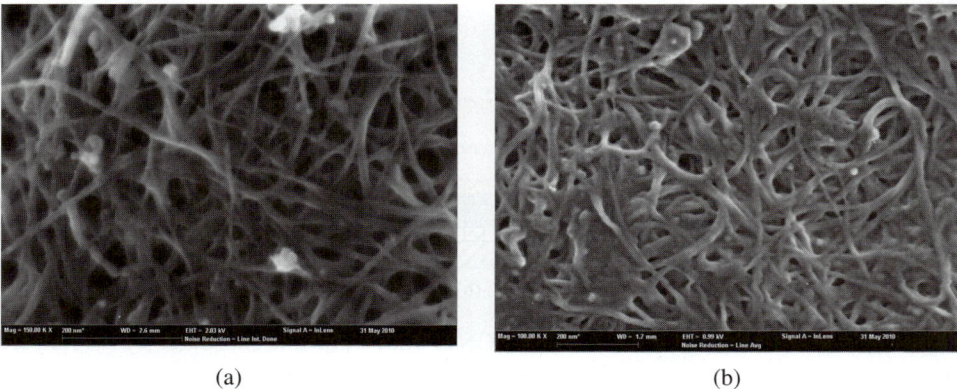

(a)   (b)

Fig. 3. SEM images of SWNTs on the carrier polymer. (a) Before ultrasonic embossing. (b) SWNTs deposited on the underside of channel by ultrasonic embossing.

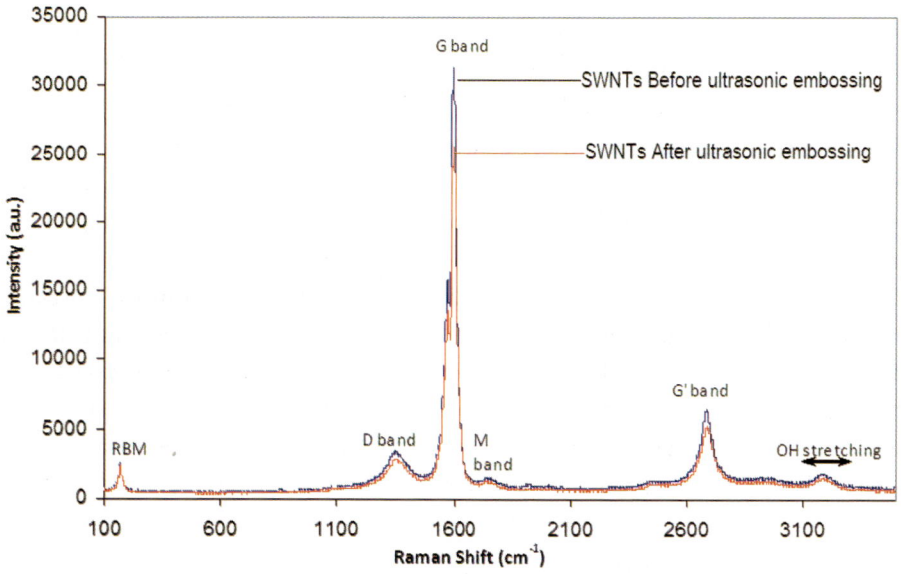

Fig. 4. Comparison of Raman spectra of SWNTs before and after ultrasonic embossing.

The SWNTs after embossing appear to be closer to one another. However, there is no change in the physical structure of the SWNTs.

The Raman spectra of SWNTs at 514 nm, 2,400 l/mm before and after ultrasonic embossing are shown in Fig. 4. The main observation of spectra peeks are on RBM, G band, and OH stretching. RBM is used to verify the diameter of the CNTs for which the results of scanning showed no shifting of the peak before and after ultrasonic embossing, which is the same with the OH stretching peak and G band. However, the G band, which represents the layer thickness of the CNTs, shows a lower magnitude of intensity. This is due to the pressure during the embossing process, which reduced the thickness of the SWNT layer.

DNA Extraction by the Carboxylic Group on Single-Walled Carbon Nanotubes

One of the excellent properties of CNTs is their large surface area-to-volume ratio. One technique used to incorporate a nanotubular surface into a biological system is through the functionalization of CNTs, which enables chemical bonding between the CNTs and the material of interest. The binding between the CNTs and the target element via chemical bonding, covalent or noncovalent, has been studied. The covalent bonding between biomolecules and functionalized CNTs may provide high stability, good selectivity, and accessibility because of its capability to control the location of the biomolecules. The carboxylic group is often the best choice as it can undergo a variety of reactions and is easily formed on CNTs via oxidizing treatment. It was reported that the presence of a carboxylic group at nanotube ends and at defects on the surface has advantages in performing acid–base chemistry and to introduce on the nanotube amide, an ester linkage. The control over reactants and/or reaction conditions may control the locations and density of the functional groups on the CNTs and the attached biomolecules.

In our work, temporary immobilization between DNA and a functional carboxylic group was used to separate the DNA from other biomolecules in the sample. We used the carboxylic group under high ionic solution to bind the phosphate group of DNA. An electrostatic attraction was formed between the functional group of CNTs and DNA at pH values ranging from 5.2 to 6.5. The separation of DNA from the rest of the biosample and chemicals from previous processes was accomplished by washing with an alcohol-based solution. As an alcohol-based solution does not change the pH value, the electrostatic force did not change during the washing process. The elution of DNA from the CNTs was performed by a low ionic strength solution at a pH value around 7.6, or it can also be achieved with DI water.

After the polymer microfluidic chip is fabricated with embedded SWNTs in the microchannels, the biosample and chemical solutions are transported into the chip by an external pump and valves. There are three steps in the DNA extraction process: the binding of DNA, washing, and elution, as shown in Fig. 5. The polymerase chain reaction (PCR) product, which contains amplified DNA fragments, primers, dNTPs, and other enzymes, is used as the starting biosample. The PCR product, mixed with a binding buffer (10 mM Tris-HCl,

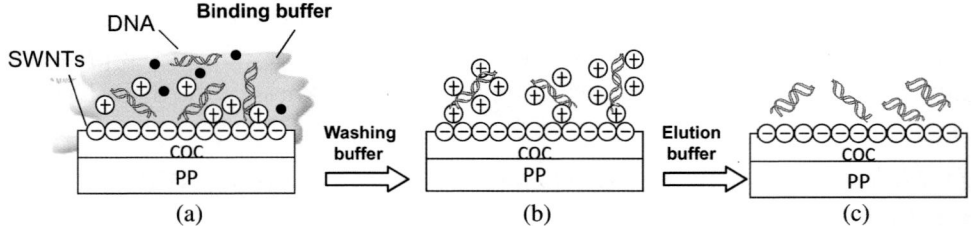

Fig. 5. DNA purification process by using SWNTs embedded in microchannels. (a) Mixed solution of DNA sample with binding buffer was flowed into the channel. At high ionic solution, DNA is bonded to SWNTs by electrostatic forces, (b) the washing buffer was used to remove the rest of sample and chemicals from the channel, (c) DNA was removed from the CNTs by changing the ionic concentration in the elution solution.

1 mM EDTA, 5 M Gu-HCl, pH 6.9) and adjusted pH with 3 M sodium acetate (pH 5.2), was then loaded into the chip. Under a high ionic solution, the DNA was immobilized onto the HO-group at surface of SWNTs (Fig. 5(a)). The sample was then washed by 500 µL 10 mM Tris-HCl, 1 mM EDTA, and 80% ethanol, at pH 7.5 (Fig. 5(b)). SWNTs were dried by air flow for 10 min, after which, the DNA was eluted from the SWNTs using 50 µL elution solution of 10 mM Tris-HCl and 1 mM EDTA at pH 7.5–8.0 (Fig. 5(c)).

Design and Fabrication of DNA Purification Microfluidic Module

In our work, a micromixer with patterned grooves with a staggered herringbone mixing (SHM) structure on the bottom channel was used to promote mixing in the channel. The channel had two inlets and one outlet; the mixer was 0.6 mm in width, 0.5 mm in height, and 45 mm in length. Slant grooves of 45° with respect to the axis were used with barriers on the side wall of the main channel to improve the mixing efficiency. The grooves were 200 µm in width and height and the barriers were 200 µm in width and 350 µm in height. The design of the micromixer is shown in Fig. 6(a). Brass was used for the mold on which protruding microstructures were milled and ground by using micromachining technique. The final surface roughness was between 10–20 nm. Figure 6(b) shows the SEM photo of the groove structure on the brass mold after machining.

Following the ultrasonic embossing process described above, the fabricated polymer chip is shown in Fig. 7. Black represents the deposited SWNTs at the underside of microchannel. The channels were enclosed by laminating a cover layer with a biocompatible adhesive film from Adhesives Research Pte. Ltd.

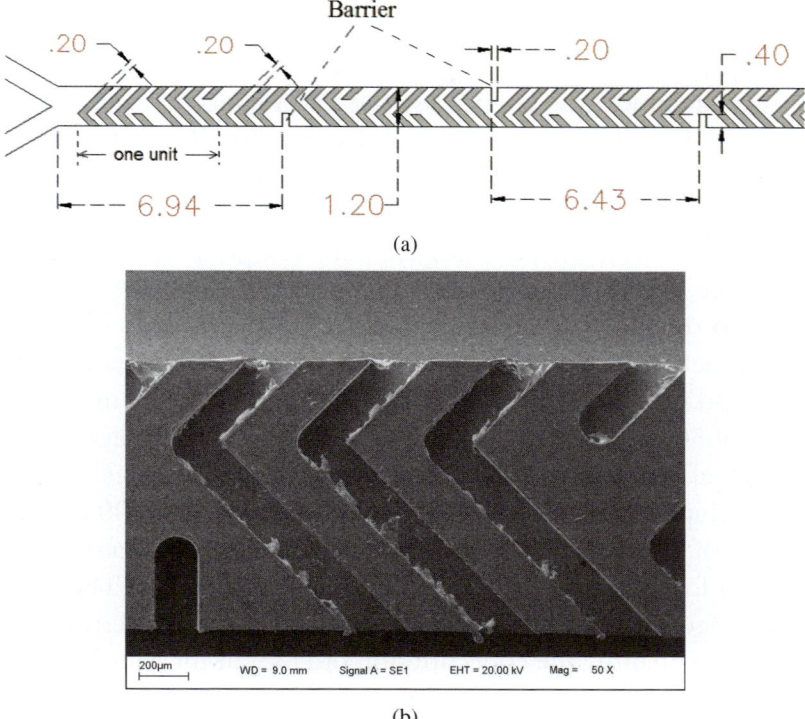

Fig. 6. Design of a microchannel with two inlets and one outlet. The grooves are 200 µm in width and height. (a) Design of groove patterns and (b) the brass mold after micromachining.

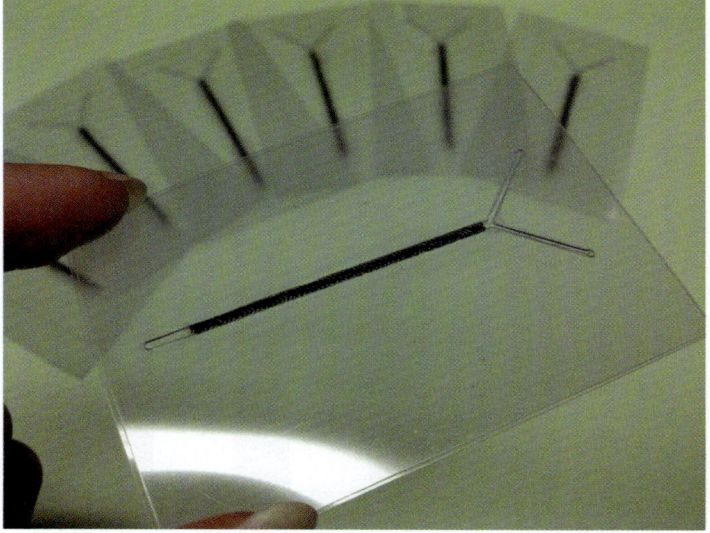

Fig. 7. Polymer chips with SWNTs embedded onto the channel. The SWNTs are located in the mixing zone and are represented in black.

Experiments and Results of DNA Extraction by SWNTs

In this study, pUC19 was used, which is a circular double stranded DNA of 2,686 base pairs in length. To increase the DNA concentration, 35 cycles of PCR were conducted. The input was a mixture of pUC19 50 ng, 2 × PCR master mix 25 µL, 25 µM pUC19_1300F 1 µL, 25 µM pUC19_1300R 1 µL, and nuclease-free water 50 µL. Each temperature cycle consists of 98°C for 45 s, 60°C for 45 s, and 72°C for 2.5 min. The DNA product after amplification process was 1,374 base pairs. The purification of DNA after PCR was performed on the polymer chip.

The channels in each polymer chip was washed by DI water at 1 mL/min for 10 min and dried out in the vacuum oven at 50°C for 10 min before use. A flow rate of 50 µL/min was use to load the pre-mix solution of DNA sample of 25 µL at 100 ng/µL concentration and binding buffer of 125 µL in the polymer chip. A washing buffer of 100 µL was flowed at 100 µL/min, and the flow rate of 50 µL/min was used for loading the elution buffer of 50 µL.

A total of four chips were used to repeat the experiment. The solution in the elution step was collected and analyzed using spectrophotometer (NanoDrop) to quantify the concentration of DNA in the solution. The purified DNA volume was measured and recorded.

Figure 8 shows the percentage of volume recovery from the elution solution with respect to the initial elution volume of 50 µL. The concentration of end results was measured and is shown in Fig. 9. From Figs. 8 and 9, it can be seen that average volume recovery is 82.6% and the average DNA concentration is 82.3%. The purity of the DNA sample was quantified by the spectroscopy method by a ratio of 260 nm/280 nm. NanoDrop was used to measure the purity ratio and it was found to be between 1.8 to 2.0. Purified

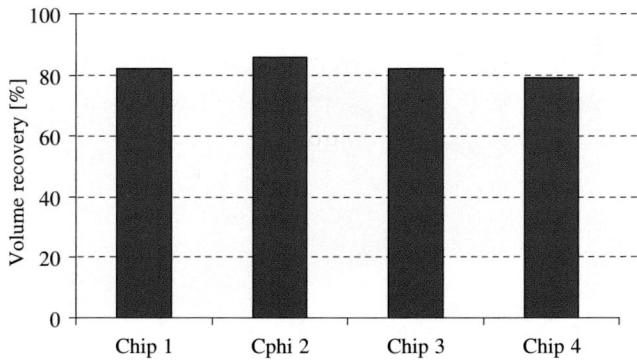

Fig. 8. Volume recovery percentage after elution from four chips with respect to the loading volume of 50 µL.

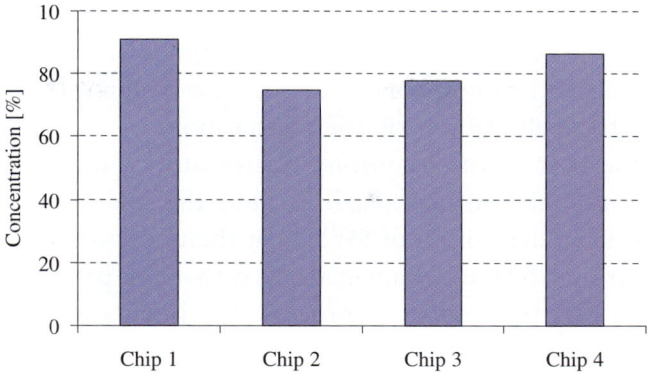

Fig. 9. Concentrations of purified DNA were measured by NanoDrop and compared to the loading concentration.

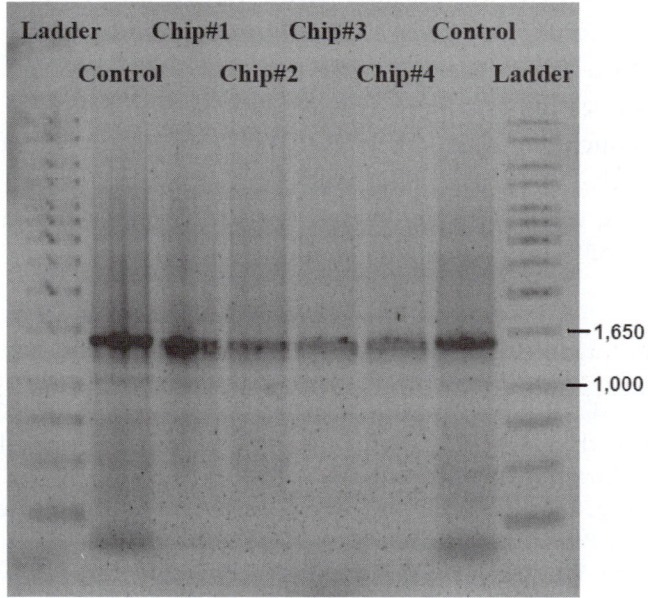

Fig. 10. Agarose gel (1%) of purified pUC19 DNA by SWNTs on polymer chips compared with control DNA and 1-kb ladder from Invitrogen.

DNA samples were run in gel electrophoresis and compared with the control DNA at the 1,374-b fragment as shown in Fig. 10.

Discussions and Conclusion

Our work demonstrated a low-cost polymer microfabrication technique by using ultrasonic embossing. Localized heat is generated within a second by ultrasonic vibration. The polymer is softened and molded into a microstructured tool. The entire process takes only a few minutes. Material shrinkage is low due to localized heating, which eases the de-molding process and minimizes damage to small features on the master mold.

In the same process of structuring microfeatures, ultrasonic embossing can simultaneously embed nanoparticles onto the microchannels. We have shown the localized deposition of SWNTs at the underside of polymer channels. Analysis from SEM and Raman showed that the properties of SWNTs did not change after the embossing process. The integration of SWNTs onto a polymer device by this method may be used for bio-sensing, filtration, flexible electronics, and other applications.

Here, SWNTs, which were functionalized by carboxylic groups, were used to extract DNA from the mixture with other biomolecules and chemicals. pUC19 after PCR was purified by the device with embedded SWNTs. The immobilization of DNA was achieved by a high ionic buffer between pH 5.2 to 6.5. Non-DNA was separated out and the elution of DNA from SWNTs was successful.

It is thus concluded that embedded SWNTs with a functionalized surface in microchannels can be used effectively for the purification of DNA; ultrasonic embossing can be simultaneously used to produce the microchannels and embed SWNTs on polymer substrates.

References

1. Zengerle, R., Microfluidics. *Proc Micromechanics Europe*, 1998. p. 111–122.
2. Heckele, M., Schomburg, W.K., Review on micro molding of thermoplastic polymers. *J Micromech Microeng*, 2004, **14**: p. R1–R14.
3. Liu, S.J., Dung, Y.T., Hot embossing precise structure onto plastic plates by ultrasonic vibration. *Proc Polym Eng Sci*, 2005, **45**: p. 915–925.
4. Liu, S.J., Dung, Y.T., Ultrasonic vibration hot embossing. *Int Polym Process XX*, 2005, **4**: p. 449–452.
5. Mekaru, H., Nakamura, O., Maruyama, O., Maeda, R., Hattori, T., Development of precision transfer technology of atmospheric hot embossing by ultrasonic vibration. *Microsyst Technol*, 2006, **13**: p. 385–391.
6. Truckenmüller, R., Ahrens, R., Cheng, Y., Fischer, G., Saile, V., An ultrasonic welding based process for building up a new class of inert fluidic microsensors and -actuators from polymers. *Sens Actuators A*, 2006, **132**: p. 385–392.
7. Mekaru, H., Goto, H., Takahashi, M., Development of ultrasonic micro hot embossing technology. *Microelectron Eng*, 2007, **84**: p. 1282–1287.
8. Khuntontong, P., Blaser, T., Schomburg, W.K., Ultrasonic micro hot embossing of thermoplastic polymers. In: *24th Annual Meeting of Conf Polymer Proc Soc*, 2008.

9. Khuntontong, P., Blaser, T., Schomburg, W.K., Ultrasonic micro hot embossing of polymers exemplified by a micro thermal flow sensor. In: *Conf Smart System Integration*, Barcelona, Spain, 2008.
10. Khuntontong, P., Blaser, T., Maas, D., Schomburg, W.K., Fabrication of a polymer micro mixer by ultrasonic hot embossing. In: *Conf 19th MicroMechanics Europe Workshop*, 2008.
11. Khuntontong, P., Blaser, T., Schomburg, W.K., Fabrication of molded interconnection devices by ultrasonic hot embossing on thin polymer films. *IEEE Trans Electron Pack Manuf*, 2009, **32**: p. 152–156.
12. Bennatar, A., Gutowski, T.G., Ultrasonic welding of PEEK graphit APC-2 composites. *Polymer Eng Sci*, 1989, **29**: p. 1705–1721.
13. Khuntontong, P., Fabrication of polymer micro devices by ultrasonic hot embossing. Dissertation at Instituition of Konstruktion und Entwicklung von Mikrosystemen, Aachen University, 2008, p. 13–14.
14. Bakajin, O., Ben-bark, N., Peng, J., Noy, A., Carbon nanotube based microfluidic elements for filtration and concentration. In: *7th Int Conf on Miniaturized Chemical and Biochemical Analysts Systems*, 2003.
15. Ghosh, S., Sood, A.K., Kumar, N., Carbon nanotube flow sensors. *Science*, 2003, **299**: p. 1042–1044.
16. Wang, J.A, Review: Carbon-nanotube based electrochemical biosensors. *Electroanalysis*, 2005, **17**(1).
17. Oh, J., Kim, G., Mattia, D., Noh, H.M., A novel technique for fabrication of micro- and nanofluidic device with embedded single carbon nanotubes. *Sens Actuators B*, 2009, 154: p. 67–72.
18. Tey, J.N., et al., Laminated, microfluidic-integrated carbon nanotube based biosensors. *Appl Phys Lett*, 2009, **94**, 013107.
19. Chen, S., Shen, W., Wu, G., Chen, D., Jiang, M., A new approach to the functionalization of single-walled carbon nanotubes with both alkyl and carboxyl groups. *Chem Phys Lett*, 2005, **402**: p. 312–317.
20. Chopra, N., Majumder, M., Hinds, B., Bifunctional carbon nanotubes by sidewall protection. *Adv Funct Mater*, 2005, **15**: 858–864.
21. Miller, S., Martin, C., Redox modulation of electroosmotic flow in a carbon nanotubes membrane. *J Am Chem Soc*, 2004, **126**(20): p. 6226–6227.
22. Barry, C.R., Lwin, N.Z., Zheng, W., Jacobs, H.O., Printing nanoparticle building blocks from the gas phase using nanoxerography. *Appl Phys Lett*, 2003, **83**: p. 5527–5529.
23. Meitl, M.A., et al., Solution casting and transfer printing single-walled carbon nanotube films. *Nano Lett*, 2004, **4**(9): pp. 1643–1647.
24. Allen, A.C., Sunden, E., Cannon, A., Graham, S., King, W., Nanomaterial transfer using hot embossing for flexible electronic devices. *Appl Phys Lett*, 2006, **88**: p. 083112.

Chapter 16

Ferrofluidics

A. Rezzan Kose and Hur Koser*

Department of Electrical Engineering
Yale University
15 Prospect Street, BCT 507
New Haven, CT 06511, USA

Overview of Ferrofluids

One of the earliest recorded studies of a magnetic material dates back to the ancient Greeks, who discovered that lodestone — naturally magnetized bulk magnetite — could attract iron and other lodestones.[1] The Chinese used the lodestone to create the first compass for navigation during the 11th century,[2] and its adoption in Europe gradually led to the age of exploration. By the mid-1800s, Faraday elucidated the link between magnetism and electricity, and Maxwell later combined electromagnetism and optics in his famous set of equations.[3] The 20th century saw the development of very large and powerful electromagnets, followed by superconducting magnets used to levitate trains,[4,5] to bend beams of charged matter around an accelerator ring, and even to capture stunning three-dimensional pictures of our bodies in magnetic resonance images (MRIs).[6]

Today, the impact of magnetism on our daily lives is immense, and we are not simply referring to large, powerful electromagnets. Advances in chemical and physical techniques over the last century have led to the synthesis and deposition of very small magnetic materials as well. For instance, nanometer-scale magnetic thin films are currently used in ultra-dense hard disk drives.[7,8]

---

*Corresponding author. E-mail: hurkoser@gmail.com

Magnetic microspheres are routinely used in immunoseparation assays,[9–11] and magnetite nanoparticles injected into patients enhance image contrast in MRI machines, saving lives by detecting tumors earlier than ever before.[12–15]

The synthesis of magnetic nanoparticles and their use within liquid media eventually inspired the development of a new class of completely manmade material — magnetic liquids, or *ferrofluids*. The coexistence of fluidic and magnetic properties in ferrofluids gives rise to a range of fascinating hydrodynamic phenomena, the study of which is known as *ferrohydrodynamics*. Working with ferrofluids requires an interdisciplinary knowledge of electromagnetism, fluid mechanics, nanotechnology, physics, and chemistry. While it is not practical to cover the entire breath of ferrohydrodynamics in a single chapter, it is quite possible and useful to cover an overview of ferrofluids, followed by a detailed look at a particular emerging application. In this chapter, we adopt that approach by illustrating existing and promising applications of ferrofluids, with an eventual focus on the physics of ferrofluids confined to microfluidic channels and their upcoming use in cellular manipulation and sorting. Ferrofluids have the potential to do for the health sciences what the introduction of the compass initiated for medieval Europe — beget a new era of exploration that eventually leads to breakthrough discoveries.

What is a ferrofluid?

Ferrofluids are colloidal mixtures of nano-sized magnetic particles (either single domain or superparamagnetic) covered by a surfactant and suspended in a compatible liquid medium.[16] The most common ferrofluids are either oil-based (where *cis*-oleic acid may be used as the surfactant) or water-based (where ionic species, such as citrate, may act as the surfactant) (Fig. 1). The surfactant layer helps overcome van der Waals forces by preventing the particles from coming too close, either by steric or electrostatic repulsion.

Typically, magnetic nanoparticles are magnetite (an iron oxide) with diameters between 5–15 nm. These particles can be obtained as precipitates of simple chemical reactions.[16] Their size is such that the thermal motion of nanoparticles at room temperature is sufficient to prevent them from agglomerating due to magnetic attraction and from settling down due to gravity.

The long-term stability of ferrofluids generally dictates a maximum magnetic particle concentration of 10% by volume (higher concentrations result in particle settling). When subjected to a strong magnetic field, magnetic moments of the particles align with the applied field and the ferrofluid is attracted toward the stronger magnetic field direction. In this manner, isolated

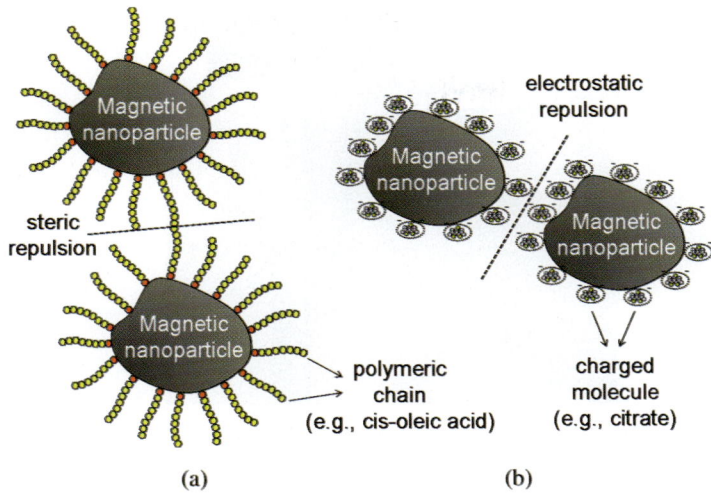

Fig. 1. Magnetic nanoparticles can be kept apart with surfactants, either through (a) steric repulsion between long polymeric chains of large organic molecules (such as *cis*-oleic acid) or (b) through electrostatic repulsion between similarly charged species (such as citrate). Steric repulsion works best in organic media such as oil, whereas electrostatic repulsion is compatible with polar liquid media such as water.

ferrofluid samples can be manipulated by a moving permanent magnet or a strong electromagnet (Fig. 2).

Existing applications

In the early 1960s, NASA scientists developed the first ferrofluids as an alternative means to move liquid fuels in a gravity-free environment.[17] Since then, ferrofluids have found many other uses within industrial, commercial, and biomedical settings.[18-22] Currently, most conventional applications of ferrofluids involve static magnetic fields from permanent magnets to hold ferrofluids for use as liquid seals,[19] bearings in rotating machinery,[23] dampers in stepper motors and shock absorbers,[24,25] as well as in grinding and polishing processes.[26] Over 50 million loudspeakers that are manufactured each year use ferrofluids to enhance voice-coil cooling and improve the speaker's frequency response through better damping.[27] Oil-based ferrofluids are also used in high-power transformers as coolants.[28]

New research focuses on the biomedical applications of very dilute ferrofluids. In most of these applications, there are two key technological capabilities that make ferrofluids useful. First is the ability to modify the surface chemistry of the nanoparticles so that they can be functionalized with various

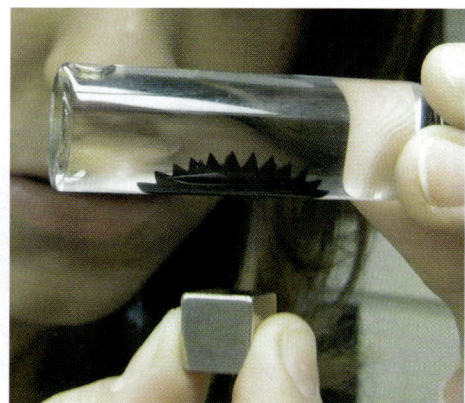

Fig. 2. Ferrofluids are attracted toward higher magnetic fields. Intense field gradients create surface instabilities (i.e., spikes) that follow the field lines.

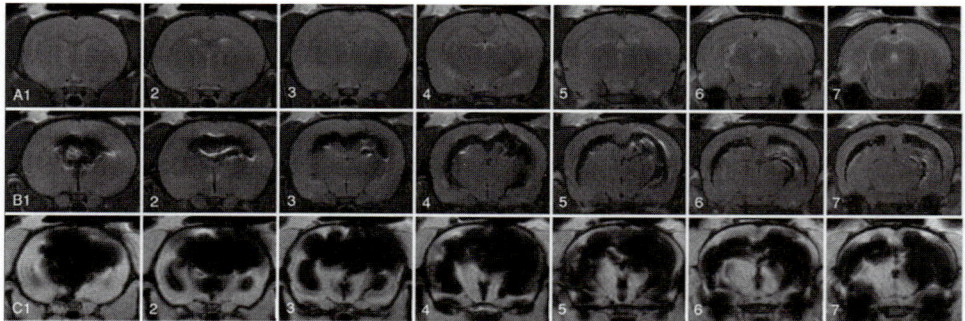

Fig. 3. Magnetic nanoparticles covered with Dextran are used as contrast agents in magnetic resonance imaging (MRI). Images shown here are from a rat brain. Areas that are progressively darker in sequences (b) and (c) indicate contrast agent localization following their injection. (Reproduced with permission from Ref. 13.)

chemicals, proteins, and antibodies. Functionalized nanoparticles can selectively attach to pathogens, live cells or tissue for use in immunoassays, immunoseparators,[9] or as contrast agents for MRI (Fig. 3).[12–15] Second, the magnetic nanoparticles can be directed to and concentrated at specific organs via external magnetic fields in a minimally invasive fashion so as to target a particular site for specific drug delivery[21,29] or hyperthermia.[30–32]

Indeed, hyperthermia is one of the most promising cancer therapies involving the *in vivo* use of nanoparticles to selectively kill tumor tissue. There exists a number of platforms that utilize different nanoparticle complexes for this purpose, such as functionalized hollow gold shells heated via infrared light from outside the body.[33] Magnetic nanoparticles can also be

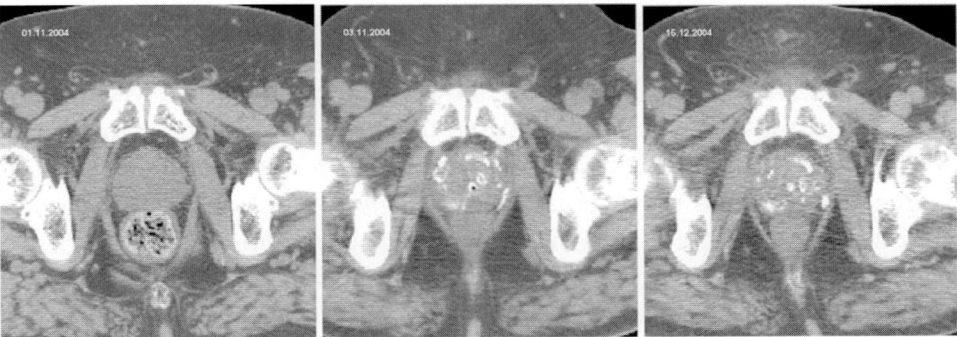

Fig. 4. Magnetic nanoparticles can be localized within or near a tumor, and they can be heated with external radiofrequency waves to kill cancerous cells in an emerging therapy commonly referred to as *hyperthermia*. These computed tomography (CT) images document a prostate tumor before the injection of magnetic nanoparticles (left), immediately following injection (center), and six weeks later (right). (Figure reproduced with permission from Ref. 35.)

heated via external radiofrequency waves,[34] and it is their "directability" that makes them attractive in the context of reducing toxicity in hyperthermia therapy (Fig. 4).[35]

Ferrofluid Synthesis

There are two general approaches for synthesizing ferrofluids from magnetic nanoparticles: (1) the reduction of larger magnetic particles into smaller nanoparticles through grinding and (2) precipitating them from chemical solutions.[16] The former was popularized by Papell,[17] although the concept is a well-known ancient technique. More recently, common synthesis methods involve the precipitation of magnetic oxides (such as magnetite) from chemical reactions of metallic salt solutions. Oil-based magnetite precipitation, first proposed by Khalafalla and Reimers in 1974,[36] remains as the foundation for many modern synthesis recipes. In 1981, Massart described an aqueous ferrofluid preparation, where the magnetic nanoparticles are stabilized by charged groups on their surfaces,[37] making water-based ferrofluids practical.

Rendering ferrofluids biocompatible makes their use in biomedicine possible. Water-based ferrofluids can be made biocompatible by encapsulating magnetic nanoparticles with a non-toxic, bio-friendly layer. Examples include coating ferrofluid particles with bovine serum albumin (BSA)[38] or salts such as citrate.[39] Citrate is a commonly used surfactant in ferrofluids and is mostly compatible in cell cultures as well. In addition, the ionic concentration and the pH level (between 7 and 7.4) of the solution must be adjusted so that the cells that come in contact with the ferrofluid can retain their viability for extended periods.[39] In this regard, determining the optimal ionic concentration is crucial; too many ions in the solution may result in magnetic particle aggregation, whereas there should be sufficient ionic species in the solution to prevent cell rupture due to the osmotic pressure on the cells.

In our laboratory, we have produced a biocompatible, water-based ferrofluid composed of cobalt–ferrite nanoparticles. Briefly, the synthesis recipe is as follows: the magnetic nanoparticles are precipitated out of a boiling solution of a strong base (such as 1 M sodium hydroxide) in which cobalt(II) chloride hexahydrate and iron(III) chloride salts are reacted. After isolation and rinsing, the magnetic precipitate is further reacted with a 0.35 M solution of iron(III) nitrate in the presence of a strong acid (such as 2 M nitric acid) at 80°C for 20 min while stirring.[36,37,39] The final precipitate comprises cobalt–ferrite nanoparticles, which can be concentrated with a magnet as the nitric acid is decanted. To achieve the right levels of surfactant, ionic concentration, and pH simultaneously, cobalt–ferrite particles within the precipitate are re-dispersed in deionized water, and the resulting ferrofluid is dialyzed for one week against 40 mM sodium citrate and citric acid solution at a pH level of 7.4. We measured cell viability within this ferrofluid using live/dead staining techniques, such as the Trypan blue staining protocol.[39] As Fig. 5 depicts, 75% of red blood cells remain viable for several hours within the resulting ferrofluid. It may be possible to increase the viability time of the cells

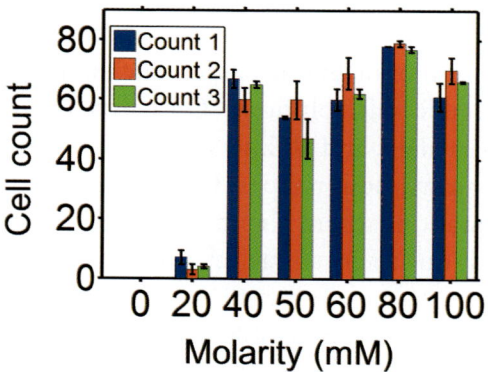

Fig. 5. Cell viability within a biocompatible ferrofluid. A citrate surfactant concentration of 40 mM was chosen as this was most compatible simultaneously with cell survival and ferrofluid stability. (Figure reproduced with permission from Ref. 39.)

even further through the partial incorporation of cellular media constituents within the final ferrofluid. However, this ferrofluid is already well suited for diagnostic and cellular manipulation applications that conclude within several hours.

Brief Introduction to Ferrohydrodynamics

In the presence of an external magnetic field, the nanoparticles within the ferrofluid tend to align their magnetic moments in the direction of the applied field. This tendency for magnetic alignment is characterized by a material property named *magnetic susceptibility*. In equilibrium, the magnetic susceptibility of a ferrofluid is given by the degree with which its magnetization (M) is influenced by the external field (H), i.e.,

$$M = \chi_0 H. \tag{1}$$

Under quasistatic conditions, this alignment tendency is offset by the rotational Brownian motion of the nanoparticles. At higher liquid temperatures, for instance, the average rotational energy of the nanoparticles is increased and the tendency for magnetic alignment is reduced.

The magnetic susceptibility is constant for low magnetic fields, typically for $|H| < 10\,\text{kA/m}$ or for a magnetic flux density ($B = \mu_0(M+H)$) magnitude below 10 mTesla. Higher field values lead to magnetic saturation, with M no longer varying proportionally to changes in H.

When the applied field is suddenly changed, the magnetization of the ferrofluid cannot respond instantaneously. Instead, M changes with a characteristic time constant (τ):

$$\frac{dM}{dt} = -\frac{1}{\tau}(M - \chi_0 H). \tag{2}$$

This is the simplest version of the magnetic relaxation equation (MRE), valid under no-flow and no-convection assumptions.[40] It is a simple, first-order differential equation whose solution to a step change in H is an exponential decay to the new steady-state magnetization value.

If the applied field is sinusoidally alternating or rotating, the ferrofluid magnetization will vary in the same fashion, as long as the ferrofluid remains in its linear, low-field regime (i.e., constant χ_0). This follows directly from the linearity of Eq. (2) and the fact that sinusoidals are eigenfunctions of linear differential equations. Assuming, for simplicity, a sinusoidal magnetic field with amplitude H_0 varying only in one dimension, i.e., $H = \text{Re}\{H_0 e^{j\omega t}\}\mathbf{i}_x$, the corresponding ferrofluid magnetization will be given by $\mathbf{M} = \{M_0 e^{j\omega t}\}\mathbf{i}_x$, where M_0 is, in general, a complex amplitude that incorporates the phase delay between $\mathbf{H}$ and $\mathbf{M}$. Substituting these field expressions into Eq. (2), we obtain the general, complex relationship between H_0 and M_0:

$$M_0 = H_0 \left(\frac{\chi_0}{1 + j\omega\tau} \right). \tag{3}$$

Equation (3) gives rise to the concept of a frequency-dependent magnetic susceptibility (also known as AC susceptibility):[41]

$$\chi_f = \left(\frac{\chi_0}{1 + j\omega\tau} \right). \tag{4}$$

Figure 6 depicts the AC susceptibility of a water-based, biocompatible ferrofluid composed of cobalt–ferrite magnetic nanoparticles.[39]

The time constant τ in Eq. (4) determines the angle of χ_f — the lag between the sinusoidal $\mathbf{H}$ and $\mathbf{M}$ fields. This lag is the result of two physical processes that control the alignment of magnetization within a given magnetic nanoparticle and the applied field. If the magnetic cores of the particles are small enough, their magnetic moment will simply rotate inside the nanoparticles (*Néel relaxation*; Fig. 7(a)) with a characteristic time constant given by[16]

$$\tau_N = \frac{1}{f_0} e^{(K_a V / k_B T)}, \tag{5}$$

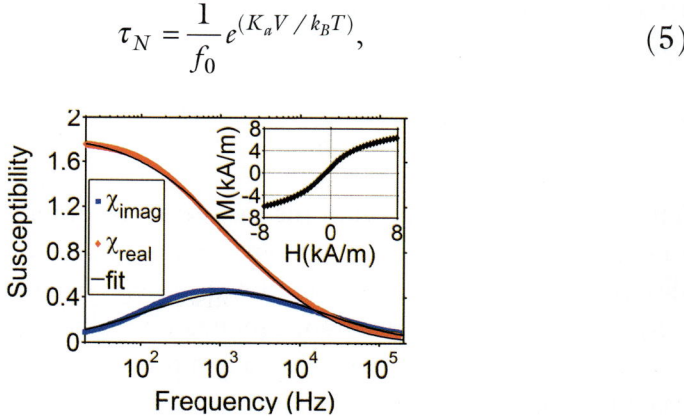

Fig. 6. AC susceptibility of the biocompatible ferrofluid in Fig. 5. The top and bottom curves are the magnitudes of the real and imaginary components of the susceptibility spectrum, respectively. Inset: Quasistatic (i.e., DC) magnetization plot of the same ferrofluid. Field values above several kA/m begin to noticeably saturate ferrofluid magnetization. (Figure reproduced with permission from Ref. 39.)

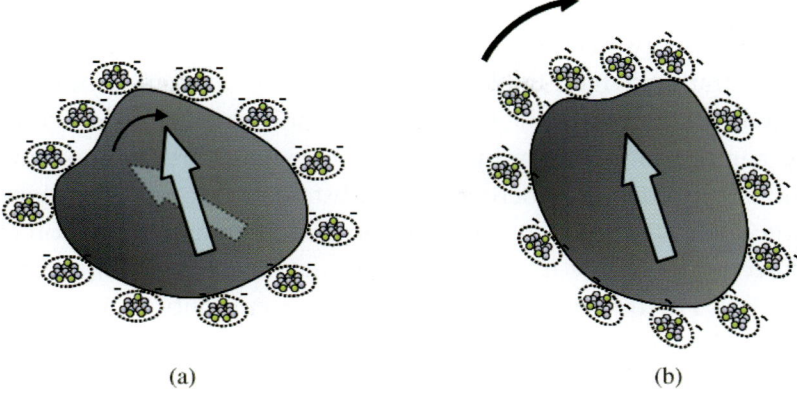

Fig. 7. Magnetic relaxation mechanisms. (a) Néel relaxation involves the magnetization vector inside the nanoparticles rotating to align with the applied field. (b) In Brownian relaxation, the crystal anisotropy within the magnetic nanoparticle is large, and the rotation of the entire nanoparticle is more favorable.

where f_0 is a precession frequency (typically in the range 10^8–10^{12} Hz), K_a is the magnetic anisotropy energy density, V is the volume of the nanoparticle core, and $k_B T$ is the thermal energy. Particles with larger cores will, in general, have a higher magnetic anisotropy energy, which fixes the magnetic moments within the cores. Here, the particles themselves will rotate in solution to orient with the applied field (*Brownian relaxation*; Fig. 7(b)), with a characteristic time constant given by

$$\tau_B = \frac{\pi D_{hyd}^3}{2k_B T} \eta, \qquad (6)$$

where η is the dynamic viscosity of the fluid, k_B is the Boltzmann constant, T is the absolute temperature (in Kelvin), and D_{hyd} is the hydrodynamic diameter of the particle, including its surfactant layer. In general, the two relaxation processes take place in parallel, and the faster of the two dominates the overall relaxation process.

The magnetic relaxation time constant depends in part on nanoparticle size. Most ferrofluids have a log-normal size distribution of nanoparticles; hence, the AC susceptibility spectrum can be fitted using Eqs. (4)–(6) to estimate the mean and standard deviation of the size distribution.

Ferrofluid Physical Characterization

Measuring the AC susceptibility of a ferrofluid can reveal a great deal about the physical state of the colloidal suspension of nanoparticles. The most direct information obtained is about the magnetic strength of the ferrofluid. The quasistatic value of susceptibility is directly related to the magnetic nanoparticle concentration, which can be inferred from a combination of susceptibility and density measurements. One can also find out if the colloid is stable or forming agglomerates simply by checking the imaginary component of the susceptibility spectrum. If there exists a substantial amount of agglomerates, their large hydrodynamic size will result in slow magnetic relaxation (Fig. 6) and a corresponding low-frequency peak in the imaginary component of AC susceptibility. The frequency-dependent AC susceptibility of a ferrofluid can be obtained by measuring the changes in the mutual inductance of an electromagnetic coil pair, with and without the presence of a ferrofluid.[42] Alternatively, commercial equipments, such as the Physical Properties Measurement System (PPMS®) allow for an automated measurement of susceptibility over a wide range of temperatures and static magnetic fields.[43]

Very often, however, it is quite useful to directly visualize and characterize the size and physical properties of magnetic nanoparticles within a ferrofluid. In this regard, transmission electron microscopy (TEM) is an indispensible tool that complements susceptibility measurements. TEM imaging reveals information about the size uniformity and distribution, and even the surface properties of the nanoparticles if atomic resolution can be obtained. Most TEM samples can be easily prepared by dipping a carbon-coated copper/rhodium grid into a ferrofluid sample diluted with ethanol.[39] Figure 8 depicts a TEM image of a ferrofluid sample and the log-normal size distribution of its nanoparticles.

Dynamic light scattering (DLS) is another measurement method used to reveal an estimate of nanoparticle hydrodynamic size.[44] For dilute colloidal

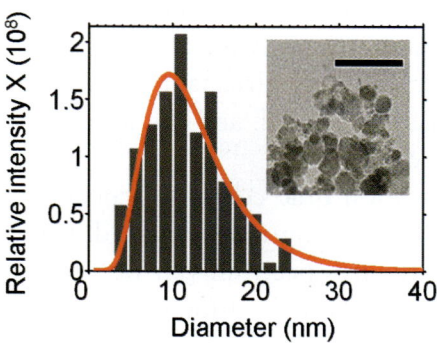

Fig. 8. Chemical synthesis of magnetic nanoparticles typically leads to a log-normal size distribution. Inset: Transmission electron microscopy (TEM) image of cobalt–ferrite nanoparticles. Scale bar: 50 nm. (Figure adopted with permission from Ref. 39.)

suspensions of most nanoparticles, it works reasonably well. However, concentrated ferrofluids need to be substantially diluted to allow for light transmission to yield an acceptable signal, and the act of dilution changes the colloidal equilibrium of the ferrofluid.[39] For instance, magnetic nanoparticles are much less likely to form chains or agglomerates when the ferrofluid is diluted. As a result, DLS measurements do not necessarily correspond to the dynamic size composition of the actual ferrofluid. A combination of TEM analysis and AC susceptibility measurements is much more reliable and accurate for the physical characterization of ferrofluids than DLS.

Ferrofluid viscosity is another material property that affects its magnetic relaxation properties (Fig. 6). Standard commercial rheometers work reasonably well to accurately characterize this physical property of ferrofluids. One should pay attention, however, to the fact that overall ferrofluid viscosity is a function of nanoparticle concentration and external magnetic fields.[45] Concentrated ferrofluids have higher overall viscosities, and if their magnetic nanoparticles are large enough to be subject to Brownian relaxation, external magnetic fields will tend to align them and make it difficult for liquid shear to rotate the nanoparticles.[45] Since viscosity, by definition, is the relationship between liquid shear forces and flow, the effective ferrofluid viscosity increases in the presence of external magnetic fields. In the presence of strong magnetic fields, the nanoparticles may even form chains, leading to dramatic (i.e., orders of magnitude) changes in viscosity.[46]

Ferro-Microfluidics — An Emerging Field

We will make a clear distinction here between the direct uses of ferrofluids within microfluidics and existing applications of dilute magnetic nanoparticles within microfluidics or other vessels. Low concentrations of functionalized magnetic nano- and microparticles have been popular in lab-on-a-chip applications for quite a while,[22] especially in the context of immunolabeling[47] and immunomagnetic separation,[9] since they can be controlled remotely and concentrated easily via an external magnet. Unlike other manipulation methods such as dielectrophoresis,[48] labeling samples with magnetic particles requires no special control of sample ionic concentrations, electrical conductivity, or dielectric properties. There are no large electric fields or intense lasers necessary (such as in optical tweezers[49]). Unlike fluorescent markers,[22] magnetic labels do not suffer from signal degradation over time as well. However, the downside of magnetic labeling is the additional incubation and wash cycles necessary, as well as the difficulty of removing the labels *a priori*.[50,51]

Bioferrofluidics

Unlike magnetic immunolabels, ferrofluids offer the potential to manipulate biological samples and individual cells without any prior labeling. The concept of bioferrofluidics involves replacing standard buffer solutions with biocompatible ferrofluids in which chemical and biological assays can be conducted. In order to make this approach practical, the ferrofluid medium has to offer a clear, easy-to-recognize advantage over traditional buffers. After all, existing buffers, such as phosphate buffered saline (PBS) are proven, clean, and biocompatible; they are also optically transparent (unlike most concentrated ferrofluids), enabling a host of standard fluorescent tagging and detection systems.

There is one aspect of traditional assays and their associated buffers that can be improved; virtually all rely on diffusion to help target moieties eventually find and bind to their complementary receptors. In this context, the most significant advantage of bioferrofluidics is their ability to magnetically manipulate and transport non-magnetic moeities, such as plastic microspheres and live cells, directly and actively toward a collection site or sensor regions within an assay. Active mass transport, as opposed to waiting for diffusion, is a very sensible strategy for cell assays, since the target moieties are large and their corresponding diffusion constants are very small.

Indeed, anything non-magnetic placed inside a ferrofluid medium will act as a "magnetic hole" and be repelled by positive magnetic field gradients,[52] in direct analogy to a hole within a semiconductor that gets repelled by a positive electric field. As the nanoparticles get attracted toward the field gradient, the non-magnetic particle gets displaced and is effectively pushed away. As long as the non-magnetic moiety is much larger than the magnetic nanoparticles, it effectively behaves as if it is inside a uniform magnetic medium. It can even assume positive buoyancy inside the ferrofluid within a vertical field gradient, just as air bubbles within a liquid have positive buoyancy under the influence of Earth's gravitational pull on the water. There is no need for any physical attachment of the magnetic nanoparticles to the non-magnetic microparticle being manipulated; hence, the approach is completely label-free. This phenomenon has been known for decades, and has even been utilized to separate valuable mineral ores within ferrogravimetric separation chambers.[16] Since the recent advent of biocompatible ferrofluids that can sustain live cells for extended periods, the magnetic hole concept can also be applied to cellular separation, manipulation, and even sorting within microfluidic channels — a realization that has led to "ferro-microfluidics."[39]

Magnetic force

In this subsection, we briefly outline a simple analytical approach that enables the estimation of non-magnetic particle velocities arising from applied magnetic fields and field gradients inside ferrofluids. We assume that non-magnetic microparticles are suspended in ferrofluids in a closed microfluidic channel and derive an analytical expression for the instantaneous force on those microparticles.

To simplify the calculations, we assume a perfectly spherical, incompressible microparticle, immersed in a magnetically linear ferrofluid. Under these assumptions, the ferrofluid magnetization in the immediate vicinity of the microparticle and the virtual magnetization (M_{eff}) within the particle's volume (V_p) can be approximated as uniform. The total instantaneous force on the microparticle is the same as that on a dipole:[16]

$$F_{ins} = \int_{V_p} \nabla(M_{eff} \times B_{in}) dV, \qquad (7)$$

where B_{in} is the magnetic flux density within the spherical microparticle. The integration is over the internal volume of the particle and hence, the instantaneous force expression can be simplified to

$$F_{ins} = V_p \nabla(M_{eff} \cdot B_{in}). \qquad (8)$$

To obtain an analytical expression for the magnetic force, we will express M_{eff} and B_{in} in terms of the external magnetic field (H_{ext}) in the absence of the microparticle. Effective magnetization in terms of the external magnetic field (H_{ext}) and ferrofluid susceptibility can be written as:[39]

$$M_{eff} = \frac{-3\chi_f}{3 + 2\chi_f} H_{ext}. \qquad (9)$$

For further calculations, the demagnetization effect within a sphere should be taken into account. Let us assume that the sphere itself possesses an effective susceptibility (χ_{eff}) that arises from being inside the magnetic medium; using the demagnetization field relationship for a sphere, we obtain:[39]

$$M_{eff} = \frac{3\chi_{eff}}{3 + \chi_{eff}} H_{ext}. \qquad (10)$$

Comparing Eq. (9) with Eq. (10) reveals the expression for χ_{eff} in terms of ferrofluid susceptibility:

$$\chi_{eff} = \frac{-\chi_f}{1 + \chi_f}. \qquad (11)$$

Note that the effective magnetic susceptibility depends on that of the ferrofluid, since the microparticle responds to magnetic forces only because it displaces the ferrofluid. The negative sign in Eq. (11) indicates that under static conditions, the effective magnetization of the microparticle is in the opposite direction of the local ferrofluid magnetization.

In the presence of AC fields, the local magnetic field varies sinusoidally, and the time-average force is half the maximum value of the instantaneous force in Eq. (8):

$$F_{ave} = \frac{1}{2} V_p \mu_0 \left(\frac{\text{Re}\{\chi_{eff}\}}{|\chi_{eff}|^2} + 1 \right) \left(\frac{9|\chi_f|^2}{9 + 6\text{Re}\{\chi_f\} + 4|\chi_f|^2} \right) \nabla |H_{ext}|^2 \qquad (12)$$

The details of this force derivation are given in Ref. 39.

One way to utilize this force on non-magnetic microparticles is to use high-resolution lithography to pattern thin-film micromagnets on a substrate.[53] As polystyrene microspheres suspended inside a ferrofluid flow through a microfluidic channel, the islands of micromagnets underneath create a periodic magnetic force landscape that traps the microspheres (Fig. 9).[53] Notice from Eq. (12) that magnetic force is proportional to microparticle size; hence, this device could potentially be used to perform size-based trapping of cells in conjunction with an externally imposed flow.

Another ferro-microfluidic device that utilizes the size dependency of magnetic force is depicted in Fig. 10.[54] Here, the idea is to use an external permanent magnet to continuously push microspheres sideways into different outlet channels in order to sort them based on size.

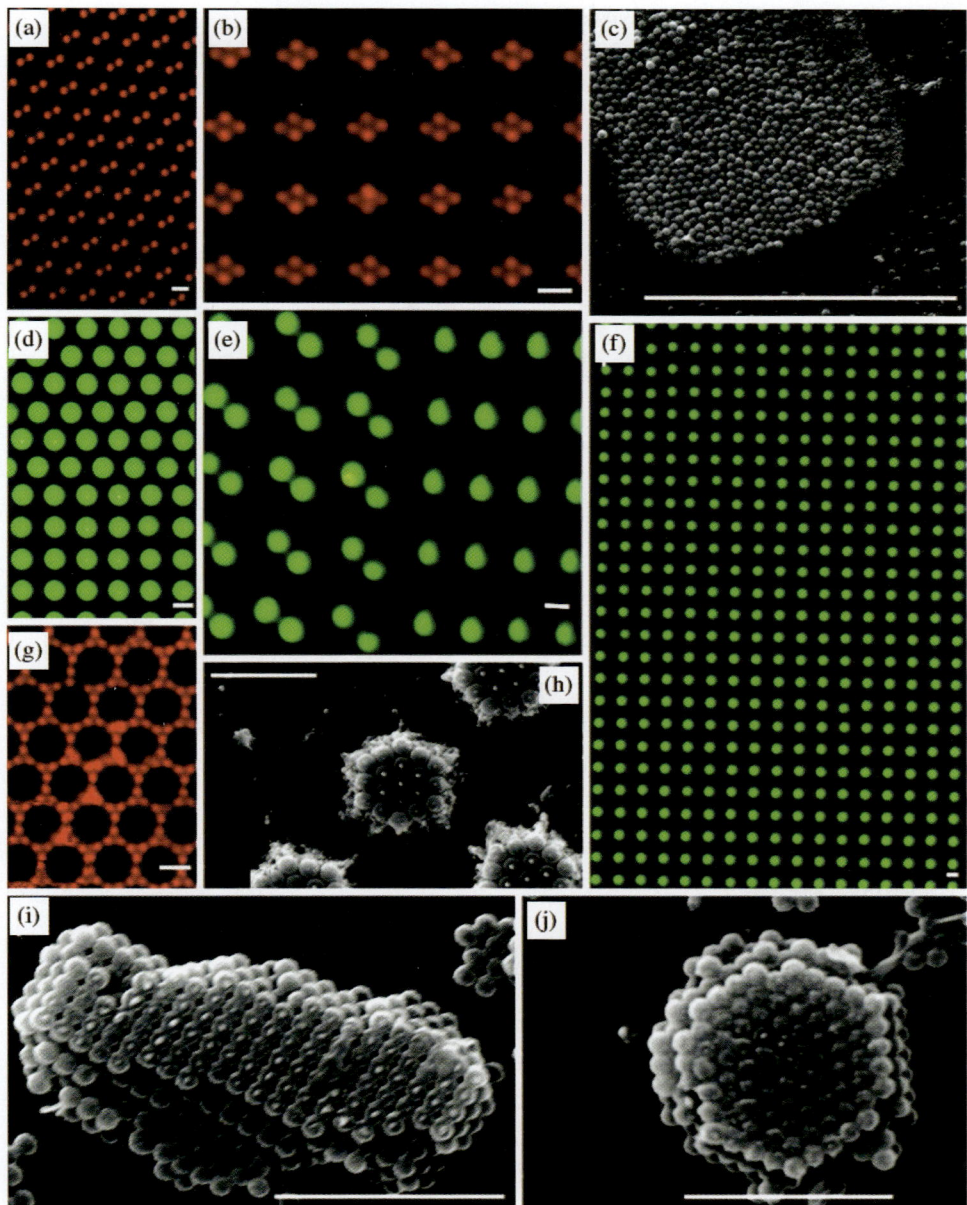

Fig. 9. Non-magnetic microbeads assembled into various configurations within a ferro-microfluidic channel via magnetic thin film islands patterned on the substrate (a)–(j). Scale bars: 5 μm. (Figure reproduced with permission from Ref. 53. Copyright 2005, National Academy of Sciences, USA.)

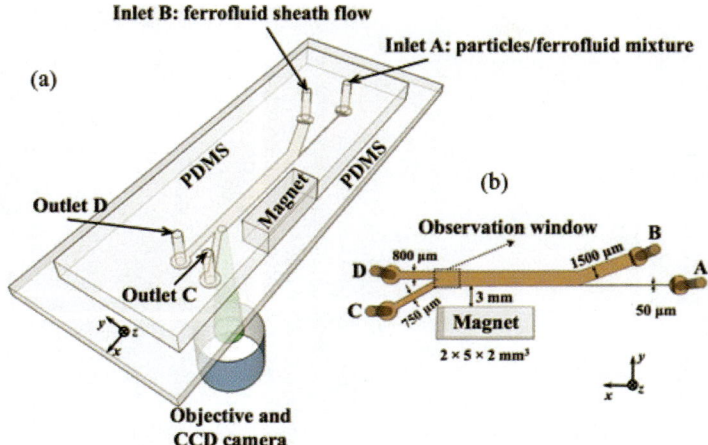

Fig. 10. A ferro-microfluidic device for the binary separation of microbeads. A permanent magnet just outside the device provides the size-dependent separation force as the microbeads flow within the channel. (Figure reproduced with permission from Ref. 54.)

In general, ferro-microfluidic devices that rely on magnetic force are ideal for rapid mass transport of microparticles or even cells towards sensor locations that can be integrated onto the same device. However, these devices typically need to utilize another actuation mechanism (such as pressure-driven flow) to induce the target microparticles to circulate within the channel.

Magnetic torque

A special field configuration — namely, traveling magnetic fields — can be used to generate a torque on a non-magnetic microparticle suspended within a ferrofluid.[55] The torque arises from a magnetic field that is locally rotating within the ferro-microfluidic channel. This field rotation, in turn, causes a rotation of the magnetic nanoparticles (assuming they relax through Brownian mechanism) and subsequently, the counter-rotation of the microparticle (Fig. 11) through a slip (s). The time-average value of this torque can be written as:[39]

$$T_{ave} = \hat{y} \frac{V_p}{2} s\mu_0 \frac{\text{Im}\{\chi_{eff}\}}{|\chi_{eff}|^2} \left(\frac{9|\chi_f|^2}{9 + 6\text{Re}\{\chi_f\} + 4|\chi_f|^2} \right) |H_{ext}|^2. \quad (13)$$

Figure 12 depicts the simplest manner to incorporate a traveling magnetic field within a ferro-microfluidic device — parallel electrodes carrying

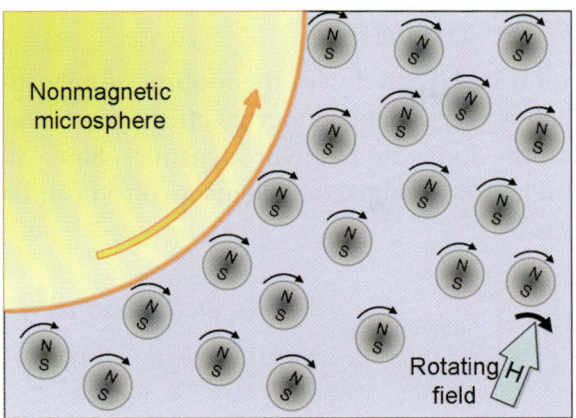

Fig. 11. AC fields that cause magnetic nanoparticle rotation can also induce the counter-rotation of non-magnetic microspheres and live cells.

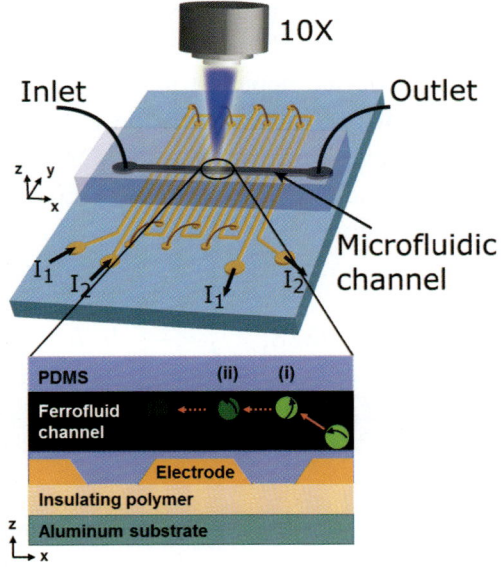

Fig. 12. Schematic for a ferro-microfluidic device designed to manipulate, separate, and sort non-magnetic microspheres and live cells via traveling magnetic waves. Microspheres and cells get pushed up toward the channel ceiling as soon as the sinusoidal currents are turned on. At low frequencies, they get trapped over the gap between the electrodes; above a critical frequency, they translate continuously. (Figure adopted with permission from Ref. 39.)

sinusoidal electric currents in quadrature (i.e., with neighboring electrodes 90° out of phase progressively in one direction). With this device, it is possible to apply a traveling magnetic field that exerts both magnetic force and torque on the microparticles.[39]

Calculating particle velocities

The Reynolds number associated with the motion of micron-scale beads in a quiescent ferrofluid is very small and inertial effects can be neglected. Hence, Stokes flow equations dominate the hydrodynamics. Since Stokes flow equations are linear, all hydrodynamic coefficients involved can be combined into a resistance matrix:[56]

$$\begin{bmatrix} F_{ave,x} \\ T_{ave,y} \end{bmatrix} = A \begin{bmatrix} v_x \\ \omega_y \end{bmatrix}, \text{ with } A = \begin{pmatrix} 6\pi\eta R f_1(h,R) & 6\pi\eta R^2 f_2(h,R) \\ 8\pi\eta R^2 f_3(h,R) & 8\pi\eta R^3 f_4(h,R) \end{pmatrix}. \quad (14)$$

Here, v_x is the linear velocity of the microparticle along the channel length, ω_y is its angular velocity, η is the ferrofluid viscosity, R is microsphere radius, and f_i is a resistance factor that depends on the particle radius and its distance (h) from a wall of the microfluidic channel. Notice that microparticle rotation will couple to its linear velocity only in the vicinity of a wall that provides some traction. Within the device depicted in Fig. 12, microparticles are all pushed up to the ceiling of the channel as soon as currents are turned on. Assuming the minimal distance of the microsphere to the ceiling (h) is much smaller than its radius (R), the resistance factors of Eq. (14) can be obtained from standard lubrication theory:[57]

$$f_1 \approx -\frac{8}{15}\ln(h/R) + 0.9588; \quad f_2 \approx -\frac{2}{15}\ln(h/R) - 0.2526; \\ f_3 \approx -\frac{1}{10}\ln(h/R) - 0.1895; \quad f_4 \approx -\frac{2}{5}\ln(h/R) + 0.3817. \quad (15)$$

Using the surface charge density on the microparticle and the channel surfaces, h can be estimated through the Derjaguin, Landau, Verwey, and Overbeek theory (DLVO theory).[58] From fits to experimental data, however, h is on the order of a few nanometers, comparable to the surface roughness of the channel ceiling, indicating that the microsphere is very strongly pushed upward.

In order to obtain an expression for particle velocities, Eq. (14) can be solved for v and ω through a simple matrix inversion:

$$\begin{bmatrix} v_x \\ \omega_y \end{bmatrix} = A^{-1} \begin{bmatrix} F_{ave,x} \\ T_{ave,y} \end{bmatrix}. \quad (16)$$

Hence, linear particle velocities due to magnetic force and torque alone can be determined by

$$v_{force,x} = \frac{f_4}{6\pi\eta RG} F_{ave,x} \quad \text{and} \tag{17}$$

$$v_{torque,x} = -\frac{sf_2}{8\pi\eta R^2 G} T_{ave,y}, \tag{18}$$

where $G \equiv f_1 f_4 - f_2 f_3$. The net particle velocity is then given by

$$v_x = v_{force,x} + v_{torque,x}. \tag{19}$$

Toward fully integrated ferro-microfluidics

The above calculations demonstrate that particle velocities in ferrofluids subjected to traveling magnetic fields have frequency and size dependences. The frequency dependence is built into the expressions for force and torque in Eqs. (12) and (13); they arise from the AC susceptibility spectrum (χ_f) of the ferrofluid. The size dependence of particle velocity is clear from the expressions in Eqs. (17) and (18). Both magnetic force and torque scale with particle volume (R^3); the hydrodynamic drag that resists linear particle motion scales with R against force and with R^2 against torque that rolls the particle. Hence, linear particle velocity due to magnetic force alone depends on R^2, whereas that due to torque scales with R. This observation indicates that torque effects on smaller particles are relatively more significant — a realization that enables size-based separation within a ferro-microfluidic device without any imposed external flow.

Figure 13 depicts the results from a binary separator that has been constructed as shown in Fig. 12. Here, each microsphere size has a distinct critical frequency above which torque effects dominate over the trapping tendency of the magnetic force, leading to the continuous roll of the microsphere along the channel ceiling. When the excitation frequency of the traveling wave is chosen to be between the two critical frequencies of two different microsphere sizes, the larger ones (9.9 µm in diameter) can be trapped, while the smaller ones (2.2 µm) can be cleared (Fig. 14).

The separation efficiency of this device is over 90%. At an excitation frequency of 400 Hz, 96.5% of the 9.9-µm microspheres within the channel can be trapped within 10 s, whereas 99.3% of the 2.2-µm particles (1,285 out of 1,294) continue to translate and clear the observation window in 45 s

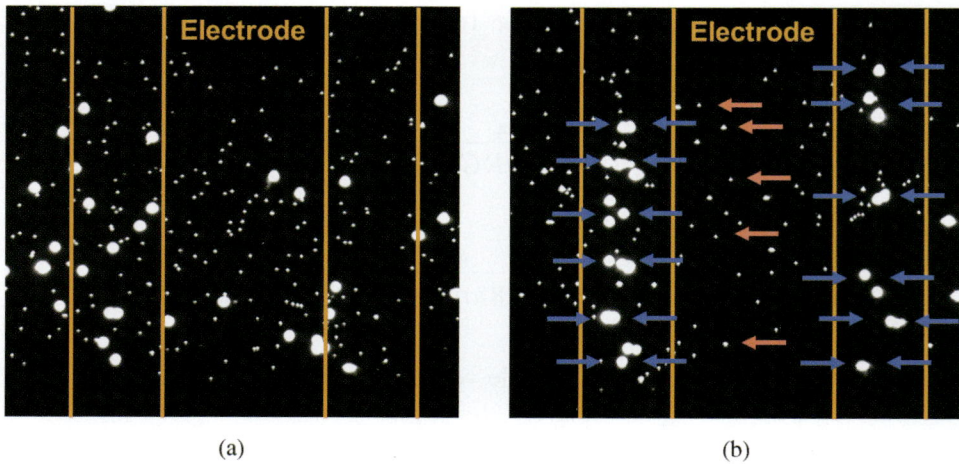

Fig. 13. Size-based separation with ferro-microfluidics. 9.9-μm and 2.2-μm microspheres — initially randomly dispersed within the channel (a) — can be rapidly separated by trapping the larger microspheres (blue arrows) and clearing the small ones (red arrows) (b). (Figure adopted with permission from Ref. 39.)

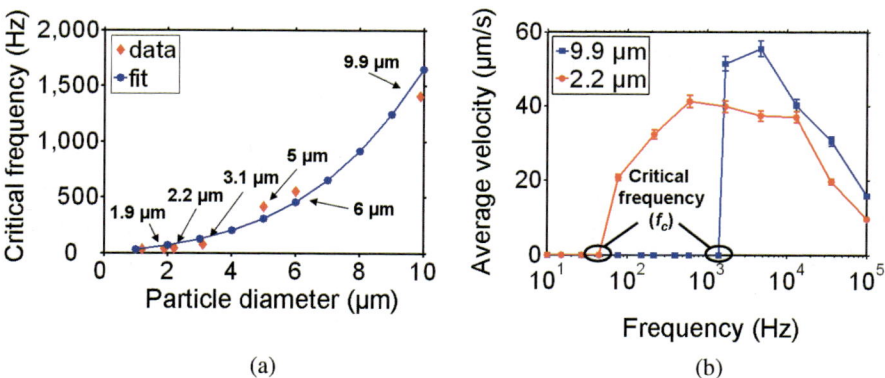

Fig. 14. The critical frequency below which a microsphere gets trapped depends on its diameter (a), getting larger with increasing size (b). (Figure adopted with permission from Ref. 39.)

(Fig. 15). The same device is capable of manipulating and separating bacteria from blood cells, achieving 95.7% bacterial separation efficiency and 93.7% blood cell trapping efficiency.

Finally, the same device can manipulate, separate, and sort live cells based on their shape and elasticity. In one experiment, healthy red blood cells were separated from sickle cells with an efficiency of 75%.[39] The sensitivity to shape and elasticity results from the way the cells are pushed up and rolled on the channel ceiling; cells that deviate more from a hard sphere are more difficult to roll and easier to trap.

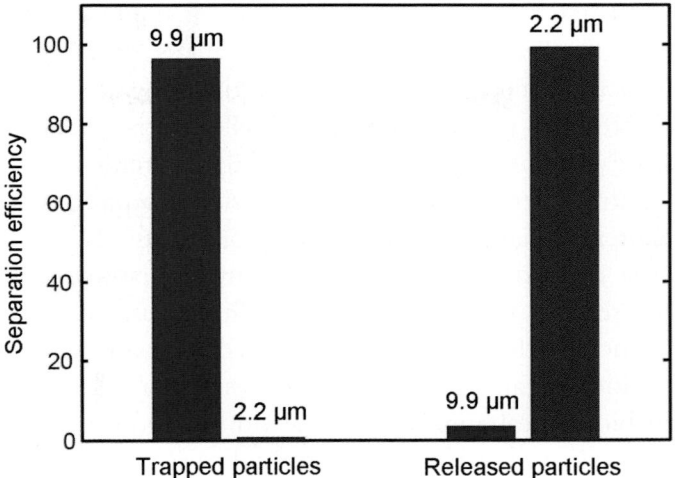

Fig. 15. The separation efficiency for 9.9-μm and 2.2-μm particles at an excitation frequency of 400 Hz. The trapping efficiency for the 9.9-μm microspheres is 96.5% within 10 s, whereas the separation efficiency for the 2.2-μm microspheres is 99.3% in 45 s.

Conclusions and Outlook

The main advantage of traveling magnetic fields is the compactness of design that they enable in ferro-microfluidics. They do away with bulky permanent magnets or large electromagnet set ups. The minimal dimensions of the electrodes and the separation between them are measured in tens of microns; as such, they make expensive, high-resolution microfabrication redundant. In fact, the substrates can be manufactured very cost-effectively from printed circuit boards.[59]

Another advantage of the device depicted in Fig. 12 is that it isolates cellular manipulation, separation, and sorting from external flow. Thus, these devices can be combined with a continuous-flow system to dramatically increase the throughput, without demanding a specific, constant flow rate. Hence, these ferro-microfluidic devices offer a level of flexibility and practicality that render them more suitable for wide deployment.

Through varying electrode geometry and input excitation frequency, these devices could be tailored for different size ranges of particles and cells. They can be integrated with lab-on-a-chip sensors and diagnostic systems, and by dramatically increasing mass transport, help existing sensor transducers achieve their theoretical limits of sensitivity and performance.

The ability to position cells in precise locations within ferrofluids can also be used toward artificial tissue generation. Patterning cells into well-organized structures is a key challenge in tissue engineering.[60] A number of techniques — from manual cell seeding to inkjet printers — have been employed since the emergence of tissue engineering as a field.[61] Recent efforts using ferrofluids to create organized cellular structures are promising in this regard.[38] Ferro-microfluidics offers the potential to dramatically increase the control over cell location and density in printed tissues.

In this chapter, we have given an overview of ferrofluids and detailed their potential in the context of on-chip particle manipulation, separation, and transport. The creation of biocompatible ferrofluids, due, in part, to advances in magnetic nanoparticle synthesis, has opened new arenas for ferrofluidic applications in biology and medical diagnostics. Note that we have focused specifically on biomedical applications that do not require any nanoparticle functionalization. Even such simple ferrofluids are capable of dramatically changing the face of microfluidics research. Ferro-microfluidic devices allow target cells to be concentrated, trapped, and simply directed toward sensor surfaces efficiently and in a label-free fashion. The approach can lead to rapid, automated, and disposable bioassays that count and estimate the concentration of target cells within minutes. The same devices could eventually be used to selectively concentrate rare cells, such as circulating tumor cells in blood samples, by exploiting their structural and elasticity differences. Applied in this manner, ferro-microfluidics could help boost the detection sensitivity of existing cellular assays dramatically.

References

1. Keithley, J.F., *The Story of Electrical and Magnetic Measurements: From 500 BC to the 1940s*, Wiley-IEEE Press, 1999.
2. Carlson, J.B., Lodestone compass: Chinese or Olmec primacy? *Science*, 1975, **189**(4205): p. 753–760.
3. Maxwell, J.C., On physical lines of force. *Philos Mag*, 2010, **90**: p. 11–23.
4. Tsuchiya, M., Ohsaki, H. Characteristics of electromagnetic force of EMS-type Maglev vehicle using bulk superconductors. *IEEE Trans Magn*, 2000, **36**(5): p. 3683–3685.
5. Polgreen, G.R., *Magnetic System of Transportation*, The General Electric Company Limited, 1959.
6. Lauterbur, P.C., Image formation by induced local interactions: Examples employing nuclear magnetic resonance. *Nature*, 1973, **242**(5394): p. 190–191.
7. Sasaki, Y.S.C., Guan, L., Ito, H., *Thin-Film Magnetic Head, A Head Gimbal Assembly and Hard Disk Drive*, Headway Technologies, Inc., Milpitas, CA, 2009.
8. Takahashi, M., Tsunoda, M., Shoji, H., Ultra-clean sputtering process for magnetic thin films on hard disk drives. *Vacuum*, 2000, **59**(2–3): p. 814–824.

9. Choi, J.-W. et al., A new magnetic bead-based, filterless bio-separator with planar electromagnet surfaces for integrated bio-detection systems. *Sens Actuators*, 2000, **68**(1–3): p. 34–39.
10. Hancock, J.P., Kemshead, J.T., A rapid and highly selective approach to cell separations using an immunomagnetic colloid. *J Immunol Methods*, 1993, **164**(1): p. 51–60.
11. Kondo, A., Kamura H., Higashitani, K., Development and application of thermo-sensitive magnetic immunomicrospheres for antibody purification. *Appl Microbiol Biotechnol*, 1994, **41**(1): p. 99–105.
12. Kim, D.K. et al., Starch-coated superparamagnetic nanoparticles as MR contrast agents. chemistry of materials, 2003, **15**(23): p. 4343–4351.
13. Kim, D.K. et al. Characterization and MRI study of surfactant-coated superparamagnetic nanoparticles administered into the rat brain. *J Magn Magn Mater*, 2001, **225**(1–2): p. 256–261.
14. Semelka, R.C., Helmberger, T.K.G., Contrast agents for MR imaging of the liver. *Radiology*, 2001, **218**(1): p. 27–38.
15. Kobayashi, H. et al., Micro-magnetic resonance lymphangiography in mice using a novel dendrimer-based magnetic resonance imaging contrast agent. *Cancer Res*, 2003, **63**(2): p. 271–276.
16. Rosensweig, R.E., *Ferrohydrodynamics*, Cambridge University Press, New York, 1985.
17. Papell, S.S., Low viscosity magnetic fluid obtained by the colloidal suspension of magnetic particles, US Patent 3215572, 1965.
18. Berkovski, B., Bashtovoy, V., *Magnetic Fluids and Applications Handbook*, Begell House, Inc., New York, 1996.
19. Bailey, R.L., Lesser known applications of ferrofluids. *J Magn Magn Mater*, 1983, **39**(1–2): p. 178–182.
20. Pankhurst, Q.A. et al., Applications of magnetic nanoparticles in biomedicine. *J Phys D*, 2003, **36**(13): p. R167.
21. Neuberger, T. et al., Superparamagnetic nanoparticles for biomedical applications: Possibilities and limitations of a new drug delivery system. *J Magn Magn Mater*, 2005, **293**(1): p. 483–496.
22. Gijs, M.A.M., Magnetic bead handling on-chip: new opportunities for analytical applications. *Microfluid Nanofluid*, 2004.
23. Schweitzer, G., Maslen, E.H., *Magnetic Bearings: Theory, Design, and Application to Rotating Machinery*, Springer, 2009.
24. Scherer, C., Neto, A.M.F., Ferrofluids: Properties and applications. *Brazilian J Phys*, 2005, **35**(3A): p. 718–727.
25. Raj, K., Moskowitz, R., Commercial applications of ferrofluids. *J Magn Magn Mater* 1990, **85**(1–3): p. 233–245.
26. Childs, T.H.C., Yoon, H.J., Magnetic fluid grinding cell design. *CIRP Ann Manu Technol*, 1992, **41**(1): p. 343–346.
27. King, J.A.M., Method for making loudspeaker with magnetic fluid enveloping the voice coil, US Patent *4017694, 1977.*
28. Raj, K., Moskowitz, R., Ferrofluid-cooled electromagnetic device and improved cooling method, US Patent 5462685, 1995.
29. Forbes, Z.G. et al., An approach to targeted drug delivery based on uniform magnetic fields. *IEEE Trans Magn*, 2003, **39**(5): p. 3372–3377.

30. Gordon, R.T., Hines, J.R., Gordon, D., Intracellular hyperthermia a biophysical approach to cancer treatment via intracellular temperature and biophysical alterations. *Med Hypotheses*, 1979, **5**(1): p. 83–102.
31. Jordan, A. et al., Presentation of a new magnetic field therapy system for the treatment of human solid tumors with magnetic fluid hyperthermia. *J Magn Magn Mater*, 2001, **225**(1–2): p. 118–126.
32. Moroz, P., Jones, S.K., Gray, B.N., Magnetically mediated hyperthermia: Current status and future directions. *Int J Hyperthermia*, 2002, **18**(4): p. 267–284.
33. Kennedy, L.C. et al., A new era for cancer treatment: Gold-nanoparticle-mediated thermal therapies. *Small*, 2011, **7**(2): p. 169–183.
34. Rosensweig, R.E., Heating magnetic fluid with alternating magnetic field. *J Magn Magn Mater*, 2002, **252**: p. 370–374.
35. Johannsen, M. et al., Clinical hyperthermia of prostate cancer using magnetic nanoparticles: Presentation of a new interstitial technique. *Int J Hyperthermia*, 2005, **21**(7): p. 637–647.
36. Khalafalla, S.E., Reimers, G.W., Production of magnetic fluids by peptization techniques, US Patent 3843540, 1974.
37. Massart, R., Preparation of aqueous magnetic liquids in alkaline and acidic media. *IEEE Trans Magn*, 1981, **17**(2): p. 1247–1248.
38. Krebs, M.D. et al., Formation of ordered cellular structures in suspension via label-free negative magnetophoresis. *Nano Lett*, 2009, **9**(5): p. 1812–1817.
39. Kose, A.R. et al., Label-free cellular manipulation and sorting via biocompatible ferrofluids. *Proc Natl Acad Sci U S A*, 2009, **106**(51): p. 21478–21483.
40. Zahn, M., Ferrohydrodynamic torque-driven flows. *J Magn Magn Mater*, 1990, **85**(1–3): p. 181–186.
41. Debye, P.J.W., *Polar Molecules*, Dover, New York, 1929.
42. Maiorov, M.M., Experimental study of the permeability of a ferrofluid in an alternating magnetic field. *Magnetohydrodynamics*, 1979, **15**: p. 5.
43. Chung, S.H. et al., Biological sensors based on Brownian relaxation of magnetic nanoparticles. *Appl Phys Lett*, 2004, **85**(14): p. 2971–2973.
44. Berne, B.J., Pecora, R., *Dynamic Light Scattering*, John Wiley & Sons, New York, 1976, p. 376.
45. Odenbach, S., Magnetoviscous and viscoelastic effects in ferrofluids. *Int J Mod Phys B*, 2000, **14**(16): p. 1615–1631.
46. Odenbach, S., *Magnetoviscous Effects in Ferrofluids*, Springer, 2002.
47. Hayes, M.A. et al., Flow-Based Microimmunoassay, 2001, p. 5896–902.
48. Pethig, R., Markx, G.H., Applications of dielectrophoresis in biotechnology. *Trends Biotechnol*, 1997, **15**(10): p. 426–432.
49. Liu, Y. et al., Evidence for localized cell heating induced by infrared optical tweezers. *Biophys J*, 1995, **68**(5): p. 2137–2144.
50. Edelstein, R.L. et al., The BARC biosensor applied to the detection of biological warfare agents. *Biosensors Bioelectron*, 2000, **14**(10–11): p. 805–813.
51. Choi, J-W. et al., An integrated microfluidic biochemical detection system for protein analysis with magnetic bead-based sampling capabilities. *Lab Chip*, 2002, **2**(1): p. 27–30.
52. Kashevsky, B.E., Nonmagnetic particles in magnetic fluid: Reversal dynamics under rotating field. *Phys Fluids*, 1997, **9**(6): p. 1811–1818.

53. Yellen, B.B., Arranging matter by magnetic nanoparticle assemblers. *Proc Natl Acad Sci*, 2005, **102**(25): p. 8860–8864.
54. Zhu, T., Marrero, F., Mao, L., Continuous separation of non-magnetic particles inside ferrofluids. *Microfluid Nanofluid*, 2010, **9**(4): p. 1003–1009.
55. Mao, L., Koser, H., Ferrohydrodynamic pumping in spatially traveling sinusoidally time-varying magnetic fields. *J Magn Magn Mater* 2005, **289**: p. 199–202.
56. Happel, J., Brenner, H., *Low Reynolds Number Hydrodynamics with Special Applications to Particulate Media*, Springer, New York, 1973.
57. Goldman, A.J., Cox, R.G., Brenner, H., Slow viscous motion of a sphere parallel to a plane wall — I. Motion through a quiescent fluid. *Chem Eng Sci*, 1967, **22**(4): p. 637–651.
58. Derjaguin, B., Landau, L., Theory of the stability of strongly charged lyophobic sols and of the adhesion of strongly charged-particles in solutions of electrolytes. *Prog Surface Sci*, 1993, **43**(1–4): p. 30–59.
59. Mao, L., Koser, H., Towards ferrofluidics for μ-TAS and lab on-a-chip applications. *Nanotechnology*, 2006, **17**(4): p. S34–S47.
60. Langer, R., Vacanti, J., Tissue engineering. *Science*, 1993, **260**(5110): p. 920–926.
61. Pirlo, R.K. *et al.*, Cell deposition system based on laser guidance. *Biotechnol J*, 2006, **1**(9): p. 1007–1013.

Chapter 17

Antibody-based Blood Bioparticle Capture and Separation Using Microfluidics for Global Health

ZhengYuan Luo[*,†], ShuQi Wang[‡], Utkan Demirci[‡,§],
TianJian Lu[†], Feng Xu[†,¶,**], and BoFeng Bai[*,†,††]

[*]*State Key Laboratory of Multiphase Flow in Power Engineering
Xi'an Jiaotong University, Xi'an 710049, China*

[†]*Biomedical Engineering and Biomechanics Center
Xi'an Jiaotong University, Xi'an 710049, China*

[‡]*Demirci Bio-Acoustic-MEMS in Medicine (BAMM) Laboratory
Division of Biomedical Engineering, Department of Medicine
Brigham and Women's Hospital, Harvard Medical School
Boston, MA 02139, USA*

[§]*Harvard–MIT Health Sciences and Technology, Cambridge, MA 02139, USA*

[¶]*The Key Laboratory of Biomedical Information Engineering
of Ministry of Education, School of Life Science and Technology
Xi'an Jiaotong University, Xi'an 710049, China*

Introduction

Progresses on global health and treatment

Global health data released by the World Health Organization (WHO) has revealed that a substantial number of people die prematurely from medical conditions every year. In 2008, the worldwide adult mortality rate was approximately 18%.[1] The extremely heavy global burden of diseases (GBDs) is

---

[**,††]Corresponding authors. Email: fengxu@mail.xjtu.edu.cn; bfbai@mail.xjtu.edu.cn

the leading cause of death worldwide (e.g., 7.9 million people died of cancers[2] and two million people died of HIV diseases[3] in 2007 alone). More than 70% of the total GBDs occur in low- and middle-income countries. The incidence and mortality of many diseases are always high even in developed countries,[4] but the burden of these diseases in developing countries remains much heavier than that in developed countries, according to the WHO report.[2] In recent times, significant progress has been made in improving global health, e.g., microfluidic devices for point-of-care (POC) diagnosis. Nevertheless, most of the new technologies and devices can be only used in developed countries because of their intrinsic limitations, such as the requirement of refrigerated storage of chemicals and highly trained technicians.[5] The improvement of global health is significantly impeded by endemic poverty,[6] hence there is an urgent need for effective and inexpensive medical technologies and devices.

For worldwide lethal diseases like HIV and cancers, efficient treatment is the most important issue, which depends on early and accurate diagnosis, which can minimize the waste of public resources on ineffective treatments and the spread of disease in the population through predicting a more prompt and proper treatment of patients.[7] This can significantly decrease deaths due to diseases, especially in developing countries. However, the cost of diagnosis is very high in resource-limited countries, resulting in low quality palliative care and sometimes, directly leading to the worsening of the condition of patients, as indicated by the high mortality in developing countries.[8] For example, cancer is often diagnosed and treated at the advanced stage, when tumor cells have already invaded and metastasized to other parts of the body.[9]

Blood tests are used to diagnose many physiological and pathological diseases and monitor treatments. However, most blood analysis methods commonly require highly skilled technicians and well-equipped, expensive laboratories.[10] More effective, inexpensive, and comprehensive approaches are critically needed for blood analysis to test for diseases and achieve early and accurate health diagnosis, especially in developing countries.

Microfluidics for the global health

To develop new diagnostic technologies and devices, scientists and engineers should consider the key factors for providing health care throughout the world, i.e., high efficiency and low cost.[11] In addition, there are other intrinsic constraints in developing countries, including limited resources, unreliable electrical power, poorly equipped hospitals, and the lack of highly skilled technicians.[12] Microfluidics-based diagnostic technologies have great potential to overcome these challenges.[13] For example, microfluidic systems integrating antibody capture for blood bioparticle separation offer efficient and speedy measurements from small volumes of reagents and samples, eliminating the dependence on highly skilled operators and demonstrating its potential to be integrated into POC diagnostic systems.[10,14]

In this chapter, we review recent experimental investigations and theoretical simulations on antibody-based blood bioparticle capture and separation using microfluidics. Next, we discuss the challenges associated with this strategy. Mathematical modeling work based on the kinetics of single adhesion bonds and the adhesion kinetic simulation of adhesive cells will also be presented. The summary and future directions are concluded in the last part of the chapter.

Experimental Studies on Antibody-based Bioparticle Capture and Separation

Bioparticles in blood as biomarkers for diseases

Blood is a very important and complex mixture that has the potential to reveal a massive amount of information about the functioning of the human body. Cells of the immune system with specific immunosurveillance functions that sort harmful bioparticles leading to specific diseases are often transported via blood.[10] For example, circulating tumor cells (CTCs), which are probably the origin of intractable metastatic diseases, have been found in the peripheral blood of cancer patients.[15,16] Accordingly, various types of multiscale bioparticles in blood bring great potential for diagnosing many physiological and pathological diseases and for monitoring treatment. Most of the bioparticles can be distinguished by their hierarchical scales (Fig. 1). Their shape, number, and characteristics have close relationships with their functions to the human body. Clinically, many bioparticles have been used as biomarkers for diseases. Through blood tests, doctors can obtain the quantities of various blood cells for disease monitoring, e.g., red blood cells (RBCs) for anemia,[17] white blood cells (WBCs) for leukemia,[18] and platelets for thrombocytopenia.[19] The count of CTCs in peripheral blood has been considered as an alternative marker for metastatic cancers.[9,20–22] The definitive diagnosis of malaria remains as the detection of malarial parasites in the blood and this inexpensive test for malaria is used widely in developing countries, which is essential for describing parasite-infected RBCs in blood.[23] Effective detection of viruses in the blood also plays a critical role in medicine and public health.[24,25] For monitoring and diagnosing HIV, the level of HIV-1 RNA in blood is widely used to indicate the viral replication condition.[26] The HIV-1 p24 antigen in blood, which is in

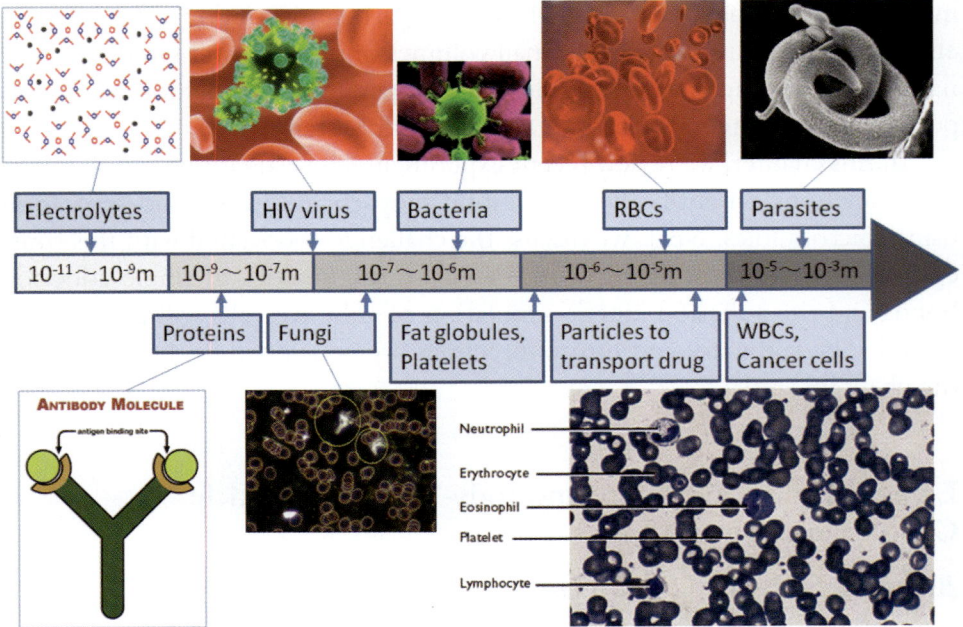

Fig. 1. Multiscale bioparticles in blood. Organic macromolecules are in the nanometer scale, cells are in the micrometer scale, and viral and bacterial particles are between the nanometer and micrometer scales. Many bioparticles are used as markers for diseases, such as RBCs for anemia, WBCs for leukemia, platelets for thrombocytopenia, HIV-1 RNA and CD4+ T cells for HIV, parasites for malaria, and CTCs for cancers.

parallel with the level of HIV-1 RNA, is also used as a marker for viral load monitoring.[27] Furthermore, the quantity of CD4+ T cells in blood has also been used as a HIV monitoring marker.[28]

Antibody-based microfluidic devices developed for global health

CD4+ T cell and viral capture devices for HIV

To decrease the number of deaths from HIV, antiretroviral treatment (ART) must be expanded rapidly in resource-limited countries. However, this has been significantly impeded due to the lack of inexpensive and easy-to-use monitoring tools. Current commercial viral load assays[29] and inexpensive alternatives (e.g., the Ultrasensitive p24 assay,[30] RT-qPCR[31]) are still not practical for resource-poor settings because of their limitations such as the need for highly skilled operators, low throughput, and high cost per test. Consequently, cost-effective and easy-to-use ART monitoring tools are urgently needed to help expand ART in developing countries.

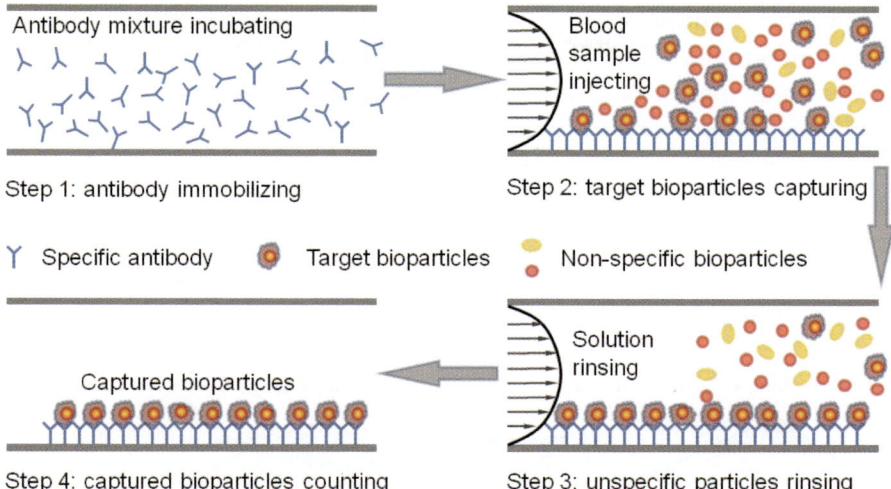

Fig. 2. Schematic illustration of the entire process of antibody-based bioparticle capture and separation using microfluidics. First, the antibody mixture is incubated to immobilize specific antibodies on the surfaces. Then, the blood sample is injected to capture target bioparticles and rinsed with solution to wash away unspecific particles. Finally, the captured bioparticles are counted using lensless CCD imaging or an optical microscope. However, if the captured cell count is based on fluorescence technology, bioparticles must be stained with specific fluorescence labels before sample injecting or after solution rinsing.

CD4+ T cell count has been used to diagnose opportunistic HIV infections and initiate ART.[32,33] Nevertheless, existing CD4+ T cell count methods (e.g., flow cytometry like FACSCount[34] and some alternative methods[35,36]) are very costly and rarely used in resource-limited countries. Therefore, CD4+ T cell count methods based on antibody capture and microfluidics have recently been developed.[37–46] Basically, these methods involve the immobilization of the CD4 antibody on microchannel surfaces through surface chemistry modification, injecting whole blood sample to capture CD4+ T cells, rinsing non-specific cells from the microchannels, and counting all the cells captured within the chip (Fig. 2). In recent years, considerable efforts have been made to simplify microfluidic devices and reduce their cost, where the involvement of fluorescence microscopes for the detection of captured cells contributes a large part to the high cost.[43,45,47] To eliminate the dependence on fluorescence microscopes, Cheng et al. achieved CD4+ T cell count using optical microscopes[40,41] and then developed a counting method through cell lysate impedance spectroscopy.[39] Thereafter, external equipment for accomplishing sample filling and rinsing is exterminated by Thorslund et al.[42] through pure capillary action. More recently, a lensless, ultrawide-field cell array based on shadow imaging has been developed, which greatly simplifies the CD4+ T cell count system and significantly increases the counting speed.[43,48] These simplified designs enhances the possibility of integrating CD4+ T cell count into a POC system. However, the low efficiency and specificity of the microchips turned out to be the greatest setbacks. Cheng et al.[40] investigated the dependence of antibody capture on shear stress and identified the differences of the sensitivity of shear stress between monocytes and CD4+ T lymphocytes. Based on their observation, a double-stage cascaded microchip was designed,[44] which included four monocyte-depletion microchannels to remove monocytes before CD4+ T cell capture (Fig. 3). Recently, Wang et al.[46] microfabricated traps coated with anti-CD4

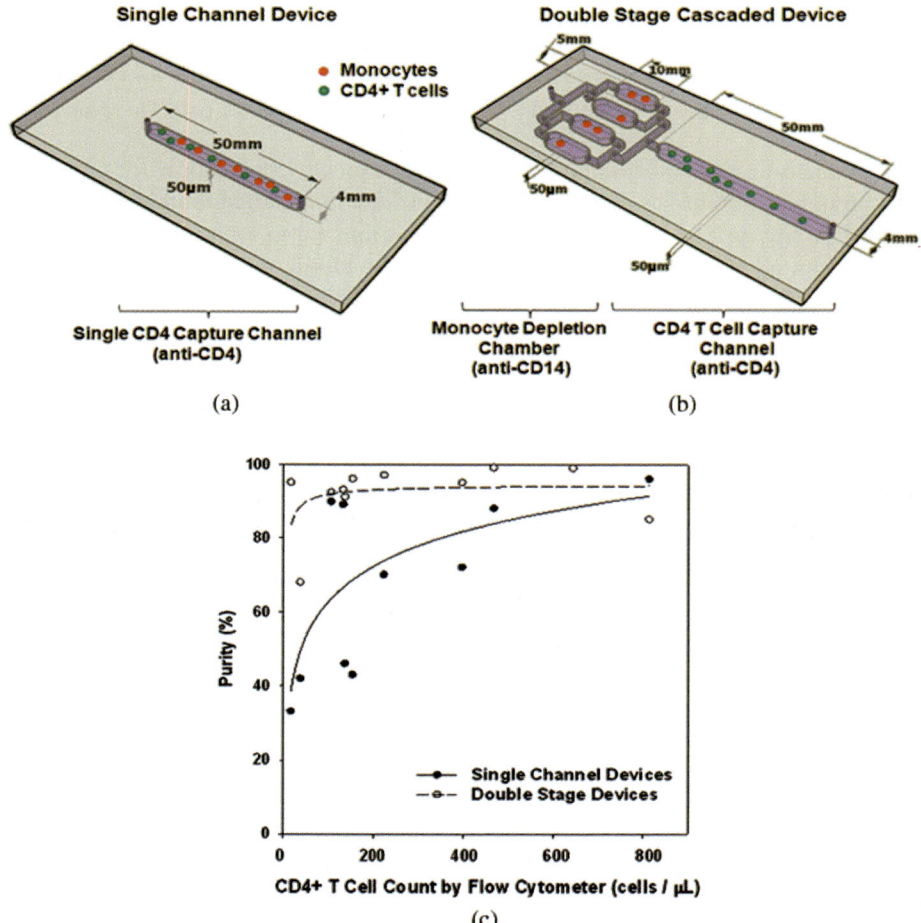

Fig. 3. A double-stage cascaded microchip for antibody-based CD4+ T cell separation and enumeration. (a) Single channel capture device used as the control group, (b) schematic illustration of the double-stage cascaded microchip, and (c) purity of the captured CD4+ T cells as a function of the absolute CD4 counts. The microchip includes four monocyte-depletion microchannels to remove monocytes before CD4+ T cell capture, and the purity was improved significantly, especially when the absolute CD4 count was below 200 cells/μL.[44]

antibody in a chamber in order to increase the binding surface in unit volume, which significantly improves the cell capture efficiency (~80%) compared to previous work (~70%[43]). However, the capture efficiency still remains low.

Though CD4[+] T cell count in the blood is an important standard for determining the extent of damage of HIV to the immune system, the count is also affected by many other factors such as genetic defects and lack of developmental control,[49] decreasing its reliability. Plasma HIV viral load count directly reflects the status of HIV patients and is more reliable for monitoring HIV.[50] Direct viral count using microfluidics and nanotechnologies has become attractive in the development of a POC viral load system.[51,52] Recently, an antibody-based microfluidic device has been applied for HIV viral particle capturing and counting from whole blood.[51] In this study, anti-gp120 antibodies were immobilized on a glass surface of a microfluidic channel made from polymethymetacrylate (PMMA) and glass, to capture HIV-1 particles when the whole blood sample was injected into the channel. Captured viral particles were imaged via quantum dots because of the small diameter of HIV (~120 nm).[53] This direct capture method for HIV presents a great potential to create a POC viral load system. However, there are still many improvements that can be made, such as the capturing of various HIV subtypes from whole blood with higher efficiency and specificity. It is also challenging to avoid using fluorescent microscope for quantum dots imaging.

Capture and separation of CTCs

Though some advances have been realized in diagnosing and treating primary tumors, cancer-related deaths remain strikingly high because of the spread of primary tumors to other parts in the body (i.e., metastasis[54]) via CTCs.[15,55] Consequently, effective capture, separation, and counting of CTCs in peripheral blood are extremely important for the diagnosis and prognosis of metastatic cancers. The capture, separation, and counting of CTCs have been studied for various cancers,[21] such as in the prostate,[16] breast,[56] and lung.[57] However, existing methods (e.g., immunomagnetic approaches, size- and deformation-based methods) are time- and reagent-consuming, and the separation efficiency and purity are also very low (e.g., < 50%).[58–60] Another major issue in the quantification of CTCs is the extremely low concentration (i.e., several CTCs per milliliter peripheral blood).[15,61,62] Hence, antibody-based CTC capture and separation using microfluidics provide a promising alternative, as indicated by the emerging CTC capture/separation microfluidic platforms.[22,63–72] Du et al.[64] used epithelial membrane antigen (EMA) and epithelial growth factor receptor (EGFR) coated on the surfaces of straight microchannels to capture breast cancer cells, and ~100% purity of captured population was achieved. However, the capture efficiency is found to be only ~30%. Nagrath et al.[22] developed an antibody-based capture

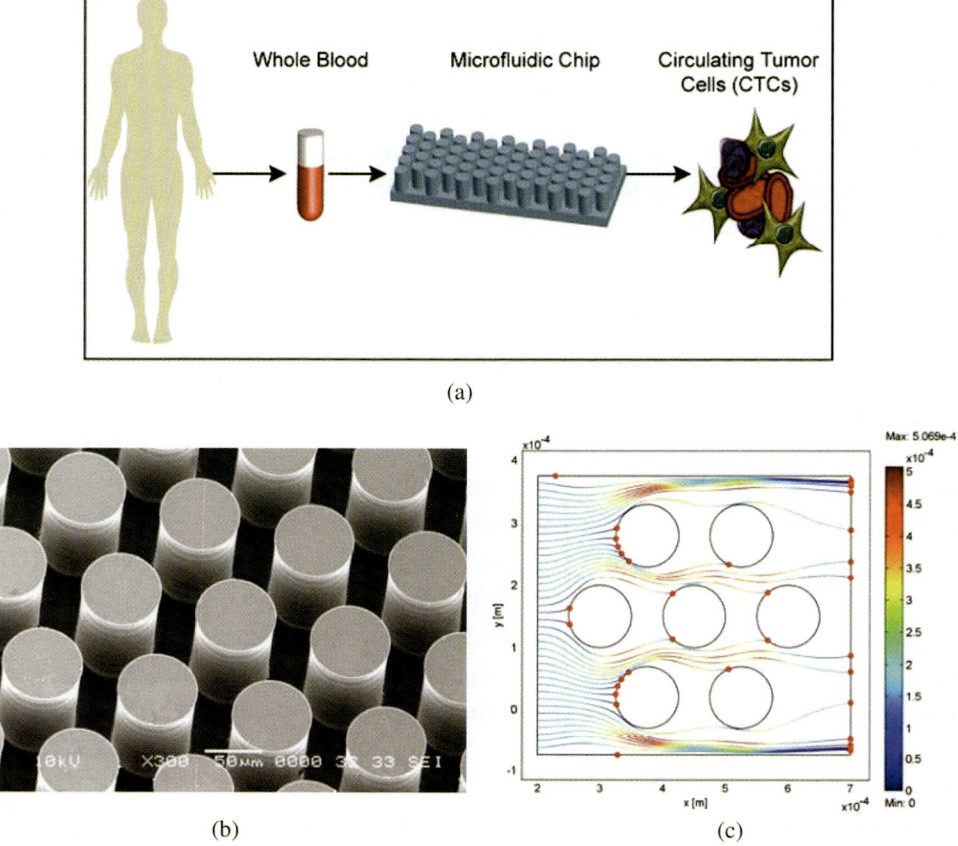

Fig. 4. A simple microfluidic array with microposts for CTC isolation. (a) One-step process for POC isolation of CTCs from peripheral blood, (b) schematic illustration of the micropost array (scanning electron micrograph), and (c) computational analysis of cells colliding with the micropost array. Solid lines are the cell trajectories and red dots indicate the end positions of the cells. By using this numerical method, the capture efficiency of CTCs using this microfluidic system can be calculated, which was about 65%.[22]

microfluidic device with high surface area-to-volume-ratio, integrating microposts arranged in an equilateral triangular format to separate cancer cells (Fig. 4). Their device achieved ~60% capture efficiency, but only ~50% purity. In order to improve both the capture efficiency and purity of a microfluidic system, great efforts have been made to optimize the structure of the capture device, such as integrating micropillars,[67] three-dimensional nanostructures,[70] obstacles,[66] and sinusoidally shaped channels.[71] Another way to enhance the antibody-based capture method is to search for more specific antibodies, e.g., prostate-specific membrane antibody (J591) for prostate cancer CTCs (PCTCs)[66] and the biomimetic combination of E-selectin and anti-EpCAM for MCF7 cells.[68] Through the abovementioned methods, the capture efficiency and purity of cancer cells were increased to >80% and >60%,[66] respectively. Aptamers immobilized on microchannels have also been used to further improve the capture efficiency (>80%) and purity (>97%) of cancer cells.[61,73,74] However, in spite of these advances, many challenges remain, including the increase of throughput, simplification of microfluidic devices, and further improvement of the capture efficiency and purity of target cells.

Capture of other bioparticles

Besides CTCs for cancers and CD4$^+$ T cells for HIV, the antibody-based capture method using microfluidics has been also used to capture and separate other bioparticles. These include leukocyte subtypes,[75–81] RBCs,[82] proteins,[83,84] bacteria,[85,86] and pathogens.[87] Except for blood checking, the techniques are also applicable to other bioparticles in the body fluid and related environments. Some innovative methods, e.g., dielectrophoresis (DEP) integrated in microfluidics[77,88,89] and nanoparticles immobilized on channel surfaces,[78,86,90] have been developed to enhance antibody-based capture and improve capture efficiency and purity. For instance, most cells behave as a dielectric particle in an external electric field, resulting in a DEP force on the cells in suspension.[91,92] Hashimoto et al.[77] integrated DEP in microfluidics to quickly concentrate cells from the flowing cell suspension to the active detection area with a specific antibody, and the sensitivity was successfully increased for the detection of a small number cells from the blood. In those studies, the innovative enumerating method also has been developed. For example, Ymeti et al.[93] reported the application of an integrated optical Young interferometer sensor for the antibody-based detection of herpes simplex virus type 1 (HSV-1). The antibody-based capture method using microfluidics has not only been used for bioparticle separation and enumeration, but also applied to other purposes such as purifying of biomarkers,[94] evaluating antibody interactions with cell surface targets,[95] analyzing protein[96,97] and cell phenotypes,[98] and progenitor cell separation for tissue engineering.[89,99–101] For more details on the DEP method, see Chapter 14.

Challenges with Current Antibody-based Bioparticle Capture Using Microfluidics

Most of current researches on the antibody-based bioparticle capture and separation using microfluidics are listed in Table 1. Although great progress has been made, several critical issues still remain to be solved. First, highly

Table 1. Several studies on antibody-based bioparticle capture and separation using microfluidics.

Bioparticle type	Capture method	Counting method	Microchannel structure	Microchannel material	Flow parameters	Efficiency	Specificity	Reference
CD4+ T cell	Monoclonal anti-CD4 antibody	Cell lysate impedance spectroscopy	Straight rectangular channel: 5 cm × 4 mm × 50 μm*	PDMS	Blood injecting: 5 μL/min; rinsing: 20 μL/min × 3 min†; cell lysing: 10 μL/min × 1 min	>75%	—	39
	Monoclonal anti-CD4 antibody	Optical microscope	Straight rectangular channel: 5 cm × 4 mm × 50 μm	PDMS	Blood injecting: 10 μL × (1~20) μL/min‡; rinsing: 20 μL/min × 3 min	>75%	60%~90%	40
	Biotinylated anti-CD4 antibody	Fluorescence microscope	Straight rectangular channel with micropillars: 4 cm × 0.5 mm × 30 μm	PDMS	Blood injecting and rinsing based on capillary filling	79%	—	42
	Anti-CD4 antibody	Lensless imaging	Straight rectangular channel: 2.4 cm × 4 mm × 50 μm	PMMA	Blood injecting: 5 μL/min × 2 min; rinsing: 20 μL/min × 3 min	70.2 ± 6.5%	88.8% ± 5.4%	43
	Anti-CD4 antibody	Chemiluminescence	Microchannel with traps based on micropillars	PDMS	Blood injecting: 1~5 μL/min; chemiluminescent solution: 5 μL/min	80%	89%	46

(Continued)

Table 1. (Continued)

Bioparticle type	Capture method	Counting method	Microchannel structure	Microchannel material	Flow parameters	Efficiency	Specificity	Reference
HIV-1 virus	Anti-gp120 antibody	Quantum dots	Straight rectangular channel: 2.4 cm × 4 mm × 50 μm	PMMA	Blood injecting: 10 μL × 0.5 μL/s	—	—	51
Breast cancer cell	EMA and EGFR	Fluorescence microscope	Straight microchannels: 2.0 cm × 0.5 mm × 500 μm	PDMS	Blood injecting: 15 μL/min × 3 min	>30%	~100%	64
Breast cancer cell	Anti-EpCAM and erbB2	Camera and microscope	Straight microchannels: 1.6 cm × 6 mm × 250 μm	Glass	Sample injecting: 90 μL/min × 10 min; rinsing: 350 μL/min × 6 min	~85%	—	65
NSCLC cell	Anti-EpCAM antibody	Fluorescence microscope	Microchannel integrating microposts	Silicon	Blood injecting: 5 mL × 1 mL/h; rinsing: 10 mL × 10 mL/h	~60%	~50%	22
Prostate cancer cell	Antibody J591	Fluorescence microscope	Microchannel (2.5 cm × 8 mm × 100 μm) integrating microposts with 80-μm diameter	Glass and silicon	Blood injecting: 1 mL × 1 mL/h; rinsing: 1 mL/h × 10 min	85 ± 5%	68% ± 6%	66

Note: *denotes microchannel length × width × height; †denotes flow rate × flow time; ‡denotes total sample volume × flow rate.

specific antibodies are needed to selectively target bioparticles. Most antibodies used to capture particular bioparticles demonstrate low specificity and capture bioparticles other than those of interest. For example, since the CD4 antibody is also expressed on monocytes, the capture specificity of CD4$^+$ T cells is very low (~60%), especially at absolute CD4 counts below 200 cells/mL due largely to monocyte contamination.[40] Low antibody specificity is the most prominent issue in cancer cell capture, limiting its application.[58,102,103] Second, expensive and complex counting (e.g., fluorescence microscope,[22] quantum dots[51]) and sample filling technologies (e.g., pressure-driven sample filling by external syringes[40]) are still used in most of the current antibody-based capture microfluidic devices. These technologies limit their application in POC systems. To address these challenges, some innovative approaches have been developed, e.g., cell lysate impedance spectroscopy,[39] lensless shadow imaging techniques,[43] optical Young interferometer sensor,[93] and pure capillary action for sample filling and rinsing.[42] However, new problems emerge with these new technologies. For example, lensless charged-coupled device (CCD) imaging further reduces the efficiency of the microfluidic device (CCD efficiency is defined as CCD count divided by all captured cells).[43] It is difficult to completely lyse captured cells and accurately determine the relationship between the number of captured cells and lysate impedance spectroscopy.[39] Furthermore, individual differences in capillary action provide different self-driven forces and flow conditions for sample injecting and solution rinsing.[42] Third, the low capture efficiency and specificity of current strategies for antibody-based capture limit their commercialization potential for replacing flow cytometry. As shown in Table 1, the capture efficiency and specificity of existing antibody-based approaches are mostly below ~80%, especially for cancer cell capture, which is less than 70%. Finally, there is no sufficient basis to guide the selection of structural parameters of capture microchannels and flow conditions of target samples (Table 1). Some researchers[22,38,40,75] have studied the effects of fluid shear stress on antibody-based capture using the Hele–Shaw microfluidic chamber,[104] which can produce linearly decreasing shear stress along the chamber length. This has been used to optimize the flow conditions of antibody-based capture. However, several other factors affecting the capture efficiency (e.g., the interaction time between target bioparticles and the active surface area, the deformation and motion of target bioparticles) are also significantly affected by the flow conditions.[40,105] In actual fact, the root cause for most of the critical issues described above is that the underlying mechanisms of antibody-based bioparticle capture and separation using microfluidics are still unknown, and there is an urgent need for theoretical studies in this respect.

Mathematical Modeling of Antibody-based Bioparticle Capture

Mathematical modeling is an effective method to study the underlying mechanisms of biological processes[106,107] and can address some issues associated with experimental studies. For experimental researches on microfluidics for bioparticle capture, the indispensable need of precise measuring technologies (e.g., fluorescence technology even with quantum dots[108]) and microstructure manufacturing technologies (e.g., soft lithography technology[109]) gives rise to many obstacles (e.g., complex operations, high cost). Besides, biological phenomena are extremely complicated and there are many factors affecting biological behaviors of cells, including adhesion, which are complex[110,111] and difficult to control.

The whole process of antibody-based microfluidic capture and separation can be divided into two key steps, i.e., (1) bioparticles being driven away from the sample mainstream and colliding with capture surfaces and (2) bioparticle adhesion and rolling on capture surfaces, possibly dissociating from surfaces and being washed away (Fig. 2). For several recent years, many researchers have made considerable efforts to model the capture process of bioparticles in microfluidics. In this section, we will review (1) studies on the kinetics of single adhesion bonds, (2) the numerical simulation of single cell adhesion and rolling on surfaces with multiple adhesion bonds, and (3) mathematical modeling of bioparticle motion and adhesion in microfluidics.

The kinetics of single adhesion bonds

Previous researches have revealed that cells adhere to a substrate via multiple receptor–ligand bonds.[112–114] Receptor–ligand interactions are mediated by kinetics including on-rate and off-rate. Sufficiently high on-rate and off-rate of bonds are prerequisites, respectively, for the formation and dissociation of receptor–ligand bonds. The hydrodynamic forces exerted by the sample flow not only prevent bioparticles from being captured, but also separate push-captured bioparticles. Adhesive forces are then generated by the receptor–ligand bonds balancing the hydrodynamic forces to promote the capture of bioparticles and prevent the captured bioparticles from being washed away. The adhesion force of single bonds and the number of adhesion bonds determine the adhesion forces applied on bioparticles and both of them depend on the on-rate and off-rate of adhesion bonds.

As selectin–ligand interactions play a critical role in the tethering and rolling of circulating cells on the vascular surface during inflammation or cancer metastasis,[115–119] they have been intensively studied and serve as an ideal system for investigating the kinetics of receptor–ligand interactions. According to previous work on selectin–ligand bond dissociation kinetics, the intrinsic off-rate in the absence of an applied force exists in selectin–ligand interactions[120–122] and varies considerably among different types of selectins, e.g., 1 s^{-1} for P-selectin, 0.7 s^{-1} for E-selectin, and 7 s^{-1} for L-selectin.[120,123] This is because the off-rate of adhesion bonds depends significantly on external forces such as hydrodynamic forces.[121,124] Since the specific dissociation kinetic behaviors depend on the applied force,[120–122,124–130] adhesion bonds have been divided into two types — slip bonds and catch bonds. The dissociation rate of slip bonds increases with increasing external force.[122,124,126,128] Recently, catch bonds, which dissociate less rapidly at higher forces than at lower forces, have been observed.[130–135] Moreover, bond rupture is regulated by the loading rate and the history of force application as revealed by the introduction of a ramp/jump model of applied force.[125,128,136–138] Compared with bond dissociation kinetics, investigations of bond formation kinetics have seen slower progress. The on-rate depends on the relative velocity and distance between interacting molecules.[139–142] The more rapidly and forcefully selectins collide with their ligands, the more easily selectin–ligand bonds form.[143] For example, the on-rate has been found to increase with increasing compressive force.[130] Recently, the formation of bonds is also found to depend on molecular length and the micromorphology of the cell surface on which they are expressed.[144–146] Although the kinetics of adhesion bonds have been studied intensively, several issues still exist and more investigations are urgently needed, such as the adhesive cooperation of multiple receptor–ligand bonds and theoretical models for the kinetics of different receptor–ligand bonds.

Numerical simulation of cell rolling and adhesion

After being captured, the bioparticles will move on the capture surfaces and may be washed away by flow. A fundamental and quantitative understanding of the adhesion of bioparticles to surfaces is needed to improve the immobilization of the captured bioparticles. During many biophysical processes such as inflammation or cancer metastasis, it has been observed *in vivo* and *in vitro* that cells adhere to and roll on vascular surfaces or substrates immobilized with adhesion molecules.[147–152] The rolling of cells on surfaces has been simulated and investigated using flow chambers with immobilized ligands or cell monolayers *in vitro*.[143,153–156] To reduce the effects of uncertain factors in experiments and better understand the mechanism of adhesion and rolling of cells on surfaces, some theoretical models have been developed.

In the early literature, cells adhered on surfaces were considered as solid parts of an adhesion plate to determine hydrodynamic forces on adhesion cells produced by the surrounding fluid.[157–159] In order to introduce the effects of adhesion bonds, cells are idealized as microvillus-coated hard spheres covered with adhesive molecules.[141,160–164] These approaches provided the possibility to simulate cell adhesion in the case of limited computing resources. However, the deformation of adhesion cells and microvilli was also identified as factors that affect adhesion behaviors[165–170] (e.g., increasing the formation of adhesion bonds[143,165,167]). Accordingly, numerical simulation methods were developed to consider cell deformation (e.g., numerical methods of tracking or capturing moving boundaries[167,171–173]). The simple liquid drop model was introduced into cell modeling and reflected several effects of cell deformation.[105,174] In addition, the capsule model, which considered cells as viscous drops surrounded by flexible membranes, simulated cells more accurately.[175–177] In recent years, more realistic models like the compound drop model were developed to include the effects of the nucleus, where the cell was considered to be made up of only the nucleus and cytoplasm.[178–182] Using these models, more realistic comparisons were realized between simulations and experimental results.

In order to investigate the adhesion behaviors and locomotion of captured particles on surfaces using theoretical models, the calculation of adhesion forces plays an essential role. Many studies have attempted to develop methods to model adhesion forces. The adhesion potential model was one of the methods used to facilitate cell adhesion, which used the adhesion potential function to describe adhesion force.[175,182] However, adhesion forces are generated by multiple adhesion bonds. The kinetic adhesion model developed by Dembo et al.[183] successfully models the formation and dissociation of adhesion bonds. In this model, a single bond is treated as a Hookean spring and the adhesion force (f_b) generated by one bond is

$$f_b = \sigma (l_x - \lambda), \tag{1}$$

where σ is the spring constant, l_x is current length of the bond, and λ is the equilibrium bond length. Microvilli on the cell membrane play a critical role in cell adhesion, because only the receptors on the microvillus tips will interact with ligands on the substrate.[162,184,185] Consequently, l_x is the distance between the microvillus tip and capture surface. Thereafter, the total adhesion force (F_b) acting on per unit area of cell membrane is deduced from

$$F_b = N_b f_b, \tag{2}$$

where N_b is the bond density. Using the reversible chemical reaction theory,[186] N_b is calculated from the adhesion kinetic relation between the formation and dissociation of bonds as follows:

$$\frac{\partial N_b}{\partial t} = k_f(N_{l0} - N_b)(N_{r0} - N_b) - k_r N_b, \tag{3}$$

where N_{r0} and N_{l0} are the initial receptor density on the cell membrane and the initial ligand density on the capture surface, respectively. k_f and k_r are the forward and reverse reaction rates.

The kinetic adhesion model conveniently calculates the value of the total adhesion force, but cannot predict the distribution of the adhesion force and the details of each bond. A stochastic simulation system using the Monte Carlo method successfully tracks the state of each bond (formation or dissociation).[187–191] In this model, P_f and P_r are used to express the probability of the formation of a new bond and the dissociation of an existing bond, respectively. Their values in a time step (Δt) are calculated by

$$P_f = 1 - \exp(-k_f \Delta t) \quad \text{and} \quad P_r = 1 - \exp(-k_r \Delta t). \tag{4}$$

During each time interval, P_f and P_r are computed for every bond and compared with two random numbers N_1 and N_2 (between 0 and 1). If $P_f > N_1$, a new bond will form; if $P_r > N_2$, an existing bond will break. The value of adhesion force (f_b) contributed by each bond is calculated by Eq. (1) and the direction is along the adhesion bond. Then, the total adhesion force (F_b), including the value of the total force and torque exerted on cell, is the vector sum of the forces generated on all bonds.

The forward and reverse reaction rates in Eqs. (3) and (4) are computed by[183]

$$\begin{aligned} k_f &= k_f^0 \exp\left(-\frac{\sigma_{ts}(l_x - \lambda)^2}{2kT}\right) \quad \text{and} \\ k_r &= k_r^0 \exp\left(-\frac{(\sigma - \sigma_{ts})(l_x - \lambda)^2}{2kT}\right), \end{aligned} \tag{5}$$

where k_f^0 and k_r^0 are the equilibrium values of the forward and reverse reaction rates, respectively; k and T are respectively the Boltzmann constant and the absolute temperature; σ_{ts} is the spring constant of the bond in the transition state. The relationship between σ and σ_{ts} determines the state of the bond — $\sigma_{ts} > \sigma$ for catch bonds and $\sigma_{ts} < \sigma$ for slip bonds. The model was modified to include the effects of many other factors (e.g., microvillus extension,[192] receptor extraction,[193] and receptor shedding[194]) on the dissociation rate of the bond, and the calculation of k_r is modified as[181,192–194]

$$k_r = k_r^0 \exp\left(\frac{(\kappa\sigma - \sigma_{ts})(l_s - l_{mv} - \lambda)^2}{2kT}\right), \qquad (6)$$

where l_s and l_{mv} are the length of the microvillus bond (that is the distance between the microvillus bottom and the capture surface) and the unstressed length of the microvillus, respectively; κ is a correction coefficient. If receptor shedding or extraction occurs, κ is a number more than unity. As described earlier, the forward and reverse reaction rates are significantly affected by external forces applied on adhesion bonds. A force-dependent model[136,195,196] was used to compute the dissociation rate using the equation

$$k_r = k_r^0 \exp\left(\frac{\gamma f_b}{kT}\right), \qquad (7)$$

where γ is the reactive compliance with length units for describing how susceptible the bond is to applied forces.

Although several models for cell adhesion have been developed and used to simulate cell rolling on capture surfaces (Table 2), they are limited to only model adhesion behavior controlled by a single type of bond and the values of parameters used for adhesion simulation vary tremendously among different studies. Further investigation is still desired to interpret much more complex phenomena such as multiple bonds cooperation, spreading and migration of cells, and cell adhesion governed by multiple types of bonds.

Mathematical modeling of bioparticle motion and adhesion in microfluidics

The range of binding, composed of the length of the adhesion bond (in nanometers[174,197,198]) and the microvilli (about 0.35 μm) (Table 2), is less than 1 μm, and it indicates that target bioparticles must approach the capture surface close enough to establish contact with the surface and then be captured. Consequently, the underlying mechanisms of bioparticle motion away from the mainstream and collision with capture surfaces are crucial to increase bioparticle capture. Previous works have identified particles moving to the center, but not close to the wall in flows between two platelets or in

Table 2. Several computer simulation studies for the adhesion and rolling of a single cell on surfaces.

Cell model	Adhesion model	Models for k_f and k_r	Computational method	Values of parameters	Reference
Simple liquid drop model; compound drop model	Eq. (3)	Eq. (5)	A multi-scale computational approach and immersed boundary technique	$\sigma = (0.5 \sim 10) \times 10^{-3}$ N/m $\sigma_{ts} = (0.48 \sim 9.5) \times 10^{-3}$ N/m $\lambda = 5.0 \times 10^{-8}$ m $N_{r0} = (2.0 \sim 5.0) \times 10^{14}$ m^{-2} $N_{l0} = (2.0 \sim 5.0) \times 10^{14}$ m^{-2} $k_r^0 = 10^{-11} \sim 10$ s^{-1} $k_f^0 = 10^{-18}$ m^2/s	174
	Eq. (4)	Eq. (5)	Immersed boundary method and Monte Carlo simulation	$\sigma = 1.0 \times 10^{-3}$ N/m $\sigma_{ts} = 0.99 \times 10^{-3}$ N/m $\lambda = 10 \times 10^{-8}$ m $l_{mv} = 3.5 \times 10^{-7}$ m $N_{r0} = 0.56 \times 10^{14}$ m^{-2} $N_{l0} = 155$ microvilli/cell × 50/microvilli $k_r^0 = 1$ s^{-1} $k_f^0 = 1$ s^{-1}	176
Capsule model	Eq. (4)	Eq. (5)	Immersed boundary method and Monte Carlo simulation	$\sigma = 1.0 \times 10^{-3}$ N/m $\sigma_{ts} = (0.98 \sim 0.99) \times 10^{-3}$ N/m $\lambda = 10 \times 10^{-8}$ m $l_{mv} = 3.5 \times 10^{-7}$ m $N_{r0} = 1.5 \times 10^{14}$ m^{-2} $N_{l0} = 252$ microvilli/cell × 50/microvilli $k_r^0 = 1$ s^{-1} $k_f^0 = 1$ s^{-1}	177
Compound drop model	Eq. (4)	Eq. (5)	Immersed boundary method and Monte Carlo simulation	$\sigma = 1.0 \times 10^{-3}$ N/m $\sigma_{ts} = 0.99 \times 10^{-3}$ N/m $\lambda = 10 \times 10^{-8}$ m $l_{mv} = 3.5 \times 10^{-7}$ m $N_{r0} = 1.44 \times 10^{14}$ m^{-2} $N_{l0} = (21$ or $155)$ microvilli/cell × 50/microvilli $k_r^0 = 3.7$ s^{-1} $k_f^0 = 3 \sim 30$ s^{-1}	178

(Continued)

Table 2. (Continued)

Cell model	Adhesion model	Models for k_f and k_r	Computational method	Values of parameters	Reference
	Modified Eq. (3)	Modified Eq. (6)	Modified volume-of-fluid method	$\sigma = 5.3 \times 10^{-3}$ N/m $\sigma_{ts} = 0.16 \times 10^{-3}$ N/m $\lambda = 1.0 \times 10^{-8}$ m $\kappa = 2.0625$ $N_{r0} = (5.0-50) \times 10^{14}$ m^{-2} $N_{l0} = (50-500) \times 10^{14}$ m^{-2} $k_r^0 = 1$ s^{-1} $k_f^0 = 10^{-11}$ m^2/s	179
	Eq. (3)	Eq. (6)	Volume-of-fluid method	$\sigma_B = 5.3 \times 10^{-3}$ N/m $\sigma_{ts} = 0.1 \times 10^{-3}$ N/m $\lambda = 1.0 \times 10^{-8}$ m $\kappa = 1.0$ $N_{r0} = 30 \times 10^{14}$ m^{-2} $N_{l0} = 15 \times 10^{14}$ m^{-2} $k_r^0 = 10$ s^{-1} $k_f^0 = 10^{-10}$ m^2/s $N_{mv} = 928$ microvilli/cell $l_{mv} = 3.0 \times 10^{-7}$ m	181
Compound drop model; drop-rigid-particle model	Adhesion potential model; Eq. (3)	Eq. (5)	The finite element method and the level set technique	$\sigma = 0.2 \times 10^{-3}$ N/m $\sigma_{ts} = 0.1 \times 10^{-3}$ N/m $\lambda = 1.0 \times 10^{-8}$ m $l_{mv} = 3.0 \times 10^{-7}$ m $N_{r0} = 3.0 \times 10^{14}$ m^{-2} $N_{l0} = 1.5 \times 10^{14}$ m^{-2} $k_r^0 = 10$ s^{-1} $k_f^0 = 10^{-10}$ m^2/s	182

Note: N_{mv} is the number of microvilli on the cell.[181] Drop-rigid-particle model considers the nucleus as a rigid particle.[182]

tubes.[199–203] RBCs were also revealed to migrate laterally to the mainstream flow and away from the wall of the vessel *in vivo* and *in vitro*, leading to the formation of the cell-free layer near the wall of the vessel.[204–209] However, WBCs are observed to drift close to the wall, which is known as the margination process, and induce effective leukocyte adhesion.[210–213] Many efforts have been devoted to leukocyte margination. RBCs are more susceptible to deformation and move to the vessel center compared to WBCs, but this would not be expected to marginate WBCs to the wall.[214] RBCs have been identified *in vivo* and *in vitro* to be critical in leukocyte margination.[215–218] Erythrocytes are found to directly affect leukocyte adhesion.[219–222] For example, the normal force imparted by erythrocytes increases the formation of new adhesion bonds, but the tangential force and torque intensify the dissociation of existing bonds.[222] The leukocyte–erythrocyte interaction has been intensively studied using numerical simulation methods.[214,215,217,222–224] Sun *et al.*[224] introduced a lattice Boltzmann approach to simulate blood flow in an expansion structure and found erythrocytes push leukocytes toward the wall. In another numerical study,[214] erythrocyte aggregation was observed to be unnecessary for leukocyte margination, but the leukocytes were also found to be pushed toward the wall by erythrocytes similar to that obtained by Sun *et al.*[224] Unfortunately, current numerical methods can only simulate the motion of few cells (e.g., 29 RBCs and one WBC in Ref. 214 and RBCs in Ref. 225) which is far from the thousands of cells in realistic blood flow. Many important factors have not been considered and their physical mechanisms are also unknown, i.e., leukocyte–erythrocyte interaction, collisions between cells, and erythrocyte aggregation. Accordingly, breakthroughs are eagerly anticipated in numerical simulation for blood flow with multiple bioparticles.

Although current computational limitations impede the use of numerical simulations for accurately modeling the entire process of multiple bioparticle capture, simulations predicting this approach coupled with *in vivo* and *in vitro* experiments have been successfully used to design and improve microfluidics for bioparticle capture and separation.[22,65,153] For example, computational fluid dynamics (CFD) analyses are used to model the motion of multiple cells in capture microchannels and obtain the calculated capture efficiency, which pose as guides for designing and improving microfluidics.[22,65] As the biomarker capture method is extensively introduced into molecular level studies like proteins analyses,[226–228] a numerical simulation method has been developed to model the whole process of surface capture.[229–232] As an example, the study by Park *et al.*[232] is introduced in this section. As shown in Fig. 5, the distribution of bond molecules in a channel with a triangular cross-section and on the channel surface is numerically studied

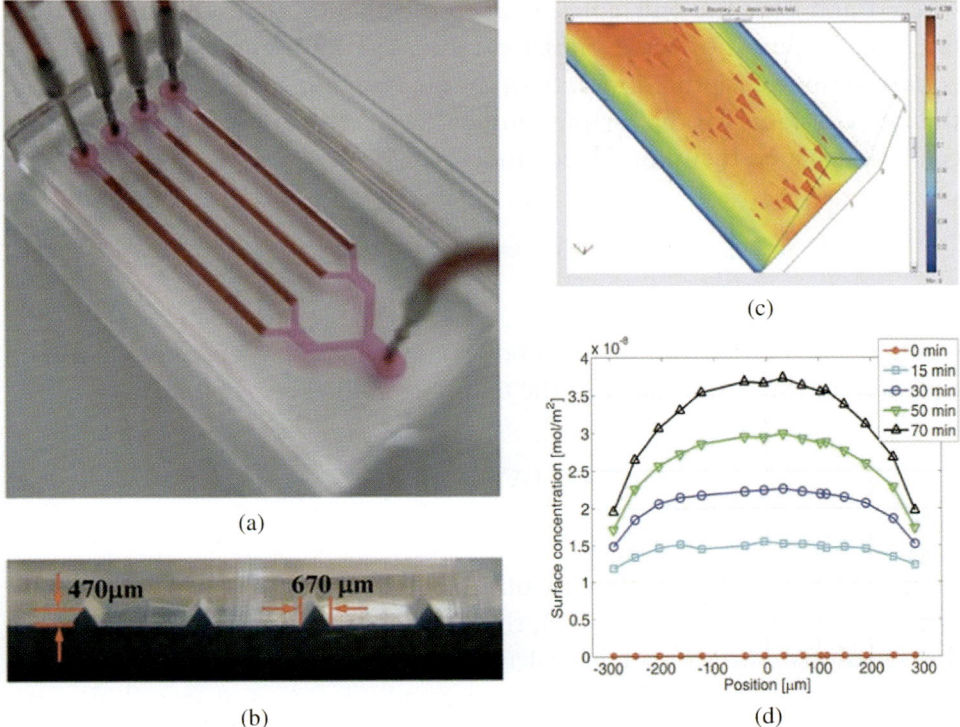

Fig. 5. Numerical study of molecule adhesion in a triangular microchannel. (a) Photographs of the fabricated microfluidic device. (b) A side view of four branched channels. (c) Computational velocity vectors (arrows) and the distribution of captured molecules on function surface (colored surface). (d) The concentration of captured molecules as functions of time and position along the cross-section. Numerical model with the Navier–Stokes equations, surface reaction equation, and convection–diffusion equation simulated the distribution of captured molecules on the functional surface.[232]

using CFD multi-physics simulation. This computational model considered the transport of molecules in the microchannel and the interaction between molecules and reacting surfaces. The fluid velocity field was obtained by using the stationary Navier–Stokes equations:

$$\rho (u \times \nabla)u = -\nabla p + \nabla \times \eta [\nabla u + (\nabla u)^T], \tag{8}$$

where ρ and η are the density and dynamic viscosity of the fluid, p represents the pressure, and u denotes the velocity vector. The interaction of adhesion molecules and the reaction surface is controlled by the equation

$$C + \theta \underset{k_{des}}{\overset{k_{ads}}{\rightleftharpoons}} C_s, \tag{9}$$

where c and θ are the bulk concentration of adhesion molecules and the surface concentration of active sites for taking the molecules; c_s is the surface concentration of captured molecules on the function surface; k_{ads} and k_{des} are the rate constants for the forward and backward reactions. Following that, c_s can be calculated by using the equation

$$\frac{\partial c_s}{\partial t} + \nabla \times (-D_s \nabla c_s) = k_{ads} c \theta - k_{des} c_s, \tag{10}$$

where D_s is the surface diffusivity and c is computed from a convection–diffusion equation for the mass balance of the transport in the bulk molecules:

$$\frac{\partial c_s}{\partial t} + \nabla \times (-D \nabla c + cu) = 0, \tag{11}$$

where D represents the diffusivity of the bulk molecules. Solving the combination of Eqs. (8), (10), and (11), the distribution of captured molecules on the functional surface can be calculated. This study provides a probable method to model antibody-based capture and separation of several bioparticles in the nanometer scale like proteins and viruses.

Computational methods have been identified by previous works to be able to predict the capture behaviors of bioparticles. However, in order to simulate the motions and adhesion of bioparticle more accurately and better understand the mechanisms of bioparticle capture, simulating models still have to be improved to overcome many limitations, such as simulating multiple bioparticles of different types, modeling much more complex geometries, and coupling many other factors like leukocyte–erythrocyte interactions, erythrocyte aggregation, collisions between bioparticles, and biochemical factors.

Conclusions

The microfluidics-based antibody capture for bioparticle separation is increasingly important in the area of disease diagnosis and monitoring for global health because of their great potential to be integrated into inexpensive and effective POC systems. This chapter reviews the progresses in experimental and theoretical studies on antibody-based bioparticle capture and separation. Four key issues hindering further developments are summarized, including the selection of antibodies with high specificity, simplification of the microfluidics device, optimization of the structural parameters and the flow conditions, and significant improvement of the capture efficiency and specificity. Theoretical approaches are enabling tools to understand the underlying mechanisms and should be paid more attention in the future. Previous theoretical simulation approaches coupled with *in vivo* and *in vitro* experiments have demonstrated the ability to design and improve the microfluidics for bioparticle capture and separation. Some theoretical models have successfully simulated the adhesion behaviors of single captured cells and a CFD multiphysics simulation was found to be able to model the entire process of protein capture. Future work should focus on overcoming computational limitations, such as simulating multiple bioparticles of different types, modeling much more complex geometries and coupling many other factors such as leukocyte–erythrocyte interactions, erythrocyte aggregation, collisions between bioparticles and biochemical factors, and then combining experimental methods to fully understand the adhesion mechanisms of multiple bioparticles and improve capture efficiency and specificity. Indeed, many efforts still remain to be devoted to simplify the current microfluidics, further reduce its cost, and establish microfluidic POC devices for disease diagnosis and monitoring, which can be used in resource-limited settings.

References

1. World Health Organization (WHO), World health statistics: Global health indicators, Retrieved June 25, 2010, from http://www.who.int/healthinfo/statistics/mortality/en/.
2. World Health Organization (WHO), Quick cancer facts, Retrieved June 25, 2010, http://www.who.int/mediacentre/factsheets/fs297/en/index.htm.
3. Bongaarts, J., Pelletier, F., Gerland, P., How many more AIDS deaths? *Lancet*, 2010, **375**: p. 103–104.
4. Jemal, A. et al., Cancer statistics, 2008. *Cancer J Clin*, 2008, **58**: p. 71–96.
5. von Lode, P., Point-of-care immunotesting: Approaching the analytical performance of central laboratory methods. *Clin Biochem*, 2005, **38**: p. 591–606.
6. Pogge, T.W., Human rights and global health: A research program. *Metaphilosophy*, 2005, **36**: p. 182–209.
7. Daar, A.S. et al., Top ten biotechnologies for improving health in developing countries. *Nat Genetics*, 2002, **32**: p. 229–232.
8. Malkin, R.A., Design of health care technologies for the developing world. *Annu Rev Biomed Eng*, 2007, **9**: p. 567–587.
9. Nie, S.M., Xing, Y., Kim, G.J., Simons, J.W., Nanotechnology applications in cancer. *Annu Rev Biomed Eng* 2007, **9**: p. 257–288.
10. Toner, M., Irimia, D., Blood-on-a-chip. *Annu Rev Biomed Eng*, 2005, **7**: p. 77–103.
11. Chin, C.D., Linder, V., Sia, S.K., Lab-on-a-chip devices for global health: Past studies and future opportunities. *Lab Chip*, 2007, **7**: p. 41–57.
12. Mabey, D., Peeling, R.W., Ustianowski, A., Perkins, M.D., Diagnostics for the developing world. *Nat Rev Microbiol*, 2004, **2**, p. 231–240.

13. Yager, P. et al., Microfluidic diagnostic technologies for global public health. *Nature*, 2006, **442**: p. 412–418.
14. Dupuy, A.M., Lehmann, S., Cristol, J.P., Protein biochip systems for the clinical laboratory. *Clin Chem Lab Med*, 2005, **43**: p. 1291–1302.
15. Cristofanilli, M. et al., Circulating tumor cells, disease progression, and survival in metastatic breast cancer. *New Engl J Med*, 2004, **351**: p. 781–791.
16. de Bono, J.S. et al., Circulating tumor cells predict survival benefit from treatment in metastatic castration-resistant prostate cancer. *Clin Cancer Res*, 2008, **14**: p. 6302–6309.
17. Gehrs, B.C., Friedberg, R.C., Autoimmune hemolytic anemia. *Am J Hematol*, 2002, **69**: p. 258–271.
18. Chiorazzi, N., Rai, K.R., Ferrarini, M., Mechanisms of disease: Chronic lymphocytic leukemia. *New Engl J Med*, 2005, **352**: p. 804–815.
19. Warkentin, T.E., Heparin-induced thrombocytopenia: Pathogenesis and management. *Br J Haematol*, 2003, **121**: p. 535–555.
20. Smerage, J.B., Hayes, D.F., The measurement and therapeutic implications of circulating tumour cells in breast cancer. *Br J Cancer*, 2006, **94**: p. 8–12.
21. Allard, W.J. et al. Tumor cells circulate in the peripheral blood of all major carcinomas but not in healthy subjects or patients with nonmalignant diseases. *Clin Cancer Res* 2004, **10**: p. 6897–6904.
22. Nagrath, S. et al., Isolation of rare circulating tumour cells in cancer patients by microchip technology. *Nature*, 2007, **450**: p. 1235–1239.
23. Gascoyne, P., Satayavivad, J., Ruchirawat, M., Microfluidic approaches to malaria detection. *Acta Tropica*, 2004, **89**: p. 357–369.
24. Cheng, X.H., Chen, G., Rodriguez, W.R., Micro- and nanotechnology for viral detection. *Anal Bioanal Chem*, 2009, **393**: p. 487–501.
25. Mao, C.B., Liu, A.H., Cao, B.R., Virus-based chemical and biological sensing. *Angew Chem*, 2009, **48**: p. 6790–6810.
26. Chang, D., Learmonth, K., Dax, E.M., HIV testing in 2006: Issues and methods. *Expert Rev Anti Infect Ther*, 2006, **4**: p. 565–82.
27. Sutthent, R. et al., p24 antigen detection assay modified with a booster step for diagnosis and monitoring of human immunodeficiency virus type 1 infection. *J Clin Microbiol*, 2003, **41**: p. 1016–1022.
28. Phillips, A.N. et al., Serial lymphocyte CD4 Counts and development of AIDS. *Lancet*, 1991, **337**: p. 389–392.
29. Fiscus, S.A. et al., HIV-1 viral load assays for resource-limited settings. *PLOS Med*, 2006, **3**: p. 1743–1750.
30. Schupbach, J. et al., Heat-mediated immune complex dissociation and enzyme-linked immunosorbent assay signal amplification render p24 antigen detection in plasma as sensitive as HIV-1 RNA detection by polymerase chain reaction. *AIDS*, 1996, **10**: p. 1085–90.
31. Rouet, F. et al., Transfer and evaluation of an automated, low-cost real-time reverse transcription-PCR test for diagnosis and monitoring of human immunodeficiency virus type 1 infection in a West African resource-limited setting. *J Clin Microbiol*, 2005, **43**: p. 2709–2717.
32. Fowler, M.G., Owor, M., Monitoring HIV Treatment in resource-limited settings: Reassuring news on the usefulness of CD4(+) cell counts. *J Infect Disease*, 2009, **199**: p. 1255–1257.

33. Mellors, J.W. et al., Plasma viral load and CD4(+) lymphocytes as prognostic markers of HIV-1 infection. *Ann Intern Med*, 1997, **126**: p. 946–954.
34. Strauss, K. et al., Performance evaluation of the FACS Count system: A dedicated system for clinical cellular analysis. *Cytometry*, 1996, **26**: p. 52–59.
35. Cassens, U. et al., Simplified volumetric flow cytometry allows feasible and accurate determination of CD4 T lymphocytes in immunodeficient patients worldwide. *Antiviral Ther*, 2004, **9**: p. 395–405.
36. Karcher, H., Bhning, D., Downing, R., Mashate, S., Harms, G., Comparison of two alternative methods for CD4(+) T-cell determination (Coulter manual CD4 count and CyFlow) against standard dual platform flow cytometry in Uganda. *Cytometry B-Clin Cytometry*, 2006, **70B**: p. 163–169.
37. Furdui, V.I., Harrison, D.J., Immunomagnetic T cell capture from blood for PCR analysis using microfluidic systems. *Lab Chip*, 2004, **4**: p. 614–618.
38. Murthy, S.K., Sin, A., Tompkins, R.G., Toner, M., Effect of flow and surface conditions on human lymphocyte isolation using microfluidic chambers. *Langmuir*, 2004, **20**: p. 11649–11655.
39. Cheng, X. et al., Cell detection and counting through cell lysate impedance spectroscopy in microfluidic devices. *Lab Chip*, 2007, **7**: p. 746–755.
40. Cheng, X.H. et al., A microfluidic device for practical label-free CD4+T cell counting of HIV-infected subjects. *Lab Chip*, 2007, **7**: p. 170–178.
41. Cheng, X.H. et al., A microchip approach for practical label-free CD4+ T-cell counting of HIV-infected subjects in resource-poor settings. *J Acquired Immune Defic Syndr*, 2007, **45**: p. 257–261.
42. Thorslund, S., Larsson, R., Bergquist, J., Nikolajeff, F., Sanchez, J., A PDMS-based disposable microfluidic sensor for CD4(+) lymphocyte counting. *Biomed Microdev*, 2008, **10**: p. 851–857.
43. Moon, S. et al., Integrating microfluidics and lensless imaging for point-of-care testing. *Biosens Bioelectron*, 2009, **24**: p. 3208–3214.
44. Xuanhong, C. et al., Enhancing the performance of a point-of-care CD4+ T-cell counting microchip through monocyte depletion for HIV/AIDS diagnostics. *Lab Chip*, 2009, **9**: p. 1357–1364.
45. Stybayeva, G. et al., Lensfree holographic imaging of antibody microarrays for high-throughput detection of leukocyte numbers and function. *Anal Chem*, 2010, **82**: p. 3736–3744.
46. Wang, Z.K. et al., Microfluidic CD4+ T-cell counting device using chemiluminescence-based detection. *Anal Chem*, 2010, **82**: p. 36–40.
47. Wang, S., Xu, F., Ipa, A., Kelesa, H.O., Demirci, U., Emerging microscale technologies for global health: CD4+ counts. 2010.
48. Alyassin, M.A. et al., Rapid automated cell quantification on HIV microfluidic devices. *Lab Chip*, 2009, **9**: p. 3364–3369.
49. Sakaguchi, S., Naturally arising CD4(+) regulatory T cells for immunologic self-tolerance and negative control of immune responses. *Annu Rev Immunol*, 2004, **22**: p. 531–562.
50. Simon, V., Ho, D.D., HIV-1 dynamics *in vivo*: Implications for therapy. *Nat Rev Microbiol*, 2003, **1**: p. 181–190.
51. Kim, Y.G., Moon, S., Kuritzkes, D.R., Demirci, U., Quantum dot-based HIV capture and imaging in a microfluidic channel. *Biosens Bioelectron*, 2009, **25**: p. 253–258.

52. Wang, S.Q., Xu, F., Demirci, U., Advances in developing HIV-1 viral load assays for resource-limited settings. *Biotechnol Adv*, 2010.
53. Zhu, P. et al., Distribution and three-dimensional structure of AIDS virus envelope spikes. *Nature*, 2006, **441**: p. 847–852.
54. Leaf, C., Why we're losing the war on cancer and how to win it. *Fortune*, 2004, **149**: p. 76.
55. Chiang, A.C., Massague, J., Molecular origins of cancer molecular basis of metastasis. *New Engl J Med*, 2008, **359**: p. 2814–2823.
56. Budd, G.T., et al. Circulating tumor cells versus imaging — predicting overall survival in metastatic breast cancer. *Clin Cancer Res*, 2006, **12**: p. 6403–6409.
57. Maheswaran, S. et al., Detection of mutations in EGFR in circulating lung-cancer cells. *New Engl J Med*, 2008, **359**: p. 366–377.
58. Tothill, I.E., Biosensors for cancer markers diagnosis. *Semin Cell Dev Biol*, 2009, **20**: p. 55–62.
59. Heath, J.R., Davis, M.E., Nanotechnology and cancer. *Annu Rev Med*, 2008, **59**: p. 251–265.
60. Kwon, K.W. et al., Label-free, microfluidic separation and enrichment of human breast cancer cells by adhesion difference. *Lab Chip*, 2007, 7: p. 1461–1468.
61. Dharmasiri, U. et al., Highly efficient capture and enumeration of low abundance prostate cancer cells using prostate-specific membrane antigen aptamers immobilized to a polymeric microfluidic device. *Electrophoresis*, 2009, **30**: p. 3289–3300.
62. Muller, V., Pantel, K., Bone marrow micrometastases and circulating tumor cells: Current aspects and future perspectives. *Breast Cancer Res*, 2004, 6: p. 258–261.
63. Feng, J., Soper, S.A., McCarley, R.L., Murphy, M.C., Separation of breast cancer cells from peripherally circulating blood using antibodies fixed in microchannels, In: Bartels, K.E. et al., eds. *Lasers in Surgery: Advanced Characterization, Therapeutics, and Systems*, Vol. XIV. SPIE Int Soc Optical Engineering, Bellingham, 2004, p. 278–293.
64. Du, Z., Cheng, K.H., Vaughn, M.W., Collie, N.L., Gollahon, L.S., Recognition and capture of breast cancer cells using an antibody-based platform in a microelectromechanical systems device. *Biomed Microdev*, 2007, 9: p. 35–42.
65. Ehrhart, J.C. et al., A new immunosensor for breast cancer cell detection using antibody-coated long alkylsilane self-assembled monolayers in a parallel plate flow chamber. *Biosens Bioelectron*, 2008, **24**: p. 467–474.
66. Gleghorn, J.P. et al., Capture of circulating tumor cells from whole blood of prostate cancer patients using geometrically enhanced differential immunocapture (GEDI) and a prostate-specific antibody. *Lab Chip*, 2010, **10**: p. 27–29.
67. Liu, Y.J. et al., A micropillar-integrated smart microfluidic device for specific capture and sorting of cells. *Electrophoresis*, 2007, **28**: p. 4713–4722.
68. Myung, J.H., Launiere, C.A., Eddington, D.T., Hong, S., Enhanced tumor cell isolation by a biomimetic combination of E-selectin and anti-EpCAM: Implications for the effective separation of circulating tumor cells (CTCs). *Langmuir*, 2010, **26**: p. 8589–8596.
69. Sivagnanam, V., Song, B., Vandevyver, C., Bunzli, J.C.G., Gijs, M.A.M., Selective breast cancer cell capture, culture, and immunocytochemical analysis using self-assembled magnetic bead patterns in a microfluidic chip. *Langmuir*, 2010, **26**: p. 6091–6096.
70. Wang, S.T. et al., Three-dimensional nanostructured substrates toward efficient capture of circulating tumor Cells. *Angew Chem* 2009, **48**: p. 8970–8973.

71. Adams, A.A. et al., Highly efficient circulating tumor cell isolation from whole blood and label-free enumeration using polymer-based microfluidics with an integrated conductivity sensor. *J Am Chem Soc,* 2008, **130**: p. 8633–8641.
72. Cheung, L.S.L. et al., Detachment of captured cancer cells under flow acceleration in a bio-functionalized microchannel. *Lab Chip,* 2009, **9**: p. 1721–1731.
73. Phillips, J.A., Xu, Y., Xia, Z., Fan, Z.H., Tan, W.H., Enrichment of cancer cells using aptamers immobilized on a microfluidic channel. *Anal Chem,* 2009, **81**: p. 1033–1039.
74. Xu, Y. et al., Aptamer-based microfluidic device for enrichment, sorting, and detection of multiple cancer cells. *Anal Chem* 2009, **81**: p. 7436–7442.
75. Sin, A., Murthy, S.K., Revzin, A., Tompkins, R.G., Toner, M., Enrichment using antibody-coated microfluidic chambers in shear flow: Model mixtures of human lymphocytes. *Biotechnol Bioeng* 2005, **91**: p. 816–826.
76. Wang, K., Cometti, B., Pappas, D., Isolation and counting of multiple cell types using an affinity separation. *Anal Chim Acta,* 2007, **601**: p. 1–9.
77. Hashimoto, M., Kaji, H., Nishizawa, M., Selective capture of a specific cell type from mixed leucocytes in an electrode-integrated microfluidic device. *Biosens Bioelectron,* 2009, **24**: p. 2892–2897.
78. Han, W.J., Allio, B.A., Foster, D.G., King, M.R., Nanoparticle coatings for enhanced capture of flowing cells in microtubes. *ACS Nano,* 2009, **4**: p. 174–180.
79. Sethu, P. et al., Microfluidic isolation of leukocytes from whole blood for phenotype and gene expression analysis. *Anal Chem,* 2006, **78**: p. 5453–5461.
80. Russom, A. et al., Microfluidic leukocyte isolation for gene expression analysis in critically ill hospitalized patients. *Clin Chem,* 2008, **54**: p. 891–900.
81. Kotz, K.T. et al., Clinical microfluidics for neutrophil genomics and proteomics. *Nat Med,* 2010, **16**: p. 1042–1047.
82. Yang, F., Gao, Y.X., Zhang, Y., Chen, J., Long, M., Developing a microfluidic based system to quantify cell capture efficiency. *Sci China C-Life Sci* 2009, **52**: p. 173–181.
83. Lee, S., Cho, N.P., Kim, J.D., Jung, H., Kang, S.H., An ultra-sensitive nanoarray chip based on single-molecule sandwich immunoassay and TIRFM for protein detection in biologic fluids. *Analyst,* 2009, **134**: p. 933–938.
84. Diercks, A.H. et al., A microfluidic device for multiplexed protein detection in nano-liter volumes. *Anal Biochem,* 2009, **386**: p. 30–35.
85. Dharmasiri, U. et al., Enrichment and detection of *Escherichia coli* O157:H7 from water samples using an antibody modified microfluidic chip. *Anal Chem,* 2010, **82**: p. 2844–2849.
86. Beyor, N., Seo, T.S., Liu, P., Mathies, R.A., Immunomagnetic bead-based cell concentration microdevice for dilute pathogen detection. *Biomed Microdev,* 2008, **10**: p. 909–917.
87. Wark, A.W., Lee, J., Kim, S., Faisal, S.N., Lee, H.J., Bioaffinity detection of pathogens on surfaces. *J Ind Eng Chem,* 2009, **16**: p. 169–177.
88. Hu, X.Y. et al., Marker-specific sorting of rare cells using dielectrophoresis. *Proc Natl Acad Sci U S A,* 2005, **102**: p. 15757–15761.
89. Ng, S.Y. et al., Label-free impedance detection of low levels of circulating endothelial progenitor cells for point-of-care diagnosis. *Biosens Bioelectron,* 2009, **25**: p. 1095–1101.
90. Rosi, N.L., Mirkin, C.A., Nanostructures in biodiagnostics. *Chem Rev,* 2005, **105**: p. 1547–1562.

91. Albrecht, D.R., Underhill, G.H., Mendelson, A., Bhatia, S.N., Multiphase electropatterning of cells and biomaterials. *Lab Chip*, 2007, **7**: p. 702–709.
92. Ho, C.T., Lin, R.Z., Chang, W.Y., Chang, H.Y., Liu, C.H., Rapid heterogeneous liver-cell on-chip patterning via the enhanced field-induced dielectrophoresis trap. *Lab Chip*, 2006, **6**: p. 724–734.
93. Ymeti, A. et al., Fast, ultrasensitive virus detection using a young interferometer sensor. *Nano Lett*, 2007, **7**: p. 394–397.
94. Stern, E. et al., Label-free biomarker detection from whole blood. *Nat Nanotechnol*, 2010, **5**: p. 138–142.
95. Sherman, D.J. et al., A differential cell capture assay for evaluating antibody interactions with cell surface targets. *Anal Biochem*, 2010, **401**: p. 173–181.
96. Rowe, C.A. et al., Array biosensor for simultaneous identification of bacterial, viral, protein analytes. *Anal Chem*, 1999, **71**: p. 3846–3852.
97. Nam, J.M., Thaxton, C.S., Mirkin, C.A., Nanoparticle-based bio-bar codes for the ultrasensitive detection of proteins. *Science*, 2003, **301**: p. 1884–1886.
98. Sun, Y., Liu, Y.Y., Qu, W.S., Jiang, X.Y., Combining nanosurface chemistry and microfluidics for molecular analysis cell biology. *Anal Chim Acta*, 2009, **650**: p. 98–105.
99. Plouffe, B.D., Kniazeva, T., Mayer, J.E., Murthy, S.K., V.L. Sales, Development of microfluidics as endothelial progenitor cell capture technology for cardiovascular tissue engineering and diagnostic medicine. *FASEB J*, 2009, **23**: p. 3309–3314.
100. Plouffe, B.D. et al., Peptide-mediated selective adhesion of smooth muscle and endothelial cells in microfluidic shear flow. *Langmuir*, 2007, **23**: p. 5050–5055.
101. Plouffe, B.D., Radisic, M., Murthy, S.K., Microfluidic depletion of endothelial cells, smooth muscle cells, and fibroblasts from heterogeneous suspensions. *Lab Chip*, 2008, **8**: p. 462–472.
102. Pestourie, C., Tavitian, B., Duconge, F., Aptamers against extracellular targets for *in vivo* applications. *Biochimie*, 2005, **87**: p. 921–930.
103. Shangguan, D. et al., Aptamers evolved from live cells as effective molecular probes for cancer study. *Proc Natl Acad Sci U S A*, 2006, **103**: p. 11838–11843.
104. Usami, S., Chen, H.H., Zhao, Y.H., Chien, S., Skalak, R., Design and construction of a linear shear-stress flow chamber. *Ann Biomed Eng*, 1993, **21**: p. 77–83.
105. Luo, Z.Y., Xu, F., Lu, T.J., Bai, B.F., Direct numerical simulation of single leukocyte deformation in microchannel flow for disease diagnosis. *J Med Syst*, 2010.
106. Verdier, C., Couzon, C., Duperray, A., Singh, P., Modeling cell interactions under flow. *J Math Biol*, 2009, **58**: p. 235–259.
107. Lim, C.T., Zhou, E.H., Quek, S.T., Mechanical models for living cells — a review. *J Biomech*, 2006, **39**: p.195–216.
108. Michalet, X. et al., Quantum dots for live cells, *in vivo* imaging, and diagnostics. *Science*, 2005, **307**: p. 538–544.
109. Whitesides, G.M., Ostuni, E., Takayama, S., Jiang, X.Y., Ingber, D.E., Soft lithography in biology and biochemistry. *Annu Rev Biomed Eng*, 2001, **3**: p. 335–373.
110. Zamir, E., Geiger, B., Molecular complexity and dynamics of cell-matrix adhesions. *J Cell Sci*, 2001, **114**: p. 3583–3590.
111. Oltvai, Z.N., Barabasi, A.L., Life's complexity pyramid. *Science*, 2002, **298**: p. 763–764.
112. Balaban, N.Q. et al., Force and focal adhesion assembly: A close relationship studied using elastic micropatterned substrates. *Nat Cell Biol*, 2001, **3**: p. 466–472.

113. Bershadsky, A.D., Balaban, N.Q., Geiger, B., Adhesion-dependent cell mechanosensitivity. *Annu Rev Cell Dev Biol*, 2003, **19**: p. 677–695.
114. Zaidel-Bar, R., Ballestrem, C., Kam, Z., Geiger, B., Early molecular events in the assembly of matrix adhesions at the leading edge of migrating cells. *J Cell Sci*, 2003, **116**: p. 4605–4613.
115. Springer, T.A., Adhesion receptors of the immune-system. *Nature*, 1990, **346**: p. 425–434.
116. Butcher, E.C., Leukocyte-endothelial cell recognition — 3(or more) steps to specificity and diversity. *Cell*, 1991, **67**: p. 1033–1036.
117. Lawrence, M.B., Springer, T.A., Leukocytes roll on a selectin at physiological flow-rates: Distinction from and prerequisite for adhesion through integrins. *Cell*, 1991, **65**: p. 859–873.
118. Carlos, T.M., Harlan, J.M., Leukocyte-endothelial ashesion molecules. *Blood*, 1994, **84**: p. 2068–2101.
119. Springer, T.A., Traffic signals for lymphocyte recirculation and leukocyte emigration — the multistep paradigm. *Cell*, 1994, **76**: p. 301–314.
120. Alon, R., Chen, S.Q., Puri, K.D., Finger, E.B., Springer, T.A., The kinetics of L-selectin tethers and the mechanics of selectin-mediated rolling. *J Cell Biol*, 1997, **138**: p. 1169–1180.
121. Evans, E., Ritchie, K., Dynamic strength of molecular adhesion bonds. *Biophys J*, 1997, **72**: p. 1541–1555.
122. Mehta, P., Cummings, R.D., McEver, R.P., Affinity and kinetic analysis of P-selectin binding to P-selectin glycoprotein ligand-1. *J Biol Chem*, 1998, **273**: p. 32506–32513.
123. Alon, R., Hammer, D.A., Springer, T.A., Lifetime of the P-selectin-carbohydrate bond and its response to tensile force in hydrodynamic flow. *Nature*, 1995, **374**: p. 539–542.
124. Evans, E., Probing the relation between force — lifetime — and chemistry in single molecular bonds. *Annu Rev Biophys Biomol Struct*, 2001, **30**: p. 105–128.
125. Fritz, J., Katopodis, A.G., Kolbinger, F., Anselmetti, D., Force-mediated kinetics of single P-selectin ligand complexes observed by atomic force microscopy. *Proc Natl Acad Sci U S A*, 1998, **95**: p. 12283–12288.
126. Evans, E., Leung, A., Hammer, D., Simon, S., Chemically distinct transition states govern rapid dissociation of single L-selectin bonds under force. *Proc Natl Acad Sci U S A*, 2001, **98**: p. 3784–3789.
127. Hanley, W. et al., Single molecule characterization of P-selectin/ligand binding. *J Biol Chem*, 2003, **278**: p. 10556–10561.
128. Evans, E., Leung, A., Heinrich, V., Zhu, C., Mechanical switching and coupling between two dissociation pathways in a P-selectin adhesion bond. *Proc Natl Acad Sci U S A*, 2004, **101**: p. 11281–11286.
129. Sarangapani, K.K. et al., Low force decelerates L-selectin dissociation from P-selectin glycoprotein ligand-1 and endoglycan. *J Biol Chem*, 2004, **279**: p. 2291–2298.
130. Snook, J.H., Guilford, W.H., The effects of load on E-selectin bond rupture and bond formation. *Cell Mol Bioeng*, 2010, **3**: p. 128–138.
131. Marshall, B.T. et al., Direct observation of catch bonds involving cell-adhesion molecules. *Nature*, 2003, **423**: p. 190–193.
132. Thomas, W.E., Nilsson, L.M., Forero, M., Sokurenko, E.V., Vogel, V., Shear-dependent 'stick-and-roll' adhesion of type 1 fimbriated *Escherichia coli*. *Mol Microbiol*, 2004, **53**: p. 1545–1557.

133. Guo, B., Guilford, W.H., Mechanics of actomyosin bonds in different nucleotide states are tuned to muscle contraction. *Proc Natl Acad Sci U S A*, 2006, **103**: p. 9844–9849.
134. Thomas, W. *et al.*, Catch-bond model derived from allostery explains force-activated bacterial adhesion. *Biophys J*, 2006,. **90**: p. 753–764.
135. Hammer, D.A., Leukocyte adhesion: What's the catch? *Current Biol*, 2005, **15**: p. R96–R99.
136. Bell, G.I., Models for the specific adhesion of cells to cells. *Science*, 1978, **200**: p. 618–627.
137. Marshall, B.T., Sarangapani, K.K., Lou, J.H., McEver, R.P., Zhu, C., Force history dependence of receptor-ligand dissociation. *Biophys J*, 2005, **88**: p. 1458–1466.
138. Zhang, Y., Sun, G.Y., Lu, S.Q., Li, N., Long, M.A., Low spring constant regulates P-selectin-PSGL-1 bond rupture. *Biophys J*, 2008, **95**: p. 5439–5448.
139. Pierres, A. *et al.*, Experimental study of the interaction range and association rate of surface-attached cadherin 11. *Proc Natl Acad Sci U S A*, 1998, **95**: p. 9256–9261.
140. Bongrand, P., Ligand-receptor interactions. *Rep Prog Phys*, 1999, **62**, 921–968.
141. Chang, K.C., Hammer, D.A., Adhesive dynamics simulations of sialyl-Lewis(x)/E-selectin-mediated rolling in a cell-free system. *Biophys J*, 2000, **79**, 1891–1902.
142. Zhu, C., Kinetics and mechanics of cell adhesion. *J Biomech*, 2000, **33**: p. 23–33.
143. Chen, S.Q., Springer, T.A., An automatic braking system that stabilizes leukocyte rolling by an increase in selectin bond number with shear. *J Cell Biol*, 1999, **144**: p. 185–200.
144. Williams, T.E., Nagarajan, S., Selvaraj, P., Zhu, C., Quantifying the impact of membrane microtopology on effective two-dimensional affinity. *J Biol Chem*, 2001, **276**: p. 13283–13288.
145. Huang, J. *et al.*, Quantifying the effects of molecular orientation and length on two-dimensional receptor-ligand binding kinetics. *J Biol Chem*, 2004, **279**: p. 44915–44923.
146. Wu, L. *et al.*, Impact of carrier stiffness and microtopology on two-dimensional kinetics of P-selectin P-selectin glycoprotein ligand-1 (PSGL-1) interactions. *J Biol Chem*, 2007, **282**: p. 9846–9854.
147. Ley, K. *et al.*, Lectin-like cell-adhesion molecule-1 mediates leukocyte rolling in mesenteric venules *in vivo*. *Blood*, 1991, 77: p. 2553–2555.
148. Jones, D.A., Abbassi, O., McIntire, L.V., McEver, R.P., Smith, C.W., P-selectin mediates neutrophil rolling on histamine-stimulated endothelial cells. *Biophys. J*, 1993, **65**: p. 1560–1569.
149. Lawrence, M.B., Kansas, G.S., Kunkel, E.J., Ley, K., Threshold levels of fluid shear promote leukocyte adhesion through selectins (CD62L,P,E). *J Cell Biol*, 1997, **136**: p. 717–727.
150. Cao, J., Donell, B., Deaver, D.R., Lawrence, M.B., Dong, C., *In vitro* side-view imaging technique and analysis of human T-leukemic cell adhesion to ICAM-1 in shear flow. *Microvascular Res*, 1998, **55**: p. 124–137.
151. Stein, J.V. *et al.*, L-selectin-mediated leukocyte adhesion *in vivo*: Microvillous distribution determines tethering efficiency, but not rolling velocity. *J Exp Med*, 1999, **189**: p. 37–49.
152. Wijnhoven, B.P.L., Dinjens, W.N.M., Pignatelli, M., E-cadherin-catenin cell-cell adhesion complex and human cancer. *Brit J Surg*, 2000, **87**: p. 992–1005.
153. Hanzlik, J., Cretekos, E., Lamkin-Kennard, K.A., Biomimetic leukocyte adhesion: A review of microfluidic and computational approaches and applications. *J Bionic Eng*, 2008, **5**: p. 317–327.

154. Chotard-Ghodsnia, R., Drochon, A., Grebe, R., New flow chamber for the study of shear stress and transmural pressure upon cells adhering to a porous biomaterial. *J Biomech Eng Trans ASME*, 2002, **124**: p. 258–261.
155. Lu, H. et al., Microfluidic shear devices for quantitative analysis of cell adhesion. *Anal Chem*, 2004, **76**: p. 5257–5264.
156. Chotard-Ghodsnia, R. *et al.*, Morphological analysis of tumor cell/endothelial cell interactions under shear flow. *J Biomech*, 2007, **40**, 335–344.
157. Brooks, S.B., Tozeren, A., Flow past an array of cells that are adherent to the bottom plate of a flow channel. *Comput Fluid*, 1996, **25**: p. 741–757.
158. Chapman, G.B., Cokelet, G.R., Flow resistance and drag forces due to multiple adherent leukocytes in postcapillary vessels. *Biophys J*, 1998, **74**: p. 3292–3301.
159. Gaver, D.P., Kute, S.M., A theoretical model study of the influence of fluid stresses on a cell adhering to a microchannel wall. *Biophys J*, 1998, **75**: p. 721–733.
160. Caputo, K.E., Hammer, D.A., Adhesive dynamics simulation of G-protein-mediated chemokine-activated neutrophil adhesion. *Biophys J*, 2009, **96**: p. 2989–3004.
161. Krasik, E.F., Caputo, K.E., Hammer, D.A., Adhesive dynamics simulation of neutrophil arrest with stochastic activation. *Biophys J*, 2008, **95**: p. 1716–1728.
162. Hammer, D.A., Apte, S.M., Simulation of cell rolling and adhesion on surfaces in shear flow general results and analysis of selectin-mediated neutrophil adhesion. *Biophys J*, 1992, **63**: p. 35–57.
163. Chang, K.C., Tees, D.F.J., Hammer, D.A., The state diagram for cell adhesion under flow: Leukocyte rolling and firm adhesion. *Proc Natl Acad Sci U S A*, 2000, **97**: p. 1262–11267.
164. Das, B., Johnson, P.C., Popel, A.S., Computational fluid dynamic studies of leukocyte adhesion effects on non-Newtonian blood flow through microvessels. *Biorheology*, 2000, **37**: p. 239–258.
165. Park, E.Y.H. *et al.*, Comparison of PSGL-1 microbead and neutrophil rolling: Microvillus elongation stabilizes P-selectin bond clusters. *Biophys J*, 2002, **82**: p. 1835–1847.
166. Yago, T. *et al.*, Distinct molecular and cellular contributions to stabilizing selectin-mediated rolling under flow. *J Cell Biol*, 2002, **158**: p. 787–799.
167. Cheng, D., Jian, C., Struble, E.J., Lipowsky, H.H., Mechanics of leukocyte deformation and adhesion to endothelium in shear flow. *Ann Biomed Eng*, 1999, **27**: p. 298–312.
168. Lei, X., Lawrence, M.R., Dong, C., Influence of cell deformation on leukocyte rolling adhesion in shear flow. *J Biomech Eng Trans ASME*, 1999, **121**: p. 636–643.
169. Cheng, D., Xiao, X.L., Biomechanics of cell rolling: shear flow, cell-surface adhesion, and cell deformability. *J Biomech*, 2000, **33**: p. 35–43.
170. Caputo, K.E., Hammer, D.A., Effect of microvillus deformability on leukocyte adhesion explored using adhesive dynamics simulations. *Biophys J*, 2005, **89**: p. 187–200.
171. Udaykumar, H.S., Kan, H.C., Shyy, W., Tran-Son-Tay, R., Multiphase dynamics in arbitrary geometries on fixed Cartesian grids. *J Comput Phys*, 1997, **137**: p. 366–405.
172. Kan, H.C., Udaykumar, H.S., Shyy, W., Tran-Son-Tay, R., Hydrodynamics of a compound drop with application to leukocyte modeling. *Phys Fluids*, 1998, **10**: p. 760–774.
173. Shyy, W., Francois, M., Udaykumar, H.S., Ndri, N., Tran-Son-Tay, R., Moving boundaries in micro-scale biofluid dynamics. *Appl Mech Rev*, 2001, **54**: p. 405–453.
174. N'Dri, N.A., Shyy, W., Tran-Soy-Tay, R., Computational modeling of cell adhesion and movement using a continuum-kinetics approach. *Biophys J*, 2003, **85**: p. 2273–2286.

175. Sukumaran, S. Seifert, U., Influence of shear flow on vesicles near a wall: A numerical study. *Phys Rev E,* 2001, **64**.
176. Pappu, V., Doddi, S.K., Bagchi, P., A computational study of leukocyte adhesion and its effect on flow pattern in microvessels. *J Theor Biol,* 2008, **254**: p. 483–498.
177. Jadhav, S., Eggleton, C.D., Konstantopoulos, K., A 3-D computational model predicts that cell deformation affects selectin-mediated leukocyte rolling. *Biophys J,* 2005, **88**: p. 96–104.
178. Pappu, V., Bagchi, P., 3D computational modeling and simulation of leukocyte rolling adhesion and deformation. *Comput Biol Med,* 2008, **38**: p. 738–753.
179. Khismatullin, D.B., Truskey, G.A., Three-dimensional numerical simulation of receptor-mediated leukocyte adhesion to surfaces: Effects of cell deformability and viscoelasticity. *Phys Fluids,* 2005, **17**: p. 21.
180. Liu, X.H., Wang, X., The deformation of an adherent leukocyte under steady shear flow: A numerical study. *J Biomech,* 2004, **37**: p. 1079–1085.
181. Khismatullin, D.B., Truskey, G.A., A 3D numerical study of the effect of channel height on leukocyte deformation adhesion in parallel-plate flow chambers. *Microvasc Res,* 2004, **68**: p. 188–202.
182. Jin, Q., et al. Migration and deformation of leukocytes in pressure driven flows. *Mech Res Commun,* 2007, **34**: p. 411–422.
183. Dembo, M., Torney, D.C., Saxman, K., Hammer, D., The reaction-limited kinetics of membrane-to-surface adhesion and detachment. *Proc R Soc Lond B Biol Sci,* 1988, **234**: p. 55–83.
184. Hammer, D.A., Lauffenburger, D.A., A dynamic-model for receptor-mediated cell-adhesion to surfaces. *Biophys J,* 1987, **52**: p. 475–487.
185. Moore, K.L. et al., P-selectin glycoprotein ligand-1 mediates rolling of human neutrophils on P-selectin. *J Cell Biol,* 1995, **128**: p. 661–671.
186. Kramers, H.A., Brownian motion in a field of force and the diffusion model of chemical reactions. *Physica,* 1940, 7: p. 284–304.
187. Lee, D., Schultz, J.B., Knauf, P.A., King, M.R., Mechanical shedding of L-selectin from the neutrophil surface during rolling on sialyl Lewis X under flow. *J Biol Chem,* 2007, **282**: p. 4812–4820.
188. Bhatia, S.K., King, M.R., Hammer, D.A., The state diagram for cell adhesion mediated by two receptors. *Biophys J,* 2003, **84**: p. 2671–2690.
189. King, M.R., Sumagin, R., Green, C.E., Simon, S.I., Rolling dynamics of a neutrophil with redistributed L-selectin. *Math Biosci,* 2005, **194**: p. 71–79.
190. Caputo, K.E., Lee, D., King, M.R., Hammer, D.A., Adhesive dynamics simulations of the shear threshold effect for leukocytes. *Biophys J,* 2007, **92**: p. 787–797.
191. King, M.R., Heinrich, V., Evans, E., Hammer, D.A., Nano-to-micro scale dynamics of P-selectin detachment from leukocyte interfaces. III. Numerical simulation of tethering under flow. *Biophys J,* 2005, **88**: p. 1676–1683.
192. Shao, J.Y., Ting-Beall, H.P., Hochmuth, R.M., Static and dynamic lengths of neutrophil microvilli. *Proc Natl Acad Sci U S A,* 1998, **95**: p. 6797–6802.
193. Shao, J.Y., Hochmuth, R.M., Mechanical anchoring strength of L-selectin, beta(2) integrins, and CD45 to neutrophil cytoskeleton and membrane. *Biophys J,* 1999, **77**: p. 587–596.
194. Hafezi-Moghadam, A., Ley, K., Relevance of L-selectin shedding for leukocyte rolling in vivo. *J Exp Med,* 1999, **189**: p. 939–947.

195. Cozensroberts, C., Lauffenburger, D.A., Quinn, J.A., Receptor-mediated cell attachment and detachment kinetics. I. Probabilistic model and analysis. *Biophys J*, 1990, **58**: p. 841–856.
196. Cozensroberts, C., Quinn, J.A., Lauffenburger, D.A., Receptor-mediated cell attachment and detachment kinetics. II. Experimental model studies with the radial-flow detachment assay. *Biophys J*, 1990, **58**: p. 857–872.
197. N'Dri, N., Shyy, W., Tran-Son-Tay, R., Udaykumar, H.S., A multi-scale model for cell adhesion and deformation. *ASME IMECE Conf FED*, 2000, **253**: p. 205–213.
198. Shyy, W., Francois, M., Udaykumar, H.S., N'Dri, N., Tran-Son-Tay, R., Moving boundaries in micro-scale biofluid dynamics. *Appl Mech Rev*, 2001, **5**: p. 405–453.
199. Staben, M.E., Zinchenko, A.Z., Davis, R.H., Motion of a particle between two parallel plane walls in low-Reynolds-number Poiseuille flow. *Phys Fluids*, 2003, **15**: p. 1711–1733.
200. Mody, N.A., King, M.R., Three-dimensional simulations of a platelet-shaped spheroid near a wall in shear flow. *Phys Fluids*, 2005, **17**: p. 12.
201. Staben, M.E., Davis, R.H., Particle transport in Poiseuille flow in narrow channels. *Int J Multiphase Flow*, 2005, **31**: p. 529–547.
202. Xuan, X.C., Li, D.Q., Particle motions in low-Reynolds number pressure-driven flows through converging-diverging microchannels. *J Micromech Microeng*, 2006, **16**: p. 62–69.
203. Griggs, A.J., Zinchenko, A.Z., Davis, R.H., Low-Reynolds-number motion of a deformable drop between two parallel plane walls. *Int J Multiphase Flow*, 2007, **33**: p. 182–206.
204. Goldsmit, Hl., Red cell motions and wall interactions in tube flow. *Fed Proc*, 1971, **30**: p. 1578.
205. Fung, Y., Stochastic flow in capillary blood-vessels. *Microvasc Res*, 1973, **5**: p. 34–48.
206. Fischer, T.M., Stohrliesen, M., Schmidschonbein, H., Red-cell as a fluid droplet-tank tread-like motion of human erythrocyte-membrane in shear-flow. *Science*, 1978, **202**: p. 894–896.
207. Yang, S., Undar, A., Zahn, J.D., A microfluidic device for continuous, real time blood plasma separation. *Lab Chip*, 2006, **6**: p. 871–880.
208. Bagchi, P., Mesoscale simulation of blood flow in small vessels. *Biophys J*, 2007, **92**: p. 1858–1877.
209. Fujiwara, H. et al., Red blood cell motions in high-hematocrit blood flowing through a stenosed microchannel. *J Biomech*, 2009, **42**: p. 838–843.
210. Jain, A., Munn, L.L., Determinants of leukocyte margination in rectangular microchannels. *PLOS One*, 2009, **4**: p. e7104.
211. Schmidschonbein, G.W., Usami, S., Skalak, R., Chien, S., Interaction of leukocytes and erythrocytes in capillary and post-capillary vessels. *Microvasc Res*, 1980. **19**: p. 45–70.
212. Firrell, J.C., Lipowsky, H.H., Leukocyte margination and deformation in mesenteric of rat. *Am J Physiol*, 1989, **256**: p. H1667–H1674.
213. Abbitt, K.B., Nash, G.B., Rheological properties of the blood influencing selectin-mediated adhesion of flowing leukocytes. *Am J Physiol* 2003, **285**: p. H229–H240.
214. Freund, J.B., Leukocyte margination in a model microvessel. *Phys Fluids*, 2007, **19**: p. 13.
215. Sun, C.H., Munn, L.L., Particulate nature of blood determines macroscopic rheology: A 2-D lattice Boltzmann analysis. *Biophys J*, 2005, **88**: p. 1635–1645.
216. Shevkoplyas, S.S., Yoshida, T., Munn, L.L., Bitensky, M.W., Biomimetic autoseparation of leukocytes from whole blood in a microfluidic device. *Anal Chem*, 2005, **77**: p. 933–937.

217. Sun, C.H., Munn, L.L., Influence of erythrocyte aggregation on leukocyte margination in postcapillary expansions: A lattice Boltzmann analysis. *Phys A* 2006, **362**: p. 191–196.
218. Pearson, M.J., Lipowsky, H.H., Influence of erythrocyte aggregation on leukocyte margination in postcapillary venules of rat mesentery. *Am J Physiol*, 2000, **279**: p. H1460–H1471.
219. Melder, R.J., Munn, L.L., Yamada, S., Ohkubo, C., Jain, R.K., Selectin- and integrin-ediated T-lymphocyte rolling and arrest on TNF-alpha-activated endothelium: Augmentation by erythrocytes. *Biophys J*, 1995, **69**: p. 2131–2138.
220. Munn, L.L., Melder, R.J., Jain, R.K., Role of erythrocytes in leukocyte-endothelial interactions: Mathematical model and experimental validation. *Biophys J*, 1996, **71**: p. 466–478.
221. Melder, R.J., Yuan, J., Munn, L.L., Jain, R.K., Erythrocytes enhance lymphocyte rolling and arrest *in vivo*. *Microvasc Res*, 2000, **59**: p. 316–322.
222. Migliorini, C. et al., Red blood cells augment leukocyte rolling in a virtual blood vessel. *Biophys J*, 2002, **83**: p. 1834–1841.
223. Hund, S.J., Antaki, J.F., An extended convection diffusion model for red blood cell-enhanced transport of thrombocytes and leukocytes. *Phys Med Biol*, 2009, **54**: p. 6415–6435.
224. Sun, C.H., Migliorini, C., Munn, L.L., Red blood cells initiate leukocyte rolling in postcapillary expansions: A lattice Boltzmann analysis. *Biophys J*, 2003, **85**: p. 208–222.
225. Doddi, S.K., Bagchi, P., Three-dimensional computational modeling of multiple deformable cells flowing in microvessels. *Phys Rev E*, 2009, **79**: p. 14.
226. Fan, R. et al., Integrated barcode chips for rapid, multiplexed analysis of proteins in microliter quantities of blood. *Nat Biotechnol*, 2008, **26**: p. 1373–1378.
227. Sorger, P.K., Microfluidics closes in on point-of-care assays. *Nat Biotechnol*, 2008, **26**, 1345–1346.
228. Xie, S.N., Moya, C., Bilgin, B., Jayaraman, A., Walton, S.P., Emerging affinity-based techniques in proteomics. *Exp Rev Proteomics*, 2009, **6**: p. 573–583.
229. Parsa, H. et al., Effect of volume- time-based constraints on capture of analytes in microfluidic heterogeneous immunoassays. *Lab Chip*, 2008, **8**: p. 2062–2070.
230. Zimmermann, M., Delamarche, E., Wolf, M., Hunziker, P., Modeling and optimization of high-sensitivity, low-volume microfluidic-based surface immunoassays. *Biomed Microdev*, 2005, **7**: p. 99–110.
231. Foley, J.O., Mashadi-Hossein, A., Fu, E., Finlayson, B.A., Yager, P., Experimental and model investigation of the time-dependent 2-dimensional distribution of binding in a herringbone microchannel. *Lab Chip*, 2008, **8**: p. 557–564.
232. Park, J. et al., Simple haptotactic gradient generation within a triangular microfluidic channel. *Lab Chip*, 2010, **10**: p. 2130–2138.

Chapter 18

Applications of Quantum Dots for Fluorescence Imaging in Biomedical Research

ShuQi Wang, Matin Esfahani*, Dusan Sarenac*,
Bettina Cheung*, Aishwarya Vasudevan*, Fatih Inci*,
and Utkan Demirci*,†,‡*

**Bio-Acoustic-MEMS in Medicine (BAMM) Laboratory
Division of Biomedical Engineering, Department of Medicine
Brigham and Women's Hospital, Harvard Medical School
Boston, MA 02139, USA
†Harvard–MIT Health Sciences and Technology
Cambridge, MA 02139, USA*

Introduction

Biomedical research is currently one of fastest growing areas with advanced technologies such as microfluidics, fluorescence imaging, and microelectromechanical systems (MEMS). For fluorescence imaging, conventional organic dyes (e.g., Rhodamine B) are extensively used.[1,2] Recently, quantum dots (QDs), due to their special optical properties,[3] are gaining momentum in their role in biomedical research such as in the development of biosensors, clinical diagnosis, and basic immunological studies.[4-9] QDs are semiconductor nanoparticles with a diameter of 5–50 nm, including a metal core and an insulator surface. Because of their nanoscale structure, QDs have strictly confined electronic wave functions,[9,10] which confer QDs with unique optical properties such as high luminescence intensity, high molar extinction coefficient, size-tunable emission

---

‡Corresponding author. Email: udemirci@rics.bwh.harvard.edu

wavelengths, broad absorption spectra, and narrow emission bands. These properties enable applications such as the labeling of cells, antibodies, proteins, and viral particles for *in vitro* and *in vivo* studies. The ability to introduce QDs into cells allows the binding of QDs to subcellular components without affecting cellular integrity and thus, facilitates the study on protein–protein interactions (e.g., cell signaling pathways) at the subcellular level.[11] Along with the fast development of QD technologies, increasing concerns on the toxicity of QDs to human beings, animals, plants, as well the environment need to be addressed. In this chapter, the special optical properties of QDs, bioconjugation methods to facilitate fluorescence imaging, and applications of QDs in biomedical research are presented. In addition, the toxicity and biosafety of QDs for further clinical applications will be discussed.

Optical properties of QDs

One of the most important features of QDs is the tunable wavelengths of emitted fluorescence.[12] QDs, as semiconductors, require an intermediate amount of energy to transport electrons in the conduction band. This energy is called the bandgap. Once the bandgap energy is overcome, electrons are excited. These electrons will revert to their initial state and emit photons in the form of fluorescence. The wavelength of emitted fluorescence largely depends on the bandgap size, which is governed by the size of QDs known as the quantum confinement effect.[13,14] The smaller the size of QDs, the shorter the wavelength of fluorescence they emit. Thus, it is possible to adjust the emission wavelengths of QDs by changing their size (Fig. 1).

Another important feature of QDs is their broad absorption and narrow emission wavelengths. The peaks of wavelength depend on the bandgap level. Since there are different values for bandgaps, the absorption spectra contain a number of overlapping peaks (Fig. 1). The energy absorbed by QDs typically decreases with increasing wavelengths of the excitation light source.[15] The maximum wavelength at which QDs absorb is called the absorption onset. The absorption spectrum of QDs can be observed at wavelengths from the ultraviolet (UV) region to the near-infrared (IR) region until the absorption onset.[15] This is particularly useful for multicolor biological labeling, where QDs with different emission wavelengths can be excited simultaneously.[5,16] The spectra of QDs can be sharp, narrow, and bell-shaped (Gaussian). This clear differentiation between the peaks of emitted fluorescence allows for multicolor staining, and the fluorescence peaks can be easily detected using different fluorescence filters. In contrast, organic dyes have overlapping fluorescence signals and the peaks are hardly distinguishable.

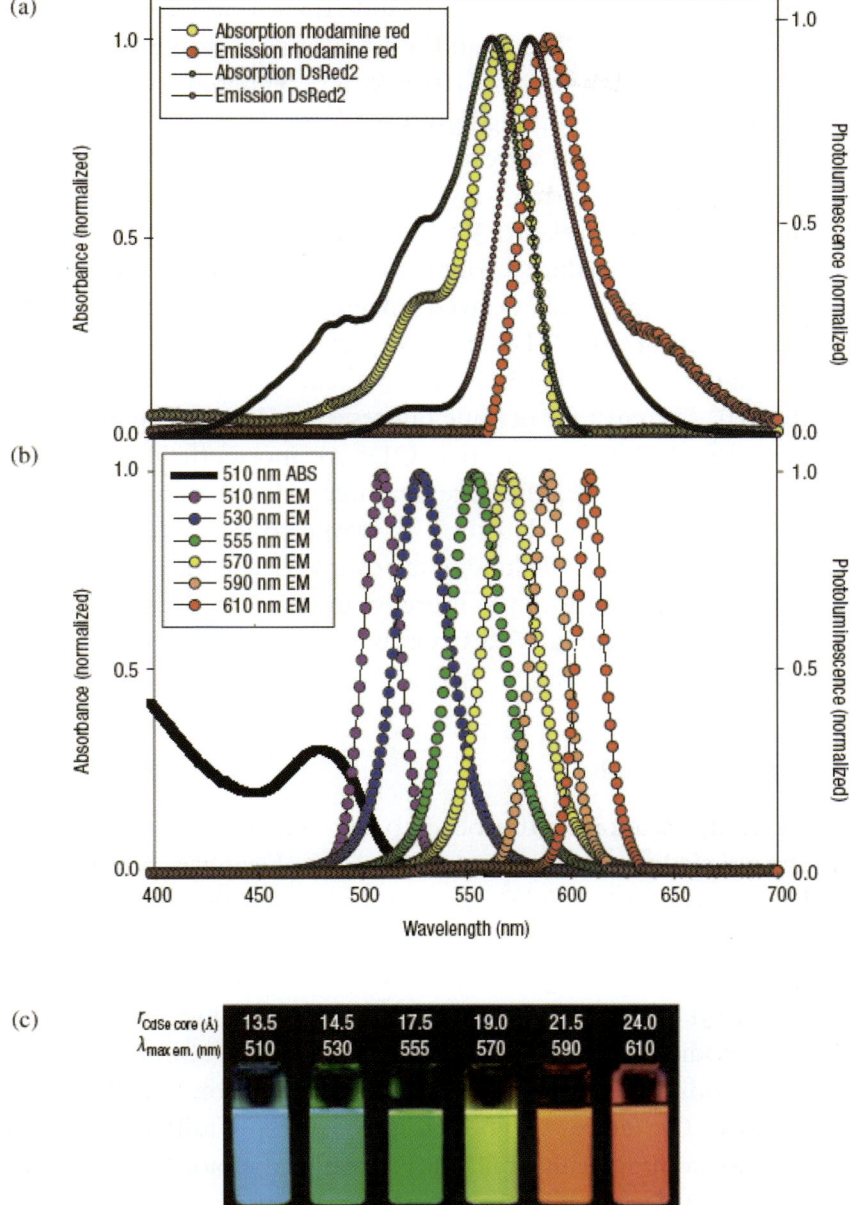

Fig. 1. Absorption and emission spectra of some commonly used fluorescent labels.[14] (a) Absorption and emission peaks of Rhodamine Red (a commonly used organic fluorescent dye) and DsRed2 (a fluorescent protein tag expressed by genetic engineering). Each dye requires a different excitation wavelength and emits fluorescence at a different wavelength. (b) Absorption and emission spectra of CdSe-core QDs indicating differences in emission peaks for CdSe-core QDs for differences in the radii of QDs for a single absorption peak. (c) Fluorescence intensities observed in actual samples.

Other important features of QDs include their long fluorescent lifetime, large molar extinction coefficient, and resistance to photobleaching.[8] In general, QDs fluoresce much longer (10–100 ns) compared to normal fluorophores (~2 ns) after discontinuing light excitation. This property can be used for the differentiation of QDs from other fluorophores. For example, fluorescent lifetime imaging microscopy (FLIM) can differentiate fluorescence caused by auto-fluorescence and QDs. This can be achieved by programming to start the imaging procedure a couple of nanoseconds after the discontinuation of light exposure. This delayed imaging procedure ensures that the fluorescence resulting from QDs can be detected over the auto-fluorescence. QDs have a molar absorption of 10^5–10^6 M^{-1} cm^{-1} compared to that of 2.5×10^4 to 2.5×10^5 M^{-1} cm^{-1} for organic fluorescence dyes.[17] The higher extinction coefficients of QDs indicate that QDs are more efficient in absorbing excitation photons. Hence, they are brighter (10–20 times stronger than organic dyes), enabling the detection of lower concentrations of analytes. Furthermore, QDs are stable under exposure to a light source, which is approximately a thousand times more than organic dyes.[6] This unique resistance to photobleaching is particularly useful for long-term monitoring of biological processes.

Bioconjugation of QDs

For biological applications, QDs need to be conjugated with biomolecules without changing the molecular structure. A number of strategies have been reported to conjugate biomolecules to QDs, including biotin–streptavidin binding,[18] electrostatic interaction,[19] mercapto exchange,[20] and cap exchange.[21] One of the simplest conjugation methods is to coat QDs with avidin.[11] Avidin-coated QDs can be further attached to biotin-tagged biomolecules, including antibodies and nucleic acids. Due to the high affinity between biotin and streptavidin, desired biomolecules can be reliably attached onto the surface of QDs. Currently, avidin-conjugated QDs of different fluorescent wavelengths are commercially available. Although the biotin–avidin conjugation method is simple and reliable, it also has some disadvantages. First, avidin molecules are bulky (~60 kDa), and can cause several steric effects. Second, avidin-coated QDs tend to aggregate. Third, cross-linking of the surface proteins also becomes a concern considering the fact that there are up to 40 biotin binding sites.[22] To overcome these disadvantages, monovalent streptavidin were conjugated with size-reduced QDs so that they bind to biotinylated biomolecules at a ratio of 1:1.[22]

Another way for bioconjugating QDs is electrostatic exchange. Biomolecules can be adsorbed to the hydrophilic shell through electrostatic interactions.[18] Mattoussi and co-workers demonstrated the conjugation of positively charged QDs to the negatively charged surfaces of the biomolecules. This method showed high stability, high yield (higher than that of non-conjugated counterparts), and little or no particle aggregation. Using the same strategy, engineered proteins can be attached to QDs for biomolecule binding. Goldman et al. demonstrated the conjugation of IgG antibodies to QDs via protein G.[19] In this method, antibodies bind to QDs through their Fc (Fragment, crystallizable) region and position their Fab sites outward, which can lead to the favorable orientation of antibodies for maximum antibody–antigen interactions.

Alternatively, biomolecules can be conjugated to the surface of QDs through a mercapto exchange process.[23] In this process, biomolecules containing thiol (–SH) groups can attach to the shell of QDs. The resulting bioconjugate, however, is not as stable as the other two bioconjugation methods mentioned earlier. The attached biomolecules do not have a strong bond with QDs and they can be easily detached from QDs. The mercapto exchange process also requires the use of strong organic solvents, e.g., DMSO, that negatively affect QD structures and biological components conjugated to QDs. To improve stability, dative thiol binding can be used to link the cysteine residues of the biomolecule and QDs.[24]

QDs can also be conjugated to biomolecules by modifying the surface QDs with functional groups.[19] This process, known as cap exchange, involves the usage of a cross-linker molecule to bind to the QD, while exposing the desired functional groups on the outer layer. To link QDs and the desired biomolecules, bifunctional cross-linkers with two main ligands — one attaching to the QDs (e.g., thiol group), and the other attaching to the biomolecule of interest (–COOH, –NH$_2$, –SH) — are used. Biomolecule–QD complexes made by this process are much more stable than those made by the mercapto exchange process.[11,25] Two commonly used cross-linking agents are 1-ethyl-3-(3-dimethylaminopropyl)carbodiimide (EDC) and N-hydroxysuccinimide ester (NHS).[26] EDC can link –NH$_2$ and –COOH groups, while SMCC links –SH and –NH$_2$ groups.

Another specific cross-linker is the nickel–nitrilotriacetic acid complex (Ni–NTA), which is used for histidine-tagged peptides and antibodies.[27] The nitriloacetic acid group covalently binds to the carboxyl group on the QD, while the histidine-tagged antibodies bind to nickel ions.[28] This method is favorable over the biotin–avidin method because of low production costs. Further, histidine-expressing molecules can be directly attached to QDs with Zn^{2+}.

Application of QDs for Fluorescence Imaging

Use of QDs for cellular imaging in vitro

Standard microscopy does not permit the direct visualization of cellular structures, as they are too small and transparent to be visualized and thus, these cells are generally labeled with a fluorescent biomarker to facilitate microscale visualization via fluorescent microscopy. It is expected that QDs will replace the conventional fluorophores for cellular imaging and multiplex tissue imaging due to their appealing optical properties. For example, QDs have successfully labeled cellular components such as nuclear proteins, mitochondria, microtubules, actin filaments, endocytic compartments, mortalin, and cytokeratin, as well as cellular membrane proteins and receptors such as serotonin transport proteins, prostate-specific membrane antigen, Her2, glycine receptors, erbB/HER, and P-glycoprotein.[11]

QDs have also been used to label molecular targets at the subcellular level. In one early study, Bruchez et al. demonstrated the labeling of cellular actin fibers with biotin-conjugated QDs.[29] More specifically, F-actin filaments were labeled by red QDs using avidin–biotin interactions for ligand–receptor binding. The preparation of this experiment involves the incubation of the fibroblasts in phalloidin–biotin and streptavidin, then labeling them with QDs with covalently conjugated biotin. The positive results of this experiment marked that QDs can be successfully used to label subcellular targets. However, the QDs bound non-specifically to the nuclear membrane in this study, but the generated signal was relatively weak.[29] Nie et al. reported the labeling of the protein transferrin and the incorporation of QDs by live cells.[8] More recently, Wu et al. performed a quantitative analysis into the labeling efficiency of QDs at the subcellular level.[30] In this study, QDs were conjugated with immunoglobulin G (IgG) and streptavidin. These QDs were used to label various cellular receptors and components at different subcellular locations; these experiments were performed with different types of specimens such as live cells, fixed cells, and tissue sections.[30] The results of this study produced high-quality multicolor labeling of different cellular structures, showing that QDs are effective in cellular imaging and multi-target cellular detection (Fig. 2).

The most direct way of using QDs for cancer cell detection as demonstrated in *in vitro* studies is the cellular labeling of cancer biomarkers. For example, it is known that the protein mortalin is located at several cellular sites and exhibits distinct staining patterns and different functional properties depending on whether the cell is normal or transformed.[31] As mortalin can be considered as a cancer cell marker, Kaul et al. compared the images of

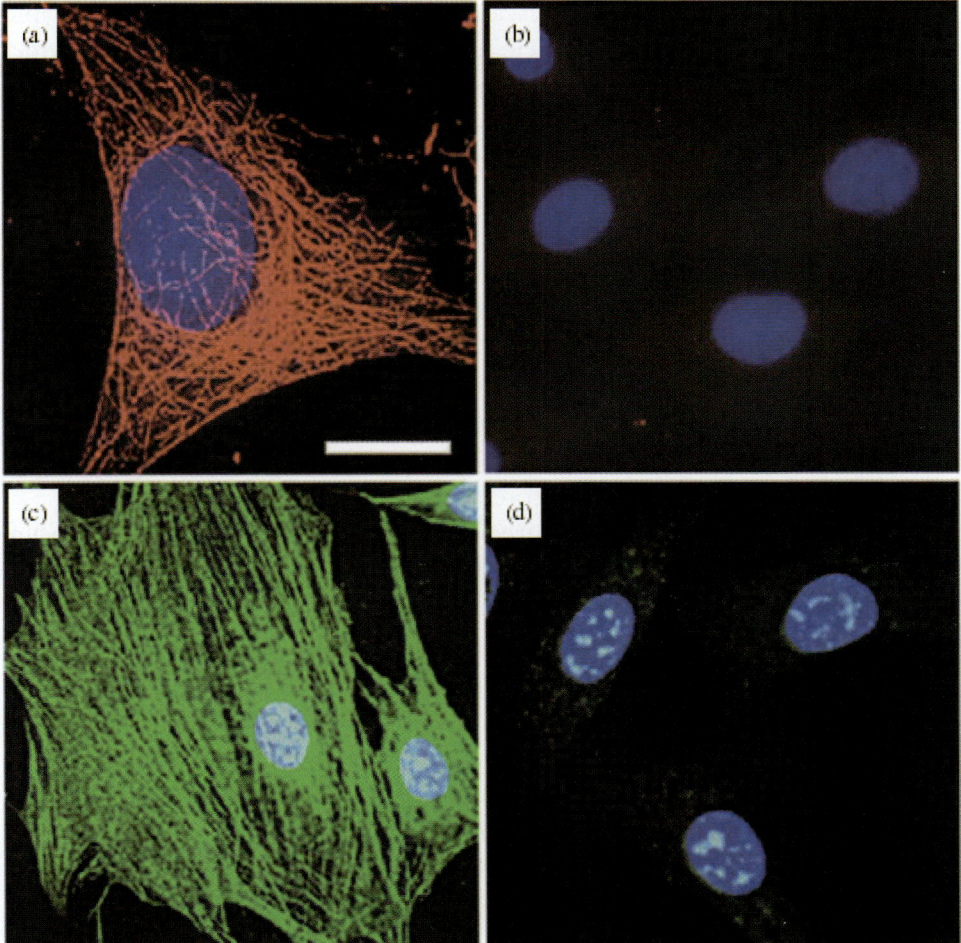

Fig. 2. Fluorescence staining of cytoskeleton fibers in 3T3 with streptavidin-coated QDs.[31] (a) Microtubules were stained with monoclonal anti-α–tubulin antibody, biotinylated anti-mouse IgG, and streptavidin-coated QDs (630 nm; red). (b) Control for (a): no primary antibody. (c) Actin filaments were stained with biotinylated phalloidin and streptavidin-coated QDs (535 nm; green). (d) Control for (c): no biotin–phalloidin. Cell nuclei were counterstained with Hoechst 33342 blue dye. Scale bar: 10 μm for (a), 24 μm for (b)–(d).

mortalin through the use of conjugated QDs.[32] Wu *et al.* used their aforementioned QDs conjugated with IgG and streptavidin to label HER2, a breast cancer biomarker.[30] The labeling of HER2 by the QD–IgG probes was successful and proved that the QD–IgG conjugates are specific toward their targets. In addition, QD–streptavidin conjugates have a low non-specific binding when there is no primary antibody in the system.[30]

QDs have also been used in research pertaining to the massive and rapid screening of proteins and nucleic acids. Han *et al.* encoded polymer microbeads with QDs at precise ratios to analyze biological molecules in parallel.[33] These results demonstrated that single-color encoded beads produced accuracies as high as 99.99%. Fluorescence staining of fixed cells with QDs allows the enhancement of fluorescence imaging because of the enlarged pore size on the cell surface and relatively free entry of QDs into the cells. For example, Lidke *et al.* stained epidermal growth factor with QDs (EGF–QDs) to image and analyze erbB/HER signal transduction.[34] EGF–QDs were prepared using size-exclusion (40 kDa) spin columns (Biorad). The significance of these experiments can be appreciated in light of the fact that the EGF–QDs bound to filopodial erbB1 and revealed a novel mechanism of retrograde transport to the cell body. Thus, QDs can be used as a fluorescence probe to explore protein–protein interactions.[34]

Further, QDs have also been used to facilitate rapid CD4$^+$ T lymphocyte counting using microfluidic devices.[35,36] In this method, anti-CD4 and anti-CD8 antibodies were conjugated to QDs via biotin–avidin interaction. When CD4$^+$ T lymphocytes were captured by antibodies coated on microchannel surface (for more information on this strategy, see Chapter 17), anti-CD4 and anti-CD8 antibody-conjugated QDs were used to achieve fluorescence imaging/counting (Fig. 3). This study demonstrated that the QD-based CD4$^+$ T lymphocyte counting system can be potentially used in resource-constrained settings, since CD4 cell count (flow cytometry) and viral load measurement (RT-qPCR) are expensive (US$50–200 per test).[37]

Use of QDs for cellular imaging in vivo

In vivo studies face the challenge of delivering the QDs through the cell membrane lipid bilayer without damaging cellular unity. Several methods have been developed that overcome this obstacle, including endocytic uptake, scrape loading, microinjection, electroporation, and mediated targeted uptake.[14,38] Through these methods, it is possible to label cells and image them for longer time intervals.[38] In addition, antibody-coated QDs improved biocompatibility for biological applications.[35,39,40] Another possibility is the encapsulation of individual QDs in phospholipid micelles.[41] These micelle-encapsulated QDs were successfully used in *in vivo* experiments of *Xenopus* embryos with better biocompatibility. However, the degradation of phospholipid micelles is still unknown. This method seems to be an attractive platform for targeting cells due to the small size and biocompatibility of lipids.

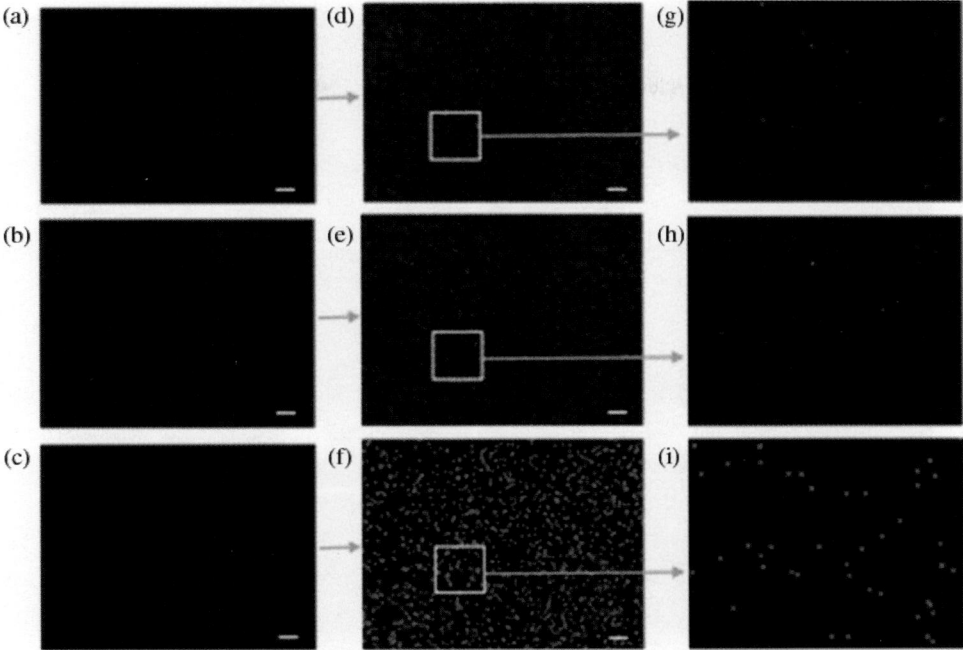

Fig. 3. Fluorescence imaging of CD4+ T lymphocytes on a microchip.[36] The captured CD4+ T lymphocytes were stained AF488-anti CD3 (a) and AF647-anti CD4 (b). To facilitate rapid counting, fluorescence images were merged for better recognition by a counting algorithm (c). Images (d)–(f) were processed and counted by the counting algorithm. Images (g)–(i) are the corresponding portions that were shown in images of (d)–(f), respectively.

Multicolor imaging of human prostate cancer cells grown in mice after systemic injection was reported by Gao *et al.*[42] The QDs used in this study contained an amphiphilic triblock copolymer, targeting ligands and multiple polyethylene glycol (PEG) molecules. The prepared QDs were capable of specific cancer cell labeling, while at the same time, possessed steady *in vivo* protection and improved biocompatibility. Due to the use of an ABC triblock polymer, QDs did not experience fluorescence loss and particle aggregation that occured in QDs in *in vivo* experiments. This study produced images of simultaneous multicolor QDs accumulating on a tumor cell growing area. To visualize and monitor tumor cells, two different mechanisms (active and passive) were performed. The active mechanism was found to be much faster and more efficient than passive targeting. It was also noted that there was no fluorescent emission available from the tumor when hydroxylated QDs were used, indicating a rapid blood clearance by the reticuloendothelial system. The use of PEGylated QDs and QDs with prostate-specific membrane antigen-specific monoclonal antibodies both showed QD accumulation at the

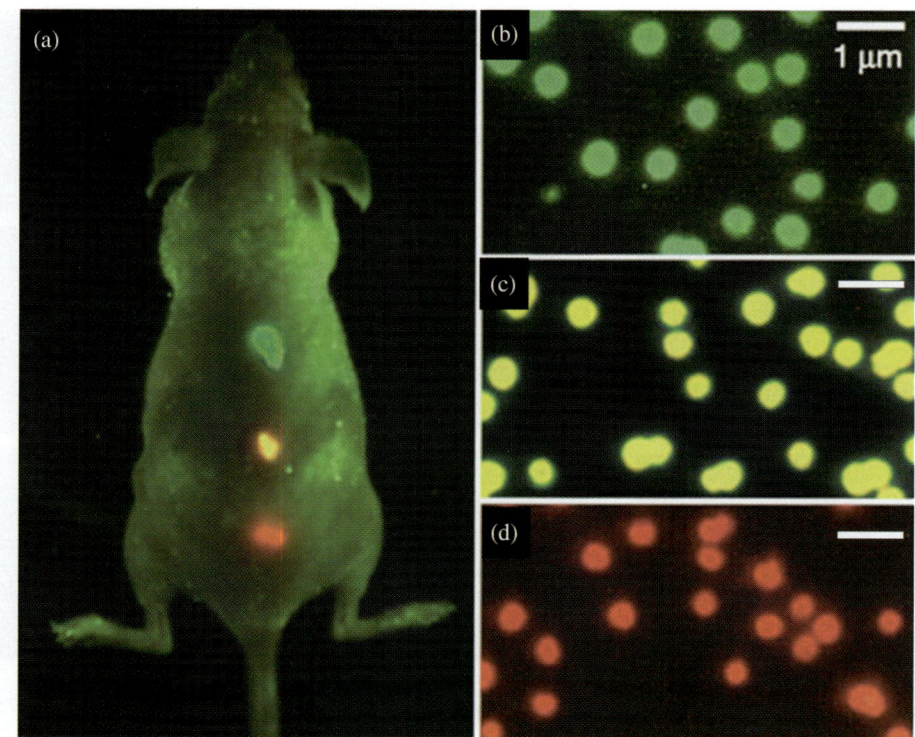

Fig. 4. Simultaneous *in vivo* imaging of multicolor QD-encoded microbeads.[42] QD-encoded microbeads (0.5 μm diameter) were injected into a single mouse at three different locations (a). They emitted green, yellow, or red light, which was observed simultaneously with a single light source in (b)–(d), respectively.

tumor site (Fig. 4). These results demonstrated the importance of considering surface charges and/or ligands on the QDs when employing them for *in vivo* experiments.

In vivo studies of cancer cells in animals showed that specifically conjugated QDs accumulate at tumor sites, which can be used to image and monitor the diseased cancer tissue. Cai *et al.* reported the successful imaging of tumor vasculatures using QDs conjugated with arginine–lysine–aspartic acid peptides.[43] Peptides attached to QDs hold more promise as target ligands over QD–antibody conjugates due to the fact that many peptides could be attached to the QD's surface simultaneously. This resulting polyvalency effect, defined as acting against or interacting with more than one valence, creates a stronger interaction between the QD and the target molecule.[44] Integrin $\alpha_v\beta_3$ was overexpressed on activated tumor cells in this study since it binds to the RGD-containing motifs of the interstitial matrix. Other studies have previously suggested that integrin $\alpha_v\beta_3$ can serve as a target for tumor imaging since it causes tumor angiogenesis and thus tumor metastasis.[45] Stroh et al. combined multiphoton microscopy and QDs in their experiment to visually differentiate tumor vessels from perivascular cells and the matrix.[46] They also succeeded in observing the activities of the precursor cells and examining the capability of particles to access the tumor sites. Similar experiments were performed by Voura et al., where QDs were used to track tumor cells in the process of metastasis.[47]

Application of QDs for viral tracking/detection

QDs have been extensively used for virological applications and viral detection. One of the applications is to continuously track the viral infection and to understand its progression. Continuous tracking of viral infection at the early phases can provide insightful information on the mechanism of the infection and offer potential opportunities for anti-viral drug and vaccine development. Long-term tracking of single viral particles was made possible by utilizing the photostability and brightness properties of QDs.[48]

Agrawal et al. first demonstrated the possibility of covalently attaching QDs to the surface of viral particles.[49] Then, Joo et al. also demonstrated that retroviruses such as the human immunodeficiency virus (HIV) can be successfully tagged with QDs through the membrane incorporation of a short acceptor peptide, which is susceptible to site-specific biotinylation and attachment of streptavidin-conjugated QDs.[50] In this approach, a biotinylated 15-amino acid peptide was first incorporated onto the surface of a virion.[50,51] This application enabled QD labeling for visualizing the dynamic interactions between viruses and target cells so as to study how different types of retroviruses enter the host cell, which helps in the understanding of viral infections. However, this method requires the careful positioning of the QDs onto the viral capsid proteins to avoid non-specific binding of QDs.[49] Dixit et al. demonstrated the incorporation of CdSe/ZnS semiconductor QDs into viral particles.[52] The encapsulation of functionalized QDs was achieved by the self-assembly of HS–PEG–COOH-tagged DNA sequence in viral capsids, which yielded a virus-like particle similar in size to native viral particles.[52]

In addition, QDs have been used to achieve the rapid detection of HIV on a chip.[40] It requires only a finger-prick volume (10 µL) of unprocessed HIV-infected whole blood and a microfluidic chip immobilizing an antibody against the viral envelope protein gp120. The use of two different-colored QDs allows the application of a dual-stain imaging technique and provides a new and effective tool for the accurate detection of HIV particles (Fig. 5).

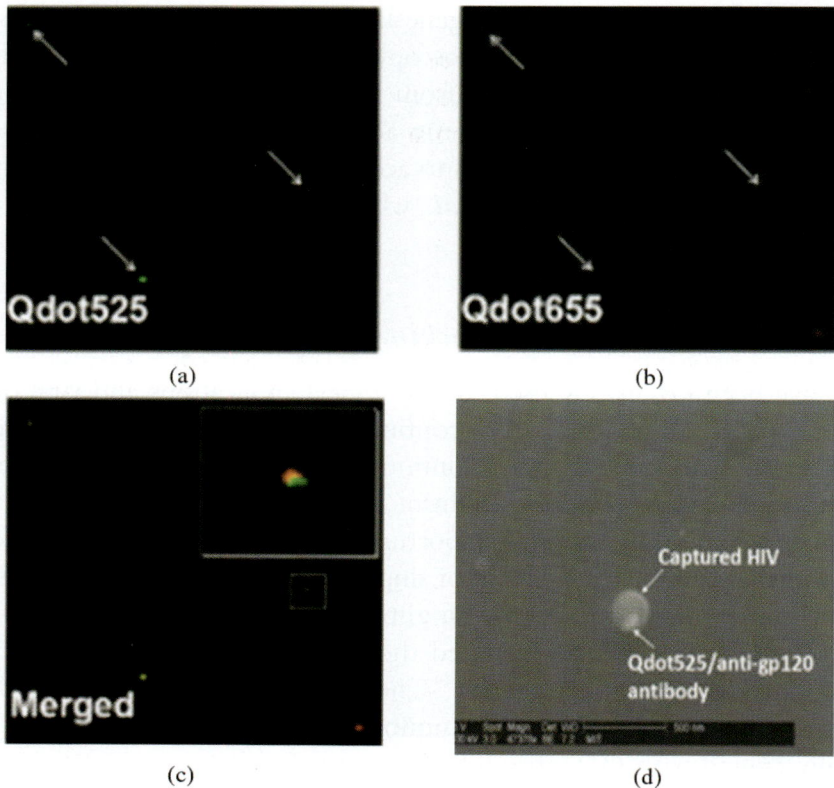

Fig. 5. Fluorescence imaging of HIV particles using QDs.[40] HIV particles were captured by anti-gp120 antibodies, which were coated on the microchannel surface. For dual-staining, streptavidin coated QDs (green QDs 525 and red QDs 655) were used to identify biotinylated ConA and anti-gp120 antibody via streptavidin–biotin interaction. These two QDs were sequentially introduced into the channel to achieve the specific detection of gp120 (a) and high-mannose oligosaccharide (b) on the viral surface. Using these two specific and independent reactions between gp120 and anti-gp120, and between mannose and ConA lectin, HIV-1 particles were specifically detected by co-recognition (c). The captured HIV particle was also confirmed via scanning electron microscopy (d).

This microfluidic device can be potentially used for HIV detection and viral load monitoring in resource-limited settings.[40]

QDs have also been successfully used for the detection of hepatitis B virus (HBV) DNA in a nanobiosensor that is simple, specific, rapid, and with high throughput.[53] This QD–DNA nanosensor is based on fluorescence resonance energy transfer (FRET). It is capable of detecting the target DNA and even a single mismatch in the HBV genome. In this application, water-soluble CdSe/ZnS QDs were first prepared by replacing the trioctylphosphine oxide on the surface with 3-mercaptoproionic acid. Then, functional QD–DNA conjugates were formed by the attachment of oligonucleotides onto the surface. To form QD–DNA-conjugated sandwich hybrids, DNA targets and Cy5-modified signal DNAs were added. The fluorescence emission resulting from Cy5 fluorophore (the acceptor) and QD (the donor) was captured by a FRET microscope. No Cy5 emission was produced due to the lack of FRET in the case of a single-base mismatch. Wang *et al.* showed that this method can be used for the high-throughput and multiplexed detection of target HBV DNA and its mutations.[53]

Another experiment on viral detection done by Liang *et al.* combined two-photon microscopy with hepatitis C virus (HCV)-specific QDs that were conjugated as fluorescent tags.[54] The study demonstrated the feasibility of detecting HCV-infected cells using QDs, and their extent and distribution within the liver of patients who have chronic HCV infection. QDs specific to HCV infected cells were used to determine the proportion of HCV-infected hepatocytes and healthy ones. Liang *et al.* showed that the QD application is able to detect the viral core and non-structural protein 3 antigens from patient liver tissues that are infected with HCV.

Toxicity and Biosafety of QDs

Although QDs have been widely used for applications in biological imaging, studies on their potential toxicity to human beings, animals, plants, as well as the environment have not been extensively carried out.[55] Recently, concerns about the potential hazards that QDs pose toward both humans and the environment are gradually gaining attention.[56,57] It is essential to understand environmental concerns and the cytotoxicity of using QDs before applying them in the clinical arena. In the following section, the potential toxicity and biosafety considerations of QDs will be discussed.

Potential toxicity of QDs

The toxicity of QDs depends on various factors derived from their inherent physicochemical properties and environmental conditions. Properties such as size, concentration, surface coating (capping material and functional groups), dispersibility, species, and exposure time are associated with QD toxicity.[58] As a result, both the intrinsic properties of QDs and the surrounding environmental conditions should be considered to evaluate QD toxicity.

Two important parameters, i.e., QD size and dose, have been associated with QD cytotoxicity. The QD size (typically ranging from 5 to 50 nm) is critical to the biological response of cells. Smaller QDs (< 2.5 nm) tend to localize in and around the cell nucleus, whereas larger QDs are typically distributed within the cytoplasm.[59,60] Lovric et al. showed that the size of QDs is related to their subcellular distribution and the severity of QD-induced cytotoxicity.[60] In addition, the dose is also an important factor due to the large influence of surface area to QD actions. A significant decrease in cell viability was observed even at a low concentration (0.1 mg/mL) of QDs, and smaller-sized QDs have been shown to have stronger effects on the death rate of exposed cells due to accumulation around the nucleus and their higher mobility inside the cell.[61] It has also been reported that the cell death rate is time-dependent,[59] indicating that long-term cytotoxicity needs to be studied.

The surface modification of QDs also plays a predominant role in determining QD toxicity. Some compounds such as mercaptoundecanoic acid (MUA) coated on QDs are responsible for genotoxicity.[62] Other surface modifications on QDs, such as the capping material, can also affect the internalization of QDs into the cells. For instance, smaller green fluorescent QDs (~13 nm) were observed inside breast cancer cells when using MPA–polymer- and polymer–silane-coated QDs. In contrast, larger red fluorescent QDs (~24 nm) were barely detected inside the cells. However, the exact opposite results were observed for PEG–silane-coated QDs. Hence, the toxicity of QDs to different cells depends on the size of the QD.[39] Although the detailed mechanisms remain to be explored, studies have shown that the toxic effects may result from the precipitation of Cd^{2+} ions on the cell surface, where the release of toxic molecules from the surface modification of QDs resulted in cellular and/or nuclear lipid membrane damage.[62,63]

The environmental conditions containing QD suspensions can also directly affect cell viability. For instance, a study indicated that cell viability dramatically decreases, if the QDs are initially exposed to air for 30 min,[64] which may be due to the oxidation of the QD surface, releasing a high level of free Cd^{2+} ions. As reported, Cd^{2+} ions cause severe kidney disorders by affecting the Na^+/glucose co-transporter on the renal cell membrane.[65] Similar toxicity results were obtained with an increase in UV radiation exposure time.[66] Evidence has shown that the release of these free radical ions is detrimental to cells. Some antioxidants were found to be able to inhibit cytotoxicity by controlling the shell structure of QDs and preventing the QDs from degradation.[60] Further studies are required to fully understand the influence of the environment on QDs. In addition, studies have shown that current two-dimensional cultures are not efficient in indicating the cytotoxicity of QDs in vivo. When a three-dimensional cell culture model of liver tissue was used, the QD's toxic effects are significantly reduced as compared to two-dimensional cultures.[67] The significant reduction of cytotoxicity in three-dimensional cultures compared to two-dimensional cultures is due to the well-developed layer of the extracellular matrix (ECM), which decreases the amount of QDs from entering inner layers of cells. In two-dimensional cultures, there is no ECM acting as a protective barrier.[67]

Several *in vivo* studies performed on rats illustrated different absorbance and clearance rates of QDs after intravenous injection, which differed in types, sizes, and surface modifications.[68–70] However, a common finding is that the liver is the organ where xenobiotics, e.g., QDs, accumulate the most. Interestingly, no significant changes of physiological or pathological parameters were observed in the target tissues.[59] Nevertheless, the current number of studies performed *in vivo* is not sufficient to draw conclusions of QD toxicity *in vivo*. Investigations should be made on QD distribution, excretion, metabolism, pharmacokinetics, and pharmacodynamics in animal models *in vivo*, which will be vital for the development of QD-based clinical applications, such as *in vivo* fluorescence imaging and monitoring drug delivery.

Biosafety of QDs

The potential routes of QD exposure to human include environmental, workplace, and therapeutic or diagnostic administration during their development, manufacture, usage, and disposal.[58,71] Most common routes of QD exposure include inhalation, dermal contact, or ingestion. Inhalation is thought to be one of the most important routes of all nanoparticle exposure, including QDs, since nanoparticles can travel great distances in air through Brownian motion.[71] The QDs' size plays a significant role in terms of their deposition in pulmonary tissues upon inhalation. For example, QDs smaller than 2.5 nm could potentially reach deeper into lungs and interact with the alveolar epithelium, whereas larger aerosolized QDs would deposit in bronchial spaces. However, little is known about the aerosolization of QDs in air, in which QDs of smaller size may form aggregates, thus increasing their size.

The interaction of QDs with the skin was also studied. One study demonstrated that QDs can barely penetrate the skin layer. In this study, a small fraction of several QDs species was shown to pass through the stratum corneum, with an even smaller fraction accumulating within the dermis. The penetration is dependent on the size, shape, and surface charge of the QDs, with smaller and spherical QDs appearing to penetrate deeper. None of the QDs were found to be able to penetrate through the entire thickness of the skin, indicating that healthy, intact skin should act as a sufficient barrier to some QDs.[72] However, further studies are needed to better understand the penetration of QDs into intact and diseased skin.

The ingestion of QDs is also possible through inhaling QDs or hand-to-mouth transfer of QDs. Like other nanoparticles, QDs may undergo limited ingestion absorption, mainly to the lymphoid system following systemic exposure. Data suggest that the absorption of QDs through ingestion is governed by the size and the surface characteristics of the particle (with increased absorption for smaller, hydrophobic, and neutral particles).[73,74]

Summary and Perspectives

QDs, owing to their superior optical properties to conventional organic dyes, have been widely used as fluorescent tags to facilitate biomedical imaging *in vitro* and *in vivo*. For a wide range of applications, a variety of bioconjugation methods such as biotin–avidin-based electrostatic interaction, as well as thiol chemistry-based mercapto exchange and cap exchange are available. These bioconjugation technologies enable versatile applications of QDs in fluorescence labeling and imaging. Through QD-based fluorescence technologies, the visualization of subcellular components, whole cells, transport processes, tumor propagation, and viruses has been achieved. Multiplex imaging of cellular components (such as nuclear proteins, mitochondria, microtubules, endocytic compartments, and serotonin transport proteins) is a distinct advantage offered by QDs. Thus, QDs hold great potential to be further utilized to probe unknown proteins or nucleic acids in basic research.

QD-based medical imaging has also been used to facilitate the diagnosis of cancer and infectious diseases. One of the future research goals is to develop QD-based biosensors to deliver point-of-care diagnostics. This aspect is particularly important to achieve the diagnosis and monitoring of infectious diseases in resource-constrained settings. The other future research direction will be investigating the cytotoxicity of QDs both *in vitro* and *in vivo*. Clinical application of QDs cannot be achieved until the pros and cons in medical imaging *in vivo* are well characterized. In addition, emphasis should be placed on the evaluation of QD toxicity to human beings, animals, plants, as well as the ecosystem. Furthermore, environmental effects on QD toxicity should be considered in the aspects of chemical modifications, electrostatic interactions, and cellular targets.

References

1. Morgan, T.T. *et al.*, Encapsulation of organic molecules in calcium phosphate nanocomposite particles for intracellular imaging and drug delivery. *Nano Lett*, 2008. **8**(12): p. 4108–4115.

2. Tosi, S. et al., Classification of deletions and identification of cryptic translocations involving 7q by fluoresence in situ hybridization (FISH). *Leukemia*, 1996. **10**(4): p. 644–649.
3. Jain, R.K., Stroh, M., Zooming in and out with quantum dots. *Nat Biotechnol*, 2004. **22**(8): p. 959–960.
4. Bruchez, M. et al., Semiconductor nanocrystals as fluorescent biological labels. *Science*, 1998. **281**(5385): p. 2013–2016.
5. Chan, W.C.W. et al., Luminescent quantum dots for multiplexed biological detection and imaging. *Curr Opin Biotechnol*, 2002. **13**(1): p. 40–46.
6. Xing, Y. et al., Bioconjugated quantum dots for multiplexed and quantitative immunohistochemistry. *Nat Protoc*, 2007. **2**(5): p. 1152–1165.
7. Bruchez, M., Jr. et al., Semiconductor nanocrystals as fluorescent biological labels. *Science* 1998. **281**(5385): p. 2013–2016.
8. Chan, W.C., Nie, S., Quantum dot bioconjugates for ultrasensitive nonisotopic detection. *Science*, 1998. **281**(5385): p. 2016–2018.
9. Murray, C.B., Kagan, C.R., Bawendi, M.G., Synthesis and characterization of monodisperse nanocrystals and close-packed nanocrystal assemblies. *Annu Rev Mater Sci*, 2000. **30**: p. 545–610.
10. Hines, M.A., Guyot-Sionnest, P., Synthesis and characterization of strongly luminescing ZnS-Capped CdSe nanocrystals. *J Phys Chem*, 1996. **100**(2): p. 468–471.
11. Medintz, I.L. et al., Quantum dot bioconjugates for imaging, labelling and sensing. *Nat Mater*, 2005. **4**(6): p. 435–446.
12. Klimov, V.I., Optical gain and stimulated emission in nanocrystal quantum dots. *Science*, 2000. **290**(5490): p. 314–317.
13. Alivisatos, A.P., Semiconductor clusters, nanocrystals, and quantum dots. *Science*, 1996. **271**(5251): p. 933–937.
14. Medintz, I.L. et al., Quantum dot bioconjugates for imaging, labelling and sensing. *Nat Mater*, 2005. **4**(6): p. 435–446.
15. Yoffe, A.D., Semiconductor quantum dots and related systems: Electronic, optical, luminescence and related properties of low dimensional systems. *Adv Phys*, 2001. **50**(1): p. 1–208.
16. Jaiswal, J.K. et al., Long-term multiple color imaging of live cells using quantum dot bioconjugates. *Nat Biotechnol*, 2003. **21**(1): p. 47–51.
17. Resch-Genger, U. et al., Quantum dots versus organic dyes as fluorescent labels. *Nat Methods*, 2008. **5**(9): p. 763–75.
18. Goldman, E.R. et al., Avidin: A natural bridge for quantum dot-antibody conjugates. *J Am Chem Soc*, 2002. **124**(22): p. 6378–6382.
19. Mattoussi, H. et al., Self-assembly of CdSe–ZnS quantum dot bioconjugates using an engineered recombinant protein. *J Am Chem Soc*, 2000. **122**(49): p. 12142–12150.
20. Willard, D.M. et al., CdSe–ZnS quantum dots as resonance energy transfer donors in a model protein–protein binding assay. *Nano Lett*, 2001. **1**(9): p. 469–474.
21. Mitchell, G.P., Mirkin, C.A., Letsinger, R.L., Programmed assembly of DNA functionalized quantum dots. *J Am Chem Soc*, 1999. **121**(35): p. 8122–8123.
22. Howarth, M. et al., Monovalent, reduced-size quantum dots for imaging receptors on living cells. *Nat Methods*, 2008. **5**(5): p. 397–399.
23. Parak, W.J. et al., Biological applications of colloidal nanocrystals. *Nanotechnology*, 2003. **14**(7): p. R15–R27.

24. Pinaud, F. et al., Bioactivation and cell targeting of semiconductor CdSe/ZnS nanocrystals with phytochelatin-related peptides. *J Am Chem Soc*, 2004. **126**(19): p. 6115–6123.
25. Uyeda, H.T. et al., Synthesis of compact multidentate ligands to prepare stable hydrophilic quantum dot fluorophores. *J Am Chem Soc*, 2005. **127**(11): p. 3870–3878.
26. Alivisatos, A.P., Gu, W., Larabell, C., Quantum dots as cellular probes. *Annu Rev Biomed Eng*, 2005. **7**: p. 55–76.
27. Hainfeld, J.F. et al., Ni–NTA–gold clusters target His-tagged proteins. *J Struct Biol*, 1999. **127**(2): p. 185–198.
28. Xu, C. et al., Nitrilotriacetic acid-modified magnetic nanoparticles as a general agent to bind histidine-tagged proteins. *J Am Chem Soc*, 2004. **126**(11): p. 3392–3393.
29. Bruchez, M., Jr. et al., Semiconductor nanocrystals as fluorescent biological labels. *Science*, 1998. **281**: p. 2013–2016.
30. Wu, X. et al., Immunofluorescent labeling of cancer marker Her2 and other cellular targets with semiconductor quantum dots. *Nat Biotechnol*, 2003. **21**(1): p. 41–46.
31. Wadhwa, R. et al., Differential subcellular distribution of mortalin in mortal and immortal mouse and human fibroblasts. *Exp Cell Res*, 1993. **207**(2): p. 442–448.
32. Kaul, Z. et al., Mortalin imaging in normal and cancer cells with quantum dot immunoconjugates. *Cell Res*, 2003. **13**(6): p. 503–507.
33. Han, M. et al., Quantum-dot-tagged microbeads for multiplexed optical coding of biomolecules. *Nat Biotechnol*, 2001. **19**(7): p. 631–635.
34. Lidke, D.S. et al., Quantum dot ligands provide new insights into erbB/HER receptor-mediated signal transduction. *Nat Biotechnol*, 2004. **22**(2): p. 198–203.
35. Jokerst, J.V. et al., Integration of semiconductor quantum dots into nano-bio-chip systems for enumeration of CD4+T cell counts at the point-of-need. *Lab Chip*, 2008. **8**(12): p. 2079–2090.
36. Alyassin, M.A. et al., Rapid automated cell quantification on HIV microfluidic devices. *Lab Chip*, 2009. **9**(23): p. 3364–3369.
37. Wang, S., Xu, F., Demirci, U., Advances in developing HIV-1 viral load assays for resource-limited settings. *Biotechnol Adv*, 2010. **28**(6): p. 770–781.
38. Jaiswal, J.K. et al., Long-term multiple color imaging of live cells using quantum dot bioconjugates. *Nat Biotechnol*, 2003. **21**(1): p. 47–51.
39. Sukhanova, A. et al., Biocompatible fluorescent nanocrystals for immunolabeling of membrane proteins and cells. *Anal Biochem*, 2004. **324**(1): p. 60–67.
40. Kim, Y.G. et al., Quantum dot-based HIV capture and imaging in a microfluidic channel. *Biosens Bioelectron*, 2009. **25**(1): p. 253–258.
41. Dubertret, B. et al., In vivo imaging of quantum dots encapsulated in phospholipid micelles. *Science*, 2002. **298**(5599): p. 1759–1762.
42. Gao, X. et al., In vivo cancer targeting and imaging with semiconductor quantum dots. *Nat Biotechnol*, 2004. **22**(8): p. 969–976.
43. Cai, W. et al., Peptide-labeled near-infrared quantum dots for imaging tumor vasculature in living subjects. *Nano Lett*, 2006. **6**(4): p. 669–76.
44. Mammen, M., Choi, S.K., Whitesides, G.M., Polyvalent interactions in biological systems: Implications for design and use of multivalent ligands and inhibitors. *Angew Chem Int Edn*, 1998. **37**(20): p. 2755–2794.
45. Liu, Z., Wang, F., Chen, X., Integrin alpha(v)beta(3)-targeted cancer therapy. *Drug Dev Res*, 2008. **69**(6): p. 329–339.

46. Stroh, M. et al., Quantum dots spectrally distinguish multiple species within the tumor milieu in vivo. Nat Med, 2005. 11(6): p. 678–682.
47. Voura, E.B. et al., Tracking metastatic tumor cell extravasation with quantum dot nanocrystals and fluorescence emission-scanning microscopy. Nat Med, 2004. 10(9): p. 993–998.
48. Michalet, X. et al., Quantum dots for live cells, in vivo imaging, and diagnostics. Science, 2005. 307(5709): p. 538–544.
49. Agrawal, A., Sathe, T., Nie, S., Single-bead immunoassays using magnetic micro-particles and spectral-shifting quantum dots. J Agric Food Chem, 2007. 55(10): p. 3778–3782.
50. Joo, K.I. et al., Site-specific labeling of enveloped viruses with quantum dots for single virus tracking. ACS Nano, 2008. 2(8): p. 1553–1562.
51. Beckett, D., Kovaleva, E., Schatz, P.J., A minimal peptide substrate in biotin holoenzyme synthetase-catalyzed biotinylation. Protein Sci, 1999. 8(4): p. 921–929.
52. Dixit, S.K. et al., Quantum dot encapsulation in viral capsids. Nano Lett, 2006. 6(9): p. 1993–1999.
53. Wang, X. et al., QDs-DNA nanosensor for the detection of hepatitis B virus DNA and the single-base mutants. Biosens Bioelectron, 2010. 25(8): p. 1934–1940.
54. Liang, Y. et al., Visualizing hepatitis C virus infections in human liver by two-photon microscopy. Gastroenterology, 2009. 137(4): p. 1448–1458.
55. Lewinski, N., Colvin, V., Drezek, R., Cytotoxicity of nanoparticles. Small, 2008. 4(1): p. 26–49.
56. Oberdorster, G., Oberdorster, E., Oberdorster, J., Nanotoxicology: An emerging discipline evolving from studies of ultrafine particles. Environ Health Perspect, 2005. 113(7): p. 823–839.
57. Nel, A. et al., Toxic potential of materials at the nanolevel. Science, 2006. 311(5761): p. 622–627.
58. Hardman, R., A toxicologic review of quantum dots: Toxicity depends on physicochemical and environmental factors. Environ Health Perspect, 2006. 114(2): p. 165–172.
59. Zhang, Y. et al., In vitro and in vivo toxicity of CdTe nanoparticles. J Nanosci Nanotechnol, 2007. 7(2): p. 497–503.
60. Lovric, J. et al., Differences in subcellular distribution and toxicity of green and red emitting CdTe quantum dots. J Mol Med, 2005. 83(5): p. 377–385.
61. Shiohara, A. et al., On the cyto-toxicity caused by quantum dots. Microbiol Immunol, 2004. 48(9): p. 669–675.
62. Hoshino, A. et al., Physicochemical properties and cellular toxicity of nanocrystal quantum dots depend on their surface modification. Nano Lett, 2004. 4(11): p. 2163–2169.
63. Kirchner, C. et al., Cytotoxicity of colloidal CdSe and CdSe/ZnS nanoparticles. Nano Lett, 2005. 5(2): p. 331–8.
64. Derfus, A.M., Chan, W.C.W., Bhatia, S.N., Probing the cytotoxicity of semiconductor quantum dots. Nano Lett, 2004. 4(1): p. 11–18.
65. Xia, X. et al., The endogenous CXXC motif governs the cadmium sensitivity of the renal Na+/glucose co-transporter. J Am Soc Nephrol, 2005. 16(5): p. 1257–65.
66. Li, J. et al., The photodynamic effect of different size ZnO nanoparticles on cancer cell proliferation in vitro. Nanoscale Res Lett, 2010. 5(6): p. 1063–71.
67. Lee, J. et al., In vitro toxicity testing of nanoparticles in 3D cell culture. Small, 2009. 5(10): p. 1213–21.

68. Ballou, B. et al., Noninvasive imaging of quantum dots in mice. *Bioconjug Chem*, 2004. **15**(1): p. 79–86.
69. Fischer, H.C. et al., Pharmacokinetics of nanoscale quantum dots: In vivo distribution, sequestration, clearance in the rat. *Adv Funct Mater*, 2006. **16**(10): p. 1299–1305.
70. Yang, R.S. et al., Persistent tissue kinetics and redistribution of nanoparticles, quantum dot 705, in mice: ICP-MS quantitative assessment. *Environ Health Perspect*, 2007. **115**(9): p. 1339–43.
71. Stern, S.T., McNeil, S.E., Nanotechnology safety concerns revisited. *Toxicol Sci*, 2008. **101**(1): p. 4–21.
72. Ryman-Rasmussen, J.P., Riviere, J.E., Monteiro-Riviere, N.A., Penetration of intact skin by quantum dots with diverse physicochemical properties. *Toxicol Sci*, 2006, **91**(1): p. 159–165.
73. Hillyer, J.F., Albrecht, R.M., Gastrointestinal persorption and tissue distribution of differently sized colloidal gold nanoparticles. *J Pharm Sci*, 2001. **90**(12): p. 1927–36.
74. Jani, P. et al., The uptake and translocation of latex nanospheres and microspheres after oral administration to rats. *J Pharm Pharmacol*, 1989. **41**(12): p. 809–12.

Biographies of the Editors

Utkan Demirci

Utkan Demirci is an Assistant Professor of Medicine and Health Sciences and Technology at the Harvard Medical School (HMS), Brigham and Women's Hospital (BWH). He received his bachelor's degree in Electrical Engineering (*Summa Cum Laude*) in 1999 from the University of Michigan, Ann Arbor, his master's degrees in Electrical Engineering in 2001 and in Management Science and Engineering in 2005 from Stanford University, and his doctorate in Electrical Engineering in 2005 also from Stanford University. His current work involves applying nano- and micro-scale technologies to manipulate cells in nanoliter volumes, with applications in infectious disease diagnostics and monitoring, cell encapsulation and assembly for cryobiology, tissue engineering, and regenerative medicine. His research interests include the applications of microelectromechanical systems (MEMS) and acoustics in medicine. Dr. Demirci has authored over 60 peer-reviewed publications in journals including *PNAS, Advanced Materials, Biomaterials,* and *Lab on a Chip*, more than 80 conference abstracts and proceedings, and 10 book chapters. He has edited a journal special issue on point-of-care diagnostics. His work has been highlighted in *Wired Magazine, Nature Photonics, MIT Technology Review Magazine, AIP News, BioTechniques,* and *Biophotonics*. He has given over 50 national and international presentations including invited keynotes at various academic, governmental, and industrial institutions.

Ali Khademhosseini

Ali Khademhosseini is an Associate Professor at the Harvard–MIT Division of Health Sciences and Technology (HST), Brigham and Women's Hospital (BWH), and Harvard Medical School (HMS), as well as an Associate Faculty at the Wyss Institute for Biologically Inspired Engineering. He is also a Junior Principal Investigator at Japan's World Premier International – Advanced Institute for Materials Research (WPI-AIMR) at Tohoku University, where he directs a satellite laboratory. In addition, he is an Adjunct Professor in the Department of Biomedical Engineering at the University of Texas at Austin. His research is based on developing micro- and nanoscale technologies to control cellular behavior, with particular emphasis on developing microscale biomaterials and engineering systems for tissue engineering. He has authored approximately 400 publications and 16 patent/disclosure applications. His work has been published in journals such as *PNAS*, *JACS*, *Advanced Materials*, *Biomaterials*, and *Lab on a Chip*, and highlighted in numerous public news media, *Nature*, and the *Scientific American and Technology Review Magazine*. As of July 2012, he has been cited over 7,000 times and invited to give nearly 150 seminars and keynote lectures.

Robert Langer

Robert Langer is a David H. Koch Institute Professor at MIT and a recipient of over 210 major awards, including the 2006 United States National Medal of Science, the Charles Stark Draper Prize (considered the equivalent of the Nobel Prize for engineers), the 2008 Millennium Prize (the world's largest technology prize), and the 2012 Priestley Medal (the highest award of the American Chemical Society). He received his bachelor's degree from Cornell University in 1970 and his Sc.D. from MIT in 1974, both in Chemical Engineering. He has also earned honorary doctorates from various universities, including Harvard University, the Mt. Sinai School of Medicine, and Yale University. He has written more than 1,175 articles and also has approximately 800 issued and pending patents worldwide. In addition to his numerous accolades, *Forbes Magazine* (1999) and *BioWorld* (1990) have named him as one of the 25 most important individuals in biotechnology in the world; *Discover Magazine* (2002) named him as one of the 20 most important people in this area; *Forbes Magazine* (2002) selected Dr. Langer as one of the 15 innovators worldwide who will reinvent our future; *Time Magazine* and CNN (2001) listed him as one of the 100 most important people in America and one of the top 18 personalities in science or medicine in America (*"America's Best"*); and he was also included in the *Parade Magazine* (2004) list of six *"Heroes Whose Research May Save Your Life."*

Jeffrey Blander

Jeffrey Blander is an experienced research scientist, non-profit foundation director, and industry executive specializing in areas of health systems strengthening, mobile health applications, operations research for point-of-care diagnostics, business plan development for start-up health care companies, and the design of strategic public private partnerships to address the emerging double burden of infectious and non-communicable diseases in sub-Saharan Africa. He has held dual research appointments at the Brigham and Women's Hospital (BWH) and Harvard School of Public Health as well as served as the co-director for courses he co-founded on global health practice, business, and medical technology within the Division of Health Science and Technology (HST) at Harvard Medical School (HMS) and MIT. In these roles, he has mentored many students and alumni on field projects in East Africa. In 2010, Dr. Blander was presented with the Olympus Emerging Education Leadership Award for his efforts. He has also served as Volunteer Country Director for the Clinton Foundation (Jamaica, 2004) and was honored with pre- and post-doctoral Fogarty International Center/NIH Fellowships in Global Health and Clinical Research (Tanzania, 2005 and 2008). He received his doctorate and master's degree from the Harvard School of Public Health and his bachelor's degree from the Wharton School of the University of Pennsylvania.

Index

acoustic 476, 478, 490
alginate 267, 270, 271
animal-on-a-chip 220
antibody-based capture 428
anti-viral antibodies 317, 319
atherosclerosis 97, 103
atherosclerotic plaque 97, 99, 107
auditory 476
axon 194

bacterial 138
BBB (blood–brain barrier) 216, 221, 229, 230
bio-ferrofluidics 402
biomimetic 475
bioparticle separation 418, 438
    CD4$^+$ T cell 420, 421
    CTCs (circulating tumor cells) 423, 424
    other bioparticles 425
    virus 420, 423, 425
biosafety 2, 13, 15
bone marrow stromal cells 70

carbon nanotubes (CNTs) 379, 383
cell-adhesive islands 71–74, 80
cell-based diagnostics 126
cell counting 126
cell culture 162, 166, 167
cell lysis 164, 171, 174–176
cell microenvironment 215, 216, 221
cell shape 71–73, 75, 81

cell size 71–73
cellular manipulation 411
CFD (computational fluid dynamics) 98, 102
CNTs (carbon nanotubes) onto polymer films 380
cochlea 476, 479, 490
cochlear 488
cochlear implant 478
cochlear-like 477
cochlear-like sensor 479
co-culture 201, 202
collagen 256, 260, 272, 273
continuous-flow PCR (CF-PCR) 136
contractility 73–75, 80, 84
coronary arteries 100, 101
    FDB (first diagonal branch of the left anterior descending artery) 100, 101, 107
    LAD (left anterior descending artery) 100, 101, 105, 107, 109
    LCX (left circumflex artery) 101, 107, 109
    LMA (left main coronary artery) 100–102, 107, 109
CT (computed tomography) angiography 99
cytoskeletal tension 73, 80, 81

detection 135
dielectrophoresis (DEP) 360, 361, 365, 367

direct detection 316, 322
DNA 348, 349
DNA extraction 383
DNA extraction by SWNTs 386
DNA/RNA 138
dose–response 37–39
droplet 29–34, 36, 37
drug development *see* Chapter 2
drug discovery 198, 199, *see* Chapter 2

electrophoresis 342, 347
embryoid bodies (EBs) 253, 255
embryonic stem cells (ESCs) 197, 241, 242, 244, 245, 249, 253, 255, 256, 259, 260, 263
emulsion 294, 296, 298, 300, 301, 304
encapsulation 293, 294, 304, 305
endothelial cells 103
enzyme-linked immunosorbent assay (ELISA) 117
extracellular matrix (ECM) 246, 248, 249, 256, 264, 269, 271

ferrofluids 392, 393, 395, 399, 401, 402
FGF (fibroblast growth factor) 256
fluid control 118
fluid–structure 490
fluorescence-activated cell sorting (FACS) 191
fluorescence imaging 1–3, 5–9, 11–13, 15, 17
focal adhesion 71, 73
focal adhesion molecules 75
frequency analysis 478
frequency analyzer 476
functionalized beads 328

genomic 223
global health 417, 418, 420, 438
glucose sensor 47, 48, 50, 52, 53, 55–60

growth factor 247, 258, 264, 267, 275

healthcare 135
hematopoietic stem cells (HSCs) 245, 247–249, 260, 262
heteroduplex analysis (HDA) 356, 357
high-throughput screening (HTS) 33
HTS (high-throughput screening) 33, 34, 36
human samples 139
hyaluronic acid (HA) 259, 265, 271
hydrogel 251, 264–268, 270, 271, 275

IC_{50} 37, 39
immunoassay 117
immunocytochemistry 177, 178
induced pluripotent stem (iPS) cells 197
inertial focusing 330
infectious disease 115, 135
inner ear 476
integrin 248, 249
interaction 490
in vitro disease model 213, 215, 219, 226
in vitro tissue model 214, 215, 219, 222
iron oxide 392

lab-on-a-chip 48, 49, 63
LDL (low-density lipoprotein) *see* Chapter 5
leaky junction 98, 102, 103
liver 214, 216, 219–221, 226, 227
liver bioreactors 226, 227

magnetic nanoparticle 392, 393, 395, 401, 406
magnetic liquids 392
magnetism 391

mathematical models 476
mathematical modeling 419, 428, 429, 433
  bioparticle motion 429, 433
  cell adhesion 429–431, 433
  cell rolling 430, 433
  single adhesion bond 429
mechanics 488
mechano-sensing 74, 75, 83
mechano-signals 71, 83, 84
mechano-transduction 78, 85
mesenchymal stem cells (MSCs) 69, 70, 72, 84, 245, 250, 251, 259–261, 266, 268
  human MSCs (hMSCs) 250
metabolic 219, 221
metabolomics 22, 24, 25
microarray 256
microcontact printing 71, 72, 80, 81
microenvironment 166, 167
microfabrication 69–70, 76, 219, 226
microfibers 302
microfluidic 69–71, 81, 83, 84, 86, 135, 241–243, 249, 252, 253, 260, 263, 401, 403, 409, 418, 419, 425, 428, 429, 433, 436
microfluidic cell arrays 223, 224
microfluidic cell culture 216, 218, 220, 232
microfluidic chips 144
microfluidic sensor 48, 49, 53
microparticle 298, 307
micropatterning 71, 78, 85
microphone 477
micro total analysis system (mTAS) 152
microvalve 307, 308
modeling 475, 488
molecular detection 221, 223
molecular diagnostics 135
morphology 73, 76, 77, 80, 81
mutation 356, 357

nanoparticles 392, 394, 395, 397
nanotopography 78
neural stem cells (NSCs) 247, 248, 258, 260, 264
neuronal heterogeneity 186
neuronal polarity 186
neuron networks 216
neuron type-specific vulnerability 190
niche 246–249
nuclear shape 73, 75
nucleic acid detection 120, 121
numerical simulation 429–431, 436
NURBS (non-uniform rational B-spline) 101, 102

patient-specific modeling 97
physical models 476
point-of-care (POC) 48, 49, 117
point-of-care testing (POCT) 135
polydimethylsiloxane (PDMS) 216–219, 222, 224, 243, 250, 251, 255, 259, 271, 273
polymerase chain reaction (PCR) 122, 135
  PCR detection 124
pore theory 98, 104
protein markers 128
proteomic 128, 221

quantum dots (QDs) 1–16, 19

sample preparation 138
sensor 475, 476, 478, 490
shear stress 98, 103, 106
signal amplification 118
signaling 162, 165, 166, 178
single-cell analysis 163, 170
single nucleotide polymorphisms (SNPs) 342, 358
solid phase extraction (SPE) 141
sound 477
spectral analysis 477

spinning   294
stem cell   241
subcellular location   171

theranostics   127
tissue engineering   241, 242, 263, 264, 272
topography   76, 86, 249, 250
topology   71, 76–78, 80, 81, 86
toxicity   2, 13–16
transducer   476–478, 488, 490

ultrafiltration   315, 327
ultrasonic embossing   377, 378

vascularization   264, 268
vascular system   216, 230
viral   141
viral antigens   317, 320
viral detection   311
viral genomes   318, 321
virion   312
viruses   311
viruses-on-a-chip   318, 319, 322

waveguide   477